Wine Faults and Flaws: A Practical Guide

Wine Faults and Flaws: A Practical Guide

Keith Grainger
Consultant in Wines and Wine Technology
Broadway, UK

WILEY Blackwell

This edition first published 2021
© 2021 John Wiley & Sons Ltd

The right of Keith Grainger to be identified as the author of this work has been asserted in accordance with law.

Registered Offices
John Wiley & Sons, Inc., 111 River Street, Hoboken, NJ 07030, USA
John Wiley & Sons Ltd, The Atrium, Southern Gate, Chichester, West Sussex, PO19 8SQ, UK

Editorial Office
The Atrium, Southern Gate, Chichester, West Sussex, PO19 8SQ, UK

For details of our global editorial offices, customer services, and more information about Wiley products visit us at www.wiley.com.

Wiley also publishes its books in a variety of electronic formats and by print-on-demand. Some content that appears in standard print versions of this book may not be available in other formats.

Limit of Liability/Disclaimer of Warranty
The contents of this work are intended to further general scientific research, understanding, and discussion only and are not intended and should not be relied upon as recommending or promoting scientific method, diagnosis, or treatment by physicians for any particular patient. In view of ongoing research, equipment modifications, changes in governmental regulations, and the constant flow of information relating to the use of medicines, equipment, and devices, the reader is urged to review and evaluate the information provided in the package insert or instructions for each medicine, equipment, or device for, among other things, any changes in the instructions or indication of usage and for added warnings and precautions. While the publisher and authors have used their best efforts in preparing this work, they make no representations or warranties with respect to the accuracy or completeness of the contents of this work and specifically disclaim all warranties, including without limitation any implied warranties of merchantability or fitness for a particular purpose. No warranty may be created or extended by sales representatives, written sales materials or promotional statements for this work. The fact that an organization, website, or product is referred to in this work as a citation and/or potential source of further information does not mean that the publisher and authors endorse the information or services the organization, website, or product may provide or recommendations it may make. This work is sold with the understanding that the publisher is not engaged in rendering professional services. The advice and strategies contained herein may not be suitable for your situation. You should consult with a specialist where appropriate. Further, readers should be aware that websites listed in this work may have changed or disappeared between when this work was written and when it is read. Neither the publisher nor authors shall be liable for any loss of profit or any other commercial damages, including but not limited to special, incidental, consequential, or other damages.

Library of Congress Cataloging-in-Publication Data

Names: Grainger, Keith, author.
Title: Wine faults and flaws : a practical guide / Keith Grainger.
Description: Hoboken, NJ, USA : Wiley-Blackwell, 2021. | Includes
    bibliographical references and index.
Identifiers: LCCN 2021001638 (print) | LCCN 2021001639 (ebook) | ISBN
    9781118979068 (cloth) | ISBN 9781118979099 (adobe pdf) | ISBN
    9781118979075 (epub)
Subjects: LCSH: Wine and wine making–Gaging and testing.
Classification: LCC TP511 .G73 2021 (print) | LCC TP511 (ebook) | DDC
    641.2/2–dc23
LC record available at https://lccn.loc.gov/2021001638
LC ebook record available at https://lccn.loc.gov/2021001639

Cover Design: Wiley
Cover Image: © RomanKozhin/Shutterstock

Set in 9.5/12.5pt STIXTwoText by SPi Global, Chennai, India

10 9 8 7 6 5 4 3 2 1

To Paula

Contents

Acknowledgements

So many people have imparted knowledge, opinions and assistance that have helped me write this work, that this list of acknowledgements is more notable for the numerous omissions than the names included. However, I would like to give particular thanks to Carlos de Jesus, Paulo Lopes and Joana Mesquita of Amorim, Matt Thomson of Kiwi-Oeno Consultancy and Blank Canvas Wines, Sebastian Beaumont, Gordon Newton Johnson, Chris Alheit, the late, great Paul Pontalier of Château Margaux, Alfred Tesseron of Château Pontet-Canet of Château Belgrave and other Dourthe properties, Jean-Luc Columbo, Nicolas Joly, Sandro Bottega, Ernie Loosen, Raimund Prüm, José and Sebastien Zuccardi, Philip Tuck MW, Gordon Burns, and Evin Morrison of ETS Laboratories, Matthias Hüttl of LANXESS, Gordon Specht and Ann Dumond of Lallemand, Pascal Chatonnet, Gevork Arakelian, and Ken Walker. Finally, thanks to Nick Catley, Trevor Elliott and Hazel Tattersall for reviewing individual chapters.

Preface

Wines are produced today in over 65 countries, and it is often stated that production standards are higher than at any time in the 8000 or more years of vinous history. The consumer rightly expects any wine purchased to be of good quality, bearing in mind the price point, and free from fault, flaw, or taint. Wine critics, writers, producers, and retailers are all in the business of selling enjoyment and entertainment, for nobody has to drink wine – it is a beverage to be savoured. However, the incidences of faulty wines reaching the consumer are greater than would be regarded as acceptable in most other industries. It is claimed that such occurrences are less than was the case in recent recorded history, and it is true that the frequency of some faults, flaws, and taints being encountered in bottle (or other packaging) has declined in the last few decades. Gone are the days when a bottle of dry white Bordeaux would have more 'struck-match' sulfur odours than the aromas of Sémillon or Sauvignon Blanc, and a white Bairrada exude the 'Oloroso Sherry-like' aromas and bitter palate resulting from oxidation. Happily too, in recent years, there has been a considerable reduction in the occurrence of haloanisole contamination (often referred to as 'cork taint'), which renders heavily affected wines undrinkable, having the musty odour of damp hessian or dry rot. However, incidences of certain faults and taints have increased, and issues that were once unheard of now affect many wines offered for sale. These include 'reduced' aromas, premature oxidation (premox), atypical ageing and, very much on the rise, smoke taint.

The reduction in the occurrence of some faults might have been expected with the increased sophistication and regulation of wine production, and the advance in scientific and technical knowledge of grape growers and winemakers. These, taken together with the utilisation of a vast array of high specification production equipment, might lead the consumer to believe that they are extremely unlikely to be faced with a faulty bottle of wine. This is sadly not the case. The increased incidence of some defects is, prima facie, surprising. However, the styles of wines that consumers want, or are perceived by the industry to want, have changed in the last 20 years or so. Grape harvests are often delayed until so-called phenolic ripeness and, together with the effects of climate change, this has resulted in higher grape sugars, higher pH, and lower acidity. Each of these presents dangers. In the case of red wines, and particularly those at so-called 'entry-level', tannins are softer, and levels of residual sugar have often increased. Inexpensive wines are usually brought to the market very quickly, sometimes within a few months of harvest. Time is money, and there may be little time or budget for all desirable stabilisation procedures to take place. Finally, the consumer and especially the serious wine lover, wants wine to be a natural, agricultural product that has been turned into something wonderful by an artisan, not a scientist. Accordingly, many producers strive to make wine as 'natural' as possible, by reducing interventionist techniques and minimising chemical and biological additions, including those that may increase stability and prevent the onset of certain faults. It is also of concern that there is an increase in the incidence of mycotoxins in wines, due to their production by some of the microbial populations on vines. This is perhaps another consequence of climate change. Related off-flavours, which were historically noted only occasionally, have been detected at a much higher frequency during the last 15 years.

This book provides a detailed examination and explanation of the causes and impact of the faults, flaws, and taints that may affect wines. As such, I believe that it will prove particularly valuable to winemakers, especially those at small, boutique wineries, wine technologists and quality control professionals. Wine critics, writers, educators, and sommeliers will also find the topics most relevant. With wine trade students and people venturing into the business of wine production in mind, the content is designed to be easily and speedily assimilated. The interested and knowledgeable wine-loving consumer, including wine collectors and investors, will also find the book highly relevant and the basis for discussion at many tastings with like-minded associates.

It is assumed that the reader has, at least, a basic knowledge of winemaking. Those who feel the need to brush up on the methods and techniques of wine production, from vine to bottle, are referred to '*Wine Production and Quality* 2nd Edition' by Keith Grainger and Hazel Tattersall, also published by Wiley. However, this book is very suitable for those with limited scientific knowledge, and I have made every effort to maximise readability, with many anecdotes and expressed opinions, including my own. The chasm between general and scientific wine publications is both wide and deep, and I have attempted to bridge this as soundly as possible. There are several excellent books on viticulture and oenology, written in scientific language. Some of these are listed in the Further Reading. Alternatively, the easy to read articles and books penned by wine writers are seen by the scientific community as overflowing with anecdotes and lacking rigour. Of course, both approaches are valid, and I have made every effort to integrate them. I have tried to speak in terms that the general reader can understand. Although the number of studies and quality of research into chemical and microbiological faults in wine has increased considerably in the last couple of decades, most of this work remains within the confines of journals, which are largely unread outside of scientific and academic communities. On many occasions, results are inconclusive, conflicting, or the focus of some studies is seen as so narrow as to be of little relevance in the real world. Indeed many of the winemakers I have spoken to whilst researching this book have not been aware of the outcomes of recent research in important areas. Accordingly, there is still much misinformation and misunderstanding of the topics by producers, students and, of course, consumers.

Sections of individual chapters cover the science behind each fault, and for these, a very basic knowledge of organic chemistry and microbiology would be valuable. However, the scientifically challenged reader will find the text in includes helpful explanations, and the Glossary contains easy to understand definitions and descriptions of many scientific terms. I have included, at the end of each chapter, references to relevant publications and research papers (over 800 in total), concentrating on those in English and published within the last 20 years. I have also included references to earlier research that was particularly ground-breaking. I have not included structural drawings, and there are very few chemical equations. I have not generally detailed isomers (isomers are where compounds have the same formula but differ from each other in the way the atoms are arranged). As the title of this work suggests, the focus is very much on practice – the book is certainly not aimed at the research scientist. In other words, I wish it to be a helpful manual for those who have little interest in the activities of the research lab, but to whom the excitement and challenges of the real world of wine are a way of life.

Keith Grainger
Saint-André-de-Lidon
19 August 2020

Introduction

This book comprises a detailed examination of faults, flaws, and taints that can affect the quality and merchantability of wines. Technically, there is a distinction between a taint and fault that will be discussed in Chapter 1. However, wine consumers, merchants, and the press rarely observe such distinction, simply referring to the affected product as faulty. Some faults render affected wines unsaleable and undrinkable. Others that have a negative effect upon quality, enjoyment, or potential for ageing.

There are no reliable figures regarding the total global financial cost of faulty and tainted wines to producers, agents, distributors, and retailers. However, there can be no doubt that it runs into many £/€ billions annually. A recent estimate of the economic costs to EU wine producers of haloanisoles taint in wines (so-called 'cork taint') is 700 million euros annually [1]. Product recalls due to matters other than incidences of physical contamination are relatively rare in the wine industry. When bottles of faulty wines do reach the consumer, the impact upon the producers' and suppliers' reputations is incalculable. Although the purchaser may not have the knowledge or skill to identify the fault in question, it is unlikely that they will repurchase any other bottles of the same wine, and may avoid the brand or producer in question. In other words, the consumer will believe the faulty wine to be very low quality. If they have no recourse to financial redress, as will most likely be the case for wines purchased many years previously and which they have been patiently nurturing in their 'cellar' in anticipation of the complex delights of full maturity, consumers may well feel somewhat cheated.

Wine is, without doubt, the most discussed food or drink in the world, and wine lovers readily communicate their experiences to friends, colleagues, and those who share a love for what can be the most exciting and individual of products. Today, such discussions are livelier and more influential than ever. With the proliferation of blogs and social media postings, the opinions of a consumer's peers are, for better or worse, as or even more important than those of professional critics and reviewers. Wine writers, authors, and journalists are generally hugely supportive of the wine industry, without which their profession would not exist. Most are 'deeply in love' with wine, or at least 'fine wine'. However, it is the nature of writing that certain topics become 'hot' and are then developed and pursued for as long as the readership retains an interest. Amongst the wine faults that have received considerable coverage in specialist consumer media in the last decade or two are so-called 'cork taint', 'reduced' aromas (often referred to as reduction or reductivity), premature oxidation (premox), and the aromas produced by

Brettanomyces yeasts. When such topics are discussed in a global context, articles can be informative, but they can also weaken confidence and influence buying habits. However, if the coverage relates to individual producers, the damage inflicted can be both instant and ongoing. Reputational damage may be done simply by naming faulty wines submitted for assessment at comparative tastings conducted by specialist magazines, or for tasting competitions. By way of example, during the early years of this century, the influential USA published magazine *Wine Spectator* revealed the identity and details of several 'high-end' Californian producers who had marketed wines tainted by 2,4,6-trichloroanisole (TCA) and 2,4,6-tribromoanisole (2,4,6-TBA). These compounds are usually, and particularly in the case of 2,4,6-TBA erroneously, referred to as 'cork taint'. The negative impact upon the reputations of producers that had taken decades to build is apparent.

On the other hand, there are 'under the radar' faults that are seldom discussed in the popular wine media or amongst professionals. A prime example is 'atypical ageing', by which white wines very rapidly lose varietal character and develop undesirable aroma and palate characteristics. This fault, often confused with premature oxidation, has been described as 'one of the most serious quality problems in white wine making in nearly all wine producing countries' [2]. It has been estimated that up to 20% of USA wines might be affected [3].

The financial impact upon producers and distributors who have sold faulty product can be immediate and direct. Supermarkets and merchants impose chargebacks upon suppliers when customers return wines. On an individual basis, this may be the cost of the bottle in question, the cost of analysis of other bottles, the cost of replacement bottles and a 'fine', or service handling fee. On a volume basis, the trade customer may demand reimbursement for pallets or even containers of affected wine, shipping and warehouse costs, handling, and possibly also excise duties which, in some countries, can amount to several times the value of the wine in question. If there are problems with subsequent shipments, the merchant may well blacklist the producer or supplier. The longer-term financial impact can be massive. For example, winery contamination with haloanisoles has, on occasions, necessitated the destruction and rebuilding of cuveries and chais, as discussed in Chapter 3.

In Chapters 3–14, I discuss in detail individual categories of wine faults. I do not claim the list of faults included to be exhaustive. The discussion of each fault generally includes:
- What it is, in basic terms;
- How it can be detected by:
  sensory recognition, including sensory detection thresholds;
  laboratory analysis;
- What the cause is;
- At which stage/s of production, maturation, or storage it can occur;
- How it might be prevented;
- Whether an affected wine is treatable, and if so how;
- The detailed science applicable to the fault.

The 'history' of the individual faults is also covered. Throughout the book, particularly Chapter 18, there is a general discussion of the implementation of what constitutes good procedures and practices in the vineyard and winery to enhance quality and minimise the likelihood of faults from occurring. Carrying out audits of premises, equipment, and inventory to identify microbial or chemical contamination can be costly and taking steps to address issues identified even more so. The cost of being unaware of problems, or doing nothing to rectify them, is incalculable. I am acutely aware of budget constraints that are an everyday challenge, particularly to the small producer, and such implementation may be generally achieved at minimum expense. Information on the identification of faults by laboratory analysis, and how faults may be rectified is given in general terms. However, producers seeking to address specific issues are advised to seek advice from any of the laboratories, consultants, and companies specialising in the identification and treatment of oenological problems. It is the responsibility of producers to check the legality of any method suggested, or the addition of any oenological products, in the country/region of production and market.

There are several challenges posed in undertaking any discussion of wine faults and flaws. These include matters of definitions, boundaries, concentrations, and the matrices of individual wines. From a sensory perspective, determining when a microbial or chemical issue is a fault is not necessarily straightforward. In addition to the issue of a taster's sensory detection thresholds, there can often be a dispute whether a particular characteristic is perceived as beneficial, harmless, a flaw or a fault. These perceptions are even subject to the vagaries of fashion. In 1982, Master of Wine and Burgundy expert Anthony Hanson wrote in the first edition of his critically acclaimed book *Burgundy:* 'great Burgundy smells of shit' [4]. If there were any raised eyebrows at the time, these were only because of Hanson's choice of language.

Indeed many Burgundies exuded the odours of stables and farmyards. By 1995, Hanson was already finding such a nose objectionable and blamed microbial activity [5]. We now know that these odours have nothing to do with Pinot Noir (the variety from which pretty much all red Burgundy is made). Nor do they stem from any of the myriads of Burgundy terroirs, but result from volatile phenols and other compounds produced by the yeast *Brettanomyces* (or to be technically correct *Dekkera* although it is rarely so-called in the wine industry). Today, *Brettanomyces* is generally regarded in the wine industry as a rogue yeast, and odours of farmyards, stables, or BAND-AID® are generally considered to be undesirable and regarded by most winemakers, oenologists, and critics to be a fault. This means that aromas in 1982 regarded by an expert taster as a sign of quality are today usually seen as a fault. However, *Brettanomyces* (often referred to as 'Brett') remains a controversial topic. Many producers, critics, and wine lovers believe it can, at low levels, add complexity to a wine. This poses the question as to where the boundary should be drawn. Purists perceive Brett always to be a fault and define it as such. Some lovers of 'natural' wines consider it to be one of nature's distinctive aroma and flavour giving yeasts. In some countries, the number of wines showing Brett characteristics is increasing, due in part to winemakers trying to satisfy perceived consumer demands. As Jokkie Bakker and Ronald J. Clarke note, 'changes in winemaking culture as a result of changes in consumer preference for the required style of wine has led to an increase in some off-flavour formation, for example volatile ethyl phenols (which are by metabolised by *Brettanomyces*) [6]'. I examine the topic of *Brettanomyces* and related volatile phenols, in Chapter 4.

We may draw another example of definition/boundary/concentration challenges when discussing volatile sulfur compounds, which can be a consequence of agronomic conditions or reductive winemaking techniques. Some of these compounds may, at modest concentrations, give notes of minerality, a hint of struck match, 'gun-flint', and of savoury 'lamb fat'. Such characteristics may be considered positive in, for example, white Burgundies and the white wines of the Central Vineyards of the Loire Valley. However, at high levels, volatile sulfur compounds can give most undesirable odours, including bad eggs, onion, garlic, skunk, town gas, and faeces. But at just what point is the desirable concentration exceeded? A risqué comedian may be very funny until a boundary is crossed, beyond which they are perceived to be obscene.

A further illustration of these boundary challenges may be made when discussing excessive volatile acidity (VA). Many Italian red wines have high levels of VA, which in wines from other countries might be considered to be at very least a flaw, but which contributes to the 'Italian' character of the wines. Whilst a high level of VA in the red wines of Bordeaux is not acceptable today, times and palates change. The 1947 vintage of Château Cheval Blanc, Saint–Émilion, made before the advent of temperature controlled fermentations and described by some critics as the greatest wine of all time, possesses such high VA that by today's thinking it would be perceived as not just flawed, but seriously faulty. The topic of excessive VA is discussed in Chapter 7.

The final example of these challenges in this Introduction relates particularly to mature Riesling wines. Many producers in Germany and Alsace have long lauded the diesel or kerosene nose that examples can exhibit after several years in the bottle. Most New World producers and wine critics regard such a nose as indicative of a flaw at least, caused by 1,1,6-trimethyl-1,2-dihydronaphthalene (TDN), a norisoprenoid. At other than very low concentrations, they consider the wine to be faulty. In common with many other European-based wine writers and authors, I disagree with the defining TDN as a flaw or fault, unless the level is overwhelming (another boundary challenge). The aroma characteristics imparted by TDN can add exhilarating notes that in wines from some regions form part of the individual, sensuous character of this most distinctive of varieties. This topic is discussed in Chapter 15. Bearing in mind that sensory characteristics change according to when in a wine's life-cycle it is assessed, and using the examples I have given, a particular compound and resulting odour or taste may be considered to comprise a fault or flaw, when in the wrong concentration, in the wrong wine, in the wrong place, and at the wrong time!

There are apparent contradictions in how we assess and define wine quality. One wine can be analysed chemically and microbiologically and be declared technically very good and free from flaws, yet may taste distinctly uninteresting. Another may show technical weaknesses or even flaws, yet when tasted, it can be so full of character and true to its origin that it sends a shiver down the spine, and must be regarded as of the very highest quality. Returning briefly to the topic of *Brettanomyces,* Château de Beaucastel, a Châteauneuf-du-Pape from France's southern Rhône valley, and Chateau Musar from the Bekar valley in Lebanon are two examples of brilliant, distinctive, and exhilarating wines that historically showed considerable 'Brett' characteristics. Many lovers of these wines are still nurturing their mature stocks, as recent vintages have been much tamer. Such characteristics are not to everybody's palate, but using analogies from the music and art worlds, a person who loves Puccini is perhaps not that fond of the works of Ed Sheeran, and somebody overwhelmed by Titian may be distinctly unmoved by Paul Klee.

Perceptions of faults and flaws are subject to the sensitivity of individual tasters, and humans vary considerably in their sensory detection thresholds and their reactions to individual taints. For most of the faults discussed in this book, detection thresholds (in wine) are stated. These sometimes split into odour detection and taste detection thresholds when researchers have quantified such information. The figures stated are not always straightforward or beyond dispute. Sensory perception thresholds are usually different for red, rosé, white, sparkling, and fortified wines. Further, the level at which a compound that may constitute a fault becomes apparent may vary according to the wine matrix, which includes grape variety or varieties, style, alcoholic content, structure and balance, and is not simply a matter of quantification of the compounds responsible. The topics of sensory perceptions, detection thresholds and consumer rejection thresholds are discussed generally in Chapters 1 and 2, and also in Chapters 3–15 that cover individual faults.

The final, and most important, dichotomy briefly considered in this Introduction is that of two contrasting and perhaps incompatible approaches to wine production: (i) 'minimum intervention' and (ii) 'technical excellence'. 'Great wine is made in the vineyard' is an oft heard expression, used in equal abundance by quality-conscious producers, critics, and serious wine lovers. In other words, if the harvested crop is of the highest quality and bears the hallmarks of a distinguished area of production, the art and science of turning this into top-quality wine is not that hard. Conversely, no amount of intervention or high-tech equipment can make a superb wine from fruit that is unripe, diseased, over-cropped or otherwise undistinguished, grown on unsuitable soils, or in adverse weather or climatic conditions. When studying the classic works on oenology and numerous research papers, it is easy to be led to believe that winemaking is all about the use of cultured yeast strains, enzymes, fining agents, sterile filtration, reverse-osmosis (RO) machines, and a thousand and one vehicles of intervention. When talking to many producers of great wines that have a sense of place, with a unique identity, a very different picture is painted, and we find that with the occasional exception of certain fining agents, many do not employ any of the above. Care, time, and 'listening to the wine' are seen as the pathways to producing distinctive wines, and any additive or cellar operation that might compromise individuality is forsaken. Indeed, numerous producers who make exhilarating wines will challenge the use of some of the methods and the necessity for the oenological products, detailed in the sections of this book where prevention of individual faults are discussed.

Whilst this book has been designed to be readable from cover to cover, each chapter is also written as a stand-alone so that the reader seeking information on a specific topic can find the required material in one place. On occasions, there is repetition as the steps to be taken to prevent an individual fault may be very similar to those undertaken to avoid others, and matters of good winery hygiene are central to producing wines free of fault or taint. To maximise readability and avoid confusion between similar sounding organisms or chemical compounds, I sometimes break from conventions used in works that are designed for academic and scientific readers, such as using only the initial of a genus after its first appearance in a chapter, and only using acronyms after the abbreviated word or phrase has been detailed.

Writing a book such as this presents numerous challenges, and it has also changed the way I assess wine, even impacting upon enjoyment. As my wife Paula said to me, 'You have stopped listening to the music and started listening to the sound-system'. I hope that readers are not affected in the same way. But as our journey proceeds through the wide territory of wine faults and flaws, and although we may find some of the 'terroir' hostile, there are also interesting vinous discoveries to be made. From time to time, we may wander down side roads, but never blind alleys. When walking any city street, we must look up and down, as well as from side to side – faults and flaws are all part of a greater picture and are relative to their surroundings. And if on occasions, we find our journey to be a little dry, we can always be uplifted by reaching for a glass of any good wine that has a story to tell.

References

1 European Commission CORDIS (2016). Electronic Nose To Detect Haloanisoles In Cork Stoppers – Final Report Summary. https://cordis.europa.eu/project/rcn/111040/reporting/en (accessed October 2019).

2 Schneider, V. (2010). Primer on atypical aging. *Wines and Vines* 4: 45–51.

3 Henick-Kling, T., Gerling, C., Martinson, T. et al. (2009). Studies on the origin and sensory aspects of atypical aging in white wines. 14–16 April 2008, Trier, Germany. *Proceedings of the 15th International Enology Symposium, International Association of Enology, Management and Wine Marketing*, Zum Kaiserstuhl 16, 79206 Breisach, Germany.

4 Hanson, A. (1982). *Burgundy*. London: Faber.

5 Hanson, A. (1995). *Burgundy*, 2e. London: Faber.

6 Bakker, J. and Clarke, R.J. (2012). *Wine Flavour Chemistry*, 2e. Chichester: Wiley–Blackwell.

1

Faults, Flaws, Off-Flavours, Taints, and Undesirable Compounds

In this chapter, the advances in wine technology and changes in the markets in the recent decades are discussed, along with the possible adverse impacts of some wine compounds on human health. The distinction, often made generally in the food industry, between 'faults' and 'taints' has limited validity in wines. The challenges of sensory detection thresholds of fault compounds are noted. Microbiological and chemical faults are distinguished, and the boundary between, 'flawed' and 'faulty' is discussed. At the end of the chapter, I consider whether faulty wines can be paradoxically excellent.

1.1 Introduction

The origin of any fault or flaw in wine may be chemical, microbiological, or physical. During the process of winemaking, thousands of biochemical reactions are taking place, and most of them contribute to the aroma and flavour profile of the wine, but some are unwanted and, if not controlled or inhibited, quality will be compromised and off-odours or faults may develop. The diverse interactions between yeasts, fungi, bacteria, and chemical compounds begin in the vineyard and continue through the production processes including maturation and even during storage after packaging. The alcohol in wine (ethanol), together with the acids, provide some stability and protection against deterioration – in fact, wine (as opposed to grape must) is a harsh environment for microbes, thereby allowing only relatively few to grow. However, some microorganisms, including unwanted yeasts and bacteria can flourish during winemaking and may remain in a wine that has finished all stages of production and cause off-odours, off-flavours, or product deterioration months, or even years, after bottling. Wine may become contaminated and develop such off-odours and flavours as a result of external factors during the production processes, and even subsequent to bottling or other packaging. Common sources of contaminants include processing aids (e.g. bentonite), the winery or cellar atmosphere and environment, packaging materials (including bottle closures), and transport or storage facilities (including shipping containers).

1.2 Advances in Wine Technology in Recent Decades

The origins of wine production date back 8000 years, but the advances in the science and technology of production since World War II have outstripped those of the previous 79

Wine Faults and Flaws: A Practical Guide, First Edition. Keith Grainger.
© 2021 John Wiley & Sons Ltd. Published 2021 by John Wiley & Sons Ltd.

centuries. There have been huge changes in the methods and improvements in standards of viticulture, although the reliance upon chemical fertilisers, pesticides, and herbicides that began in the 1960s declined only recently and was certainly both damaging to the environment and wine quality. The recent advent of geolocating and remote and proximal sensing and soil mapping, so-called 'precision viticulture', and techniques to determine the hydric stress of vines have enabled growers to fine tune site-specific management practices, including the precise addition of desirable nutrients and, where permitted, measured quantities of adjusted irrigation water to individual vineyard blocks, or even precise parts of blocks.

In the winery there have been numerous advances in equipment design and quality, and winemaking procedures. Some of the most important of these are:

- Grape sorting and selection systems to exclude unripe, rotten, or damaged fruit: spectacular innovations have been made in the technology of sorting equipment in the last 15 years, including the use of sophisticated optical sorters;
- The widespread utilisation from the 1970s of temperature-controlled fermentation vessels (usually constructed of easily cleanable AISI 316 or 304 grade stainless steel), although the move today is 'back' to vats made of wood or concrete;
- Development of numerous individual strains of cultured yeasts aiding the control of fermentations and development of required flavours;
- A detailed understanding of malolactic fermentation (MLF) and the development of suitable inoculums for use when deemed necessary;
- The utilisation of a wider range of extraction techniques including pre-fermentation cold soaks, rack, and return (délestage), flash détente and thermo détente;
- The use of gases namely carbon dioxide, nitrogen, and argon, primarily to avoid unwanted oxidation;
- The development of programmable enclosed pneumatic presses that can be gas-flushed;
- The utilisation of in-tank micro-oxygenation, particularly for inexpensive red wines, that helps polymerise long-chain tannins and is a valuable tool if costly barrel ageing is not to take place;
- Pre-bottling cold stabilisation, often also using the 'contact process', to precipitate crystals of potassium bi-tartrate or calcium tartrate. Alternatively, membrane electrodialysis may now be employed for this purpose;
- The availability of systems employing membrane technology, for must and wine correction, and filtration, including front end microfiltration (MF), cross-flow (tangential) MF, ultrafiltration (UF), reverse osmosis (RO), and pervaporation;
- the improvement in cork closure quality, particularly regarding the cleansing of corks of 2,4,6-trichloroanisole (TCA), and the introduction of effective alternative closures, including screw-caps and synthetic closures with determined oxygen transmission rates (OTRs).

Of course, small- and medium-scale producers, many of whom are focussed upon the individuality and quality of their wines, may choose not to utilise 'advanced' techniques in the vineyard or winery, including oenological additives, processing aids, or sophisticated (and expensive) technical equipment.

1.3   Changes in Markets and the Pattern of Wine Consumption in Recent Decades

The changes in the pattern of wine consumption of wine during the last 40 years have been dramatic. There has been spectacular growth in sales in many countries that have limited domestic production and accordingly rely upon imports to satisfy demand: for example, in the United Kingdom, per capita wine consumption rose from just 2 l in the early 1960s to 19.7 l in 2010–2016 [1]. In 2018, UK per capita consumption was approximately 24.6 l [2]. However, during the same period, the per capita consumption of many traditional European wine production countries declined by some 50%, albeit mostly of very low quality wines. Prior to the 1970s only 10% of all wines were exported from producing countries, including intra-Europe 'exports'; by 2016 over one-third of wine consumed globally was produced in another country [1]. The journey undertaken to market wine in both bottle and bulk may be many thousands of miles, from the New World to the Old or vice versa, or from the West to the East. This massive rise in exports has only been possible because of the considerable improvements in the technical quality and particularly the stability of the product sold. 'It's a great wine to drink locally but it doesn't travel', an often heard expression when visiting wine regions in the 1970s and 1980s, is but a distant memory today. Microfiltration and ultrafiltration using membrane technology have been important tools in ensuring clarity and stability of the product, particularly in the case of the output of large producers. Other membrane processes, including reverse osmosis (RO) have enabled producers to achieve product of reasonably consistent quality and in styles that the marketers and gatekeepers believe to be popular with a wide audience, including those who were traditionally not wine drinkers.

Several 'new' markets have emerged in the last decades, particularly Asian countries: in the last few years, sales in China have surpassed the wine industry's expectations, especially for so-called 'fine' wines, although, at the time of writing, this market has been contracting. The interest amongst wine lovers in the qualities and differences in styles in wines produced around the world is also high, as evidenced by the proliferation of wine societies and attendance at tutored tasting and other wine events, the growth in wine 'tourism', and the plethora of blogs and other Internet and media discussions. Annual global wine sales are estimated to amount to 246 million hectolitres, the equivalent of nearly 33 billion 75 cl bottles [3]. The diversity of retailers selling wine, declining for many years, has recently increased. The typical consumer of everyday wines is most likely to make their purchase in a supermarket rather than in a specialist outlet, but the Internet has been the salvation of independent merchants, providing a relatively inexpensive marketing vehicle and access to a national or even international customer base. Many of these outlets offer high quality wines from small, individualistic producers and, in some cases, the production operations are 'crowd-funded' by enthusiastic and loyal customers.

The quality and style of wines from any region, or any producer, however small or large, will not be *totally* consistent from vintage to vintage – indeed it is these variations that makes individual wines so exciting to wine lovers. In other words, the quality properties of the finished product are not precisely predictable, unlike the product of other drink production technologies [4]. There is constant debate how much winemaking is considered to be an art or science. However, there is no doubt that the use of modern winemaking techniques has led to wines moving closer in style [5], and this is decried by both passionate small producers and serious and educated wine lovers. Certainly the use of technological equipment has led to less incidences of some faults and flaws but paradoxically has also contributed to the increase in others.

Although almost all 'fine wines' are bottled at source (usually at the winery but sometimes elsewhere in the region of production), there has been an increase in the last decade in shipping 'everyday wines' in bulk and bottling at destination. In fact this was commonplace until the 1970: cheaper wines were often transported in 'SAFRAP' (lined mild-steel containers) or even ships' tanks, and of course wine had been transported in barrels for many centuries. The return to bulk shipping for inexpensive wines has been largely driven by economic factors, together with the need to be seen to reduce the environmental impact of transportation. There are other pros and cons to bulk transportation, particularly from a quality perspective. International Organization for Standardization (ISO) tanks and flexitanks are the two most widely used transport containers, and wines may become tainted from poorly maintained or cleaned ISO tanks, or the ethylene vinyl alcohol (EVOH) copolymer liner of flexitanks, although such instances are thankfully very rare. There are historic incidences of wines being tainted with naphthalene on journeys from Australia to Europe. Flexitanks may also permit some undesirable oxygen ingression, albeit at a very low level – the permeability of the material has decreased considerably since the 1990s. As flexitanks are generally 'single-use', contamination from previous contents is not usually a problem, but the disposable nature perhaps does not sit well the purported environmental advantages of bulk wine shipping. A defective seal on an ISO tank or the use of a flexitank material that is highly permeable could allow oxygen ingress leading to degradation of the entire contents. Wines to be transported in bulk will require adjustment and stabilisation before their journey, and often again prior to packaging at destination – the latter operations being outside of the control of the producer, who may nevertheless bear the brunt of any fallout resulting from product deterioration, or the manifestation of faults or flaws. On the positive side, temperature variations during transport are very often less for wines in large tanks, and the standards of bottling at a dedicated plant at destination may be higher than those in some wineries.

1.4 The Possible Impact of Some Fault Compounds Upon Human Health

Most of the faults discussed in this book, however unpleasant they may be from an organoleptic point of view, are not generally harmful to human health, at least at the concentrations in which they may be found in wine. Wine has a low pH and pathogens harmful to humans will not generally grow in the product (although we should remember that every year somewhere in the world people die as a result of consuming 'fake' wine). There are, however, some compounds of microbiological origin that may be found in wine that have been shown to be potentially harmful to humans. The most important of these are biogenic amines, ochratoxin A and ethyl carbamate. Ethyl carbamate is classified as 'a probable human carcinogen' (Group 2A) by the International Agency for Research on Cancer (IARC) [6], and several countries have set limits for its concentration in wine.

Biogenic amines are toxic at high concentrations, and some people have an intolerance at the levels very often found in wine. Ochratoxin A is a known carcinogen in some animals and a suspected human carcinogen by IARC (Group 2B) [6]; accordingly limits to its concentration in wine have been set by the European Commission (EC) for member states of the European Union (EU) at $2\,\mu g/l$ [7]. In addition, some chemical compounds which may be found in wine can be harmful, particularly phthalates that can be hormone disruptors in humans. Their presence in wine is always due to contamination from an external source, such as the epoxy resin lining of concrete fermentation or storage vats. In a research paper published in 2014, Pascal Chatonnet et al. revealed that significant quantities of dibutyl phthalate (DBP) had been found in 59% of the (French) wines analysed [8]. Although, with the possible exception of biogenic amines that can have a 'blood-like' odour and taste, the above compounds do not influence aroma or taste, their presence in wine other than at very low levels should be regarded as a fault; more of this will be discussed in Chapter 14.

Of course ethanol, the alcohol of all fermented drinks, is a known carcinogen and is classified as 'a human carcinogen' (Group 1) by the IARC [6]. It is toxic if consumed in excess, and there are reported cases of death due to alcohol poisoning and other issues related to single acts of excessive consumption. Long-term regular consumption of ethanol, other than at low levels, is also a causal factor in several carcinomas, liver diseases, and other health problems such as obesity, as wine lovers and imbibers of other alcoholic beverages are regularly made aware. Acetaldehyde is considered a fault in wine only if present in excessive amounts, which generally means having a negative impact upon aroma, but when associated with alcohol consumption is also classified as 'a human carcinogen' (Group 1) by the IARC [6].

1.5 Sulfur Dioxide and Other Possible Allergens

Some individuals are allergic to grapes or alcohol. However, some compounds may be present in wine that may cause allergic reactions. The most important of these is sulfur dioxide (SO_2), which is generally added at several stages of the winemaking process for its antimicrobial and antioxidant properties. Even if none is added by the winemaker, most wine contains SO_2 as it is naturally formed by yeast during the alcoholic fermentation. Allergic reactions to this compound, which is also used as a preservative in a wide range of foods and drinks, are not uncommon, and individuals with asthma may suffer particularly adverse reactions. Many other people show an intolerance. The total SO_2 content of wine is regulated in the European Union (EU) and all major markets and any wine marketed that exceeds this must be considered as both faulty and illegal.

Certain processing aids, particularly fining agents, contain milk or egg products, which are allergens. For wines marketed in the EU, allergen labelling was made compulsory from 25 November 2005 under a European Commission Directive. Initially only the presence of sulfites/sulfur dioxide had to be declared on the label, for concentrations at or exceeding $10\,mg/l$. This threshold remains applicable. Most wines will contain in excess of this figure even if no SO_2 is added during the production process. In 2007 European Directive EC 2007/68 was issued, which provided for the mandatory labelling of further allergens – this directive has since been incorporated into EU Regulation No. 1169/2011 [9]. Due to objections from the wine industry this did not generally come into force until 2012. Insofar as wine is concerned, the only further allergens to be declared on the label are milk or egg products, if the concentration of either exceeds 0.25 mg/l. The current EU wine labelling regulations (EC) No: 2019/33 state the wording and form of allergen labelling information [10]. Wine marketed in the European Union remains exempt from the compulsory ingredients listing that was introduced for food products in 2011: EU Regulation No.: 1169/2011 [9].

1.6 Faults and Taints

Generally in the food and drink industries, a distinction is conventionally made between the terms 'fault' and 'taint'. A taint may be defined as an 'unwanted and unacceptable odour or flavour; a contaminant derived from an external source including the environment and packaging'. Equipment used in wine production processes, additives, and processing aids (e.g. bentonite), the winery atmosphere, transport containers, and packing materials including cork bottle closures can be sources of taints. Rather simplistically, the ISO defines a taint as a 'taste or odour foreign to the product originating from external contamination' [11]. Conversely, a fault may be considered to be an internal chemically or microbiologically produced off-odour, off-flavour, or cause of product deterioration. Oxidation and the off-flavours from fermentative sulfur compounds are examples of faults. Biological and enzymatic degradation of compounds such as fatty acids frequently result in off-flavours. Whilst faults may be due to poor or careless winemaking, care and diligence throughout the production, and packaging processes are also necessary to minimise the risk of taints.

However, the technical distinction between 'faults' and a 'taints' is not always clear, and restricting the use of the words in a such narrow manner makes little sense. For example, so-called 'smoke taint' affects grapes, being adsorbed onto the skin of berries and translocated to grapes via leaves. However, it is the compounds that are created (metabolised) during winemaking that give smoke-taint associated off-odours and flavours. The so-called 'rogue' yeast *Brettanomyces* produces compounds that may give off-odours and flavours, and the words contaminant and taint are often used in this regard. For example Manuel Malfeito-Ferreira talks of the 'horse sweat' taint when reviewing the impact of *Brettanomyces* [12]. In a restaurant, the diner rejecting a wine that exudes the damp, musty odours of the TCA compound, always derived from an external source such as a cork closure or even the cellar atmosphere, will inform the sommelier or waiter that it is faulty (or corked), and is most unlikely to use the descriptor 'tainted'. In fact, at low levels, TCA compounds will not actually taint a wine, but will strip it of fruitiness and result in tasting very flat. Accordingly, a wine contaminated with TCA, other than at miniscule levels of concentration, will *not* always be tainted, but will *always* be faulty. Further, many researchers use the word 'taint' in papers when discussing what are technically faults, as do many professional wine writers and authors. David Bird MW, speaks of 'reductive taint' in his book *Understanding Wine Technology* (3rd Edition), which is used by many wine students worldwide studying for the internationally recognised Diploma examination of the *Wine and Spirit Education Trust* [5]. Dr Eric Wilkes of the Australian Wine Research Institute (AWRI) refers to reduced aromas, oxidation, and *Brettanomyces* as taints [13]. Accordingly, I consider that the word 'fault' should be regarded as an all-encompassing descriptor for something that is wrong with a wine (to include any taint or other contamination) that results in a significant off-odour or flavour, a marked reduction of quality, or is adverse to human health. For the sake of both clarity and simplicity, it is generally used as such in this book.

1.7 Distinguishing Between Faults and Flaws

In a tasting assessment, the condition of a wine may be described as

- Fault-free (sound);
- Flawed (showing minor defects);
- Faulty (showing one or more serious defects).

As stated, the use of the word 'fault' should be reserved for major defects, including those off-odours and off-flavours that have a significant organoleptic impact upon a wine, or for compounds that may cause accelerated deterioration, or are harmful to human health. Where the impact is minor, including defects that result in a reduction of typicality (the word tipicité is often used amongst wine lovers and critics), or a modest reduction in quality or ageing potential the term 'flaw' is generally more appropriate, although some authors do regard a lack of typicality of style as a fault [14]. The level, or concentration, of causal compounds is obviously key here, but whilst the physical level is generally relatively easy to quantify by laboratory analysis, the sensory impact upon wines of differing styles and aroma and flavour matrices may be less easy to qualify: much will depend upon the style and matrix of the wine. So the boundaries between faulty, flawed, and 'in good condition' may, on occasions, be somewhat blurred. However, there are some compounds which, even if present at a low level, cause a such a reduction in organoleptic attributes and a loss in quality, that affected wines must always be regarded a faulty. TCA and other haloanisoles (see Chapter 3) are classic examples.

Whilst there is little doubt that a wine contaminated from an external source should always be regarded as faulty, the internally produced 'off'-aromas and flavours may be subject to dissent as to the concentration at which they become unacceptable and are considered to be flaws or faults. When tasting a wine, the judgments made are, to a large degree, subjective. Individuals have varying sensitivities, responses and reactions to aromas, odours and flavour compounds based on their culture and education [15], experience, and age [14]. Members of a panel of professional tasters and critics may be unanimous in their judgements when assessing a wine or there may, on occasions, be out and out dissent. There can also be disagreement between professionals and consumers as to what constitutes a fault or flaw, for example, the acceptability or otherwise of sediments. Precipitated crystals of potassium bi-tartrate or calcium tartrate are sometimes found in the bottom of wine bottles and, of course, these will often appear in glasses of wines when poured. They are most likely in high acidity wines from cool climates, but can appear in reds too. The crystals are harmless and have no negative organoleptic impact, but to the consumer they may be cause for concern and even rejection of the bottles in question. Accordingly, although many industry professionals are unconcerned by tartrates, some do consider them to be an 'appearance' fault (see Chapter 15).

1.8 Sensory Detection (Perception) Thresholds and Sensory Recognition Thresholds

1.8.1 Sensory Detection Thresholds

There are many compounds that may give rise to faults but, with the notable exception of TCA and other haloanisoles, their presence usually only becomes important when their concentration is at or above, or has the potential to reach, their individual sensory detection thresholds. To complicate matters, there are also some compounds which, even if below their individual sensory detection thresholds, may produce off-odours or-off-tastes if present with other compounds, which may also be below their individual sensory detection thresholds.

Sensory perception thresholds may be divided into odour detection thresholds (ODTs) and taste detection thresholds (TDTs), but many authors and researchers do not distinguish between these. A simple definition of a sensory perception threshold is 'the level at which the aroma, flavour, taint, or fault will be detected by the 50% of the general public'. An alternative definition is 'the lowest concentration at which individuals can reliably perceive a difference between a sample and its corresponding control, with 50% performance above chance' [13]. Of course, individuals vary markedly in their perception thresholds to aromas and tastes. Threshold values may be stated for detection in air, water, wines generally, red, white, or sparkling wines, and even wines made from individual grape varieties or in different styles, such as light- or full-bodied. However, the determination of such thresholds is perhaps far from a precise science, and individual research papers will often state markedly differing figures. The 'Triangle Test' is one widely used method of determining the sensory detection thresholds of compounds in wine and foodstuffs generally and is included in the methodology of the *Society of Sensory Professionals*. In this method members of a panel are presented with three samples: one different from the two identical. All three samples should be presented to each panel member at the same time, and the samples are tasted from left to right. There are six possible order combinations, and these should be randomised across panel members [16]. There is an ISO standard for measuring sensory detection thresholds by a three-alternative forced-choice (3-AFC) procedure: ISO 13301 (2018) [17]. ASTM International, formerly known as the American Society for Testing and Materials (ASTM) has produced document E679-04 (2011): Standard Practice for Determination of Odor and Taste Thresholds by a Forced-Choice Ascending Concentration Series Method of Limits [18].

1.8.2 Sensory Identification (Recognition) Thresholds

Whilst an ODT is the lowest concentration at which a particular odour is perceivable upon nosing, the identification of the odorant responsible may not be achieved at that level. The 'odour identification (or recognition) threshold' is the concentration at which a particular odorant is not only detected, but is also recognised and can be named. ISO 5492 defines and odour identification threshold as the 'minimum physical intensity of a stimulus for which an assessor will assign the same descriptor each time it is presented' [11]. However, different compounds can give arise to aromas which may have a similar or possibly the same descriptor, or be difficult to distinguish, even by trained, expert tasters. As with detection thresholds, recognition thresholds may vary according to the matrix of the wine and the interplay between numerous compounds. Most research papers only refer to detection thresholds when discussing individual odorants, and accordingly the thresholds detailed in this book are generally those of detection, not identification.

1.8.3 Odour Activity Values

The levels at which a fault compound in wine is detected and identified will very much depend upon the wine matrix. Additionally, there may be multiple compounds responsible for an off-aroma or flavour in a wine, and the threshold for determining the multiplicity of these may differ from that of the individual components. This is particularly so in the case of the compounds metabolised by yeasts of the genus *Brettanomyces*, as discussed in Chapter 4. Depending upon the fault in question, sensory detection thresholds for fault compounds may be measured in concentrations of milligrammes per litre (mg/l), which equates to parts per million (ppm), microgrammes per litre (µg/l), which equates to parts per billion (ppb), or even nanogrammes per litre (ng/l), which equates to parts per trillion (ppt). The human nose can actually detect some odours at levels of picogrammes per litre (pg/l) equivalent to parts per 1000 trillion, and possibly and incredibly at concentrations measured in femtogrammes per litre (fg/l). A femtogramme is 0.000 000 000 000 001 of a gramme! Only volatiles smell – we cannot smell most wine acids, with the exception of the volatile acids (mainly acetic acid).

Usually the greater the concentration of a compound, particularly if above detection threshold, the more it will impact upon odour. Accordingly some researchers consider 'odour activity values' (OAVs), which may be defined as the concentration of a compound present in a matrix divided by the ODT for that compound in that specific matrix [13]. Of course, the higher the OAV the more the compound will usually also impact the palate or taste profile, bearing in mind that most of the taste of a wine is derived from retronasal sensations. Of course, the strength of an odour is not simply a matter of is OAV, and strong odours can mask weaker odours.

1.9 Consumer Rejection Thresholds (CRTs)

There is often a marked difference between the concentration of a compound at which a consumer may be able to perceive a fault – the consumer detection threshold (CDT), and the level that would lead them to reject the wine – the consumer rejection threshold (CRT). With all consumer goods, the buyer is often prepared to accept minor blemishes and per-formance idiosyncrasies as long as the goods remain fit for purpose – in the case of wine, the consumer may accept it as fit for purpose as long as it is drinkable and has an element of enjoyment. Of course, consumers frequently reject wines that are simply not to their taste, or are of general low quality, and are often confused as to whether an individual aroma or taste characteristic should, or should not, be present. Visible faults such as hazes and the aforementioned tartrate crystals are very likely to lead to rejection, and on occasions sound wines may be rejected simply due to careless or inappropriate handling. Mature red wines very often throw sediment in the bottle which, in the absence of decanting, may become suspended in the wine or fall to the bottom of the glass when the wine is poured.

1.10   Basic Categories of Wine Faults

1.10.1   The Origin of Wine Faults

There are three basic categories of wine faults, according to their nature:

- Microbiological
- Chemical
- Physical

However, the defining of some faults as being of microbiological or chemical origin or nature is not necessarily straightforward: chemical reactions consequential to microbiological synthesis may produce fault compounds, and some faults, e.g. excess acetaldehyde and oxidation, may be formed by microbiological activity and/or chemical reactions.

I will briefly examine each category, and consider factors that may lead to, or help prevent related faults.

1.10.2   Microbiological Faults

1.10.2.1   Types of Microorganisms Involved

This group comprises faults that result from the actions of microorganisms; these may be grouped as (i) yeasts, (ii) bacteria, and (iii) moulds. It is perhaps worth looking briefly and very generally at each of these.

Yeasts   Yeasts are single celled microbes, belonging to the fungi kingdom. The genetic content of a yeast cell is contained within a nucleus which is enclosed within a nuclear membrane – this classifies them as eukaryotic organisms, unlike their single-celled counterparts, bacteria, which do not have a nucleus and are considered prokaryotes. There are approximately 1500 species of yeast. Yeasts mostly reproduce by budding, usually multilateral budding where buds appear from different points in the shoulder of the cell, but polar budding, where the buds repeatedly grow from the same site are common for some 'wild' yeasts, including the genus *Kloeckera* [19].

The main species responsible for fermentation of grape must into wine is *Saccharomyces cerevisiae*. Most non-*Saccharomyces* yeasts do not ferment to a high alcoholic degree, perhaps 4–5% abv. Although present in relatively low quantities on grape skins, together with many other yeasts and microflora, *S. cerevisiae* becomes dominant in the vast majority of fermentations. It is the main 'non-spoilage' yeast species that is able to produce high levels of alcohol and survive in such a hostile environment. It has the ability to consume all the sugars generally present in grape must (other than must from late harvested or botrytised grapes destined for sweet wines) and its predominance minimises the risk of sluggish or stuck fermentations. It is also particularly associated with enzymatic activities involved in the transformation of aromatic precursors contained in grapes [20]. Many winemakers use laboratory cultured strains of *S. cerevisiae* for the alcoholic fermentation: there are approximately 700 strains, each with different behavioural and flavour characteristics. However, there are a great many (indeed a growing number of) winemakers particularly at smaller properties who prefer to leave their fermentations to the various natural yeasts present on the grapes and in the winery. Of course, this as was always the case until the last 30 or 40 years. Many artisan producers and wine lovers regard natural yeasts as being an extension of 'terroir'. Whichever is their choice, winemakers usually try to ensure that *S. cerevisiae* is the species that dominates in the alcoholic fermentation process, and other species of yeasts and other microorganisms (with the exception of certain lactic acid bacteria) may be suppressed. However, for some wines, there are other *Saccharomyces* species that may be encouraged or even inoculated, including *Saccharomyces bayanus* (which is also often used for the 'prise de mousse' or second [bottle] fermentation of Champagne and other quality sparkling wines) and *Saccharomyces beticus*, being one of the key yeasts that produces the 'flor' essential for the production of *Fino* Sherries. Very occasionally, and particularly with producers who are thinking and working 'outside of the box', there are other genera and species that may be encouraged or co-inoculated, including *Lachancea kluyveri*, *Lachancea thermotolerans*, *Metschnikowia pulcherrima*, and *Torulaspora delbrueckii* [21]. However, non-*Saccharomyces* species can rise to off-odours and flavours and produce compounds that may lead to faults or spoilage, so extreme care must be taken when using these, whether they are inoculated or naturally present on grapes or in the winery. In particular, *Brettanomyces bruxellensis* and *Brettanomyces anomala* are regarded by most oenologists as major spoilage organisms in wine, although the aromas and flavours metabolised by these yeasts can, at low levels and in some wine matrices, add interest and complexity. Several 'film-forming' yeasts may grow on the surface of wine in the presence of oxygen, such as will be the case in ullaged vats or barrels. These too can result in off-odours and flavours and even spoilage. Figure 1.1 shows film-forming yeasts on the surface of wine in a small vat, when the 'floating' lid had been carelessly fitted a few centimetres above the surface.

Figure 1.1 Film forming yeasts on wine surface.

Moulds   Moulds are filamentous fungi that grow as hyphae, which are multicellular filaments. A mould is a member of one of two distinct groups of fungi:

- *Ascomycota* (sac fungi) – there are some 64 000 species;
- *Zygomycota* (bread moulds) – there are some 1000 species.

In order for moulds to grow, they require food in the form of organic matter, moisture, warmth, and oxygen. Certain moulds are implicated in spoilage and product deterioration, and filamentous fungi have a major role in the formation of haloanisoles, which give rise to so-called 'corkiness' that can render affected wines undrinkable.

Bacteria   There are more species of bacteria than any other form of life. Bacteria can be single or multi-celled microbes. Unlike yeasts, they do not have a nucleus and are classified as prokaryotes. A simple way of describing bacteria is by their morphology, i.e. their shapes. There are five groups of shapes; the two most important groups that may be found in wine are cocci that are round or oval shaped and bacilli that are rod shaped. These descriptors often appear as part of genus names. Some species of bacteria exist as individual cells; others group themselves together in pairs, rods, chains, or clusters.

Many species of bacteria grow only in aerobic conditions – these are termed obligate aerobes, and include the dreaded *Acetobacter*, which can turn wine into vinegar. There are also species, termed obligate anaerobes, which grow in anaerobic conditions. The presence of oxygen poisons some of the key enzymes of obligate anaerobes, so they will not grow in aerobic conditions. There are some obligate anaerobes which require a high level of carbon dioxide for growth, as will be the case during fermentation. Facultative anaerobes are organisms that manufacture adenosine triphosphate (ATP) by aerobic respiration in the presence of oxygen, but are able to change to fermentation or anaerobic respiration in the absence of oxygen. Bacteria that grow in an acid environment are termed acidophiles, but these are very rarely present in grape musts and wines as the pH is invariably higher than they can tolerate.

The pH of musts and wines varies considerably between extremes of pH 2.7 and 4.3 depending, inter alia, upon growing conditions including soil type, aspect, climate, and weather in the growing season. As grapes ripen, pH increases. A high pH will not only have a negative influence upon the taste profile, but increase the risk of bacterial growth. Although there are exceptions, the ideal pH of must for white wines lies in the range of pH 2.9–3.4, and for red wines pH 3.3–3.7, although in recent years this latter figure is often exceeded. Generally speaking the lower the pH the less the risk of the growth of unwanted microorganisms, but there are some bacteria, including acetic acid bacteria, that may grow in wines with a low pH.

Whilst the different species of bacteria vary considerably in the other conditions in which they thrive, they all require food, water, and suitable environmental conditions. During grape growing and winemaking bacteria have numerous sources of nutrients, including sulfur, carbon (particularly six carbon sugars), nitrogen, and phosphorus. Although, with some notable exceptions, sugars are largely consumed by yeasts during fermentation, wines that have completed all production processes will contain other nutrients, including typically between 80 and 230 mg/l of phosphorous. Whilst generally bacterial growth is slower at low temperatures, most species flourish in the wide range of 5 °C (41 °F) and 60 °C (140 °F).

Pre-fermentation heat treatments of red musts, including pre-fermentation hot maceration, flash détente, and thermo détente which, depending upon the individual process, involve heating the must to between 75 and 85 °C (158–185 °F) will kill any bacteria present at that stage, as will pasteurisation, which may be also undertaken when the finished wine is bottled.

Although many species of bacteria are implicated in microbiological wine faults some, particularly certain lactic acid bacteria of the genus *Oenococcus*, perform a most useful role in the winemaking process, being responsible for the MLF, which leads to textural changes that are almost always desirable for red wines. The MLF is often undertaken for certain styles of white and sparkling wines too, but whether or not it is desired it is crucial that it never spontaneously takes place subsequent to bottling – see Chapter 9.

1.10.2.2 Examples of Microbiological Faults

Faults of microbiological nature that may be found in wine include

- Contamination with haloanisoles – see Chapter 3;
- *Brettanomyces* related faults – see Chapter 4;
- Excess volatile acidity – see Chapter 7;
- Lactic bacteria and *Pediococcus* related faults – see Chapter 11.

1.10.2.3 Minimising the Occurrence Microbiological Faults

The key tools in minimising the risk of the occurrence of microbiological faults are as follows.

In the vineyard:

- Creating and maintaining open leaf canopies to help air-flow;
- Controlling pests such as European grapevine moth (*Lobesia botrana*), and vinegar fly (*Drosophila melanogaster*);
- Preventing or controlling vine diseases such powdery mildew and downy mildew;
- Preventing or controlling grape rots, such as *Botrytis cinerea*;
- Picking only healthy and undamaged fruit;
- Picking at an appropriate pH;
- Harvesting as cool as possible and transporting to the winery without delay.

In the winery and cellar:

- Sorting fruit to exclude rotten and damaged berries and materials other than grapes (MOGS);
- Maintaining an appropriate pH in must and wine;
- Scrupulous winery and cellar hygiene;
- Using commercial preparations of *S. cerevisiae* for alcoholic fermentations and lactic acid bacteria strains for the MLF.
- Careful oxygen management in wine – e.g. avoiding ullage in vats and barrels;
- Controlled temperatures for fermentation, maturation, and storage;
- Creating and maintaining a nutrient desert [22];
- Controlled humidity in the winery (maximum 75%) and barrel store (maximum 80%);
- Maintaining an appropriate level of molecular sulfur in wine.

1.10.3 Chemical Nature Faults

1.10.3.1 Examples of Chemical Faults
Chemical faults result from unwanted chemical changes in wine and may be due to internal or external factors. Two of the most common faults of chemical origin are chemical oxidation and, conversely, reduced aromas or reduction. There are thousands of chemical reactions that take place during the winemaking processes; some of these can result in the synthesis of compounds noted for their off-odours and flavours, whilst others give simple changes, the impact of which may include immediate or rapid product deterioration.

Faults that are generally regarded as being primarily of a chemical nature include

- Excessive acetaldehyde – see Chapter 5;
- Chemical oxidation – see Chapter 5;
- Reduced aromas/reduction – see Chapter 6;
- Iron haze and copper haze – see Chapter 10;
- Eucalyptol – 1,8-cineole – see Chapter 12;
- Smoke taint related compounds, including guaiacol, 4-methyl-guaiacol, 4-methyl-syringol, *m*-cresol, *o*-cresol, and *p*-cresol – see Chapter 12;
- Brown marmorated stink bug related compounds, including *trans*-2-decenal – see Chapter 13.

1.10.3.2 Minimising the Occurrence Chemical Faults
The key tools in minimising the risk of the occurrence of chemical faults are as follows.

In the vineyard:

- Maintaining sufficient, but not excessive, soil nitrogen levels;
- Discontinuing the use of all chemical contact and systemic treatments well before harvesting;
- Avoiding using agrochemicals that may break down into unwanted odours compounds such as 2,4-dichlorophenol and store all agrochemicals away from possible sources of contamination;
- Minimising sources of aerial pollution, including the burning of winery rubbish;
- Minimising the vineyard presence of pests, particularly ladybug (*Coccinellidae*) and brown marmorated stink bug (*Halyomorpha halys*).

In the winery and cellar:

- Careful oxygen management – e.g. avoiding ullage in vats and barrels and using gases to sparge equipment, but ensuring that adequate levels of dissolved oxygen are maintained;
- Avoiding pick-up of metals, particularly copper and iron – these metals can be contaminate wine from inappropriate or damaged equipment, particularly valves and fittings;
- Maintaining adequate levels of molecular SO_2 in wine;
- Avoiding possible contact with sources of chemical contamination, e.g. damaged tank linings.

1.10.4 Physical Faults, Contamination, and Packaging Damage

Some physical faults are as a result of chemical or microbiological factors. Packaging damage or failure, sometimes due to poor storage conditions, can have devastating consequences. By way of a very simple example, a loss of bubbles (probably accompanied by oxidation) in a Champagne or other sparkling wine due to shrinkage or other failure of the cork closure would be regarded as a most serious fault. In fact a cork 'champagne' stopper does not provide a complete hermetic seal. Some 30% of CO_2 may be lost in Champagne aged for 75 months at 12 °C (less in the case of magnum [1.5 l] bottles and more in the case of half [37.5 cl] bottles) [23]. Wine may be contaminated with a variety of objects, including flies and other insects, pieces of metal, filter materials, and oil. Whilst some of the contaminants are likely to affect bottles on an individual basis, others may require the recall of an entire batch. Occasionally pieces of glass may enter the bottle, due to problems with filling heads, the jaws of the corking or closure unit, or bottle manufacture. Wine may also become contaminated with paint, resins (including epoxy resin), brine, glycol, metals, and other substances due to damage in tanks, pipes, hoses, or other production equipment. Physical faults, physical contamination, and packaging damage are issues that affect the food, drink, and many other industries at large, and as such are not discussed in detail in this book. However, there are faults that are, or may be, related to packaging, e.g. chloroanisole contamination, that have a major organoleptic impact and these will be covered in depth.

1.11 Flaws

1.11.1 Poor Wines, as a Consequence of Adverse Weather, Sub-standard Viticulture, Careless Winemaking, or Inappropriate Additives

Of course, there remains the question as to wines that are simply poor. A wine that is made from unripe grapes will have few attributes and may well be thin, highly acidic, lacking in fruit or varietal character and show no hints of complexity. Overripe grapes may result in a wine that is highly alcoholic, lacking in acidity, flabby, and showing baked, burnt and dried-out tones and with reduced potential for ageing. Yet, poor as these wines may be they cannot be regarded as faulty, and it is appropriate to consider them to be flawed. In fact, any individual factor or any combination of factors that contributes to a wine being unbalanced is without doubt a flaw. The topic of 'balance' in wine is discussed briefly in Chapter 14. However, yet again the boundary between what is a fault or a flaw may often be soft. A wine made from *grapes* affected by downy mildew, commonly known as Peronospera, (taxonomic name: *Plasmopara viticola*), or powdery mildew, commonly known as Oïdium, (taxonomic name: *Erysiphe necator* [also known as *Uncinula necator*]), or by *B. cinerea* in its unwelcome form of 'grey rot', may result in unwanted off-odours and off-flavours and may be regarded as flawed, or if the off-odours are totally unacceptable, as faulty. *U. necator* can impart unacceptably high levels of the compounds 1-octen-3-ol and/or 1-octen-3-one in wine (see Chapter 3), but fortunately most of this is enzymatically reduced during the alcoholic fermentation process. Of course, in the case of many of the world's great sweet wines both over-ripeness and *B. cinerea* (in this case in the form of 'noble rot') are sought after, and the aroma and taste characteristics imparted as a result of the chemical changes consequential to berry infection by the fungus can send a delightful shiver down the spine. The careless or heavy-handed use of some additives, for example, oak products in the form of powder, chips, or beans, can result in unwanted aromas, flavours, and even textural changes that can render a wine to be flawed. Wines showing a lack of typicality and, particularly in the case of protected designation of origin (PDO) wines, an absence of 'the sense of place', may also be regarded as flawed, although such a judgement is somewhat controversial. The 2012 vintage of 'Les Hauts de Pontet-Canet', the 'second' wine of the flagship biodynamic Pauillac cru classé estate Château Pontet-Canet, was denied its Appellation Controlée (AC) status by the authorities, allegedly on the grounds of lack of typicality, which led to the assessment panel considering the wine to be flawed. The owners had to accept the humble 'Vin de France' designation for the wine, although knowledgeable consumers were more than happy to purchase the 'declassified' product which, by no stretch of the imagination of anyone with an in-depth knowledge of wine could be regarded as even slightly flawed.

1.11.2   The Presence of 'Fault' Compounds at Low Concentrations

The presence in wine of certain compounds at anything above trace levels, even if the concentration is below ODT or TDT, will always result in a wine a wine being faulty. TCA is a classic example. However, particularly at low concentrations, some 'fault' compounds may have a minor negative impact upon the nose or palate, but this may be balanced out or even outweighed by other positive attributes they exhibit or that are present in the wine, which remains eminently drinkable. The compounds metabolised by *Brettanomyces* yeasts are one example, and their impact at differing concentrations very much depends upon the structure and matrix of the wine.

1.12   The Incidence of Wine Faults

Should there be any illusion that wine faults are a rare occurrence, at least insofar as faulty bottles reaching the consumer is concerned, let us look at the incidence as evidenced by wines entered into a major wine tasting competition. The 'International Wine Challenge' is one of the world's best known annual wine tasting competitions that now attracts over 10 000 entries a year. The AWRI has reviewed wines submitted for this competition but rejected as faulty during a period of 10 years from 2007 to 2016, and quantified these as to the category of fault, and also type of closure [24]. The total number of entries reviewed was 106 627. The figures regarding faults make shocking reading. Some 3.7% of the entries were faulty. Three-quarters of the faulty wines (2.8% of the entries) showed one of three most common faults – oxidation (1.0% of the entries), so-called 'cork taint' (0.9%) and reduction/sulfides (0.9%). Alarmingly, for the lovers of aged fine wines, over 7% of the entries that were over six years old exhibited a fault, with oxidation being the most common, although excessive volatile acidity is also noted. Whilst it is obviously not possible to extrapolate these figures onto a global scale, one might estimate that, based on OIV figures for the years 2007–2018, when global wine sales annually averaged 243 million hectolitres [25], each year

Table 1.1 Percentage (if over 2%) of faulty wines submitted for panel tastings in two leading wine publications.

Category of wines tasted	Date	No. tasted	No. faulty	%
Publication: Decanter				
Affordable Rioja	March 2020	183	4	2.2
Priorat	November 2019	82	2	2.4
Rosso di Montalcino	September 2019	93	2	2.2
Value Douro Reds	July 2019	76	2	2.6
South American Premium Red Blends	June 2019	98	3	3.1
Chilean Sauvignon Blanc	October 2018	62	2	3.2
South American Cabernet Franc	October 2018	44	1	2.3
Gualtallary and Altamira Reds	October 2017	99	2	2.0
New Zealand Sauvignon Blanc	August 2017	93	4	4.3
Southern Rhône Whites	August 2017	96	2	2.1
Piedmont Nebbiolo	July 2017	72	2	2.8
Muscadet with extended lees ageing	June 2017	113	3	2.7
Bordeaux Crus Bourgeois	April 2017	203	5	2.3
Publication: The Word of Fine Wine				
Historic Non-Vintage Champagne	Q3 2019	44	3	6.8
Brunello di Montalcino 2012	Q2 2017	34	1	2.9
White Châteauneuf-du-Pape	Q1 2017	44	1	2.3
Champagne 2008 and 2009	Q4 2016	53	2	3.8
Savennières	Q2 2016	27	1	3.7

the equivalent of some 1.2 billion 75 cl bottles of faulty wine are sold, and presumably the vast majority of these are consumed – a particularly sobering thought!

Let us also glance at some figures for wines ascertained to be faulty wines when sent for ranking in tasting assessment features in two UK published wine magazines: 'Decanter' and 'The World of Fine Wine'. As detailed in Table 1.1, there were many instances when over 2% of the wines submitted were detected as being faulty. The publications named the wines they found to be faulty, and one can only speculate upon the impact that such naming had upon the producers.

It should be noted that there were also several wine categories assessed when none, or less than 2%, of the submitted samples were found to be faulty. However, the price points for the wine types detailed are generally relatively high, and if consumers find their purchases to be faulty they might well feel particularly aggrieved when discovering that the product in their glass is not worth anything, let alone the price paid.

1.13 'Faulty' Wines that Exude Excellence

There are many wines, particularly those at so-called 'entry level' that are technically correct, fault and flaw-free but which are distinctly uninteresting, dull, and by any set of values cannot be regarded as high quality. Such wines usually do not claim to express any regional identity, although they should be largely true to the characteristics of the grape variety or varieties named on the label. But how should we regard wines, often from small, artisan producers that are technically faulty, but which are complex, highly exciting and exude style and class and individuality? Our judgements are usually framework dependent, but excellence in wine, as perhaps in music, art, comedy, or literature can take us outside of our comfort zones where all aspects adhere to the rules. Breaking the 'you can't do that' norm may lead us into new territories that prove as exhilarating as they are disturbing. Nicolas Joly is a famous, perhaps even infamous, biodynamic producer in the Anjou district of France's Loire Valley. His family are the sole owners of La Coulée de Serrant, a seven hectare vineyard in the Savennières commune, which has its own Appellation Controlée (Protegée). He produces two other wines: 'Le Clos de la Bergerie', appellation Savennières-Roche aux Moines, and 'Les Vieux Clos', appellation Savennières. He is noted for breaking many wine-making rules but produces award-winning wines that have been described as amongst the world's best dry white wines, and yet which are regarded by some experts as faulty. He is revered by many wine lovers and reviled by many oenologists. Of course he believes great wine is 'made in the vineyard', and it is a straightforward task to turn wonderful grapes into wine. 'What is a winemaker?', he rhetorically asked me – 'I don't understand the meaning of the word' [personal discussion]. When describing a Nicolas Joly wine, the 2005 vintage of Clos de la Bergerie, Savennières-Roche-aux-Moines, Lisa Perrotti-Brown MW noted (in 2015) 'ethyl acetate, acetaldehyde and high volatile acidity' along with 'beautiful, rich, mature fruit' – she stated that 'the wine tasted delicious' [26].

1.14 Final Reflections

Many wine lovers express unhappiness because, coinciding with the technological and scientific advances of recent years, wines have become more standardised and globalised. Regional characteristics that were once so distinctive have very often become blurred. As Clark Smith states in his book *Post Modern Winemaking*, 'replacing empirical systems with theoretical methods devalues hundreds of years of specific knowledge and practice, tending to bring a squeaky-clean sameness to all wine' [20]. To many in the industry, quality in wine may be defined as 'an objective standard of excellence with the absence of any faults'. The drive of producers to produce 'high-end', consistent quality wines has perhaps paradoxically resulted in real quality being compromised. Wine is an agricultural product, subject to the vagaries of all that the natural world can present which, of course, gives the varying styles and qualities of each vintage. So is total consistency really desirable or possible? Achieving consistency may require a good deal of blending and technical intervention which can negate the sense of place and individuality of the wine in the glass. In other words, to be consistent the peaks of excellence achieved in some vintages would have to be flattened or rounded down to a repeatedly achievable standard. The words 'authentic', 'local', 'artisan', and 'natural' can be music to the ears of those who seek aromas and flavours that are as exciting as they are unpredictable. Perhaps the odd flaw may add to this excitement, and the acceptance of the occasional fault is a price worth paying for such individuality. If we examine a work of art, or read a book, and judge it to be perfect, the chances are that on each subsequent visit we perceive a flaw or two. Maybe these do not detract, but raise questions and make us want to come back again.

References

1 Anderson, K., Nelgen, S., and Pinilla, V. (2017). *Global Wine Markets, 1860 to 2016: A Statistical Compendium*. Adelaide: University of Adelaide Press. https://doi.org/10.20851/global-wine-markets License: CC-BY 4.0.
2 Statista Research Department (2015). Global per capita wine consumption 2014/2018, by country. https://www.statista.com/statistics/232754/leading-20-countries-of-wine-consumption/ (accessed 23 November 2019).
3 International Organisation of Vine and Wine (2019). State of the Vitiviniculture World Market. http://www.oiv.int/public/medias/6679/en-oiv-state-of-the-vitiviniculture-world-market-2019.pdf (accessed 24 January 2020).
4 Urkiage, A., de las Fuentes, L., Acilu, M., and Uriarte, J. (2002). Membrane comparison for wine clarification by microfiltration. *Desalination* 148: 115–120. https://doi.org/10.1016/S0011-9164(02)00663-X.
5 Bird, D. (2010). *Understanding Wine Technology*, 3e. Nottingham: DBQA Publishing.
6 International Agency for Research on Cancer [IARC], World Health Organisation (2019). List of classification. https://monographs.iarc.fr/list-of-classifications (accessed 1 June 2020).
7 European Commission (2006). Commission Regulation (EC) No 1881/2006 of 19 December 2006 setting maximum levels for certain contaminants in foodstuffs (Text with EEA relevance). EUR-Lex - 02006R1881-20140701 - EN - EUR-Lex (europa.eu) (accessed 20 March 2020).
8 Chatonnet, P., Boutou, S., and Plana, A. (2014). Contamination of wines and spirits by phthalates: types of contaminants present, contamination sources and means of prevention. *Food Additives and Contaminants A* 1 https://doi.org/10.1080/19440049.2014.941947.
9 EUR-Lex (2011). Regulation (EU) No 1169/2011 of the European Parliament And Of The Council of 25 October 2011 on the provision of food information to consumers, amending Regulations (EC) No 1924/2006 and (EC) No 1925/2006 of the European Parliament and of the Council, and repealing Commission Directive 87/250/EEC, Council Directive 90/496/EEC, Commission Directive 1999/10/EC, Directive 2000/13/EC of the European Parliament and of the Council, Commission Directives 2002/67/EC and 2008/5/EC and Commission Regulation (EC) No 608/2004. *Official Journal of the European Union*. L 304: 18-63. https://eur-lex.europa.eu/legal-content/EN/TXT/PDF/?uri=CELEX:32011R1169&qid=1461072342439&from=EN (accessed 24 January 2020).
10 EUR-Lex (2019). Commission Delegated Regulation (EU) 2019/33 of 17 October 2018 Supplementing Regulation (EU) No 1308/2013 of the European Parliament and of the

Council as regards applications for protection of designations of origin, geographical indications and traditional terms in the wine sector, the objection procedure, restrictions of use, amendments to product specifications, cancellation of protection, and labelling and presentation. *Official Journal of the European Union*, L9/2. Available at: https://eur-lex.europa.eu/legal-content/EN/TXT/?uri=CELEX%3A32019R0033 (accessed 24 January 2020).

11 ISO 5492 (2008). *Sensory analysis – vocabulary*. Switzerland: The International Organization for Standardization. https://www.iso.org/obp/ui/#iso:std:iso:5492:ed-2:v1:en (accessed 5 September 2019).

12 Malfeito-Ferreira, M. (2018). Two decades of "horse sweat" taint and *Brettanomyces* yeasts in wine: where do we stand now? *Beverages* 4 (32) https://doi.org/10.3390/beverages4020032.

13 The Australian Wine Research Institute (AWRI) (2017) AWRI APEC Wine Regulatory Forum, May 11–12, 2017. Slides available at: http://mddb.apec.org/Documents/2017/SCSC/WRF/17_scsc_wrf_013.pdf (accessed 14 December 2019).

14 McKay, M., Bauer, F.F., Panzeri, V., and Buica, A. (2018). Testing the sensitivity of potential panelists for wine taint compounds using a simplified sensory strategy. *Foods* 7 (11) https://doi.org/10.3390/foods7110176.

15 Grainger, K. and Tattersall, H. (2016). *Wine Production and Quality*, 2e. Chichester: Wiley Blackwell.

16 Society of Sensory Professionals (2020). Triangle test. https://www.sensorysociety.org/knowledge/sspwiki/Pages/Triangle%20Test.aspx (accessed 13 June 2019).

17 ISO 13301 (2018). *Sensory analysis - methodology - general guidance for measuring odour, flavour and taste detection thresholds by a three-alternative forced-choice (3-AFC) procedure*. Switzerland: The International Organization for Standardization. https://www.iso.org/obp/ui/#iso:std:iso:13301:ed-2:v1:en (accessed 5 June 2019).

18 E679-04 (2011). *Standard practice for determination of odor and taste thresholds by a forced-choice ascending concentration series method of limits*. West Conshohocken, PA: ASTM International. Available online (paywall): https://www.astm.org/Standards/E679.htm (accessed on 5 June 2019).

19 Fugelsang, K.C. and Edwards, C.G. (2007). *Wine Microbiology – Practical Applications and Procedures*, 2e. New York: Springer.

20 Avramova, M. (2017). Population genetics and diversity of the species *Brettanomyces bruxellensis*: a focus on sulphite tolerance. PhD thesis: Agricultural sciences. Université de Bordeaux.

21 Jolly, N.P., Varela, C., and Pretorius, I.S. (2014). Not your ordinary yeast: non-*Saccharomyces* yeasts in wine production uncovered. *FEMS Yeast Research* 14 (2): 215–237. https://doi.org/10.1111/1567-1364.12111.

22 Smith, C. (2014). *Postmodern Winemaking*. Berkeley: University of California Press.

23 Liger-Belair, G. and Villaume, S. (2011). Losses of dissolved CO_2 through the cork stopper during Champagne aging: toward a multiparameter modelling. *Journal of Agricultural and Food Chemistry* 59: 4051–4056. https://doi.org/10.1021/jf104675s.

24 Wilkes, E. (2017). Is it the closure or the wine? *Wine & Viticulture Journal*: 22–25. https://www.awri.com.au/wp-content/uploads/2015/10/s1879-Wilkes-WVJ-31-6-2016 .pdf.

25 International Organisation of Vine and Wine (2018). OIV statistical report on world vitiviniculture. http://www.oiv.int/public/medias/6371/oiv-statistical-report-on-world-vitiviniculture-2018.pdf (accessed 9 July 2019).

26 Perrotti-Brown, L. (2015). *Taste Like a Wine Critic: A Guide to Understanding Wine Quality*. The Wine Advocate.

2

Wine Tasting

In this chapter, I consider wine tasting in general and how to detect faults and flaws using a structured tasting technique. The concept of primary, secondary, and tertiary aromas is explained. The faults and flaws that are indicated by individual characteristics perceived during a tasting assessment are noted, with pointers to Chapters 3–15 in which they are discussed. Some important odour and flavour compounds and their sensory detection thresholds are detailed, noting that sensory such thresholds vary from wine to wine according to the matrix and, of course, from taster to taster.

2.1 Introduction

There are two ways a fault or flaw in a wine may be detected: by laboratory analysis, or by the organoleptic or sensory method, i.e. tasting. It may be argued that, as wine is made for tasting and drinking, a 'technical' fault, that is not apparent to a trained and experienced taster, is unimportant unless the fault were to develop and manifest itself on a subsequent occasion or lead to product deterioration. Sensory detection thresholds for important 'fault' compounds are discussed in the succeeding chapters in this book. It should be pointed out that published literature and research papers are far from being in consensus on the sensory detection threshold of individual compounds. Some of the disparities may be due to the nature of the panels of tasters: trained, untrained, experienced, inexperienced, age, gender, cultural background, etc. In fact, although one might expect that wine tasting 'experts' would be very sensitive to common aromas and flavours, it has been noted that they can show high olfactory detection thresholds for some key wine compounds, which is far from ideal when assurance of wine quality depends on fault detection at low levels [1]. Another challenge lies in the use and understanding of tasting lexicons, aroma, and taste descriptors. Whether described odour categories are innate or learned depends on the influence of language on odour processing. The question may be raised as to whether olfactory description systems are basic linguistic arrangements based on the taster's experience and exposure to expert lexicons [2].

Even if the primary purpose of a tasting assessment is to determine possible faults, a taster must be able to undertake a complete and thorough assessment of the wine, and the techniques and structure of doing this to ensure that all characteristics are examined, as detailed in this chapter. Some fault compounds, particularly haloanisoles, should result in rejection by a professional taster, at any level. However, there are other 'fault' compounds whose presence, particularly at a low level, may not lead to the rejection of a wine. It will depend upon the matrix of the wine and the relationship with all the sensory characteristics, including the wine's structure and balance. Some characteristics are sought after as part of the identity of certain wines but may be unacceptable in others. For example, the 'struck flint' and smoky notes of benzenemethanethiol, also known as benzene mercaptan, are almost a trademark of a classic Pouilly-Fumé (Sauvignon Blanc) from the Central Vineyards of the Loire valley. Such aromas would be unacceptable in a Vouvray (Chenin Blanc) from Touraine, further down the river, so context is crucial. Of course, wine tasting encompasses a degree, sometimes a large degree, of subjectivity but following a rigorous tasting structure, together with subsequent comparing notes and opinions with colleagues, helps the taster reach an honest and balanced assessment, including whether some characteristics are indeed faults or flaws.

Some of the information in this chapter is based upon content from *Wine Production and Quality* 2nd edn. by Keith Grainger and Hazel Tattersall [3]. This material has been revised and expanded. Many additional topics related to wine tasting are discussed in further detail in that work.

Every wine drinker is, to a greater or lesser extent, a wine taster. However, most wine consumers taste and drink the product without much thought other than whether or not it is enjoyable and meets their expectations. The wine lover gives the wine a more detailed appraisal and considers balance, complexity, and typicality (in the wine world very often referred to as *tipicité*). Neither is looking to find a fault, although if present, it may scream out and prohibit all enjoyment of the wine. The trade student, oenologist, winemaking consultant, or quality control manager have a much more detailed agenda in their tasting assessments and will follow a consistent and rigorous structure. The detection of any fault or flaw is an essential part of such tasting analysis.

The selection of tasters, for making professional judgements will encompass many criteria, including expertise in the types of wines to be assessed. There is an International Organisation for Standardization (ISO) standard for such selection: ISO 8586/2012: General guidelines for the selection, training, and monitoring of selected assessors and expert sensory assessors [4]. This standard is under review at the time of writing and is due to be replaced by ISO/CD 8586. ISO 8586/2012 defines expert sensory assessors as 'selected assessors with a demonstrated sensory sensitivity and with considerable training and experience in sensory testing, who can make consistent and repeatable sensory assessments . . . . .' Key to an individual's suitability is being able to perform with precise accuracy and, crucially, reproducibility. Of course, a long-term sensory memory is essential to make reliable comparative judgements.

2.2 Anosimics, Fatigue Effect, and Supertasters

2.2.1 Anosmics and Fatigue Effect

An anosmic has no sense of smell, and somebody with this condition is sadly precluded from being a capable wine taster. However, even amongst wine lovers and professionals, some people are specific anosmics, i.e. lacking the ability to detect one or more individual odorants. Further, tasters can become victims of fatigue or the adaption effect. When regularly or continuously exposed to particular odours, the sensitivity to them is reduced, sometimes completely lost. Professional tasters should be regularly monitored to ensure they remain sensitive to odorants commonly found in wines, and in particular, to fault compounds.

2.2.2 Supertasters

Some people have tongues with a high density of fungiform papillae and other papillae, which contain the taste buds, making them particularly sensitive to bitter sensations. This is probably on account of genetics. The psychologist Linda Bartoshuk, formerly of Yale University and latterly Presidential Endowed Professor at the University of Florida, defines these people, who perhaps comprise 25% of the population, as 'supertasters'. The work done by Ann Noble, the inventor of the 'wine aroma wheel', at UC Davis also established that there are no 'supertasters in general'. An individual who is a supertaster with one bitter compound, e.g. naringin(e), might be a non-taster with another, e.g. 6-*N*-propylthiouracil (PROP), or caffeine. It should be noted that supertasters do not necessarily make the best wine tasters, for the intense sensations they perceive from bitterness and astringency, impacts on other sensations and perceptions of the balance of a wine.

2.3 Tasting Conditions, Equipment, and Glassware

2.3.1 The Tasting Room

For any detailed and professional assessment, and particularly when checking for faults and flaws, it is important to taste in suitable conditions. The ideal tasting room will have the following characteristics:

- *Large*: Plenty of room is necessary to give the taster his or her personal space and help concentrate on the tasting;
- *Light*: Good daylight is ideal, and the room (if situated in the northern hemisphere) should have large, north-facing windows. If artificial light is required, the tubes/bulbs should be colour-corrected so that the true appearance of the wines may be ascertained;
- *White tables/surfaces*: Holding the glasses over a white background is necessary to assess the appearance and show the true colour and depth of colour of the wine, uncorrupted by surrounding surfaces;
- *Free from distractions*: Extraneous noises are undesirable, and smells can severely impact on the perceived nose of the wines. Tasting rooms should be sited away from the winery, avoiding all the smells of the vat room and cellar, together with kitchens and restaurants. This is vital that when tasting wines for suspected faults, as ambient odours will have a severe impact on the assessment process and sensory detection thresholds. There is no doubt that building materials, decorations, furnishings, and people all exude odours. Indeed, identical wines can be perceived differently according to the surroundings in

which they are assessed. Tasters should avoid wearing aftershaves or perfumes, and smoking should not take place in the vicinity;

- *Sinks and spittoons*: Spittoons, regularly emptied, are essential (see Section 2.3.2.4) and sinks for emptying and rinsing glasses are desirable.

There is an ISO standard for the design of tasting rooms: ISO 8589:2007 Sensory analysis – General guidance for the design of test rooms [5].

2.3.2 Appropriate Equipment

Having appropriate equipment for the tasting is most important. This includes an adequate supply of tasting glasses, water, spittoons, tasting sheets for recording notes, and, at a formal sit-down session, tasting mats.

2.3.2.1 Tasting Glasses
It is important to taste wines using appropriate glasses. Experts do not universally agree as to the detailed design of the ideal tasting glass, but certain criteria are essential. Two of the key characteristics are:

- *Fine rim*: A fine rim glass will roll the wine over the tip of the tongue, whilst an inexpensive glass with a beaded rim will throw the wine more to the centre. The tip of the tongue is the part of the mouth where we most detect sweetness, with other parts of the tongue being less sensitive to this;
- *Cup tapering inwards*: The cup of the glass must taper inwards towards the top. This will develop, concentrate, and retain the nose of the wine in the headspace above the liquid, and also facilitate tilting the glass and swirling the wine. Cut glass is not appropriate for wine tasting, as it is impossible to ascertain the true depth of colour and brightness.

Glasses manufactured to the ISO tasting glass specification ISO 3591:1977 [6] remain popular among many wine tasters, both professional and amateur. The ISO tasting glass is particularly good at revealing those faults perceptible on the nose. The International Organisation of Vine and Wine (OIV) recommends the use of tasting glasses that comply with the requirements of ISO 3591, and that glasses are rinsed in ultra-filtered and deionised water [7]. Whether ISO glasses are the best glass for tasting particular wine types is subject to much dispute. The nose of full-bodied and complex red wines certainly develops more in a larger glass. Wine glass manufacturers, particularly *Riedel* and *Zwiesel*, have designs to bring out the best of individual wine types, so perhaps the only real advantage of the ISO glass is that it is a standard reference.

An appropriate tasting sample is 3–4 cl, which will be sufficient for two or three tastes. Pouring 5 cl into the glasses allows tasters to return for further assessment should this be required. If glasses of a larger size than the standard ISO tasting glass are used, it is appropriate to pour correspondingly more wine.

2.3.2.2 Glass Washing and Storage
Ideally, wine glasses should be washed by hand just in hot water. If the glasses show signs of grease or lipstick, a little detergent may be used. The glasses should be well rinsed with hot water, briefly drained then dried using a clean, dry, glass cloth. The glass cloth must

be previously washed without the use of fabric conditioner in the washing cycle, which can leave the glass with aromas, and a film on the surface. Glass cloths should be changed regularly – perhaps after drying as few as six glasses. The odour of a damp or dirty glass cloth stays in the glass and impacts on the content. Glasses always need to be nosed to check for basic cleanliness and the absence of 'off' aromas before use.

Glasses should not be stored bowl down on shelves, for they may pick up the smell of the shelf and develop mustiness. Standing glasses upright on shelves may lead them to collect dust, so a rack in which glasses are held upside down by the base on pegs is ideal. To be sure that no taint from the glass is transmitted to the wine, it is a good idea to rinse the glass with a little of the wine to be tasted. This is also useful if tasting several wines from the same glass.

2.3.2.3   Water

There should be a supply of pure, still, mineral, or spring water for the taster to refresh the palate between wines, if necessary, for drinking and perhaps rinsing glasses. The variable amount of chlorine contained in most tap water usually makes it unsuitable. Sparkling water is best avoided because the carbonic acid (H_2CO_3) content will impact upon, the assessment of the wine's acidity, sweetness, and balance. Plain, salt-free biscuits such as water biscuits may also be provided, but some tasters believe that these corrupt the palate a little. Plain, low-salt content, soft bread is a possible alternative. Cheese, although sometimes provided at tasting events, should be avoided. The fat it contains will coat the tongue. The protein content combines with and can soften the perception of wine tannins. Any salt content will fight the tannins.

2.3.2.4   Spittoons

Spittoons, placed within easy reach of the participants, are essential at any wine tasting. Depending on the number of attendees and the capacity required, there are many possibilities. In the absence of purpose-made spittoons, wine-cooling buckets will suffice, perhaps with some sawdust or shredded paper in the bottom, to reduce splashing. There are many designs of spittoons suitable for placing on tables and larger units for standing on the floor. Consideration should be given to the construction material. Plastic, stainless steel, and aluminium are all good. Unlined galvanised metal should be avoided at all cost as wine acids can react with the galvanisation and create disgusting aromas. The importance of spitting at wine-tastings cannot be overemphasised. The taster needs to keep a clear head and generally avoid unnecessary ingression of alcohol. Even when wines are spat out, a tiny amount will still make its way to the stomach, and indeed a minute amount will also enter the body via the act of nosing the wines. The reader is advised not to drive after wine tasting, even if all the wines have been spat out.

2.3.3   When to Taste

As to when to taste, the decision is unfortunately often dictated by matters beyond the taster's control. However, the ideal time is when the taster is most alert and the appetite stimulated – namely in the late morning. After a meal is certainly not the best time, for not only is the taster replete and perhaps drowsier (as all early afternoon seminar presenters know), but also the palate is jaded and confused. The constituents of many foods will have an impact, positive or negative, upon taste perceptions.

2.4   The Use of a Structured Tasting Technique and Detection of Faults

A structured tasting technique for wine assessment is valuable in ensuring all the characteristics of a wine are considered by a taster, the members of a panel of tasters, and groups of panels, are using the same criteria in their assessments. The usual sequence of tasting is to examine the appearance of the wine, followed by the nose, then the palate. Conclusions about the wine may be summarised, and these will include considerations as to quality. A fault or flaw may be revealed at any or all stages of the tasting assessment, depending upon the defect in question. For example, oxidation may manifest itself on appearance, nose, and palate. Taints by haloanisoles cannot be seen, but will certainly be smelt and tasted. The flaw of excessive acidity, other than volatile acidity, will not be seen or smelt but will be disclosed upon the palate. It is possible for wines to exhibit multiple faults. Some faults, particularly if at a low level, may not reveal themselves upon first assessment, but become apparent upon returning to the wine after some minutes. In other cases, a characteristic that initially appears to represent a fault may dissipate after the wine has been oxygenated and kept in the glass for a few minutes. The taster should note the indicators of each fault or flaw when suspected or confirmed at each stage of the organoleptic assessment.

2.5   Appearance

2.5.1   The Appearance Assessment

The first stage in a tasting assessment is to examine the appearance. This can reveal much about the wine and give indications as to the origin, style, quality, and maturity, as well as revealing some possible faults. Faults that are or may be detectable on the appearance are shown in Table 2.1. This list is not exhaustive.

Wine appearance should be examined in several ways, including holding the glass at approximately 30° from the horizontal over a white background, a tablecloth or sheet of white paper, as illustrated in Figures 2.1 and 2.2. This will enable the clarity, the intensity

Table 2.1   Some faults detectable on appearance of wine.

Visual indicator	Possible fault	Refer to:
Browning, loss of colour (red wines), 'flat' or dull appearance	Oxidation	Chapter 5
Deepening of colour (white or sparkling wine)	Oxidation Lightstrike	Chapter 5 Chapter 6
Unexpectedly light colour	Atypical ageing	Chapter 8
Unexpected bubbles, sludgy sediments	Fermentation in bottle	Chapter 9
Haze	Presence of yeast or bacteria. Unstable proteins	Chapter 10
Cloudy appearance with brown tinges	Heat damage	Chapter 14

Figure 2.1 A cask sample Margaux – Ch. Margaux 2017.

Figure 2.2 A 40 years old Margaux – Ch. Lascombes 1979.

of colour, and the true colour of the wine to be seen, uncorrupted by other colours in the room. Figure 2.1 shows a vibrant purple cask sample – Château Margaux 2017 (photograph taken in April 2018). Figure 2.2 shows a 40 years old Margaux – Château Lascombes 1979 (photo taken in February 2020).

2.5.2   Clarity and Brightness

A wine that has finished fermentation and been stabilised should be clear and bright. Brightness is a sign of healthiness. Brightness is how light is reflected off the surface of the wine and is related to the pH – the lower the pH, the brighter the wine usually is, as the charge on suspended particles is higher. Clarity is how light is scattered as it passes through the body of the wine, and this is related to turbidity, i.e. suspended particles. If a mature wine that has thrown sediment has been carelessly handled, then there may be fine or larger particles in suspension. This is not a fault per sé, but a natural consequence of the tertiary development of the wine. However, any packaged wine that appears cloudy, oily, milky, or otherwise murky is suffering from one or more faults, for example, refermentation in the bottle – see Chapter 9 – or ropiness – see Chapter 11.

The clarity of wine may vary from *clear* to *hazy*, and the brightness from *bright* to *dull*. Dullness may well indicate a tired or faulty wine. Wines from the New World often appear brighter than those from Europe. Wines that have high acidity, particularly if tartaric acid has been added in the winemaking process, may appear to be especially bright. On occasions, such brilliance might be an early indicator of excessive acidity, but the taster is cautioned against drawing such a conclusion at this early stage. A young wine that appears dull probably has a high pH (low acidity), which is generally indicative of poor quality and total lack of ageing potential. As wines mature, they lose brightness, and with over-maturity, they become dull. Dullness may be one of the first indicators of oxidation.

2.5.3 Intensity

The intensity, i.e. the depth of colour, should be noted. As well as holding the glass at a 30° angle as discussed, the depth of colour can be observed by looking directly down into the wine in the glass that is standing on a white background, as illustrated in Figure 2.3.

The intensity of wine can range from *very pale* to very *deep*. When writing a note, several steps between these extremes that might be detailed, e.g. medium-deep. For novice tasters, it is perhaps useful to know that examining a standard tasting sample (3–4 cl) of a red wine, by holding the glass at 30° over a sheet of white paper containing printed type and the print is clearly visible, the wine may be described as pale or medium-pale. If the print cannot be seen, then deep or very deep might be the appropriate descriptors.

Very often, wines that are pale in intensity will be lighter in flavours and body than those that are deep, but this is not always the case, and sometimes pale wines can often be overtly aromatic. Whites that have had extensive barrel ageing will have a greater visual intensity, although red wines may lose intensity after a long period in barrel. Some red grape varieties such as Merlot and Syrah (also known as Shiraz) can produce deep-coloured wines, especially when from a hot climate, or made with small berries which have a high skin to pulp ratio, or low yielding vines. Conversely, several red grape varieties, including Pinot Noir and, to a lesser extent, Nebbiolo produce wines that are usually not deep-coloured. However, there are some notable exceptions to this, e.g. some Pinot Noirs from New Zealand's Central Otago region. The extraction techniques used, and particularly pre-fermentation maceration (cold soaking) if undertaken, impact upon the colour intensity. However, paleness in red wine, especially a young wine, is often indicative of a lack of concentration, which might be the consequence of many various factors such as a difficult growing season or inadequate extraction. Full-bodied and concentrated reds, especially those from hot climates, will be deep or opaque in youth, and this suggests intense flavours and may be indicative of good quality. The colour intensity of red wines decreases with maturity, but

Figure 2.3 Looking down to see depth of colour.

the opposite is the case with aged whites. However, occasionally extremely deep-coloured red wines can be surprisingly lacking in flavours. There are a few red varieties that have coloured flesh as well as black skins (e.g. Alicante Bouschet) but are surprisingly lightly flavoured. These *teinturier* grapes may, very occasionally, be used in a blend to deepen the colour of a wine that might be perceived as too pale.

Further, intense colours can sometimes indicate winemaking methods that have focused on colour, rather than flavour, extraction. Mindful of the association in the eyes of drinkers (and critics) of deep red wines and perceived high-quality, winemakers can manipulate intensity, for example, by the addition of *8000 color* or *Mega Purple*, concentrates made from grape skins and seeds. Thus, the taster is cautioned against drawing conclusions as to style or quality at this stage. However, a red wine that is paler than expected or white wine deeper than expected may be suffering from oxidation, which will be further evidenced by the colour, nose, and palate.

2.5.4 Colour

Many factors affect the actual colour of a wine, as well as its intensity. These include the climate and region of production, grape ripeness, the grape variety or blend of varieties, vinification techniques including any barrel ageing, and the wine's state of maturity.

2.5.4.1 White Wines

White wines can vary in colour from almost *water clear* to *deep gold* or even *amber.* Words used to describe colour might also include *green, lemon, straw,* and (almost always a signal of alarm) *brown.* Some white varieties such as Sauvignon Blanc generally produce wines at the green-lemon end of the colour range. In contrast, others such as Gewürztraminer (which has a more heavily pigmented skin) usually give straw or gold colours. Whites from cooler climates often appear lemon-green or lemon, and those from warmer areas may veer towards gold. Sweet wines such as Sauternes are usually gold, even in youth, and will deepen, perhaps even to copper with age. The colour of all whites will darken with time. The rate at which this takes place varies considerably depending on several factors, particularly pH and acidity (higher acidity slows down the process) and how well the bottle has been stored. Brown colours are certainly a warning sign of faults, either lightstrike or severe oxidation. Lightstrike, discussed in Chapter 6, is most likely in white wines that have been bottled in clear glass or pale glass and exposed to ultra-violet light sources.

Of course, the taster will be aware that many wines made from white grapes are deliberately oxidised during the production process, e.g. Amontillado and Oloroso Sherries, and the tuilé and rancio wines from Rivesaltes in the Roussillon region of France. Oxidation, as a fault, is discussed in Chapter 5.

2.5.4.2 Rosé Wines

Of all wine types, rosé is the category that is made for visual appeal. Winemakers and marketing departments know that an attractive appearance is crucial to the drinker's perceptions of style and quality. The colour of rosé wines depends on several factors, particularly the production techniques, e.g. whether the colour is the result of skin contact in the press, a short maceration in the tank, or if the wine has been made by the *saignée* method. This technique involves draining juice from a vat of fermenting crushed red grapes before heavy colour extraction, perhaps after 6–24 hours or so. If the vat is bled after six hours the colour will usually be pale, whilst 24 hours will give a deeper coloured rosé, but the actual colour and intensity will depend on many factors, including climatic conditions and grape variety. There is a very wide range of possible colours and intensities for rosé wines from *onion-skin* to *pink* or even *orange.* Any orange or brown tints should be regarded as a danger sign, indicating oxidation which even in its early stages will result in a loss of fruit, and bitterness on the palate.

2.5.4.3 Red Wines

Red wines can vary in colour from *purple* through *ruby* to *mahogany.* Purple is indicative of a young red wine, some being so intensely coloured that they appear almost to be blue-purple. As the wine begins to age, the purple tones lighten to ruby, with further ageing to a warm brick red colour, and with full maturity possibly to garnet or even tawny. We may consider the colour of red wine on a scale that runs from *purple,* through *ruby* and *garnet,* to *tawny.* Red wines that have undergone lengthy barrel ageing, in which controlled oxygenation has been taking place, change colour and also lighten in intensity faster than those bottled early and aged in bottle. A good illustration of this is the contrast between a Vintage Port (not Late-Bottled Vintage – LBV) of about 12 years old, following two years or so in vat, aged in

Figure 2.4 Vintage Port and Tawny Port.

bottle for 10 years and a 10-Year-Old Tawny Port, which has been matured in pipe (cask) and bottled when ready for drinking. This contrast is illustrated in Figure 2.4.

A brown wine is tired, oxidised, and probably undrinkable. Red and white wines that are heavily oxidised are pretty much indistinguishable in colour.

2.5.4.4 The Rim and Core

The colour gradation from the heart or core of the wine to the rim, where it touches the glass should be noted. The greatest colour intensity is at the heart, but in the area approaching the rim, the colour is paler and different. For example, a wine that is ruby coloured at its core may gradate to brick red or garnet tones towards the rim, indicating maturity. As the rim of the wine touches the glass, the last millimetre or two will be water clear. The distance of the colour gradation will vary from just a couple of millimetres in a young wine to perhaps a centimetre or more in a mature example. Mature white wines too will have considerable gradation in colour approaching the final few millimetres of the rim, which again will be water clear. The colour of the rim should be observed (any hints of brown being a sign of possible oxidation), and the width of the rim noted, e.g. *broad* or *narrow*.

2.5.5 Other Observations

These may include, among other things, **bubbles, petillance, legs/tears, and deposits.**

2.5.5.1 Bubbles or Petillance

Bubbles are a key feature of sparkling wines, but a small amount may also be observed on still wines.

Still Wines Occasionally, the presence of bubbles on a still wine could be indicative of a fault – either an alcoholic or malolactic fermentation (MLF) is taking place or has taken place in the bottle. This issue is discussed in Chapter 9. Cloudiness and/or presence of sludgy sediments might be a further indication of this. However, still wines in good condition may contain bubbles or petillance. Gases, namely carbon dioxide (CO_2), nitrogen (N), and argon (Ar) may be used as a blanket at various stages of winemaking to prevent oxidation or other spoilage. Particularly if a very fresh style of wine is desired, it is common to flush the vats with one of these gasses. Bottles too may be pre-evacuated of oxygen and gas sparged immediately before filling. Some of the gas (particularly CO_2 which is highly soluble) may become dissolved in the wine – this does not generally detract from the quality, and can often add a sensation of freshness. Some wines, e.g. wines from the Mosel region of Germany may naturally retain some CO_2 from the alcoholic fermentation. In the case of still wines, a brief observation of the size and quantity of bubbles should be made. The bubbles may appear on the tasting glass surface, in which case they are likely to be large, on the rim, or in the heart of the wine.

Sparkling Wines The quality of the mousse is considered to be an essential part of the overall quality of sparkling wines. The size, quantity, and consistency of the bubbles should be noted. The bubbles may rise from the base of the cup or a seemingly random point in the heart of the wine. Generally speaking, small bubbles are indicative of a desirable cool, slow second fermentation, especially when this has taken place in the bottle as in Champagne and other high-quality sparkling wines made by the traditional method. The character and volume of bubbles will vary somewhat according to the type and washing of the tasting glass. Of course, a lack of bubbles in 'sparkling' wine is a physical fault, almost certainly indicating failure of the cork or other closure. Such a wine will almost certainly show other faults resulting from oxygen ingress, including oxidation.

2.5.5.2 Legs

One of the most misunderstood visual aspects of wine tasting is the presence or otherwise of legs, often referred to as tears. The wine should be swirled in the glass, held to eye-level and, after waiting for several seconds, viewed horizontally and observed as to how the swirled wine runs back down the glass. If the liquid congeals into little tears, arches, or rivers running down the glass, these are called legs. The legs may be *broad* or *narrow* (thin), *short* or *long,* and run slowly or more quickly down the glass. Wines that contain a high degree of alcohol will normally show broad legs, formed by the difference in surface tension between water and alcohol and the differential evaporation of alcohol, influenced by sugar and glycol. Several authors and critics claim that legs are purely a sign of high-alcohol, glycerol or very high residual sugar. This is refuted by examples of their presence in high-quality wines which are relatively light in alcohol, for example, fine Riesling Kabinetts from the Mosel region in Germany. The amount of dry extract also contributes to legs. Further, the amount and type of legs, if any, can be very dependent on the condition, and particularly the washing and drying of the tasting glass.

2.5.5.3 Deposits

Any deposits in the glass should be noted. These may comprise tannin sediments (which are simply the coagulation of phenolic substances) in the case of red wines or tartrate crystals in either reds or whites. Phenolic sediments naturally occur in fine red wines following bottle maturation, and such wines should be decanted before service. Although some might disagree, tartrate crystals should not be regarded as a wine fault, and are often visible in wines of the very highest quality. Thick deposits in red wines may also be tannin-stained tartrate deposits (especially in low pH wines). However, in white wines, they can look alarmingly like pieces of broken glass and worry the consumer, but they are completely harmless. The crystals are most likely a precipitate of potassium bitartrate ($KC_4H_5O_6$) or occasionally calcium tartrate ($C_4H_4CaO_6$). They are often found in bottles of German or other wines that have a high level of tartaric acid. They may precipitate if the wine is subjected to cold conditions – perhaps in a cold cellar or refrigerator. Many winemakers go to great lengths to try to ensure that the crystals do not appear in the bottle. This topic is discussed in detail in Chapter 15. The money and time spent by the industry on such treatments might be better invested in consumer wine education.

Grey or pale brown deposits are certainly a warning sign – they are most likely to be dead yeast cells or bacteria. Whilst it is possible that these are a consequence of inadequate racking and filtration, they are more likely an indication that the wine has undergone an alcoholic or MLF in the bottle, as discussed in Chapter 9. In any event, the aroma and flavour profiles together with the texture, will have been modified.

2.6 Nose

The second stage in a tasting assessment is to nose the wine. The olfactory epithelium, situated at the top of the nasal cavity, is a very sensitive organ. The tongue reveals only a very limited number of tastes, and most of the 'taste' sensations are detected by the receptor cells of the olfactory epithelium, received either via the nasal or retro-nasal passage. The information is turned into electrical signals and sent via the olfactory bulb to the olfactory cortex in the brain. There is a difference between the impressions of aromas obtained nasally and retro-nasally, and the sensations transmitted via the retro-nasal passage will be discussed later in this chapter. There is no doubt that repeated and overexposure to particular smells reduces sensitivity to them, and this can be an issue for winemakers who are regularly exposed to odours such as sulfur dioxide (SO_2), or whose cellars are contaminated with haloanisoles (see Chapter 3).

When nosing a wine, we are smelling the headspace in the glass, above the surface of the liquid. There must be plenty of headspace for the aromas, the volatile compounds in the wine, to develop. The inwards tapering bowl of a well-designed tasting glass (narrower at the top than the bottom) enables aromas to be retained in the headspace.

The nose of the wine should be assessed in the following stages:

- Condition;
- Intensity;
- Aroma characteristics;
- Development.

As discussed below, the wine should be first nosed without swirling, then swirled around the glass to vaporise the volatile compounds, and then given a comprehensive nosing.

2.6.1 Condition

The initial nosing of the wine will assess the condition: is the wine clean, or does it show a fault? The wine should not be swirled before this, and just one or two short sniffs are all that is required. Long and deep sniffs of a wine affected by 2,4,6-trichloroanisole (TCA) or 2,4,6-tribromoanisole (2,4,6-TBA), or possessing high volatile acidity, would result in the taster's nose being desensitised for some time! Basically, at this point, the light nosing is to check if we want to go any further in the tasting procedure. Most faults and flaws are detectable upon the nose to a trained and experienced taster. Depending on the nature and severity of any fault revealed, a decision has to be made whether to proceed with the wine tasting or not. A nose that is free from faults is described as *clean*. It should be borne in mind that on occasions wines, particularly after some years in bottle, may smell somewhat unclean when first opened, a phenomenon sometimes referred to as 'bottle stink'. This should dissipate after aeration. If there is doubt on this matter, the wine should be nosed again, having being swirled around the glass and allowed to 'breathe' for a few minutes.

Faults that are or may be detectable on the nose are shown in Table 2.2. This list is not exhaustive.

2.6.2 Intensity

This is simply assessing how strong or 'loud' the nose of the wine is, and can give an indication of quality. A nose light in intensity may be expected from a simple, inexpensive wine; a more pronounced nose is generally indicative of higher quality. However, high-quality reds, in particular, can be very closed in youth. Conversely, there are some grape varieties that nearly always give a very intense nose even in a wine of modest quality, especially aromatic whites. These include members of the Muscat family, Gewürztraminer, and Argentina's three Torrontés varieties. The wine should be aerated by swirling around the glass to reveal the full intensity of the nose.

2.6.3 Aroma Types, Development, and Characteristics

It is perhaps convenient if we consider here the topic of aroma types and development before that of aroma characteristics. However, when making a tasting note, the aroma characteristics might be noted first. Development is assessing, on the nose, the state of maturity of the wine. Wines have a lifespan that depends on many factors, and maturity should not be confused with age. Generally speaking, the higher the quality of the wine, the longer is the lifespan. The finest quality reds including, by way of examples, *cru classés* from Bordeaux, classic wines from the Rhône Valley, Toscana's Brunello di Montalcino and Piemonte's Barolo may require 10 years or more even to begin to approach their peak. Conversely, a simple branded Bordeaux or Côtes du Rhône or Chianti may be past its best at the age of three or four years.

Table 2.2 Faults detectable on nose of wine.

Olfactory indicator	Possible fault	Refer to:
Damp sack, wet cardboard, dusty, musty. Muted fruit	Chloroanisoles or Bromoanisoles	Chapter 3
Earthy, beetroot	Geosmin	Chapter 14
BAND-AID® stables, faecal, animal, smoky, spicy	*Brettanomyces* related faults	Chapter 4
Burnt, dried-fruitcake, metallic	Oxidation	Chapter 5
Struck match, burning on nose or back of throat	Excessive sulfur dioxide	Chapter 6
Rotten egg, drains	Hydrogen sulfide	Chapter 6
Garlic, gas, cabbage, sweetcorn, wet wool	Volatile sulfur compounds, lightstrike	Chapter 6
Vinegar, balsamic	Excessive volatile acidity, ethyl acetate	Chapter 7
Nail varnish, solvent	Ethyl acetate	Chapter 7
Wax, loss of varietal character, wet-dishcloth, mothballs	Atypical ageing	Chapter 8
Vomit, rancid butter	Lactic acid bacteria associated faults. Diacetyl	Chapter 11
Ash, charcoal, bacon	Smoke taint	Chapter 12
Jammy 'sweet' nose	Heat damage	Chapter 14
Beetroot	Geosmin	Chapter 14

To understand the concept of nose development, we need to consider the basic types and sources of wine aromas. These are classified into three main groups.

- Primary;
- Secondary;
- Tertiary.

2.6.3.1 Primary Aromas

Primary aromas are those derived from grapes and are generally fruity or floral. A predominance of these on the nose normally indicates a wine that is in a youthful stage of development. Varietal aromas usually belong in this group. Many of the compounds that give varietal aromas remain largely unchanged by the fermentation process. Accordingly, the pronounced blackcurrant or cassis aromas contributing to the nose of a young Cabernet Sauvignon, the overt gooseberry characteristics of a Sauvignon Blanc or the elderflower fragrance of a young Seyval Blanc are considered as primary. Grape compounds contributing to individual varietal characteristics may include methoxypyrazines, norisoprenoids, furan derivatives, lipoxygenase pathway products, and phenylpropanoid pathway products [8]. Terpenes, both mono- and sesquiterpenes, are present in grapes and are contributors to primary aromas – they are largely unchanged by the fermentation process and have little impact upon secondary aromas. The concentration of monoterpenes declines as wines age, thus reducing the primary aromas. Maturation and ageing of the wine may also result in them being modified and accordingly contributing to tertiary aromas. Young wines made from grapes that have high levels of monoterpenes, such as the Muscat family, Gewürztraminer and, to a lesser extent, Riesling can have a nose that screams of primary fruit and show overt grape-like aromas. By way of comparison, Chardonnay, a neutral variety may have less than 1 mg/l of monoterpenes, Riesling up to 4 mg/l and varieties of the Muscat family up to 6 mg/l.

It is worth noting, at this point, that the varieties we all know and love are only a few generations removed from their wild ancestors [8]. During the last 150 years, viticulturists have generally focused more and more on relatively few 'elite' varieties, much to the loss of aroma diversity, and possibly also disease resistance.

2.6.3.2 Secondary Aromas

Secondary aromas are the 'vinous' aromas resulting from the fermentation – put simply, the smells that are different in wine to those of unfermented grape juice. Many compounds in grapes are precursors of secondary aromas, including free amino acids, phospholipids, glycolipids, aldehydes, and phenols. During the fermentation process, numerous chemical changes and enzyme-catalysed modifications take place producing secondary aromas. Numerous esters are generated in the process, and the aromas of these are often assertive on the nose of young wines, sometimes imparting pear, banana or even boiled sweet or bubble gum characteristics. Several fatty acids are also generated during fermentation, and some of these, if produced at high concentrations, can be distinctly unpleasant and render the wine flawed or even faulty. These include acetic, isobutyric, isovaleric, butyric, hexanoic, and decanoic acids. Several volatile sulfur compounds may also be produced from their precursors. Some of these are desirable contributors to a wine's aromatics, but others, if they remain in the wine, may also contribute to faults. Also in the secondary aroma group are the by-products of MLF, and bâtonnage (if undertaken), which may contribute to aromas of butter and cream. Compounds extracted from oak, giving aromas of vanilla, coconut, or toast may also be considered secondary aromas.

2.6.3.3 Tertiary Aromas

Tertiary aromas result from the maturation and ageing process of the wine, particularly in bottle. During this time, there will be many chemical reactions. Following fermentation the wine will already contain some dissolved oxygen (DO), and for wines that are barrel-matured, more oxygen will be absorbed through the cask. Provided the barrels are kept topped-up, there will only be a small amount of beneficial oxygenation taking place, which will increase the aldehyde content of the wine. Wines matured in new barrels will absorb the greatest amount of oxygen, together with oak compounds including vanillin, lignin, and tannin. Those matured in second and third fill barrels will also pick up these, but to a lesser extent, for the pores in the wood become blocked with repeated use. Wines that are not matured in barrels may be micro-oxygenated, and 'oaked' in other ways, including chips, beans, and powder.

Wine maturation in bottle results in continuous changes to the volatile compounds of the wine. At bottling time, there will be some DO in the wine, and typically some in the headspace of the bottle. This oxygen present at bottling is rapidly consumed by the wine [9]. During ageing and storage more may be absorbed from the closure itself (particularly if this is cork), through the closure depending upon type, and the interface between the closure and bottle. However, many of the changes that take place during bottle maturation are reductive. These changes take place as the oxygen content of the wine is gradually reduced. It is this process that makes bottle ageing a necessity for the maturation of many fine wines, especially full-bodied reds. A wine that is not able to be reduced is a dead wine. However, *reduced aromas* are usually regarded as a fault, and this topic will be discussed in Chapter 6. As fine wines mature the tertiary aromas develop, and these can exhibit a complex array of seamless, interwoven characteristics, at best all in total harmony. Tertiary aromas are those that often evoke the most descriptive comments. Examples that might be found on red wines include shoe-leather, Havana cigar, woodland floor, truffle, autumnal gardens, game – the list is almost endless. Tertiary aromas on white wines might include hay, honey, nuts, nutmeg, and kerosene (this might be regarded as a fault, as discussed in Chapter 15). Many vegetal characteristics that may be present in wines of any colour also belong to the tertiary group. Some of these, if present in excess, may be distinctly unpleasant.

When wines are aged beyond their projected lifespan, or suffer premature decline (perhaps due to poor storage), all of the primary and many of the secondary aromas will be lost. Of the tertiary aromas that remain, some such as caramel, toffee or soaked fruitcake may be pleasant, but many will not. The smell of rotting cabbage, stale sweat, old trainers and burnt saucepans may come to the fore, and every hint of complexity will have been drained out, rendering the wine flawed.

The descriptors for the development of the nose may range from *youthful* to *fully developed*. When desirable aromatic compounds are waning and unwelcome characteristics of a tertiary nature, e.g. rotting vegetables, are assertive, the wine may be described as *tired*.

Some wines go from youthful to tired without ever passing through the fully developed stage. Examples include much Beaujolais and inexpensive Valpolicella. However, there are also types of wine that are bottled fully developed, such as Sherry, Tawny Port and some sparkling wines.

2.6.3.4 Aroma Characteristics

Some older tasters, and wine writers, distinguish between the terms 'aroma' and 'bouquet'. They use the term 'aroma' when referring to the primary aromas derived from the grapes. 'Bouquet' would encompass the assembled characteristics resulting from changes that have taken place during fermentation (secondary aromas), and particularly maturation and ageing (tertiary aromas). The distinction is far from clear-cut, and as such, the words 'aroma' and 'odour' are used in this book as all-encompassing terms.

There are some 800 or more volatile compounds that have been identified in wines, in concentrations ranging from a few nanogrammes to hundreds of milligrammes per litre. The olfactory perception thresholds of odorous compounds vary from 4 pg/l, to 100 mg/l, but the sensory threshold for any individual compound may differ when present with other compounds. It is even possible that some compounds may be detected at concentrations of a few femtogrammes per litre (fg/l).

There has been considerable research over the last couple of decades into the identification of the individual compounds (and their precursors) that contribute to wine aromas. These activities are still in their infancy. Table 2.3 lists some of the compounds contributing to individual aromas, including their origins and typical odour detection thresholds. The Sauvignon Blanc variety has been particularly well studied in respect of the precursors,

Table 2.3 Some compounds responsible for wine aromas, their origins, aroma descriptors, and typical sensory detection thresholds.

Compound	Aroma descriptors	Sensory detection threshold[a]
Primary aromas		
Terpenes		
Wine lactone	Lime, coconut, wood, sweat	10 ng/l
Linalool	Floral, rose, lavender 'Muscat' grape	25 µg/l
Geraniol	Geranium, rose, citronella	30–130 µg/l
Nerol	Rose	200–300 µg/l
α-Terpineol	Lily of the Valley, spice	250–400 µg/l
Citonellol	Citronella	18–100 µg/l
Cis-Rose oxide	Lychee, rose	100–200 ng/l
Trans-rose oxide	Rose	80 µg/l
1,8-Cineole (eucalyptol)	Eucalyptus	3.2 µg/l
N.B. Over 50 terpenes have been identified in grapes		
Secondary (fermentation) aromas		
Esters		
Ethyl acetate	Pear drops, solvent, nail varnish remover (levels above 100 mg/l may regarded as a fault)	3.6–13 mg/l
Ethyl propanoate	Cherry	2.1 mg/l
Ethyl-2-Methylpropanoate	Strawberry, banana	1.8 mg/l
Ethyl hexanoate	Green apple, aniseed	14–140 µg/l
Ethyl octanoate	Meat fat	5 µg/l
Ethyl Decanoate	Floral, grape	200 µg/l – 1.5 mg/l
Ethyl butanoate (ethyl butyrate)	Pineapple	20–400 µg/l
Ethyl isovalerate	Fruits	3 µg/l
Isoamyl acetate	Banana	30 µg/l – 2.7 mg/l
Hexyl acetate	Pear	2.4 mg/l
Phenylethyl acetate	Rose, honey	250 µg/l – 6 mg/l
2-Phenylethanol	Floral	20 mg/l

(Continued)

Table 2.3 (Continued)

Compound	Aroma descriptors	Sensory detection threshold[a]
Alcohols		
Ethanol	Alcohol!	100 mg/l
2-Phenylethyl alcohol	Rose, lilac	14 mg/l
Isoamyl alcohol	Malted barley, Whisky	30 mg/l
1-Hexanol	Cut grass	4.5 µg/l
Wood (maturation) aromas		
2-Furfurylthiol	Roasted coffee	4 pg/l
Furfural	Almond	20 mg/l
Eugenol	Clove, spice	500 µg/l
Trans-Nonenal	Sawn wood	250 ng/l
cis/trans-Whiskylactone	Coconut	125 µg/l
Guaicol	Smoke	75 µg/l
Vanillin	Vanilla	200–320 µg/l
Maltol	Caramel	30 µg/l

a) Sensory detection thresholds depend upon many criteria, including the presence of other compounds and the wine matrix. The figures stated have been extrapolated from numerous sources.

development, and reactions of volatile thiols in the grapes, must, and wine, as reviewed in 2012 by Coetzee and du Toit [10]. The aromas commonly found in a young wine made from Sauvignon Blanc, and the volatile thiols that responsible for these aromas, are shown in Table 2.4.

It may sometimes be helpful for the taster to consider wine aromas in basic groups:

- Fruits (primary aromas);
- Flowers (primary aromas);
- Spices (primary, secondary, and tertiary aromas);
- Vegetables (primary and tertiary aromas);
- Oak aromas (secondary aromas);
- Other aromas.

Of course, it is possible to divide each of these basic groups into subgroups. For example, the fruits group might be divided into citrus, green, red, stone, and tropical fruits. Dried fruits may also be included. Each subgroup contains individual aromas and flavours.

All aromas detected during nosing should be noted, and when detailing individual descriptors, these may be linked to known varietal characteristics. For example, green apple, lime, peach, and mango are just some of the aromas that may be associated with Riesling. Strawberry, raspberry, red cherry, green leaf, and mushroom are typical aromas associated with Pinot Noir. The detection of any oak related aromas including vanilla, toast, nuts, and coconut may warrant interpretation (barrels/beans/chips?).

Table 2.4 Compounds contributing to the varietal aromas of Sauvignon Blanc wines.

Aroma	Thiol	Sensory detection threshold[a]
Boxwood – broom, cat's pee	4-Mercapto-4-methylpentan-2-one	0.8–3.3 ng/l
Citrus zest and peel	4-Mercapto-4-methylpentan-2-ol	55 ng/l
Grapefruit, passion fruit	3-Mercaptohexan-1-ol	60 ng/l
Mango, guava, passion fruit	3-Mercaptohexyl acetate	4 ng/l
Cut grass	1-hexanol	4.5 µg/l
Smoke, gunflint	Benzenemethanethiol	0.3 ng/l
Aroma	**Methoxypyrazine**	**Sensory detection threshold[a]**
Green pepper, grass	2-methoxy-3-isobutyl pyrazine (iBMP)	1–2 ng/l

a) Sensory detection thresholds depend upon many criteria, including the presence of other compounds and the wine matrix. The figures stated have been extrapolated from numerous sources.

2.7 Palate

2.7.1 Palate Sensations

Palate is a convenient expression to describe the sensations a wine gives once in the mouth. Palate may comprise a 'complex combination of the olfactory, gustatory, and trigeminal sensations' [7]. In this part of the tasting process, we assess the taste and tactile sensations detected in the mouth, particularly on the tongue, and the flavour characteristics detected as a result of the wine's volatile compounds being breathed through the retro-nasal passage at the back of the mouth and transmitted to the olfactory bulb. The impressions of aromas sensed retro-nasally are often different from those sensed nasally. The sensory cells within the mouth are mostly contained within the 5000 or more taste buds on the tongue – young people may have up to 10 000 active taste buds. There are also taste buds on the roof of the mouth and the back of the throat, which is why some people claim to only get 100% of what the wine has to offer if they swallow rather than spit. Interestingly, a recent study has indicated that olfactory receptors are functionally expressed in taste papillae [11]. In other words, it may be possible for the tongue to 'smell'.

A reasonable quantity of wine should be taken into the mouth. If it is too little, the wine will be diluted and modified by saliva. It is important to breathe air through the wine in the mouth to vaporise the volatile compounds. Accordingly, a free passage is needed to and from the nose to enable transmission of these aromas. If a person has a blocked nose, it is not just the sense of smell that disappears, but most of the sense of taste. The taste sensations can be increased if we breathe out via the nose after we have spat out the wine, and this action will help determine the 'finish' of the wine.

As we taste, chew, and dissect the wine in the mouth, before finally spitting, numerous sensations develop. This evolution is considered in stages. The initial **attack** is the sensation as the wine taken into the mouth. This is followed by the **development** on the palate

Table 2.5 Faults detectable on palate of wine.

Retro-nasal, gustatory or trigeminal indicator	Possible fault	Refer to:
Musty, damp sack, wet cardboard	Chloroanisoles and Bromoanisoles	Chapter 3
BAND-AID® stables, animal, spicy	Brettanomyces (Dekkera) related faults	Chapter 4
Burnt, bitter, dried-out	Oxidation	Chapter 5
Rotten egg, garlic, cooked cabbage, skunk	Excessive sulfur dioxide, volatile sulfur compounds, reduced aromas	Chapter 6
Marmite (Vegemite), wet wool, wet cardboard	Light strike	Chapter 6
Vinegar, nail varnish, solvent	Excessive volatile acidity, ethyl acetate	Chapter 7
Thin body, metallic, non-tannic bitter finish	Atypical ageing	Chapter 8
Vomit	Lactic acid bacteria associated faults, (including mousiness)	Chapter 11
Bacon, smoky bacon crisps	Smoke taint	Chapter 12
Peanut, earthy, bell pepper (unexpected)	Ladybug (*Coccinellidae*) taint	Chapter 13
	Brown marmorated stink bug (*Halyomorpha halys*) taint	
Prunes, burnt, overcooked fruitcake	Heat damage	Chapter 14

where we perceive the flavour characteristics and intensity, and then the **finish** which comprises the final impressions of the wine including the balance. The **length** is a measurement of how long the sensations of the finish and aftertaste last. Tasters sometimes refer to the progressive sensations as 'front-palate', 'mid-palate', and 'back-palate'.

When assessing the palate of wine, we may consider the following headings:

- Dryness/Sweetness;
- Acidity;
- Tannin;
- Alcohol;
- Body;
- Flavour intensity;
- Flavour characteristics;
- Other observations;
- Finish.

Faults that are or may be detectable on the palate are shown in Table 2.5. This list is not exhaustive.

2.7.2 Sweetness/Bitterness/Acidity/Saltiness/Umami/Trigeminal Sensations

Although highly sensitive, the receptors of taste buds can only detect five basic tastes: sweetness, bitterness, saltiness, acidity, and umami (the savoury taste of some amino acids).

These are the non-volatile compounds present in wine (although acetic acid is volatile). The fifth basic taste, umami, has only been recognised in the western world in 1985 and has only appeared in wine tasting notes in very recent years. There are claims of a sixth basic taste, the bitter, chalky taste of calcium, and other basic 'tastes', but such claims remain highly controversial. Trigeminal sensations, i.e. those detected by the trigeminal nerve, have recently come to be regarded in some quarters as yet another basic 'taste'. However, these are not 'tastes', but sensations, e.g. the cooling effect of menthol/mint, or astringency. Of the five basic tastes, saltiness (comprising mainly sodium chloride) is usually not important in wine. The sensory cells of the tongue convert the detected tastes into electrical signals and send them to the brain's taste cortex. Until the late twentieth century, it was generally accepted that different parts of the tongue detect these basic tastes, and many wine-tasting books and human biology texts still illustrate a diagram of the tongue detailing these areas. However, this concept has been discredited, largely by the work of Linda Bartoshuk when at UC Davis. For this chapter, I will rely on the approach that defined areas of the tongue are more sensitive to the individual basic tastes. It is not disputed that the 'traditional' areas of detection identify the tastes, only that the other areas do not. It is also accepted that the centre part of the tongue is considerably less sensitive to the basic tastes. There are tactile sensations of the wine that are also detected in the mouth, on the cheeks, teeth, and gums. These include tannin, body and alcohol, and the trigeminal sensations.

2.7.3 Dryness/Sweetness

Before discussing perceptions of sweetness, it is pertinent to visit the topic of grape sugars briefly. Grapes contain glucose (grape sugar) and fructose (fruit sugar) which will be completely or partially converted by the action of yeasts to ethanol and carbon dioxide during the fermentation process. If there is insufficient natural sugar in grapes to produce a balanced wine with the required alcoholic degree, in some countries the winemaker may add sucrose to the must, a process generally known as must enrichment or chaptalisation. There can be no doubt that the process reduces a wine's concentration. In theory, any added sucrose should be fermented to dryness. The purpose of chaptalisation is not to produce wines with sweetness. However, in practice, most fermentations cease before total dryness, as discussed below. At the time of writing, chaptalisation remains permitted in the more northerly zones of the European Union (EU), and incredibly is sometimes undertaken even in 'good' years when grape ripeness should not be a problem. For example, 2016 is generally regarded as an excellent year in Bordeaux, but chaptalisation of vats of Cabernet Sauvignon, Cabernet Franc, and Petit Verdot was permitted, and undertaken, in many districts. An alternative to chaptalisation is the addition of concentrated grape must which is comprised largely of glucose and fructose.

Sweetness, if any, in a wine will be particularly detected on the tip of the tongue. It is important to remember than we cannot smell sweetness (sugar is not volatile), although the nose of some wines may lead us to expect that they will taste sweet. This may or may not be the case. For example, a wine made from one of the families of Muscat varieties may have a fragrant and aromatic nose reminiscent of sweet table grapes, but the wine may be bone dry when tasted. Other characteristics can also give an illusion of sweetness, in particular, high-alcohol levels (although sometimes too much alcohol can lead to a bitter taste), and vanillin oak. Thus, a high-alcohol wine that has undergone oak treatment can mislead the taster into perceiving that it is sweeter than the actual level of residual sugar. Pinching the nose whilst rolling the wine over the tip of the tongue can help the novice overcome any distortions that the nose may be giving. However, the acidity of the wine also impacts on the taster's perception of sweetness. The higher the acidity, the less sweet a wine containing residual sugar may appear to be.

Thresholds for detecting sweetness vary according to the individual. Nearly 50% of tasters can detect sugar at a concentration of 1 g/l or less, with just 5% unable to detect sugars at more than 4 g/l. Residual sugar in a wine is due to fructose remaining after the fermentation. The level of residual sugar in white wine can range from 0.3 to 300 g/l. Most high-quality red wines are fermented to dryness or 'off'-dryness, i.e. between 0.2 and 3 g/l of residual sugar. However, because dry wines are very fashionable, some wines are labelled or described as 'dry' when they are not. It is common for many New World branded whites, particularly Chardonnays, to have between 5 and 10 g/l of sugar, and popular branded New World reds may contain up to 8 g/l. A small amount of residual sugar in white wine helps to soften any bitterness, and a little sweetness in red wine can serve to balance any phenolic astringency.

2.7.4 Acidity

Acidity is detected, particularly on the sides of the tongue and cheeks, as a sharp, lively, tingling sensation. Medium and high levels of acidity encourage the mouth to salivate.

Of course, all wines contain acids: whites generally more than reds, and those from cooler climates more than those from hotter regions. In the grape ripening process, as sugar levels increase, acidity levels fall (mostly due to a reduction in malic acid), and pH increases. Thus, a cool climate white wine might have a pH of 2.8, whilst in a hot climate, the pH of red wine might be as high as 4. Uniquely amongst fruits of European origin, grapes contain tartaric acid: this is the main wine acid, although malic and citric acids are also important. These three acids account for over 90% of the total level of acidity. The other acids present may include lactic, ascorbic, sorbic, succinic, gluconic, and acetic acids. A high level of volatile acetic acid is most undesirable. At an extreme, it will impart a nose and taste of vinegar. Accordingly, excessive volatile acidity is generally regarded as a fault. If the grape must is deficient in acidity, the winemaker may be allowed to add acid, usually in the form of tartaric acid. Within the EU, such additions are only permitted in warmer, southern regions.

Perception thresholds for various acids vary according to the individual, with under 50% of tasters detecting tartaric acid in concentrations of 0.1 g/l or less, and the remainder between 0.1 and 0.2 g/l. However, sweetness negates the impact of acidity, and vice versa, and the relationship between these is one of the considerations when considering *balance*, as discussed below.

2.7.5 Tannins

Tannins are mostly detected by tactile sensations, particularly on the teeth and gums, making them feel dry, furry, and gritty. The sensations can be mouth-puckering, and after tasting wines high in tannin, you want to run your tongue across the teeth to clean them. Hard, unripe tannins also taste bitter and 'green'. Tannins are a key component of the structure of classic red wines and give 'grip' and solidity.

Tannins are polyphenols, the primary source in wine being the skins of the grapes. Stalks also contain tannins, which are generally of a green, hard nature and nowadays, with some notable exceptions, are generally excluded in the winemaking process. Oak is another source of tannin, and wines matured in new barrels or those that have been used only once or twice will absorb tannin from the wood, which depletes with each use of the barrels. Also, with repeated use, the pores in barrels become blocked by tartrates and other compounds further reducing the uptake amount of tannin, as well as oxygen. Oak products such as chips, beans, and powder are often used as a low-cost alternative to barrel ageing.

Tannin binds and precipitates protein. This is one of the reasons why most red wines match red meats and cheeses successfully. This combination causes wines containing tannin to congeal into strings or chains as it combines with protein in the mouth, and thus our perception of tannin in wine will change if we keep it in the mouth too long. Novice tasters often confuse the sensations of acidity and tannin. A classic traditionally made Barolo, which is high in both, is a good example to taste to distinguish between them. The tannin gives the dry, astringent sensations on the teeth, gums, and even hard palate. The acidity produces the tingling sensations on the sides of the tongue and cheeks.

It is often written that white wines contain no tannin. This is not true, although generally, the levels are low compared with red wines. The grapes for white wines are pressed pre-fermentation, the solids are settled or the must otherwise clarified, and reasonably clear juice is fermented. Unless there is any period of skin contact (macération pelliculaire), post-crusher and pre-press, the phenolics in the skins will have limited impact. In the case of whole cluster pressing the presence of grape phenols is minimal. White wines that have been fermented or matured in oak barrels (or otherwise oaked) may contain significant oak tannins.

The quantity of tannins in white wines ranges from 40 to 1300 mg/l, with an average of 360 mg/l. Red wines contain from 190 to 3900 mg/l, with an average of 2000 mg/l. Thus, it will be seen that whilst the average tannin level of red wines is six times that of whites, many white wines contain considerably more tannins than some reds. Grape tannin, usually in powder form, may be added during the fermentation process, to give an over-soft red a little more 'grip'. Grape tannin is an authorised oenological product in the *OIV CODEX* [12], and its addition is legal in most countries worldwide, including member states of the EU.

Wine tannins should be assessed and noted for both level and nature. We may assess the level on a scale that runs from *low*, to *high*. The nature of the tannins may be described as *ripe/soft* or conversely *unripe/green/stalky*. The texture of the tannins should also be noted. Tannins that are assertive, rough, and very gritty may be described as *coarse*, whilst those with a smooth, velvety texture as *fine-grained*. Wines with high or very high levels of tannins are coarse-textured. Those with very unripe tannins may be considered to be flawed, as will be discussed in Chapter 14.

2.7.6 Alcohol

Alcohol is the most abundant volatile compound in wine. It is detected on the palate as a warming sensation, perceived at the back of the tongue, the cheeks and in the mouth generally. The higher the level of alcohol, the more of a warm 'glow' will be perceived by the taster. Alcohol also gives a combination of sweet and bitter tastes, and bitterness increases with higher alcohol levels [13]. Over-alcoholic wines will give burning sensations, and this may be considered to be a flaw (see Chapter 14). The 'weight' of the wine in the mouth also increases with the concentration of alcohol.

With an increase in the level of alcohol, there is a decrease in the volatility of wine aroma compounds.

The alcoholic content of light, i.e. unfortified wines, ranges from 7.5% to 16% abv (alcohol by volume). As grapes ripen, the levels of fructose and glucose increase, thus increasing the potential amount of alcohol. Generally speaking, wines from hotter climates contain more alcohol than those from cooler regions. However, the average alcohol level of wines has increased during the last two or three decades. This is due to many reasons, including changes in canopy management, growers delaying harvesting until so-called phenolic ripeness is reached (especially as the market now demands a softer style of reds than in the past), the impact of climate change and the use of cultured, alcohol tolerant yeasts. In 1989, referring to Australian Cabernet Sauvignons, Bryce Rankine wrote in the reference work *Making Good Wine*: 'A ripeness of 10°–12° Baumé (18–21.6° Brix) is usual, which results in a wine containing between about 10% and 12% alcohol by volume' [14]. Today, any Australian Cabernet Sauvignon with less than 13% abv would be regarded as atypical – probably over-cropped and certainly under-ripe. Alcohol levels in white wines can range from as low 7.5% abv (e.g. some Riesling wines from Germany's Mosel region), up to 14.5% abv (e.g. some Californian Chardonnays). Red wines may range from 11% abv (e.g. some Bardolinos from Italy's Veneto region) up to 15% or more (e.g. some Chilean Cabernet Sauvignons). There is not much point in the taster attempting to estimate the actual alcoholic degree, (or the actual level of acidity) – it is the impression rather than quantification that is relevant. There are methods of removing alcohol from over-alcoholic wines, including the use of reverse osmosis (RO) machines or spinning cone columns, but these remain controversial.

If a fortified wine is being assessed, the alcohol will be in the range of 15–22% abv. We consider whether the wine has been fortified to a *low level* (15–16% abv), e.g. Fino Sherry or Muscat de Beaumes de Venise, a *medium level* (17–19% abv), e.g. Sherries other than Fino/Manzanilla, some *vin doux naturels* (VDNs), or a *high level* (20% abv or more), e.g. Port. The presence of the 'hot' aromas or palate that shows spirit tones indicate a flaw, as the spirit addition to fortified wines should be seamlessly integrated.

2.7.7 Body

Body, sometimes referred to as weight or mouthfeel, is more of a tactile than a taste sensation. It is a loose term to describe the lightness or fullness of the wine in the mouth. Body should not be confused with alcohol, although it is unlikely that a wine low in alcohol will be full-bodied. Generally, wines from cooler climates tend to be lighter-bodied than those from hotter areas. Certain grape varieties usually produce light-bodied wines, whilst others make full-bodied ones. Although it is a huge generalisation, wines made from Sauvignon Blanc or Riesling tend to be fairly light in body, whilst those made from Chardonnay or

Viognier maybe medium to full-bodied. Of red grape varieties, Pinot Noir usually produces a lighter-bodied wine than Cabernet Sauvignon or Syrah.

The body of a wine is supported by its *structure*, made up of a combination of acidity, alcohol, tannin (red wines), and any sweetness. The structure may be thought of as the architecture of a wine.

2.7.8 Flavour Intensity

Flavour intensity should not be confused with body. A wine can be light-bodied but with a very pronounced intensity of flavour, for example, a fine Riesling from Germany's Mosel region. However, as with the other aspects of style and quality, the flavour intensity of a wine will depend upon many factors in the viticulture and winemaking processes. Of particular significance is the yield in the vineyard, both per hectare and per vine. It is generally accepted that flavours of wines from high-yielding vines are often more dilute and lack the concentration of those from vines with a low yield. However, this view has been challenged by some viticulturists including Mark Matthews, a professor of viticulture at the Robert Mondavi Institute at the University of California at Davis [15]. It should be noted that a low yield for one grape variety, e.g. Sauvignon Blanc would be considered high for another, e.g. Pinot Noir. Flavour intensity is one of the key considerations when assessing quality.

2.7.9 Flavour Characteristics

The flavour characteristics perceived on the palate may closely replicate the aromas identified on the nose, but new flavours may emerge, and some characteristics previously identified may be muted. When noting individual terms, these may be linked to known varietal characters. Table 2.6 lists some of the flavours that may be perceived in white wines and the grape varieties or other wine components commonly associated with them. Table 2.7 lists some of the flavours found in red wines and their associated varieties and other wine components.

2.7.10 Other Observations

2.7.10.1 Texture

The *texture* of a wine should be considered and crucially the *balance* of the characteristics already discussed. The easiest way to understand texture is by imagining running the tips of your fingers over the skin of various parts of the body of people of different ages and professions: the smooth, soft face of a model, the hands of a cashier, the weathered face of a deep-sea-fisherman, the pre-shave chin of a builder. The texture of a wine might be described as *silky*, *velvety*, *smooth*, or *coarse*. Several wine faults impact upon texture, including some faults resulting from the actions of lactic acid bacteria – see Chapter 11.

Bubbles or spritz, if present, give tactile sensations on the tongue. In a poor quality sparkling wine, they are very aggressive, whilst a creamy feeling mousse is indicative of the well-integrated carbon dioxide in a good quality example. Thus in the case of sparkling wines, the mousse might be described as *delicate*, *creamy*, or *aggressive*.

Table 2.6 Some white wine flavours and the grape varieties or other wine components commonly associated with them.

Apple	Chardonnay (cool climate), Riesling
Apricot	Riesling, Viognier
Asparagus	Sauvignon Blanc
Banana	Chardonnay (hot climate)
Butter	Malolactic fermentation completed
'Catty'	Sauvignon Blanc
Citrus	Chardonnay (cool climate), Riesling
Coconut	Oak ageing
Cream	Malolactic fermentation completed
Creamy texture	Lees ageing
Elderflower	Sauvignon Blanc
Gooseberry	Sauvignon Blanc
Grapefruit	Chardonnay, Sémillon
'Herbaceous'	Sauvignon Blanc
Herbs	Pinot Grigio
Honey	Chenin Blanc, Riesling, Viognier
Kerosene	Riesling (aged)
Kiwi	Pinot Grigio, Sauvignon Blanc
Lanolin	Sémillon
Lemon	Chardonnay, Pinot Grigio
Lime	Riesling (moderate climate), Sauvignon Blanc
Lychee	Gewürztraminer
Mandarin	Sémillon
Mango	Chardonnay (hot climate)
Melon	Chardonnay (moderate climate)
Nectarine	Sémillon
Nettles	Sauvignon Blanc
Nuts	Chenin Blanc, Oak ageing
Passion fruit	Sauvignon Blanc
Peach	Chardonnay (moderate climate), Riesling, Chenin Blanc
Pear	Chardonnay (cool climate), Pinot Grigio
Pepper – bell (green)	Sauvignon Blanc
Petrol	Riesling (aged)
Pineapple	Chardonnay (hot climate)
Roses	Gewürztraminer
TCP	Noble rot (*Botrytis cinerea*)
Toast	Oak ageing
Vanilla	Oak ageing (especially American oak)
'Wet wool'	Chenin Blanc

Table 2.7 Some red wine flavours and the grape varieties or other wine components commonly associated with them.

Animal	Pinot Noir, high level of *Brettanomyces*
Aniseed	Malbec
Banana	Gamay, carbonic maceration
Blackberry	Grenache, Merlot, Shiraz (Syrah)
Blackcurrant	Cabernet Sauvignon
Bramble	Zinfandel
Cedar	Oak ageing, Cabernet Sauvignon
Cherry – black	Cabernet Sauvignon, Merlot, Pinot Noir (very ripe)
Cherry – red	Pinot Noir (fully ripe), Sangiovese, Tempranillo
Chocolate – dark	Cabernet Sauvignon, Shiraz
Cinnamon	Cabernet Sauvignon
Clove	Grenache
Coconut	Oak ageing
Game	Pinot Noir
Grass	Unripe grapes
Herb (mixed)	Grenache, Merlot, Sangiovese
Leafy	Pinot Noir
Leather	Shiraz (Syrah), aged wines
Liquorice	Grenache, Malbec, Cabernet Sauvignon
Meat	Pinotage
Metal	Cabernet Franc
Mint	Cabernet Sauvignon (especially cool climate)
Pencil shavings	Cabernet Sauvignon
Pepper – bell	Cabernet Sauvignon, Carmenère
Pepper – ground black	Shiraz (Syrah)
Pepper – white	Grenache
Plum – black	Merlot
Plum – red	Merlot, Pinot Noir (overripe)
Redcurrant	Barbera
Raspberry	Cabernet Franc, Grenache, Pinot Noir (ripe)
Roses	Nebbiolo
Smoke	Oak ageing
Soy sauce	Carmenère (very ripe)
Stalky	Wet vintage, unripe grapes, inclusion of stalks
Strawberry	Grenache, Pinot Noir (just ripe), Merlot, Tempranillo
Tar	Nebbiolo, Shiraz
Tea	Merlot
Toffee	Merlot
Toast	Oak ageing
Tobacco	Cabernet Sauvignon
Truffle	Nebbiolo

2.7.10.2 Balance

Balance is the interrelationship between all the taste and tactile sensations and the components that create them. If anyone or a small number of them dominate, or if there is a deficiency of any of them, the wine is unbalanced. An easy to understand example is that a white wine described as sweet or luscious but with a low acidity will be flabby and cloying. In other words, it is unbalanced. A red wine with a light body, light flavour intensity, low to medium alcohol but high tannin will feel very hard astringent and unbalanced. A balanced wine has all the sensations in proportions that make the wine a harmonious whole. In a well-balanced wine, the sensations are seamlessly integrated.

If any component or several components are making the wine unbalanced, these should be noted. However, the state of maturity of a wine is an important consideration. Whilst a very low quality wine may never be in balance at any stage in its life-cycle, a high-quality wine, particularly reds, will often only achieve balance when approaching maturity. Tannins and acidity may dominate in youth. The taster needs to evaluate all the components and the structure of the wine to anticipate how these will be interrelated at maturity. Balance is a major consideration when assessing wine quality, and wines that are substantially 'out-of-balance' may be considered as flawed, as will be briefly discussed in Chapter 14.

2.7.11 Finish – Length

Put simply, the length of the finish and aftertaste is the best indicator of wine quality. The terms 'finish', 'aftertaste', and 'length' sometimes give rise to confusion. 'Finish' refers to the final taste sensations of the wine as it is swallowed or spat. 'Aftertaste' encompasses the sensations that remain and develop as we breathe out, whilst 'length' is the measure of time for which finish and aftertaste last. To determine length, after the taster has spat the wine, they should breathe out slowly, concentrate on the sensations observing any changes or development, and count the number of seconds that the taste sensations last. The sensations delivered by poor quality and inexpensive wines will disappear after 5–10 seconds (short length), and any remaining sensations are likely to be unpleasant. Acceptable quality wines will have a length of 11–20 seconds (medium length), good wine 20–30 seconds (long length) and outstanding wines a length of 30 seconds or more (very long length). Truly excellent wines may have a length that runs into minutes. It is important that throughout this test of length, the sensations remain in tune with the actual taste of the wine, and also that everything remains in balance. The taster may wish to adjust the number of seconds timing given above for the various lengths to their own, individual perceptions.

Finish may be considered on a scale that runs from *short*, to *very long*. If any unpleasant characteristics dominate the length, they will affect the quality and should be noted accordingly. For example, a wine with unripe tannins and other bitter compounds might have a medium or even longer length. However, the bitterness will dominate, and the nature of the length becomes increasingly unpleasant. Sometimes faults that might not have manifested themselves previously become apparent on the finish of the wine. On occasions, these may include haloanisoles at a medium level – see Chapter 3 and smoke taint at a low level – see Chapter 12.

2.8   Assessment of Quality

2.8.1   Quality Level

Quality judgements are framework dependent. This poses a dilemma. Do we consider the quality of a wine only within the context of its peer group or against the entire wine world? Can a wine such as Beaujolais, which is usually made for early drinking in a soft, immediately approachable style, be described as outstanding quality even though it is carefully crafted, exquisitely perfumed, expressive of its origin and superior to most others of its type? The key to answering such questions is to be as objective as possible in the assessment, noting the quality of the wine, as perceived by the taster, according to the origin and price levels of the wine.

2.8.2   Reasons for Assessment of Quality

It is, of course, very possible that many wines of outstanding quality are not to our palate and, on occasions, simple wines may be very appealing. The reasons for our quality judgements should be logical and as possible. When reviewing the tasting assessment, consideration should be given to the intensity, concentration and complexity of nose and flavours on the palate, the structure of the wine, the balance, and length of finish. The following guidelines form a framework for quality assessments:

- *Faulty*: showing one or more faults, at a level that makes the wine unpalatable;
- *Poor*: a wine that is, and will always be, unbalanced and poorly structured. A wine with light intensity, very simple one-dimensional fruit flavours, maybe some flaws, and short finish;
- *Acceptable*: straightforward wine with simple fruit, of medium intensity, somewhat lacking in complexity, with a medium finish;
- *Good*: an absence of faults or flaws, well-balanced, medium or pronounced intensity and with a smooth texture, complexity, layers of flavours, and development on palate, fairly long or long finish;
- *Very good*: a complex wine, concentrated fruit, very good structure, well-balanced, long length of finish;
- *Outstanding*: intense fruit, and or tertiary flavours, perfect balance, very expressive and complex, classic typicity of its origin and very long length of finish.

A wine of very good or outstanding quality will present the taster with an unbroken 'line', i.e. a continuity from the sensations of attack, when the wine first enters the mouth, through the mid-palate and on to the finish. It will develop and change in the glass and gain complexity. In other words, it will not say all it has to say within a few seconds of the initial nose and taste. An outstanding wine will also exude a clear, definable, and individual personality, true to its origin, making a confident statement of time and place. It will excite in a way that seems to go beyond the organoleptic sensations. In other words, it will have the ability to move the taster in a similar way to a work of literature, art, or music.

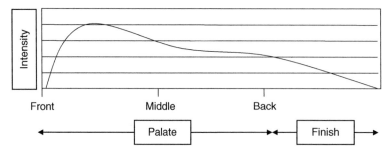

Figure 2.5 Palate profile.

The 'line' of a wine, as detailed above, may be depicted visually in the form of a palate profile. This is a graph that illustrates the intensity and the texture of a wine from the attack (front-palate) through to the mid-palate, to the back-palate and to the finish. An example of a palate profile is shown in Figure 2.5.

It will be noted that a criterion for a wine to be assessed as good, very good, or outstanding quality is the absence of any fault or flaw. As previously discussed, this can be a contentious issue. There are technical faults which, at low levels, and dependent upon their relationship with the multiflorous compounds giving aromas, tastes, structure, and texture to a wine may be acceptable or even add complexity and an extra quality dimension.

2.9 Assessment of Readiness for Drinking/Potential for Ageing

2.9.1 The Life-Cycle of a Wine

The topic of when wines are ready for drinking and the assessment of this during tasting is, by nature, complex. The life-cycle of wines depends on several factors: origin, colour, style, structure, and particularly quality. Inexpensive wines, be they red, rosé, or white, are made to be drunk immediately. The reds will generally have been made without, or with little, post-fermentation skin maceration that would give tannic structure. They will have been highly processed, including fine-filtration and technical stabilisation before bottling. A period of three or four years in the bottle is the maximum keepability, and after this time they will have lost fruit and become 'dried-out'. The further we move up the price and quality scales, the more wines benefit from some bottle ageing. Fine red wines are designed for bottle maturation. The time taken for them to reach their peak, and just how long they will remain there, will vary according to the quality and style of the vintage, the origin of the wine, and the winemaking techniques. The intensity of fruit on the nose and flavour on the palate are important indicators, but these may be 'closed' in youth. The components of solid structure, including medium to high levels of tannin, medium to high acidity and appropriate alcohol content, are the key indicators to a red that will improve in bottle. In youth, these will be fragmented, and to the novice, the wine will appear hard and unbalanced. Considerable bottle ageing will be required for them to evolve and integrate. High acidity, in particular, is a great preservative, but balance is crucial.

2.9.2   Level of Readiness for Drinking/Potential for Ageing

All wines have a window of drinkability, beyond which they will be in a state of decline, either slow or rapid, the latter particularly in the case of low quality wines. Within the window of drinkability, there will, in the case of high-quality wines, be a period when a wine is at its peak, in perfect balance and expressing complex tones that almost defy description. The tertiary characteristics will be fully developed, and the length of finish will be at a maximum. Unfortunately, there can be a disparity in the peak of drinkability between individual bottles of the same wine, sometimes due to the efficiency of the closure, but also if the storage temperature and other conditions have been inconsistent. As wines decline, the fruit will start to dry out. Richness will fade, oxidative and vegetal characteristics may overwhelm, bitterness comes to the fore, and the length of finish diminish. A decrepit wine may show high levels of oxidation and other faults. There are numerous factors the taster might consider in reaching an assessment as to where a wine is in its life-cycle including development on the nose and palate, structure, balance, concentration, complexity, and length.

2.10   Grading Wine – The Award of Points

2.10.1   Is the Awarding of Points Appropriate for Wines?

The grading of wine is a controversial topic. Some claim that wines cannot be assessed by scoring points or on star scale and that the whole tasting process should be qualitative, not quantitative. Wine crosses the boundaries of art and science, and the exciting and complex characteristics of quality wines cannot be reduced to mere numbers. Countering this argument, many critics point out that to show which wines are superior to others, they have to be rated on some scale. Critics of musical performances or theatre often give star ratings, as do restaurant inspectors, reviewers of cars, washing machines and pretty much everything marketable. However, an opera critic would not dream of rating Puccini's La Traviata 99 points, and Tosca 98, or rate an individual performance on such a fine scale. Of course, for the less knowledgeable consumer (or investor), knowing how well a wine has been scored makes the buying decision easier.

2.10.2   The Systems in Use

There are several different grading systems in use. Historically, a 0–7 scoring scale was often used, and of course, star ratings, from 0 to 5 stars remain a popular system. To many critics, the use of simple scales such as these means that expression cannot be given to the perhaps significant differences in quality between wines within one of the grading bands. Accordingly, the preferred systems now mark the wines out of 100 or out of 20 (possibly including fractions). The marks can, if deemed appropriate, be translated into a star rating, or descriptive band, e.g. 'Outstanding' or 'Highly Recommended'. Both systems have strengths and weaknesses. The scores given can be a sum of the individual scores for appearance, nose, palate, conclusions, etc., and in which case, the assessor will mark within a framework or can be an overall score. Critics of the former method point out that simply adding the scores means that glaring weaknesses or flaws in one section can result in a wine still being scored rather highly. If scores are useful at all, it is only to people who understand the system. As much as low scores are not used in marketing materials, to the uninitiated, a score of 75 out of 100 would equate to 15 out of 20, and both would seem very good!

2.10.3 Nothing but a Snapshot Reflecting a Moment in Time

A couple of personal anecdotes will illustrate that the award of points or medals are little more than an illustration of how a wine is perceived at the time it was assessed. In April 2020, I escaped COVID-19 lockdown to buy essential goods. The bottles of 2016 'Côtes de Bourg' from a Petit Château look appealing on the shelf of my local Super-U supermarket. The price is right, and hey – they bear a sticker proclaiming that the wine scored 90 points in the 2018 Decanter World Wine Awards: a Silver Medal. I buy three bottles, taken from an unopened box. I give them a few days to rest, open a bottle, and even before I can pour a glass, I can smell the unmistakeable odours of oxidation. I check the closure – a Diam technical cork, and this seems fine. Let's open another bottle – just the same. And so is the third. There are no signs of the wine being poorly stored. Assuming the Decanter assessment panel was competent, they made snapshot judgements of the wine. At the time in its life-cycle that I would have expected the wine to be in top order, it was undrinkable. A similar situation had happened to me in the UK a couple of months earlier, with a 2017 South Australian Shiraz, a Trophy winner in the 2019 International Wine Challenge. When I twisted the screw-cap, on this occasion, it was the 'reduced' nose of garlic that screamed out and did not dissipate with aeration in the glass. I could only reflect with sadness that the unwitting consumer was excitedly buying award-winning, faulty wines.

2.11 Blind Tasting

2.11.1 Why Taste Blind?

Tasting wines 'blind' without the taster being given some or all the information about their identity is regarded as the most 'objective' way to assess wine. Blind tasting is also the best way of improving tasting technique, making the tasters rely on their individual perceptions and apply their individual descriptors. It is a valuable means of expanding the memory bank, particularly about the relationship between the descriptors and the type of wine tasted. Depending on the objectives, the taster may have absolutely no information, or maybe given certain relevant details, e.g. the wines are all Burgundies, are all made from one particular variety or are all in a certain price range. In other words, the wines are semi-specified. An alternative approach, sometimes called 'single blind', is when the details of the wines to be tasted are revealed beforehand, but not the order of tasting. Such frameworks can help concentrate the mind in evaluating wines for their quality, typicity, style, and maturity.

2.11.2 Blind or Sighted?

There is no doubt that knowing the identity of the wine to be tasted impacts a taster's perception. There are good reasons why the taster might wish to know the details of the wine being tasted. First, the wine is immediately placed in context, including location within a region, i.e. the sense of place, and the characteristics of the vintage in question. What may be considered a flaw may be explained by the context, including geographical location. For example, many Italian red wines possess high levels of volatile acidity, which is part of their Italian identity. Second, knowledge of the region and/or producer, e.g. their production methods and 'philosophy', can help the taster understand the wine. As discussed in the Introduction, wines from some regions and producers display characteristics that might be regarded as faults or flaws in other wines. Third, knowledge of a wine will help the taster place it at a particular point in its life-cycle. Is it at its best? What is the potential to age based on the reputation of the producer, region, and vintage?

2.11.3 Tasting for Quality

It can be argued that only by tasting blind can the taster come anywhere near to an objective assessment. This is especially important when, considering the factors that reveal quality, including complexity, balance, and length, as the taster is divorced from being influenced by preconceptions. If the purpose of the tasting is to judge relative qualities, the wines chosen for the event should be comparable from this point of view, and in the broadest sense be stylistically similar. As we have seen, there is no point in trying to judge the quality of a Beaujolais against a *cru classé* Bordeaux.

2.12 Final Reflections

I have used the word *objective* a few times in this chapter, referring to the approach that a taster should use when assessing a wine. However, a taster should recognise that objectivity is a noble ideal, but one that is impossible to achieve. We are all influenced by our history, education, training, culture, and a hundred and one other factors. We may try to eliminate bias and prejudices and to avoid stereotyping, but try is all we can do. Several of my professional colleagues claim to be 'objective' tasters, and I am sure they make every effort in this regard. When assessing a wine, with all the art and science that is in the glass, we are making judgements. Can there really be such a thing as an objective judgement?

References

1 Tempere, S., Cuzange, E., Bougeant, J.C. et al. (2012). Explicit sensory training improves the olfactory sensitivity of wine experts. *Chemosensory Perception* 5: 205–213. https://doi .org/10.1007/s12078-012-9120-1.
2 Kaeppler, K. and Mueller, F. (2013). Odor classification: a review of factors influencing perception-based odor arrangements. *Chemical Senses* 38 (3): 189–209. https://doi.org/10 .1093/chemse/bjs141.
3 Grainger, K. and Tattersall, H. (2016). *Wine Production and Quality*, 2e. Chichester: Wiley Blackwell.

4 ISO 8586/2012 (2012). *General guidelines for the selection, training and monitoring of selected assessors and expert sensory assessors.* Geneva: International Organization for Standardization.

5 ISO 8589:2007 (2007). *Sensory analysis — General guidance for the design of test rooms.* Geneva: The International Organization for Standardization.

6 ISO 3591:1977 (1977). *Sensory analysis – Apparatus – Wine-tasting glass.* Geneva: The International Organization for Standardization.

7 International Organisation of Vine and Wine (OIV) (2015). *Review Document on Sensory Analysis of Wine.* Paris: OIV. http://www.oiv.int/public/medias/3307/review-on-sensory-analysis-of-wine.pdf.

8 Lin, J., Massonnet, M., and Cantu, D. (2019). The genetic basis of grape and wine aroma. *Horticulture Research* 6 (81) https://doi.org/10.1038/s41438-019-0163-1.

9 Ugliano, M. (2013). Oxygen contribution to wine aroma evolution during bottle aging. *Journal of Agricultural and Food Chemistry* 61 (26): 6125–6136. https://doi.org/10.1021/jf400810v.

10 Coetzee, C. and du Toit, W.J. (2012). A comprehensive review on Sauvignon Blanc aroma with a focus on certain positive volatile thiols. *Food Research International* 45 (1): 287–298. https://doi.org/10.1016/j.foodres.2011.09.017.

11 Malik, B., Elkaddi, N., Turkistani, J. et al. (2021). Mammalian taste cells express functional olfactory receptors. *Chemical Senses* 44 (5): 289–301. https://doi.org/10.1093/chemse/bjz019.

12 International Organisation of Vine and Wine (OIV) (2019). *International Oenological Codex.* Paris: OIV. http://www.oiv.int/public/medias/6550/codex-2019-en.pdf https://www.oiv.int/public/medias/7790/codex-2021-en.pdf

13 Noble, A.C. (1998). Why do wines taste bitter and feel astringent? In: *Chemistry of Wine Flavour* (eds. A.L. Waterhouse and S.E. Ebeler), 156–165. Washington, DC: American Chemical Society.

14 Rankine, B. (1989). *Making Good Wine.* Sydney: Macmillan Australia.

15 Matthews, M.A. (2015). *Terroir and Other Myths of Winegrowing.* Oakland: University of California Press.

3

Chloroanisoles, Bromoanisoles, and Halophenols

In this chapter, the longest in the book, I discuss one of the most feared wine faults/taints: contamination by haloanisoles. Contaminated wines are very often undrinkable, exhibiting a distinctly musty nose and palate. Even at concentrations below sensory thresholds, the compounds mute fruity aromas. Haloanisole-affected wines are usually, and sometimes erroneously, described as 'corked'.

3.1 Introduction

Contamination of wine by haloanisoles, i.e. chloroanisoles and bromoanisoles, is not an uncommon fault, although incidences of related taints have decreased since the early 2000s. Haloanisoles and their halophenol precursors (that can also taint wines in their own right) do not naturally occur in wine, their presence is always a result of external contamination. Minute concentrations of chloroanisole and bromoanisole compounds negatively impact wine: sensory perception thresholds range from 0.13 to 50 ng/l depending upon the compound(s) involved and the wine matrix, but their presence even at levels below sensory perception threshold can be damaging.

If a wine is heavily contaminated by haloanisoles, it will be instantly recognised by nosing – there is usually no need (or certainly any desire) to taste. Upon raising the glass to the nose, the smell is immediately of damp cellar, a heavy mustiness, reeking of wet sack, musty old newspapers, and perhaps tones of mushrooms or dry rot. If the wine is tasted, it will be dirty, fusty, and earthy – rather like biting into a rotten apple or other mouldy fruit. The presence is particularly noticeable on the back of the mouth. However, at low levels of contamination, none of these may be apparent, but the aroma and taste will be masked and muted; fruitiness, identity, and complexity are reduced, and the taste is flat and perhaps a little dusty. The haloanisole most widely found in wine is 2,4,6-trichloroanisole (2,4,6-TCA) that suppresses the transduction currents of olfactory receptor cells (ORC) [1]. Low levels of contamination are of particular concern to wine producers, as the consumer is most unlikely to recognise the wine as faulty, but will simply consider it to be of low quality. Accordingly, a wine must always be regarded as faulty if it contains haloanisole compounds at anything other than trace levels, even if the trademark musty odours are not apparent.

There continues to be considerable research into the role of haloanisoles and their impact on affected wines affected wines. Cork closures have long been identified as a major source

Wine Faults and Flaws: A Practical Guide, First Edition. Keith Grainger.
© 2021 John Wiley & Sons Ltd. Published 2021 by John Wiley & Sons Ltd.

of chloroanisole contamination, and the term 'corked' is often used in relation to badly contaminated wines, whether or not the source of the taint is a cork closure. In the last 20 years, the major manufacturers of wine corks have changed their working practices and introduced new technologies aimed at reducing and eventually eliminating instances of contaminated corks. However, related taints are often found in bottled wines not closed with any type of cork and in wine pre-packaging. There can be many other sources of haloanisole compounds in wine, including production equipment, processing aids, water, and the environment, atmosphere, and conditions of the cellar. Whilst the occasional presence of haloanisoles in used oak barrels and their consequential migration into the wine stored in them has long been known, of particular current concern is the fairly recent discovery of the compounds in new oak barrels purchased directly from certain coopers.

The main haloanisoles that taint wine are 2,4,6-TCA (causing over 80% of incidences); 2,3,4,6-tetrachloroanisole (2,3,4,6-TeCA); pentachloroanisole (also known as 2,3,4,5,6-pentachloroanisole [2,3,4,5,6-PCA]); and 2,4,6-tribromoanisole (2,4,6-TBA). There are other related taint compounds that may be found in wine, including halophenols, which may impart a plastic-like odour: halophenols are erroneously usually considered only as the precursors of haloanisole taints. Musty taints in wine can also be caused by other non-related compounds.

For the avoidance of confusion, it is worth noting at this point that many wine related publications and research papers give 2,4,6-tribromoanisole (chemical formula $C_7H_5Br_3O$) the acronym of TBA. However, TBA is also an acronym of tertiary butyl alcohol aka terbuthylazine (chemical formula $C_9H_{16}ClN_5$) used in Europe as a herbicide and pesticide (it is not registered for use in United States). This compound is unrelated to 2,4,6-tribromoanisole, which will be abbreviated as 2,4,6-TBA in this chapter. Additionally, for clarity, when discussing the various chlorophenols, the full name, rather than just the acronym, will often be used.

3.2 Haloanisole Contaminations in the Food, Drinks, Water, and Pharmaceutical Industries

To most people involved in the world of wine, the problems of TCA, 2,4,6-TBA and related compounds are considered to be an industry-specific issue. However, haloanisole taints can also affect many foods and drinks other than wine, and cause a particular concern to the pharmaceutical industry. Food and beverages that may be affected include beer, sake, whisk(e)y, tea, coffee, cocoa powder, soft drinks, milk, water, potatoes, potato crisps (chips), apples, chestnuts, flour, raisins, rice, sultanas, bananas, baked buns and cakes, eggs, broiler chickens, shell-fish, pre-packaged vegetables (particularly carrots), and pre-packaged washed salads.

3.2.1 Haloanisole Contamination of Pharmaceutical Products

Taints by haloanisole has been detected in pharmaceutical products in recent years. Contamination has frequently resulted in the recall of several branded over-the-counter (OTC) products, for example, nearly 43 million bottles of the painkiller Tylenol[*], the allergy relief medicine Benadryl[*], and other products from the same manufacturer in 2010 due to contamination with 2,4,6-TBA [2, 3], as well as 52 lots of the diabetes drug Glumetza® by Depomed and 360 000 bottles of the cholesterol-lowering drug Lipitor® by Pfizer [4]. Haloanisole contamination contributed in the enforced closure of a pharmaceutical manufacturing plant at Fort Washington in that year [4, 5]. More drugs were recalled in 2011, such as risperidone (Risperdal) [5] and at least 11 700 bottles of HIV/AIDS drug Prezista® manufactured in the United Kingdom [4]. In all these cases, 2,4,6-TBA contamination of wooden pallets from South America was found to be the major source of the taints. Although at concentrations that they may be present in wine and other products, haloanisoles are generally regarded as harmless to human health (as emphasised by the wine and cork industries), some of the above contaminated products are reported to cause nausea, vomiting, and diarrhoea.

3.2.2 Haloanisole Contamination of Food and Non-Alcoholic Drinks

The main haloanisole to taint wine, 2,4,6-trichloroanisole (TCA), is also a known taint of coffee – as such it is often referred to as the 'Rio' defect. Coffee is, of course, highly flavoured, but in coffee that has been brewed, the odour threshold for TCA is just 8 parts per trillion (ppt), and the taste threshold as low as 1–2 ppt. The main origins of taints by haloanisoles in foods have been found to be wooden pallets and wooden shipping container floors that have been chemically treated with preservatives or bleached with hypochlorite (common bleach). Packaging materials, such as multi-walled paper sacks, jute sacks, plastics, and cardboard, are also significant sources. In broiler chickens and eggs, related taints including 2,3,4,6-tetrachloroanisole (TeCA), have been found to derive from preservative-treated wood shavings used a component of cage litter. There are many possible sources of haloanisole contamination in both draught and packaged beers. An interesting incidence in Australia some years ago was the tainting of canned beers, due to TCA being absorbed into the internal lacquer of cans and also the sealing compounds of the can ends the source was found to be the contaminated floor of a shipping container [6].

3.2.3 Haloanisoles in Tap Water and Bottled Water

Haloanisoles have been found in tap waters [7, 8]. A study published in 2016 showed the occurrences of four kinds of haloanisoles in the municipal water in 22 cities of China, with Beijing representing the most serious occurrence [9]. In areas where there is persistent contamination, resident consumers become desensitised to the taints to a large degree. A rather amusing and ironic incident that took place in early 2000 at a dinner at the Stanford Plaza Hotel in Adelaide, Australia, is related by George Taber in his book *To Cork Or Not To Cork* [10]. António Amorim, CEO of the closure division of Amorim, the largest cork manufacturer in the world, was hosting a dinner attended by critics of cork wine closures from the Australian wine industry and the Australian Wine Research Institute (AWRI). Having spent over four hours fielding hostile questions from the attendees, António was suffering from a dry mouth as a result of all the talking. He reached for a glass of local tap water, took a taste, and passed it around the table for all to smell. Everybody recognised that the water was contaminated with TCA [10]. Some 19 years later, in 2019 at the 'Australia Redefined' trade and press tasting in London, I was tasting alongside a colleague, wine educator Richard

Bost, and we were alternating white and red wines at the exhibitors' stands. Upon rinsing my glass with London tap water, I was immediately struck with the haloanisole odour – a judgement immediately confirmed by Richard much to his amazement and winemaker Larry Cherubino whose wines we were tasting. Importantly, it is not just tap water that can be affected: haloanisoles have also been known to taint bottled water, flavoured water, and soft drinks.

3.3   Haloanisole Contamination of Wines

Whilst there can be discussion or even dissention as to the concentration at which many of the compounds that are implicated in the wine faults and taints discussed in this book are, or become, unacceptable, there is no doubt that contamination by haloanisoles at any level above trace is a serious fault. Concentrations of the various compounds at a level of just a few ppt are sufficient to have a severe impact upon affected wines. By way of often quoted analogies, 1 ppt is equivalent to 1 grain of wheat in 100 000 tonnes or, from an alternative viewpoint, a single second out of 32 000 years. Parts per trillion in liquids are usually scientifically expressed as nanogrammes per litre (ng/l) – both terms will be used throughout this book. Haloanisole contamination of wines has resulted in many product recalls. Winery production facilities and maturation cellars have been destroyed and rebuilt to eliminate sources of related taints (Section 3.4). Haloanisoles are very volatile and migrate easily, and ensuring a winery and the wines produced therein remain unaffected requires constant diligence and regular screening and monitoring of the winery environment. Whilst it is sometimes possible to remove or reduce the taints in bulk wine without loss of flavour by the methods discussed in Section 3.15, such treatments are expensive and may compromise quality. Standard filtration, using sheet, membrane, or crossflow filters, is totally ineffective, as is conventional fining, although historically several producers in Bordeaux did claim that the taints were reduced following fining with skimmed milk (casein is the principal protein found in milk and it is commonly used as a fining agent, particularly for white wines). Further, there is no way of neutralising haloanisoles in wine – they must be removed.

3.4   The Economic and Reputational Costs to Wine Producers and the Wine Industry

Contamination of wines by haloanisoles has resulted in substantial economic costs and reputational damage to the wine industry, generally to individual wine producers, and also to the cork industry.

3.4.1   The Economic Cost to the Industry at Large

There are no totally reliable statistics available as to either the current or the historic worldwide economic cost of haloanisole contamination to the wine industry. In the period 1995 – 2003, estimates of the financial cost each year due to 'cork taint' caused by TCA varied from €1000 million (€1 billion) [11] to as much as US$10 billion (approximately €9 billion)

[12], this latter figure being often cited. The final report summary, published in 2016, for the 'ENCORK' (Electronic Nose To Detect Haloanisoles In Cork Stoppers) project, which was largely funded with a grant of €971 000 from the European Union, suggested that the annual cost of TCA to European Union (EU) wine producers is €700 million [13]. There continues to be a massive financial impact on the cork industry, from research (over €500 million in the 10 years up to 2019), investment in new equipment and processes aimed at eradicating the problems, and the loss of market to the alternative closure industries. Additional expenses for PR campaigns have been deemed necessary to restore the confidence of consumers in cork closures and the cork industry.

3.4.2 The Economic and Reputational Cost to Individual Wine Producers

The economic and reputational cost to individual wine producers of haloanisole contamination of wines and cellars can be huge. Whilst some producers have been open with regard to discussing discovered contaminations, others have preferred not to publish information for commercial and perceived reputational reasons. Wine is, without doubt, the most discussed and written about drink in the world. The possibility of having bottles contaminated by TCA and associated taints (together with the chance of buying counterfeit wine) are the main fears amongst aficionados and collectors. If word gets out that there is a problem with a batch, or even a producer, the effect on the company can be devastating. James Laube, who writes about Californian wines for the hugely influential *Wine Spectator* magazine is known for being highly sensitive to haloanisole taints, and has on several occasions 'outed' producers whose wines exhibited contamination. With the proliferation of wine blogs and forums, information (and sometimes misinformation) on faulty bottles and tainted cellars quickly reaches wine buyers. When they feel they have been cheated, consumers spread the word, and memories are long. On the *Bordeaux Wine Enthusiasts* forum in 2012, a consumer reported that he bought four undrinkable bottles of a 1997 Cabernet Sauvignon from a Sonoma winery, opened shortly after bottling and asked the winery if they knew of a possible problem. 'There was a flat denial and they became accusatory.... I have never had (the winery's) product since. Amazing how a little bad PR can turn someone off for ever!' [14].

3.4.3 The Nightmare of Haloanisole Contaminated Wineries and Cellars

There are many reported cases of cuveries and barrel cellars at wineries becoming contaminated with TCA, TeCA, 2,4,6-TBA, PCA and/or halophenols that have necessitated considerable expenditure for remedial works. As is the case with constant exposure to sulfur dioxide (SO_2), cellar staff can become accustomed to haloanisoles compounds in the atmosphere (the so-called fatigue or adaption effect) and regard it as part of the winery's ambient 'smell'. In other words they become partially anosmic, and it is only when the contamination is revealed by third parties that the problem is revealed and action taken. In the case of severe and widespread contamination, there are several instances of entire cellar buildings having to be demolished and replacement facilities built.

3.4.3.1 California Pays a High Price

In California several high-end producers have historically been found to have a haloanisole-related problem, including the flagship estates Beaulieu Vineyard, Pillar

Rock [15], Hanzell Vineyards [16], and Gallo Sonoma [17]. Hanzell, having become aware of a problem with wines showing (low level) TCA contamination, including the 1999 vintage Pinot Noir and 2000 vintage Chardonnay, contracted St. Helena-based ETS Laboratories in 2003 to conduct a thorough investigation of the production areas. They found TCA in multiple locations, including hoses, barrels, water tanks, drains, and the air conditioning system [18]. Hanzell chose to build an entirely new building for wine production and decided to withdraw its wines from the market for nine months whilst it assessed the problem. In 1976 Château Montelena, a Napa Valley winery, stunned the wine world when its 1973 Chardonnay was declared the Number 1 white wine in the infamous 'Judgement of Paris' tasting, beating several prestigious white Burgundies. The events leading up to that remarkable event are depicted, with a great deal of artistic licence, in the film 'Bottle Shock'. If 1976 was a high for the producer, then from 2002 to 2004 was perhaps a low, due to TCA contamination in some parts of the winery and cellar, albeit at a low level [18, 19]. James Laube of *Wine Spectator* investigated and the magazine had seven samples of the winery's Cabernet Sauvignon analysed by ETS Laboratories. These showed TCA, and also low levels of TeCA and PCA [19]. The eventual confirmation of winery contamination necessitated the replacement of plumbing, several tanks, barrels, barrel racks, wooden catwalks and ladders (replaced with aluminium), and other items. Every stone surface was hand cleaned. Haloanisoles can contaminate all parts of a winery, but pipes, drains, and air conditioning systems are particular danger areas as they can quickly spread contamination to other parts of a building. However, good ventilation is essential to help reduce the risk of cellar contamination, and air conditioning systems, regularly monitored and cleaned, can actually lower the level of existing contamination in production and storage areas (Section 3.14).

3.4.3.2 Affluence and Crisis in Bordeaux

In France, particularly in Bordeaux, there have been many cases of cellar contamination with haloanisoles, with a very high incidence in the mid to late 1980s and the 1990s. The early 1980s were generally very good for the Bordeaux châteaux, with the top wines selling for what at the time were unprecedented high prices following the excellent 1982 and the very good 1983 vintages. The United States had recently turned on to fine French wines, aided by articles and reviews in *Wine Spectator* magazine, founded in 1976, and particularly the critical assessments in *Robert Parker Wine Advocate,* the first issue published in 1978. The importance of the rise of the critic Robert Parker Jr. cannot be overstated – his influence was to become so great that he would be nicknamed the 'Emperor of Wine', and even flagship Bordeaux and Rhône Valley producers would change their styles of wine to suit his palate and hopefully gain high scores in his evaluations. Although Parker has recently announced his retirement, the work of his team continues unabated. Affluent times since the 1980s speared many top Bordeaux properties into renovating their wineries and chais. As I well recall from visits made at the time, some of the châteaux had little work on the fabric of the buildings for more than a century, although the installation of the best 316 grade stainless steel vats with temperature control had become *de rigueur*. Included in many of the renovations were new timbers, including joists, purlins, and rafters. Disastrously the new timbers were treated with wood preservatives based on chlorophenol or bromophenol compounds, which are the direct precursors of chloroanisoles and bromoanisoles. There were certainly many other sources of cellar taints, but the timber suppliers were to bear the brunt of the blame. Contamination of cellars with haloanisoles and their halophenol precursors was far from uncommon, and although affected châteaux embarked on remedial programmes, at enormous expense, many denied that they had a problem to people who had bought their wines. Communication issues were perhaps compounded with regard to the prestigious Bordeaux châteaux, as the properties generally do not sell their wine directly to wholesalers or consumers, but trade via third parties on the so-called 'Place de Bordeaux'. Until the early 2000s, it was sometimes joked that the consumer was little more than an inconvenience who got in the ways of the smooth running of the Bordeaux châteaux and their wine business.

Perhaps the most widely discussed historic case of cellar contamination in Bordeaux is the disaster that befell Château Ducru-Beaucaillou. From 1986 to 1990, the chai of this illustrious Saint-Julien property, a Deuxième Cru Classé in the infamous 1855 Bordeaux Classification, suffered haloanisole contamination [20]. The problem was probably TCA and TeCA combined, and possibly even 2,4,6-TBA. However, all early reports referred to just TCA (the identification of 2,4,6-TBA as a wine contaminant that also imparts musty odours and flavours was not made until 2004), and many commentators continue to do so. Whilst only an unquantifiable percentage of the bottles produced in that period were affected, information about the problem, both true and false, circulated amongst lovers and collectors of fine wines. Prior to this period, the reputation of the property and its wine had been very high – it was regarded by writers and critics as a so-called 'super-second', i.e. along with a few other properties including Châteaux Léoville-Las Cases and Pichon Longueville Comtesse de Lalande, performing at a level superior to other Deuxième Cru Classés. As a consequence of the severe contamination, the chai was demolished and a new concrete facility was constructed. The quality of the wines and the reputation of the property recovered and is back with the super seconds, but even in 2018 the historic problems were still causing concern to collectors. In April of that year, on the *Wine Berserkers* forum, claiming to be 'the world's largest and most active online wine community', a series of 30 posts began under the heading '*Buyer beware – Ducru Beaucaillou 1988*'. The poster stated: 'I noticed on an auction site here in France that Ducru-Beaucaillou 1988 is still being sold and bought, at very high prices .... I was served one by a friend a few weeks back - like all the others I've tried, it was undrinkable' [21]. Also writing in 2018 on the *SpitBucket* website, Amber LeBeau notes, 'from 1986 to 1995, the estate was plagued with systematic cork taint issues that required significant investment to eradicate. Many of the bottles from this period had to be recorked, with those demonstrating noticeable TCA destroyed' [22]. That the descriptor of 'cork taint' was still given to what was clearly a cellar issue so many years after its occurrence is perhaps more than a little surprising.

Several other famous Bordeaux properties also suffered contamination of their cuveries and/or chais in the 1980s or 1990s. These included Château Latour, Premier Grand Cru Classé in Pauillac, with documented incidences affecting the 1994 and 1996 vintages; Château Gruaud-Larose a Deuxième Cru Classé in Saint-Julien with the 1994 vintage particularly affected; and Château Canon, Premier Grand Cru Classé (B) in Saint-Émilion, which suffered problems in the period 1992–1995 [23]. The contamination necessitated the rebuilding of cellars at Château Latour and extensive remedial works at Canon that had only had new fir timbers installed as recently as 1987. The renovations to rid the contaminations included replacing the entire framework with eighteenth century oak beams, planing the surfaces of the vats, burning the barrels, and sandblasting the walls followed by rendering with lime. For Château Canon, the cost to the house of Chanel and its owners the Wertheimer brothers (who had only purchased the property in 1993) was some US\$ 500 000. Another Bordeaux property, with a new chai installed in 1992, suffered an almost immediate contamination and the chai had to be destroyed in 1996. The odour and taste thresholds for the various haloanisoles are just a few ppt, as detailed in Section 3.6. However, figures for the level of chloroanisoles in the 1994 Château Latour were a staggering 360 ppt, the 1996 Les Forts de Latour (sometimes erroneously called the second wine of Latour) 340 ppt, and the 1989 Château Ducru-Beaucaillou 149 ppt.

Although the crisis years for the illustrious properties have now passed, I have no doubt that at the time of writing there remains the existence of haloanisole cellar contamination at several, mostly small, Bordeaux châteaux together with at least one 'left bank' cru classé, and in my mind there can be little doubt too that owners and winemakers have become desensitised to the odours. On several occasions in the last decade when tasting such wines, I, together with colleagues, have identified a problem, only for this to be vehemently denied by the makers and cellar staff. Of course, Bordeaux is not the only region in France where the properties have suffered related contaminations. However, due to the closeness of the Atlantic Ocean, the Gironde estuary and the Garonne and Dordogne rivers there is high humidity in many areas, as evidenced by frequent outbreaks of downy mildew, which thrives in humid conditions. In 2018 this fungal disease, which often breaks out after periods of rain, hit many areas badly resulting in considerably reduced yields at some properties. Due to a high water table, wines are generally matured in chais that are above ground and are often less cool than would be the case with underground cellars. In the absence of adequate ventilation, air circulation, and temperature control, fungi and moulds can grow. The role of filamentous fungi and moulds in the production of haloanisoles is discussed in Section 3.7.

3.4.3.3   Cellar Contaminations in South America

In South America too, there were many examples of wines suffering haloanisole odours resulting from contaminated cellars, with 2,4,6-TBA often found to be the compound involved. From 2004 until 2006 Viña Errázuriz, a highly regarded producer based in the Aconcagua Valley in Chile, realised that parts of the beautiful old cellar at Panquehue were contaminated with 2,4,6-TBA [24], prompting a strident eradication programme. I visited this cellar several times in the early 2000s, and the musty odours struck me on each occasion. Another Chilean winery, Viña Santa Laura recalled 300 cases of the 2001 vintage of Laura Hartwig Cabernet Sauvignon after the owner found samples that showed a mustiness, also caused by 2,4,6-TBA. The source was the wooden beams on which barrels were stacked [24].

3.5   Sensory Characteristics and Detection of Haloanisoles in Wine

The presence in wine of haloanisoles at high levels (which are in fact just a few ng/l as detailed below) will be detected as musty odours on the nose and palate, and the wines usually rejected, even by most untrained or relatively inexperienced tasters. Even at low levels, untrained consumers can often detect related odours, although they will not necessarily reject the wines [25].

3.5.1 The Odours and Tastes of Haloanisoles in Wine

Descriptors for both nose and palate of affected wines include:

- Musty;
- Damp;
- Earthy;
- Wet sack;
- Dry rot;
- Mushrooms;
- Mould;
- Wet concrete – this may be apparent at haloanisole concentrations below those at which 'musty'-like aromas are detected.

On the palate, especially the back palate, there can be the taste of biting into a rotten apple, or other decaying fruit. At low levels of contamination, the above odours and flavours may not be apparent but wines may be described as

- Acrid
- Dusty

However, contamination also has a strong masking effect on fruity notes [25] and result in the wine tasting dull and flat. Low level contamination, i.e. below accepted sensory detection thresholds, is of major concern to producers as consumers may not consider an affected wine to be faulty, but simply to be of poor quality. As such they are less likely to purchase the wine in question again and may even avoid the brand completely. It should be noted that experienced tasters, trained tasters, and those with knowledge of related taints are generally more sensitive to haloanisole contamination and will usually detect and reject wines at lower concentrations than inexperienced, untrained, or unaware consumers.

3.5.2 Variance in Sensory Detection Thresholds and Consumer Rejection Thresholds

The detection thresholds (DT) of haloanisoles in wine, categorised into odour detection thresholds (ODT) and taste detection thresholds (TDT), are very low (although not as low as in water, where they can be as low as 0.03 ng/l). TCA has the lowest DT in wine, perhaps in some sparkling wines, as low as 0.13 ng/l. The carbon dioxide in sparkling wine rapidly volatises the compound making the musty character overwhelming at minute levels of concentration. By way of comparison, the odour detection threshold for alcohol (ethanol) in wine is 100 mg/l. However, there is considerable variance in the DT figures stated in published literature, and a great deal depends upon the wine matrix as well as the skills, experience, and training of the assessors. The level of taint at which a consumer will reject a wine is generally much higher than that at which it is detected, and there is usually a different odour rejection threshold (ORT) and taste rejection threshold (TRT). A 2006 study, with a panel of mostly regular but untrained wine consumers, students and laboratory staff tasting, nosing and tasting unaffected white wines and wines 'spiked' with various levels of TCA, found an ODT of 1.5 ng/l and TDT of 1.0 ng/l in white wines, but the consumer ORT was 10.4 ng/l [26]. In the case of red wines, the ODT was 0.9 ng/l and the TDT 1.7 ng/l, but the ORT was 16 ng/l [26]. In this study, TRTs were not recorded, but the difference in reported ODTs and ORTs is staggering. Another study in 2005, using regular consumers of wine and evaluating responses to tasting white wines only, found a TCA Consumer Detection Threshold of 2.1 ppt, but a Consumer Rejection Threshold of 3.1 ppt [27]. At concentrations of TCA at 8 ppt and higher, between 80% and 90% of the panellists in this study rejected the wines.

Wines can sometimes smell musty for reasons other than haloanisoles or related compound contamination, for example, when produced from fruit affected by powdery mildew (often *called Oïdium tuckeri*, but correctly named *Uncinula necator* [synomym *Erysiphe necator*]) [28]. Many wineries that purchase grapes reject batches if more than 3% of the berries or clusters show signs of this mildew. However, many or even most of the musty off-odours due to powdery mildew will disappear during the fermentation. Sometimes wines can pick up musty notes from old, damaged barrels or un-sanitised barrels. If, in general wine drinking circumstances such as at a restaurant, a taster has doubt as to whether or not a haloanisole is the reason for a musty nose, a simple test may assist. Some pure, non-chlorinated, water should be added to the suspect wine in the glass, in the proportion of two-thirds wine to one-third water, and the wine is nosed again. If the musty aroma is reduced, then it is unlikely that a haloanisole is the cause; conversely if the musty aroma is increased, the taster can be reasonably confident that a haloanisole is the culprit.

It has been suggested that people are becoming desensitised to haloanisole taints, perhaps due to their occasional presence in many products, as discussed in Section 3.2. Speaking to the San Francisco Chronicle's wine critic Esther Mobley in 2019 [29], Lindon Bisson, professor emeritus at UC Davis gave the observation that when she started teaching some 30 years ago maybe one student in a 100, or even 200, could not detect TCA. 'But then, over time, it started to be higher and higher percentages of the students'. Bisson suggested that the reason was the contaminated bags of ready-to-eat carrots that were a student favourite. 'To them it wasn't negative', she says, 'they were perfectly happy drinking this wine that would turn our stomachs [29]'.

3.6 The Haloanisoles Responsible and Their Detection Thresholds

There are 11 haloanisoles that are known to contaminate wine.

3.6.1 The Main Haloanisoles Responsible

The four main haloanisoles responsible for the contamination of wine are

- 2,4,6-TCA
- 2,3,4,6-TeCA
- PCA
- 2,4,6-TBA

By far the most important of these contaminants is TCA, being responsible for over 80% of haloanisole tainted wines.

Other haloanisoles are known to taint wines, but in practical terms are not significant contaminants they include

- 2-Bromoanisole
- 2-Chloroanisole
- 4-Bromoanisole
- 4-Chloroanisole
- 2,4-Dibromoanisole
- 2,4-Dichloroanisole (2,4-DCA)
- 2,6-dichloroanisole (2,6-DCA).

It is often the case that more than one of the haloanisoles will co-occur; i.e. they are simultaneously present in wines (and/or corks) [30, 31]. Mark Sefton and Robert Simpson of AWRI note that 2,4-DCA, 2,6-DCA, TeCA, and PCA were frequently found together with TCA in wines sent for analysis on account of suspected 'cork taint' [31].

Other haloanisoles that are not known to taint wines include 2,3,6-TCA, and 2,3,4-TCA.

3.6.2 2,4,6-Trichloroanisole (TCA)

This compound is also known as 2,4,6-trichloro-1-methoxybenzene (molecular formula, $C_7H_5Cl_3O$). The precursor is 2,4,6-trichlorophenol (TCP). TCA is by far the most common cause of musty and associated aromas in wine, with estimates varying from 80% to 85% of tainted wines and as such it has been by far the most widely researched. TCA was the first compound to be identified as a cause of 'cork taint' in wine as long ago as 1981 by the Swiss scientist Hans Tanner et al., but the paper was only published in German [32]. The first paper to be published in English was in 1982 [33], but the importance of the findings still took some years to be recognised. As the first compound identified as responsible for associated musty aromas, and with much research, particularly prior to 2004 either targeting or concentrating on this compound, it is possible that TCA's overall contribution to musty wines has, at times, been overstated. However, it does have the lowest odour and taste detection thresholds of all the haloanisoles. TCA is chemically stable, and as such does not significantly degrade or diminish in wine over a period of time *per se* [31], although the concentration in bottled wine can reduce on account of absorption by the cork closure. Fortunately, TCA cannot be synthesised in bottled wine.

TCA can be formed anywhere there is the following combination:

- Material containing cellulose – e.g. cork, wood, cardboard, and polymers;
- Mould – particularly filamentous fungi (moulds grow under damp conditions);
- 2,4,6-trichlorophenol (2,4,6-TCP) – e.g. where chlorinated water or hypochlorite-based cleaning/sanitising materials have been used.

The other haloanisoles that may affect wine are formed by similar pathways, from the relevant halophenol precursor.

Sensory odour detection threshold of TCA in still wine is now generally accepted as 1.4–3 ng/l, though other studies have reported values beyond this range. The sensory threshold of 3 ng/l is quoted by Andrew Waterhouse of University of California at Davis and colleagues [34]. As we have seen, the perception threshold of TCA, other haloanisoles, and unrelated musty taints is usually lower in sparkling wines and white wines than in reds: the TCA odour detection threshold in sparkling wines is perhaps as low as 0.13 ng/l. Some commentators state higher thresholds, particularly for red wines, often between 4 and 6 ng/l, although even higher figures have at times been brandished. It is generally accepted that higher levels are needed to detect the compound in a red wine matrix due to its greater complexity. Greater concentrations are also needed to detect TCA in wines with a high alcohol content [26], and fortified wines. At low levels, less than 2 ng/l, the compound may not be picked up as a taint even by a trained taster, but the wine aromas and flavours will be muted and the wine will taste flat, being to a greater or lesser extent stripped of fruit flavours. Accordingly, the impact of the compound at any level other than trace, always renders the wine faulty. A level of taint above 4 ng/l might paradoxically be regarded as a 'high level' and will be usually detected on the nose and palate of a trained taster. At 6 ng/l the mouldy, musty, damp sack aromas are almost always apparent. However, in a study by McKay et al. [35] using a spiked, de-aromatised Shiraz, only 14 judges (41%) of the 34 judges were able to detect a difference between a control and samples spiked with TCA at the 4 ng/l level.

The wide variation in the reported detection thresholds is perhaps unsurprising, bearing in mind the huge diversity in wine styles and matrixes (including grape variety) and, of course, the diverse constitutions and characteristics of the panel, and particularly if the tasters are trained or untrained. Examples of the variation in reported DT in research papers include 0.9 ng/l (odour) and 1.7 ng/l (taste) in red wine and 1.5 ng/l (odour) and 1 ng/l (taste) in white wine [26], 1.4 ng/l in a Pinot Noir [36], to a staggeringly high 17.4 ng/l in a Sauvignon Blanc (surprisingly in a panel using experienced tasters) [37]. Statistician Dr Russell Gerrard notes that there is a 50% chance of expert tasters detecting TCA in white wine at a level above 1.2 ng/l, and for red wines 50% would detect at a level above 2.5 ng/l [38]. These figures differ markedly with those from the Cork Quality Council USA (QCC). The council has conducted several studies to identify sensitivity to TCA in a variety of wines (and also in cork soaks – see Section 3.14.4), with panellists drawn from the wine and cork industries. The results obtained consistently show that 50% of panellists recognise TCA at a concentration of approximately 6 ppt, and 25% at a concentration of 2.5 ppt [39]. The council states that studies with consumer panellists indicate higher thresholds. Although the above examples show a wide variance, it is perhaps worth reminding ourselves that the highest of the stated figures (17.4 ng/l) [37] might be compared with one sugar cube in 6 Olympic-sized swimming pools!

3.6.3 2,3,4,6-Tetrachloroanisole (TeCA)

The molecular formula of TeCA is $C_7H_4Cl_4O$. The precursor is 2,3,4,6-tetrachlorophenol, which may be found as an impurity in fungicides, herbicides, and wood preservatives that are based on pentachlorophenol, although these are now restricted in many countries (see Section 3.7.3.3). The detection threshold for TeCA in still wines is 10–15 ng/l, but perhaps as low as 4 ng/l in sparkling wines. The odour characteristics are generally indistinguishable from TCA.

TeCA has been implicated in many cases of musty odours in wines, the source of which has generally been found to be cellars contaminated with the compound. The percentage of wines tainted by TeCA is perhaps higher than has often been considered. A study published in 2009 examined 966 Spanish red wines that had been aged for six months or more in oak barrels. A staggering 155 (16.1%) were contaminated with one or more haloanisoles or halophenols. Surprisingly, the most common contaminant was TeCA with some 6.8% of the wines contaminated with the compound, followed by TCA with 5.3% of wines showing contamination [40].

3.6.4 Pentachloroanisole (PCA)

Pentachloroanisole (PCA) is sometimes referred to as 2,3,4,5,6-pentachloroanisole. The molecular formula is $C_7H_3Cl_5O$. The precursor is pentachlorophenol. The perception threshold in wine is high compared with the other haloanisoles at over 4000 ng/l [41], and any presence is usually below this figure.

As is the case with other haloanisoles, wood treated with pentachlorophenol-based preservatives, including structural timbers in cellars has been (and in some instances remains) a major source of PCA formation.

3.6.5 2,4,6-Tribromoanisole (2,4,6-TBA)

2,4,6-TBA is also known as 2,4,6-tribromo-1-methoxybenzene. The molecular formula is $C_7H_5Br_3O$. The precursor is 2,4,6-tribromphenol (molecular formula $Br_3C_6H_2OH$).

2,4,6-TBA was identified in wine by Pascal Chatonnet and his team as recently as 2004 [42], although it had earlier been known to affect various foods. Chatonnet et al. determined the sensory detection threshold in wine as 4 ng/l by nosing, and a lower level of 3.4 ng/l with retro-olfaction [42]. Indeed, for many years following the discovery of the compounds as a cause of musty taint in wine the sensory perception threshold was accepted as 3.4 ng/l, and this figure is widely quoted. However, the perception thresholds in wine are now regarded as lower by some commentators, and a level of 2 ng/l has been reported by several commercial laboratory analysts [personal discussion]. The threshold is much lower in water, perhaps 20 pg/l! In addition to the 'musty' aromas and tastes common to other haloanisole contamination, wines contaminated with a high level (over 20 ng/l) of 2,4,6-TBA can have an iodised or phenolic character that is mainly noted on the back palate and finish [42].

2,4,6-TBA accounts for 5–13% of the total of haloanisole tainted wines and is often stated as the second most likely cause of musty odours, after TCA, although 2-methoxy-3,5-dimethylpyrazine (MDMP) may also lay claim to this dubious honour. There are no totally reliable global figures available as to the number of bottles affected by 2,4,6-TBA, and there can be little doubt that prior to 2004 (and even after) many 2,4,6-TBA taints wines were misdiagnosed as TCA. Over a 10 years period up until 2009, the AWRI commercial services division analysed over 2000 wines subject to musty aromas and found 861 of these were affected by TCA, and 109 by 2,4,6-TBA [43]. Bearing in mind that 2,4,6-TBA was not identified until 2004, these figures would seem to indicate that the compound is an important source of musty odours.

2,4,6-TBA pickup in wine is consequential to production or storage in 2,4,6-TBA contaminated environments and containers, including barrels, the use of contaminated processing aids or pick up from contaminated plastic closures. Winery and cellar contamination have been noted in many countries but have been a particular problem in South America with many documented instances, including those discussed in Section 3.4.3.3. As with other haloanisoles, 2,4,6-TBA is known to taint many products other than wine. Other drinks include beer, water (including flavoured water), and tea.

2,4,6-TBA is not naturally found in corks and any presence in or on the closures results from absorption or adsorption from an external contaminated source, including the storage conditions of closures pre-use, or contact with tainted wines in bottle. Also any presence of 2,4,6-TBA in synthetic closures, including plastic stoppers and the Saranex liners of screw caps and the liners of crown corks (which are usually used to seal bottles of traditional method sparkling wines prior to the second fermentation) will also only result from such external contamination. Of course, if closures become contaminated, they can then be the source of the compound in wine.

3.7   The Formation Pathways of Haloanisoles from Halophenols

3.7.1   Origins

Organohalogens are a class of chemical compounds whose structure contains at least one atom of a halogen (astatine, bromine, chlorine, fluorine, or iodine). Halophenols are phenol derived groups of these compounds, chlorophenols and bromophenols. Chlorophenols may be mono-, di-, tri, tetra-, or pentachlorophenols, depending upon the number of chlorine atoms in their structure. The numbers preceding the name of each chlorophenol indicate the position of the chlorine atoms on the six points of benzene ring hexagon. Thus 2,4,6-TCP has three chlorine atoms situated at positions 2, 4, and 6 on the benzene ring. Other halophenols and anisoles, including bromophenols, chloroanisoles, and bromoanisoles are detailed in a similar manner.

3.7.1.1   Chlorophenols

Chlorophenols are aromatic organic compounds that are only present in the natural world owing to anthropogenic sources, i.e. resulting from the actions of humans [44] – they are synthetic, not natural compounds. In all, there are 19 chlorophenols. They are regarded as environmental pollutants introduced into the environment (including the atmosphere, but particularly ground and surface water) consequential to the activities of the chemical, petroleum, tinctorial, and pharmaceutical industries [45, 46]. They are largely recalcitrant to biodegradation, as there are very few microorganisms that are capable of such degradation. Many derivatives of chlorophenols are toxic, mutagenic, and carcinogenic [47]. Phenols generally are highly soluble and often found in natural expanses of water, surface, river, and ground waters and also municipal tap water. The chlorination of municipal water and wastewater and incineration of organic and municipal waste results in chlorophenols, including 2,4,6-TCP, with consequential environmental pollution. However, the synthesised production of chlorophenol-based products, including biocides, by the chemical industries and their subsequent usage has been the major source of environmental pollution.

The chlorophenols that are of particular concern to us as far as their role as precursors of related chloroanisole taints are: trichlorophenols (particularly 2,4,6-TCP), tetrachlorophenols (particularly 2,3,4,6 tetrachlorophenol), and pentachlorophenols. 2,4,6-TCP (together with the other chlorophenols except tetrachlorophenols and pentchlorophenols) can be formed when two-and-a-half equivalents of chlorine react with phenol in aqueous solution [48]. Historically cork bark and corks may have become contaminated post-harvesting with 2,4,6-trichlorophenol (but not 2,3,4,6-tetrachlorophenol or pentcahlorophenol) during the now obsolete processing of boiling cork bark in chlorinated water and/or bleaching of individual corks in a calcium hypochlorite solution.

3.7.1.2 Bromophenols

Of the 3200 or so naturally occurring organohalogens, some 1600 contain bromine. There are 19 bromophenols, including 6 tribromophenols (TBP); 2,4,6-TBP is the only bromophenol of concern to us with regard to its being a precursor of the 2,4,6-TBA wine contamination compound. Bromophenols are produced both naturally in water bodies by brown algae to remove excess bromine from their environment [49] and anthropogenically in municipal water and wastewater treated with chlorine in the presence of bromine ions and organic phenols; specifically, 2,4,6-TBP has often been found in chlorinated tap water. However, since the 1930s, the compound has also been extensively industrially synthesised using the reaction of elemental bromine with phenol, for usage as fungicides, herbicides, and flame retardants.

3.7.2 The Biological Transformation of Halophenols into Haloanisoles

Haloanisoles are transformed from halophenols by the action of microorganisms, mainly filamentous fungi. All the pathways to the existence of chloroanisoles require fungal methylation of chlorophenols [44]. Only one enzyme, chlorophenol-O-methyltransferase (CPOMT) is responsible for the microbiological biomethylation reaction (O-methylation) that produces haloanisoles from halophenols. Biomethylation is a biochemical defensive reaction that occurs in filamentous fungi (and some other microorganisms) when they come in contact with halophenols, which are toxic to them. Haloanisoles are not toxic to these fungi, so the biomethylation process enables the fungi to survive without serious damage. Biomethylation of 2,4,6-TCP is the pathway to the production of TCA. As well as chlorophenols, the CPOMT enzyme can methylate bromophenols and iodophenols to produce bromoanisoles and iodoanisoles [50]. 2,4,6-TBA is formed by biomethylation from 2,4,6-TBP. In addition to filamentous fungi, *Paecilomyces variotii*, a xerophilic fungus (i.e. preferring dry conditions) can also convert 2,4,6-TBP into 2,4,6-TBA. Xerophilic fungi only need a moisture content between 12% and 16% and a temperature between 25 and 35 °C to germinate and grow. Subsequent to the biomethylation process, chloroanisoles and bromoanisoles are secreted from the cells of the fungi. They are highly volatile and rapidly absorbed by the material, upon which the fungi are growing. Of course, cork, wood, and other materials containing cellulose all support the growth of filamentous and other fungi and thus become haloanisole contaminated.

3.7.3 The Role of the Pesticides, Herbicides, and Fungicides

3.7.3.1 Chlorophenol-Based Products

The industrial use of chlorophenols began in 1936 when pentachlorophenol was introduced as a wood preservative by the companies Dow and Monsanto [51, 52]. Pentachlorophenol and 2,4,6-trichlorophenol and 2,3,4,6-tetrachlorophenol were also very widely used as cheap pesticides, herbicides, fungicides, bactericides, and mould inhibitors over the next 60 years or so. From the chemical industries' point of view, chlorophenols are easy to synthesise and very effective owing to their liposolubility, thus allowing them to transverse membranes and enter into cells, destroying proteins and the DNA of cells [44]. Of course, they are highly toxic to their targets and other organisms. It is this very toxicity to filamentous fungi (and some other microorganisms) that necessitates their defence mechanism, the biomethylation detoxification process that converts chlorophenols into harmless (for the organisms) chloroanisoles. Thus the formation of PCA by biomethylation of pentachlorophenol in the biocide products was, with current knowledge, to be expected. However, to add to the problem, pentachlorophenol-based products were very often contaminated with a small percentage of 2,3,4,6-tetrachlorophenol, 2,6-dichlorophenol, and 2,4,6-trichlorophenol which, of course, are the respective precursors for TeCA, 2,6-DCA, and TCA.

For many years post World War II, most soft wood timber was treated with pentachlorophenol (PCP) to prevent, inter alia, the growth of fungi which impart a blue or purple stain to the wood, commonly known as sapstain. In 1968 and 1969 over 11 000 tonnes of chlorophenols were used in the United States for timber preservation. As with the pesticide and herbicide preparations, these PCP-based wood treatments also contained 2,3,4,6-tetrachlorophenol and 2,4,6-TCP. The TCP-based products were marketed, inter alia, under the brand Dowicide 2S™ and incredibly, at the time of writing, are still available for global sale from China [53]. TCP can also be produced from the aquatic degradation of another pesticide, triclosan, in the presence of low concentrations of free chlorine [54]. Triclosan was first registered for use in USA in 1969 – EPA completed its re-registration eligibility decision (RED) for its pesticide uses in 2008 [55].

There is a host of evidence as to the significant pollutant effects of chlorophenols, including to the atmosphere, soils, and ecosystems [46, 47, 51]. The resulting environmental changes are worldwide. Pentachlorophenol, 2,4,6-TCP, and some other chlorophenols are included the priority pollutant list of the United States Environmental Protection Agency [56] and in the Agency for Toxic Substances and Disease Registry 2017 Substance Priority List [57].

3.7.3.2 Bromophenol-Based Products

It was estimated that in 2001 the global production of 2,4,6-TBP amounted to 9500 tonnes, over a quarter of this being in Japan [58]. A study published in 2018 noted 2,4,6-TBP as the most widely produced brominated phenol [59]. As with chlorophenols, brominated phenols are not easily biodegradable, although some adapted and specialist communities of microorganisms can carry out some degradation of the compounds.

2,4,6-TBP has been and remains widely used as an intermediate in the preparation of flame retardants, timber preservatives, and fungicides: such usage increased following the restrictions placed on the use of preparations containing pentachlorophenols from the end of the 1980s [60]. It has been incorporated into the composition of a vast array of materials, including polyvinyl chloride (PVC), polyethylene terephthalate (PET), polybutylene terephthalate (PBT), high impact polystyrene (HIPS), and polystyrene generally, paints, leather, textiles, paper, and cardboard. It has been very commonly used as a treatment for wooden pallets. In Section 3.2, I detailed many instances of 2,4,6-TBA contamination of food and pharmaceutical products due to pallets that had been treated with 2,4,6-TBP, and the wooden floors of shipping containers are a common source of both 2,4,6-TBP and 2,4,6-TBA contamination. As a fungicide, 2,4,6-TBP has been added to many products, including paper, cardboard, and adhesives. Of crucial importance in a winery environment is its use as a wood preservative. Structural timbers including floors, beams, purlins, and voliges (the wooden planks commonly used across the rafters in rooves in France and elsewhere) have been and in some cases continue to be sources of taints. Its use as a flame retardant intermediate and fungicide has resulted in its presence not only in winery structures (including insulation materials and paints) but also in numerous items of equipment and consumables. These may include: hoses, rubber gaskets, tank linings (epoxy resins were often used to line mild-steel and other tanks), small tanks manufactured from plastics, rakes and shovels (used for removing grape solids from tanks prior to pressing), other polyethylene- or polyester-based equipment, silicon items (including barrel bungs) filter pads, rotary vacuum filters, fining materials (particularly bentonite), barrels, oak pieces, chips and powder, and packaging materials. Many cleaning and sanitising agents contain bromine. Tap water, of which copious amounts are used in every winery, may also contain 2,4,6-TBP or 2,4,6-TBA.

3.7.3.3 Restrictions on Usage of Chlorophenols and Bromophenols

By 1991 pentachlorophenols were banned in Australia, India, Indonesia, New Zealand, Sweden, and Switzerland. Pentachlorophenol usage has also been restricted in the EU since 1991 by Directive 91/173/EEC of 21 March 1991 [61] (although this was not incorporated in French law until 27 July 1994, by Decree No. 94-647) [62]. The legislation (which has since been subject to modifications, derogations, and repeals in part) included a total prohibition of PCPs to treat wood in buildings to be used for the manufacture or packaging of food products. Incredibly, the legislation allows continued use for building frames, provided they are covered with a varnish – I suppose there have never been instances of poorly or incompletely applied varnish coatings, and they never become damaged! Chlorophenol usage continues in some countries in Africa, Asia, and South America. The countries that are signatories to the Stockholm Convention on Persistent Organic Pollutants [63] voted in 2015 to ban the use of pentachlorophenols (allowing a five year phase-out period for the treatment of utility poles). The United States, Italy, and Israel are not signatories and retain restricted usage. However, even in countries that have long ago enacted bans or restrictions, chlorophenols remain in ecosystems for many years or even centuries, as they are not easily degraded. The manufacture and usage of bromophenols, including 2,4,6-TBP increased subsequent to the general banning of pentachlorophenols by the early 1990s, continues in many countries. It is the most widely produced brominated phenol, used mainly as flame retardant intermediates, fungicides, and wood preservative treatments. It is not a registered pesticide with the US Environmental Protection Agency, and its use as a fungicide has also been banned in the United States, Canada, and the EU, although certain other uses are permitted.

3.7.3.4 The Advent of Chloroanisole and Bromoanisole Wine Contamination

It is a sobering thought that TCA and other chloroanisole contamination of corks and wines almost certainly did not exist before World War II, although bromoanisole contamination of wine possibly did, for as we have seen bromophenols can be naturally present in the environment. The precursor of 2,4,6-TBA (2,4,6-TBP) can be formed in nature by marine brown algae. Although Professor George Saintsbury refers to a corked wine in his classic work *Notes on a Cellar Book* published in 1920 [64], we cannot know the precise fault noted, but it is most unlikely that it was TCA. Of course it is also possible that the fault in question was not haloanisole related, but due to a non-related compound, such as MDMP, which is sometimes referred to a 'fungal must taint', or perhaps another issue due to mould, or just the degradation of the cork closure. Even as recently as the 1970's, very few bottles of wine were perceived as 'corked'. Christian Vannnequé, former head sommelier at Paris's famed La Tour d'Argent Restaurant recalls, that in those years he and staff would upon between 800 and 1000 bottles a week, but rarely found more than four or five of them 'corked'. He says that if there were problems it was almost always with wines from the 1960s, rather than those from the 1930s, 1940s, or 1950s: 'If you had a problem with a 1929 wine, for example, it was never cork taint. It might have been a crumbly old cork…But it wasn't corked' [10]. However, the 'corked wine' crisis was just a decade or so away.

3.7.3.5 Other Halophenol Formation Pathways

2,4,6-TCP, 2,3,4,6-tetrachlorophenol, pentachlorophenol, and 2,4,6-TBP can also be formed when phenols present in wood from the decomposition of lignin react with a source of chlorine or, in the case of 2,4,6-TBP, bromine. Lignin is present in cork, being the semi-rigid, woody walls surrounding the lenticels. Cardboards and plastics also contain phenols and any contact with active chlorine can result in chlorophenol formation. Another formation pathway, again resulting from the manufacture of chemicals, is the simple biodegradation of herbicides and pesticides that include chlorobenzenes [65] and chlorinated cyclohexanes [66]. Lindane (gamma-hexachlorocyclohexane), which began use as a pesticide from 1942, degrades very slowly into a large number of metabolites, the main ones being tri- and tetrachlorophenols. Its use was finally banned in agriculture (in signature countries) as recently as 2009 under the Stockholm Convention on Persistent Organic Pollutants (POPs) [63].

3.7.4 The Fungi that Convert Halophenols into Haloanisoles

Halophenols are converted into haloanisoles by the biomethylation actions of filamentous fungi. There is a wealth of research into this process, and I will mention just a few key points from a couple of important papers. In 2002, a study found that of 14 fungal strains examined, 11 isolates of filamentous fungi could produce significant amounts of TCA when grown directly on cork that contained sub-lethal concentrations (1 µg/ml) of TCP as the precursor [67]. The species of fungi able to perform the biomethylation from halophenols to haloanisoles include *Acremonium* spp. (previously known as *Cephalosporium* spp.), *Actinomyces* spp., *Aspergillus* spp., *Chrysonilia sitophila, Cladosporium* spp., *Fusarium* spp. (an abundant soil microbe), *Mortierella alpine, Mucor plumbeus, Paecilomyces variotii, Penicillium chrysogenum, Penicillium citregenum, Penicillium decumbens, Penicillium purpurogenum, Talaromyces pinophilus, Trichoderma viride*, and *Trichoderma longibrachiatum* (present in all soils and which can be catalysed by the bacterium *Streptomyces* that often dwells in wineries) and *Verticillium psallitae*. Another interesting research project was undertaken in 2008 by Maggi et al. [68] when the ability of eight fungal strains to transform 2,4,6-TCP to 2,4,6-TCA was studied. The fungi were isolated from both cork and grapes: the isolations from cork belonged to the genera *Penicillium*, *Aspergillus*, *Trichoderma*, and *Chrysonilia*, and the isolations from grapes were of the species *Botrytis cinerea*. In this study all the isolates except *Chrysonilia* produced TCA when grown directly on cork in the presence of 2,4,6-TCP; *Aspergillus* and *B. cinerea* with the highest level of production. This study showed for the first time that *B. cinerea*, a microorganism often present on grapes and also in winery environments, transformed 2,4,6-TCP into TCA. The authors noted that their research partially explained so-called 'cork taint' being present in wine prior to bottling (although there are many other possible explanations, as discussed generally in this chapter). If present on grapes, *B. cinerea* is usually highly damaging – always so in the case of red grapes. However, in certain climates it can affect white grapes positively and many of the very finest sweet white wines in the world are made from *Botrytis* infected grapes. In my experience in tasting and judging wines in many countries, there is higher than average incidence of TCA taint in 'nobly sweet' wines, i.e. those made from grapes affected by *B. cinerea* in its beneficial form of 'noble rot'. When I and two colleagues tasted the 2016 vintage during the 'primeurs week' in Bordeaux in April 2017, three of the 23 Sauternes and Barsac wines assessed (some 13%) clearly exhibited a haloanisole-related taint. In the April 2018 primeurs week there were further instances with two wines of the 2017 vintage. Yet again in 2019, two wines of the 2018 vintage showed related taints. These wines are shown before the main bottling run has taken place, and I would consider it unlikely that the cork closures were responsible in every case, or even at all. Upon raising the issue with the owners or winemakers, none were aware of any problem and declared the wines to be sound, perhaps examples of the adaption effect. However, Didier Frechinet of the Sauternes Premier Cru Classé Château La Tour Blanche, believes that the taints have nothing to do with TCP being transformed into TCA by *B. cinerea*, but arise due to 'cellar conditions' at the properties in question [personal discussion].

3.8 Contamination of Cork with TCA and Other Chloroanisoles

TCA is the haloanisole most commonly found in wine, and the most likely, but far from the only, source of the taint compound is the cork closure. Accordingly, I will discuss briefly how cork may become contaminated by this compound. The pathways to cork contamination by other chloroanisoles are generally similar.

3.8.1 Cork Production and How Cork Becomes Contaminated

Corks are made from the bark of an evergreen species of oak: *Quercus suber*. It takes about 27 years following the planting of an oak for bark of sufficient thickness to be produced. It may then be stripped – the tree does not die as there is an inner cork bark layer that is not penetrated – and subsequent strippings will take place every 9 or 10 years. The harvested cork bark is stacked for seasoning, before being transported to a processing plant. After the planks of the oak bark have been boiled, corks are punched out along the vertical axis. Today, the production process is highly technical, at least as far as the larger cork producers are concerned. Cork production is discussed in more detail in Chapter 17.

There are several pathways to the formation of TCA that account for its presence in cork [69]. Bark to be used for the production of cork wine stoppers, may become contaminated with chlorophenols in the forest, either from the atmosphere or from rainwater. Historically, chlorophenols were used as fungicides in cork oak forests, such use continuing until the 1980s. The activity of filamentous fungi, as described above, would then convert the chlorophenols into chloroanisoles. Some authors report that the formation of anisoles and thioanisoles in cork may also be a result of the methylation of halogenated phenols due to the presence of bacteria in the cork oaks, in particular *Rhodococcus* spp., *Acinetobacter* spp., and *Pseudomonas* spp. Historically an insecticide comprising pentachlorophenol, together with 2,3,4,6-tetrachlorophenol and 2,4,6-trichlorophenol, which was marketed in Europe as Raco, was applied to the lower part of the trunk of cork oaks as an insecticide. It is likely that high levels of contamination in the first metre or so of trunk closest to the ground was on account of this biocide usage. Of course, chlorophenols are retained in soils (as well as in water), and the part of the bark closest to the ground often showed higher levels of contamination, even if Raco had not been used. In the last 15 years or so the major manufacturers have excluded this portion of the bark from use for wine closures. Large producers store planks of cork barks in the forests on metal (not wooden) pallets. The boiling of corks in municipal water containing chlorine, which was very often recirculated, and the calcium hypochlorite (chlorine) bleaching process used historically, were both without doubt major contributors to the formation of chlorophenols and chloroanisoles. The use of municipal water in the boiling process generally ceased in the 1990s and is not now permitted by the 'SYSTECODE' quality assurance system for the cork industry to which are accredited cork manufacturers in Portugal, Spain, France, Italy, Germany, Morocco, and Tunisia [70]. Hypochlorite bleaching compounds are no longer used either, replaced with baths of hydrogen peroxide and ammonia. Today, the major cork closure producers claim that no TCA is formed during the production processes and that the presence of any TCA in corks results from it being the raw material from the cork forests. As the use of chlorophenol-based insecticides ceased many years ago, contaminations of the bark appear to be lessening in frequency.

Of course, corks may become contaminated post production, during shipping, or otherwise prior to use. This problem was recognised as long ago as 1990 when a shipment of 'Champagne corks' to Australia was found to be heavily contaminated. The corks were packed in polyethylene bags inside fibreboard cartons and transported in a shipping container already contaminated [71].

3.8.2 The Transmission of TCA from Corks to Wine

Generally, only a small amount of TCA contained in or on a cork will be transmitted to wine. TCA can, of course, be present on any part of the cork surfaces, as well as in any part of the body. It does not migrate easily through cork, so it is only the content from the part of the stopper nearest to the wine that will contaminate the liquid. The extent to which this might happen depends upon the solubility of the compound in the particular wine, its infinity with the cork, its rate of migration through the cork (which is limited) and the extent to which the wine and headspace gases penetrate the cork. Some of the TCA content in the cork may be releasable into the wine over a period of months or even years (releasable TCA) and some may be released after just a few hours (rapidly releasable TCA). Group cork soaks, as discussed in Section 3.14 below can give a producer an indication of assimilation in the wine of the rapidly releasable TCA content in the corks.

3.8.3 The Historic Incidences of So-Called 'Cork Taint' and the Role of the Cork Industry

Natural cork became the closure of choice for wine bottles in the early eighteenth century, and until the last two or three decades of the twentieth century, despite the occasional physical failure of the closure or musty impacted wine, its use remained almost entirely unchallenged. The seeds of what was to become a crisis in the industry were sown in the years after World War II. Portugal was (and remains) the largest cork producer in the world, but in many ways in the post-World War II years Portugal was a backward and feudal country with much agricultural land owned by a few wealthy families and worked by peasants. Cork was the country's most important economic product. Then in 1974 there was a national uprising and the government toppled. Following this left-wing 'carnation revolution' many experienced owners, together with their skilled workers, were driven from the cork forests they had owned for generations and their land expropriated [10]. Many fled abroad. Their place was taken by inexperienced workers and 'labour committees'. A quick return from the forests was sought, although cork production had always been a long-term business, with harvests taking place every nine or ten years. Standards of farming fell, nitrogen based fertilisers were applied, harvesting methods were compromised, with bark being harvested before the nine years needed for satisfactory growth and cork taken from parts of the trees that had previously been off-limits [10]. Disastrously the uses of pentachlorophenol-based herbicides and fungicides became commonplace. Cork barks would be left lying on the ground until taken for processing. Standards in every aspect of the industry left much to be desired, but the market was rising and the industry pretty much had a monopoly. The 1970's was the start of a huge growth in worldwide wine production and sales. For example, in the United Kingdom from 1960 to 1969, the average person drank 2.2 l of wine per annum – by 2000 to 2009, this had risen to 18.1 l [72]. In the 1960s Australia was producing only 174 millions of litres of a year and the figure had increased to 1196 million litres by the year 2000 [72]. The demand for wine corks was unprecedented, and the cork industry would fulfil that demand. There were few if any controls in the industry and many producers would take short-cuts. The municipal water used for boiling would often not be changed between batches, building up contaminants. Even the small producers, numbered in the hundreds, who took care, were lacking in knowledge and technology.

3.8.4 The Rise in Incidences of So-Called 'Cork Taint'

Ever since cork was introduced as a wine bottle closure in the seventeenth century, there have been instances of closure related spoilage, but so-called cork taint only became a major issue post World War II, as discussed above. The late Professor Denis Dubourdieu of Université Victor Segalen Bordeaux 2, who was also a winemaker and owner of Bordeaux châteaux, speaking to author George Taber in 2006 said that the problem 'started maybe 30 years ago' [10]. There was a significant difference in the quoted figures regarding the incidence of affected bottles at the height of the TCA 'crisis' (the period 1985–2005) with figures ranging from 0.1% to 10% of bottles being quoted. The range of 2–5% was largely accepted by researchers [12, 30, 73]. Specialised trade and consumer wine publications during the 'crisis years' generally suggested that between 2% and 7% of bottles were affected. A large-scale study was carried out by Soleas et al. in 2002, using expert assessors. Over 2400 different wines were examined and it was found that TCA taint affected 6.1% of the wines generally, with white wines showing the greater incidence [74]. If those figures were bad enough, the most horrifying incidence was for wines sealed with an agglomerated cork where 31.9% of bottles were stated to be TCA tainted. However, 49% of the wines that were affected by 'musty taint' were not TCA contaminated, so the cause of the taint in the wines was attributed to other compounds. Whatever the TCA figure suggested, even if it was at the bottom of the 2–7% range, it was hotly disputed by the cork industry. The world's largest manufacturer of wine corks, Amorim said that in 2007 random sensory testing of 24 000 of their corks revealed just 0.48% were defective [personal discussion with Carlos de Jesus of Amorim]. Such a 'low' figure was to be disputed by many wine writers and journalists. An interesting study examining the incidence of taint in large format corks (those that would be used for bottling of wines in the size of double magnum or larger) found that of 2296 corks analysed, 1.3% had releasable TCA levels of between 1 and 5 ppt and 0.57% had releasable TCA at more than 6 ppt [75]. Bearing in mind that large format bottles are usually collectable examples of very highly priced fine wines, these figures, though not high, make

Table 3.1 Suspected cork taint rate in *Wine Spectator*'s Blind California wine tastings.

Year	Percentage of wines suspected to be tainted
2016	3.27
2015	2.63
2014	4.53
2013	4.28
2012	3.73
2011	3.87
2010	4.76
2009	6.9
2008	7.5
2007	9.5
2006	7.0
2005	7.5

depressing reading. Writing in 2011 Jamie Goode and Sam Harrop MW suggested that surveys indicated 3.5% [76]. The magazine *Wine Spectator* records the number and percentage of wines suspected to be 'cork tainted' in its blind tastings. Their figures of suspected taint in Californian wines (stoppered with cork) for the years 2005–2016 are shown in Table 3.1, indicating a considerable decline in the percentage suspected during the last 10 years, from 9.5% in 2007 to 3.27% in 2016, although the latter figure will still be alarming to the magazine's readers and collectors of fine wines.

It must be stressed that these figures are of suspected 'cork taints' revealed at sensory evaluation, and neither the wines nor the corks were analysed in a laboratory, so it is very possible that some of the taints were derived from sources other than the corks.

Of course, companies with vested interests are likely to quote taint figures that are favourable to them. Talking about natural cork in 2015, Malcolm Thompson, vice president for strategy and innovation at the alternative closure manufacturer Nomacorc, says, 'We'd put TCA levels in general at 2–3%, which is where levels are so high it is apparent wines are corked' [77]. Several studies have been contested especially those funded or conducted by interested parties, including cork and alternative closure manufacturers.

3.9 The Cork Industry in the Dock

3.9.1 The Industry in Denial

There has been much debate, at times very heated, about the implication of cork closures in musty taints in wine in the last 30 years, the adversaries being the producers of cork and the manufacturers of alternative closures, and the pro- and anti-cork lobbies amongst journalists and critics. There was (and to a large extent remains) an Old World/New World divide as to whether cork was an appropriate and reliable closure. High incidence of so-called 'cork taint' resulted in a loss of reputation of cork manufacturers. It might be argued that this was deservedly so for the cork industry's practices and procedures, including using chlorine as a biocide in the boiling of cork planks, and for bleaching, had been major contributors to the problem. It was not just New World winemakers and winery owners that were up in arms during the 1980s and 1990s – all wine producers saw cork price increase heaped upon price increase (partially on account of small harvests). Between 1982 and 1989 cork export volume increased by 11% but value increased by an incredible 340% [10], and this at a time when there was a high incidence of tainted corks. The cork industry was viewed as not only arrogant but in denial, with cork producers often blaming the wineries and their production processes, including alleged dirty conditions and mistreatment of corks. Particularly in 'faraway' countries, including Australia, New Zealand, and the United States, many wine producers felt that the Iberian cork producers were dumping inferior and tainted product into their burgeoning wine industries. Of course, the transport of the product to these markets was a recipe for disaster. The bags of cork had to undergo long sea journeys in the warm and humid holds of ships, where moulds flourished, and the cartons often stacked on timbers that had been treated with PCP or 2,4,6-TBP. Even those European wine producers who were situated far away from Portugal and Spain, the epicentres of cork production, felt that cork producers were sending them low quality or tainted corks that would be rejected by wineries closer to home. There are numerous anecdotes of wineries having identified and returned suspect bales, only to find that the same bales were then resold to others in their region. To put it bluntly, by the 1990s and the early years of this century, faith in the cork industry was at an all-time low.

3.9.2   Hostile Coverage in the Wine Press

Many wine writers, and critics regarded the cork industry totally out of touch and disinterested in the problem, and this situation was perhaps not helped by seemingly endless inappropriate and 'off-topic' PR campaigns that often backfired. Hostile press had become commonplace, and even in recent years it would appear that the heat has not gone out of the debate. In an interview for *The Drinks Business*, published in September 2017, Patrick Spencer, executive director of the Cork Forest Conservation Alliance, asserted that the claim that one in every 10 wines bottled under cork is tainted was a 'myth' [78]. The article elicited numerous comments, with several slamming Spencer's statements. Peter Scudamore Smith MW did not hold back: 'misinformed…inaccurate information; this exposition touches on political diatribe…Look forward to future retractions and apologies to industries in this area who have solved the failures by natural cork'. Spencer considered it appropriate to provide four responses, some in great detail, to the comments [78]. Writing in FT Magazine in August 2017 and referring to the growth of alternative closures in the first decade of this century, Jancis Robinson MW notes that the problem was 'the shockingly high incidence of TCA…in natural corks' [79].

3.9.3   Alternative Closures to the Fore

The state of affairs in the cork industry from 1980s to the early 2000s helped build the industries of alternative closures, mainly synthetic cork-shaped stoppers and screw caps. During the first decade of this century, such became the confidence in screw caps that many consumers began to consider it unnecessary to taste a sample of wine from a screw cap closed bottle, when offered by the sommelier or waiter in a restaurant. Of course, wines sealed with a screw cap or other synthetic closure can exhibit many types of fault, including haloanisole contamination. The fact that there were numerous instances of screw cap sealed bottles exhibiting' post-bottling reductivity' (see Chapter 6) was regarded as a minor distraction, particularly by Australasian producers, and much of the wine press. Of course, a reduced wine is usually drinkable (and may improve with aeration), whilst a badly haloanisole-affected wine is not. Although many incidences of haloanisole cellar contamination were exposed, as discussed in Section 3.4, many more instances in several countries were kept under wraps by the producers, and invariably cork closures were blamed for any and all bottles illustrating musty related faults, whether or not they were in fact to blame.

   It is generally accepted that if there is significant variation in the level of TCA or other chloroanisole compounds between several individual bottles of the same wine, the cork closures are most likely the source of the contamination. However, it is interesting to note that cork closures can act like blotting paper and actually absorb haloanisoles from contaminated wines, thus reducing the concentration. Cork closures can reduce the level of TCA in a contaminated wine can by 50%, and most of the TeCA and PCA can also be absorbed, resulting in the wine no longer being significantly tainted [80]. Plastic-based 'bag in box' liners (wine bladders) absorb haloanisoles from affected wines. A relatively simple test as to whether the source of the haloanisoles is the cork or the wine is to analyse the presence of the compounds along the length of the entire corks of tainted bottles. If the haloanisole content of the cork is greatest in the section that has been nearest the wine, the cork has probably absorbed the taint from the wine. Conversely, if the level of contamination is uniform throughout the cork, or greater at the end furthest from the wine, then the source of the taint is almost certainly the cork.

TCA or other haloanisole contamination of wine storage cellars, both producer and private, can result in finished product becoming tainted. Bottles sealed with cork are most likely to prevent the wine absorbing TCA from aerial contamination, due to the material's low rates of permeation, particularly compared to Saranex-lined screw caps. The crown cap, as used for champagne and many bottle-fermented sparkling wines before disgorgement, is perhaps a particularly vulnerable closure in this regard. Over the last 10 years, I have noted a much higher incidence of haloanisole-related taints in sparkling wines than in other white or rosé wines, and the use of the crown cap may have had a part to play, although, as we have noted, the sensory detection threshold is lower in sparkling wines.

3.10 The Cork Industry Begins to Address the Issues

3.10.1 Cork Producers Fight Back

In the early 2000s, major manufacturers of corks began to undertake considerable research, and make substantial investments and institute new working practices in order to try to eradicate cork contamination. They have also implemented many initiatives to cleanse TCA and related compounds from contaminated cork closures, as discussed in Section 3.11. There can be no doubt that all the initiatives have resulted in a substantial reduction in the incidence of the problem since the early 2000s. The cork company Amorim now individually analyses each of their 'top-of-the range' corks, branded 'NDTech', and offers a 'non-detectable TCA' guarantee. Some other manufacturers are also beginning to offer a similar service as discussed in Section 3.11. However, this comes at a price, and contaminated cork closures not subject to such stringent quality control remain one of the likely reasons for chloroanisole affected bottles. The source of 2,4,6-TBA in wine, which from an odour and taste perspective is largely indistinguishable from TCA (other than a possible iodine taste on the back plate and finish), is not cork related but due to winery or cellar conditions. Figures from the Cork Quality Council USA (CQC) state that from 2001 to 2019, there was a 95% reduction in releasable TCA in incoming cork – it should be noted that the first two quarters of 2001 showed particularly high levels, so the subsequent reduction figures may be distorted by this. In 2019, 93% of incoming corks analysed had releasable TCA of less than 1 ppt, and 98% less than 2 ppt. These figures are below what is the generally accepted consumer sensory detection threshold. The figures are detailed in Table 3.2.

The table shows figures for all incoming natural cork lots analysed. CQC members normally reject 3–5% of incoming lots, mostly as a result of these tests, and take samples from

Table 3.2 CQC incoming TCA analysis.

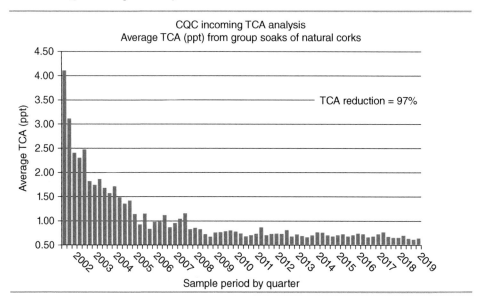

multiple bales from each cork lot that is received prior to acceptance into their inventory. Peter Weber of CQC comments, 'If a sample suggests that TCA might be present, we do not reject that bale, we reject the entire lot' [personal communication]. The CQC methodology has been consistently applied for 18 years and the number of data points is very rich. CQC test over 30 000 cork samples each year, and more than half of these consist of incoming natural cork shipments. Peter Weber admits that the data do not establish a percentage of cork taint in the US market, but it does offer a valuable picture into the industry's reduction of TCA [personal communication].

3.10.2 Many Wine Professionals Remain Hostile to Cork

However, there are highly qualified professionals who still dismiss these efforts and results. Lisa Perrotti-Brown MW is the Editor-in-Chief of Robert Parker's *Wine Advocate*. The historic influence of Robert Parker and *Wine Advocate* cannot be overstated. In November 2017 she penned an article, re-published on the Michelin Guide website, entitled *Industry Insider: Corked Wine Reality Check: Why are cork manufacturers not making some headway in the battle against cork taint?* [81]. In this article Perrotti-Brown notes, 'During a seven-day tasting of more than 900 Sonoma County wines last month, my helper and I had to discard around 7–8% of the wine samples because the wines were corked' [81]. If we accept that her highly trained palate was correct with its haloanisole analysis (and without knowing her personal perception thresholds), these figures would seem to fly in the face of all recently published statistics, including those from the CQC, as noted above. The question has to be asked as to what percentage of the wines were tainted as a result of the cork closure, and how many were possibly due to other sources of contamination? Writing in *Wine Enthusiast* in February 2018, Sean P. Sullivan also thought that the cork industry has not gone nearly far enough: 'To me, cork taint remains a significant issue, certainly in the wines that I sample. It is a problem that I won't consider to be fully addressed until all wines don't have taint caused by cork. To aim for anything less is shooting too low' [82]. Tim Atkin, writing in the UK wine trade publication *Harpers Wine & Spirit* in 2019 notes, 'Cork taint, like random oxidation in Burgundy (see Chapter 5), is less prevalent than it was, but there's still too much of it about' [83].

There are many in the wine industry who believe that cork is an antiquated product with which to seal a bottle and that wine is a highly technical product which warrants more than a piece of tree-bark as a closure. It is perhaps interesting to note that the same people do not question the use of oak barrels for the maturation and storage of wines. Barrels, as we now know, have been used for wine storage since the first century BCE (BC) and that the advantages of making them from oak and the positive impact this would have upon wine quality were realised not long thereafter. Of course, the species of oak used for barrel manufacture, mainly *Quercus petraea* and *Quercus robur*, are very different to the cork oak (*Q. suber*).

Therefore, discussions about so-called 'cork taint' and the plusses and minuses of cork as a wine closure amongst wine industry professionals and in the press, having been 'hot' for over 20 years, have subsided a little but are far from dead. In the last 10 years these discussions have been widened to include considerations of cork and alternative closures' oxygen transmission rates (OTRs), post-bottling reduction, as discussed in Chapters 6 and 17, and oxidation as discussed in Chapter 5. On these topics there remains divided opinion amongst producers, trade buyers, writers, journalists, and critics, and also an Old World/New World divide. The vast majority of wines produced in Australia and New Zealand are sealed with screw caps (often referred to as 'twist-offs' in the United States). In the United Kingdom many supermarkets and other large retailers now insist on screw caps or other synthetic closures when detailing the product specification with their supplier wineries, at least for so-called 'entry level' and other inexpensive wines. However, many countries, particularly in Europe, have remained largely loyal to cork as a closure, and it is exceptional for alternative closures to be used on fine red wines from France, Italy, Spain, or Portugal.

3.11 The Cork Industry's Recent Initiatives for Haloanisole Prevention and Extraction

3.11.1 Investment and Initiatives by the Stakeholders

During the last 20 years there has been a plethora of research and initiatives funded and/or undertaken by the cork industry generally, individual cork-producing companies, the Cork Technology Centre (CTCOR) which is a not-for-profit research laboratory, and APCOR (the Portuguese Cork Association). The cork manufacturer Amorim spends €7.5 million annually on research, development, and innovation alone [84]. The main aims have been to prevent haloanisole formation in corks and to extract any that may be present in the planks and stoppers. Of course, individual manufacturers wish to protect their own commercial interests, and some of the processes have been patented by them, which perhaps does not help the challenge of global eradication of haloanisoles from cork. I will briefly consider some of the most effective of these initiatives.

3.11.2 Symbios

Symbios is a process developed by Cork Technology Centre (CTCOR) that prevents the formation of chloroanisoles in cork. It is a biological process of a preventive nature that promotes the development of 'benign' microorganisms naturally occurring in cork, to the detriment of microbial species that may potentially form unwanted metabolites, and it promotes the inhibition of chloroanisoles biosynthesis during the cork processing stages.

3.11.3 Methods to Extract TCA During Cork Production

Each major cork manufacturer has introduced systems and processes at its factories with the aim of removing chloroanisoles during the cork production chain. The following initiatives are notable.

3.11.3.1 Analysis and Rejection of Bark Affected by Chloroanisoles

Samples of incoming cork bark are analysed upon receipt for chloroanisoles, and batches that are out of parameters are returned to source.

3.11.3.2 Changes in Boiling Systems

New boiling systems have been introduced in the industry, which are dynamic system processes. The water is circulating continuously and it is treated before it is re-introduced to the boiling system. Uniform boiling of all the planks as high temperatures are achieved. These systems enable not only an increase in the extraction of soluble compounds but also the extraction of volatile organic compounds such as TCA whilst also preventing the possibility of cross-contamination Figure 3.1 shows the boiling of cork planks at an Amorim factory.

3.11.3.3 Chloroanisole Removal by Steam Vaporisation

There are several systems of chloroanisoles removal using steam:

- *Controlled steam distillation*: Introduced in 2003, the Rate of Optimal Steam Application (ROSA) system is a process developed by and patented by Amorim used to extract volatiles, particularly TCA. ROSA is based on a steam distillation whereby steam and water under pressure force volatile compounds out from the cork cells. Initially the system was only suitable for cork granules as used for the manufacture of agglomerated or technical corks, and the cork discs positioned at each end of technical corks (Amorim's Twin Top brand) and at one end of sparkling wine closures. However, subsequent generations ROSA Evolution and the latest SUPERROSA treat whole, natural corks, with 80% of any TCA content removed by the process. For a cork that contains 1 ppt TCA, the 80% removal figure is excellent, but should a cork have 100 ppt TCA, then following processing this would still leave 20 ppt which is, of course, an unacceptable figure.
- *Volatilisation by entrainment at controlled temperature and humidity*: This patented process takes advantage of the fact that TCA has a volatilisation temperature of 60 °C (140 °F). In a constantly renewed atmosphere of high humidity and temperatures above 60 °C, a significant extraction of TCA from cork stoppers is achieved. The process is used on natural cork stoppers since, besides being highly effective in reducing TCA, it does not deform the stoppers.

Figure 3.1 Boiling of cork planks.

- *Volatilisation by distillation in the gaseous phase of adjusted polarity, under controlled temperature and humidity*: This is a patented process which introduces ethanol in the distillation phase, based on the principles of distillation and steam entrainment and looking for a polarity adjusted to the extraction of molecules such as TCA. The process allows for the effective treatment of natural cork stoppers through the optimised combination of temperatures close to 60 °C, concentration of ethanol in the steam phase, and continuous inflow of hot air. The process simulates the transfer of molecules from the cork to the wine in the bottle, through the solvent effect of ethanol, and does not result in any deformation of the cork stopper, and mechanical properties remain unaltered.

3.11.3.4 Extraction with Supercritical Carbon Dioxide

The process is only suitable for use with granulated corks and is at the heart of the manufacturing process of DIAM closures. A stream of carbon dioxide in a supercritical state is used to remove volatiles, including TCA. More information on this process is given in Chapter 17.

3.11.3.5 NDtech

NDtech is a system developed by Amorim that analyses each and every cork for traces of TCA using rapid gas chromatography. It was launched following five years of research and development and a spend in excess of €12 million [84]. Using the technology any cork with more 0.5 ng/l of releasable TCA is removed during the production process. The company offers a *non-detectable* TCA guarantee on corks in the NDtech range. The use of NDtech stoppers, particularly for fine wines has been gaining ground rapidly since their launch.

Figure 3.2 Selecting corks by gas chromatography. Source: Courtesy of Amorim.

A major boost for brand came in 2016 when the Chablis producer, Domaine Laroche, having moved to screw-cap closures for a period of 10 years, announced it was reverting to natural cork and using NDtech for wines (initially the Grands and Premier Crus) from the 2015 vintage onwards [85]. The first generation of gas chromatography analysis machines could analyse 10 corks per minute, the second generation currently in use has speeded up the process to 14 corks per minute, and a third generation, under trial at the time of writing, will be even faster. A bank of gas chromatography machines undertaking selection at Amorim is shown in Figure 3.2.

The individual analysis of each cork results in a more expensive end product, but with a guarantee of non-detectable TCA, many high-end producers in particular consider the price to be well worth paying. Of course, for so-called 'entry level' wines the product is, at present, beyond reach. However, the company's Chairman Antonio Amorim, speaking to *The Drinks Business* in July 2018, says the Company is addressing the challenge: 'We will have a non-detectible TCA guarantee for everything, it is in our company strategy for 2020; we are working as fast as ever' [86]. In 2019 Amorim introduced the NDtech Sparkling cork, for sparkling wines, which has two end discs that have been subjected to gas chromatograph (GC) analysis and rejected if containing more than 0.5 ng/l releasable TCA.

Clearly Amorim believes it has finally cracked the TCA challenge. During a visit to the Amorim factories in February 2019, director of marketing Carlos de Jesus commented, 'TCA will continue to be an issue for wines for many years. But the discussion will move away from cork' [personal discussion].

3.11.3.6 Other Companies' Non-Detectable TCA' Guarantees
Other cork companies have developed their own systems and guarantees to give customers confidence regarding the absence of detectable TCA. Lafitte Cork Portugal, and Lafitte Cork and Capsule, based in Napa, California, are now offering for their highest grade corks (Fleurs de Lafitte) an option for undergoing individual sensory and machine testing using GC/MS with the 'Electvs' programme, with a 100% bottle buy back guarantee of non-detectable TCA. M A Silva has developed a 'onebyone' system using gas phase spectroscopy, again with a rejection of corks containing detectable TCA. The USA Company Cork Supply has developed a DS100+ system that analyses every cork and rejects those with TCA levels above 1 ng/l. The company claims that the detection process is a proprietary one developed by them by which a cork enters a chamber where it is warmed. The process will release any volatile compounds which are collected in the chamber and concentrated in order to provide a sample for analysis. They also offer a 'buy back guarantee' [87].

3.12 Winery and Cooperage Sources of Haloanisole Contamination in Wines

There can be little doubt that instances of chloroanisole taints consequential to contaminated corks have reduced considerably in the last decade or so. However, there remain several other possible sources of haloanisole contamination in wines. Whilst in recent years some of these have also reduced in likelihood, particularly due to awareness of the dangers of the presence of chlorophenols and bromophenols in the structure of buildings and the use of hypochlorite cleaning agents, others are perhaps on the increase.

3.12.1 Winery Environment, Structure, Sundry Items, and Routine Processes as Sources of Contamination

The presence of haloanisoles or halophenols of any areas of a winery, including the cuverie or the barrel store, can result in the contamination of each and every wine produced therein. As we have seen, haloanisoles are very volatile, rapidly becoming airborne, and localised areas of contamination can quickly spread to other areas and wines. Wines can become contaminated at many stages of the production processes, being particularly exposed during rackings, other cellar operations, and storage.

Several haloanisoles can co-occur in affected wines, and this is also the case with cellar contamination. By way of example, a sample of white wine together with its DIAM (technical) cork stopper sent in May 2017 to Laboratoire Pure Environnement in Perpignan showed contamination with various haloanisoles: in the wine TCA was at a concentration of up to 21.8 ng/l (considerably above sensory detection threshold), TeCa at 14.7 ng/l, PCA at 7.1 ng/l and 2,4,6-TBA at up to 14.8 ng/l (also considerably above sensory detection threshold). The TCA in the closure was less than 0.3 ng/l, the TeCA and PCA each less than 1 ng/l and 2,4,6-TBA less than 2 ng/l [report seen by me]. The most likely conclusions to be drawn from this analysis is that the migration into the wine of the haloanisoles is a consequence of cellar contamination, or the use of contaminated processing aids.

The use of hypochlorite beaches (which thankfully has largely ceased) and even chlorinated tap water in the winery can be a source of chloroanisole contamination. Many wineries, including Hanzell (as discussed in Section 3.4.3.1) have attributed cellar contamination with TCA to historic uses of chlorine as a cleaner/sanitiser.

Aerial (atmospheric) contamination can be a very serious issue. In Section 3.4.3, I detailed several cases where wines were contaminated due to haloanisole presence in the winery structure and atmosphere. Whilst wines are particularly at risk during barrel maturation, even short exposure such as may take place during open top fermentations, pump-overs, punch downs, pressing, and rackings can result in pickup. Aerial presence may result in wine contamination above sensory detection thresholds *per se*, but it may also impart a low level of TCA to the wine, i.e. below sensory detection threshold. However, if there is also releasable TCA in a cork closure, also below sensory threshold, then this additional contribution could take the total level in the wine above sensor detection threshold.

Contamination with TeCA has often been identified prior to bottling in wines that have been stored in vats. The main source of TeCA in wineries and cellars is fungicide treatment (anti-blue) of wood that has been treated with pentachlorophenol (which also contain small quantities of 2,3,5,6-tetrachlorophenol, the precursor of TeCA) and its sodium derivatives. 2,4,6-TBA wine taints are invariably derived from sources within the winery or transport. Possible sources of contamination by 2,4,6-TBA include oak barrels, barrel bungs (especially those made of silicon), filters, pallets, plastics (including shrink-wrap as commonly used on pallets of cases of wine), wood (including wooden wine cases), bottling lines, winery structures, and the atmosphere within wineries and cellars.

3.12.2   Oak Barrels as a Source of Haloanisole Contamination of Wine

Oak barrels can become contaminated by the atmosphere in the cellar, even at some distance from the original source. Re-using oak barrels always poses risks of contamination of the newly filled wine with many unwanted compounds, yeasts and bacteria that may be present in the wood, including chloroanisoles, bromoanisoles, and *Brettanomyces*. However, haloanisoles can also be present in new oak barrels, and it would appear that a small number of coopers are particularly implicated. Between 2004 and 2009, Pascal Chatonnet et al. studied 11 different instances of wines becoming contaminated with TCA that were related to the use of new oak barrels [88]. The wines, mostly red, from France, Italy, Spain,

Australia, Austria, South Africa, and the United States showed TCA contamination at levels that varied from 1.5 to 15.5 ng/l. The contamination was clearly attributed to the barrels used. When examining a red Tempranillo wine from Rioja, samples drawn from some of the new barrels from a certain cooper contained levels of TCA detectable upon sensory analysis, whilst wines from the same lot but in barrels from two other coopers and stored in the same cellar, did not. From Cooper A, four out of 12 barrels contained detectable levels of TCA: the wine from one barrel contained 12.1 ng/l, a second barrel 5.2 ng/l, a third barrel 4.1 ng/l, and a fourth barrel 2.4 ng/l of TCA. Some of the barrels also contained TeCA and/or PCA, but at a concentration below perception threshold. From coopers B and C no barrels contained TCA, but four out of 12 barrels from each cooper contained TeCA and/or PCA but again below perception threshold. Upon detailed examination of one barrel, the TCA was found to be located in just one small area. There have been several other instances. In 2015 in the United States, a wine that had tested negative for haloanisoles prior to barrelling and then aged in a French oak barrel was found to have 11 ng/l of TCA and 82 ng/l of 2,4,6-TCP [89].

It has been suggested that the flour paste (made with approximately 1 kg of flour and 1 l of water) traditionally used by coopers to seal the groove near the ends of the staves in which the barrel head is seated, can be a source of moulds. The flagship Napa Valley producer Shafer Vineyards identified the paste to be a potential source of contamination, and persuaded the barrel makers Demptos, which has cooperages in France, Spain, Australia and the United States to develop a system of making the seal with food-grade paraffin instead of flour paste [90]. Research undertaken by independent laboratories has noted that the growth of moulds, and the ensuing production of TCA, TeCA and 2,4,6-TBA is favoured by the residue of the flour sealant [personal discussions].

In December 2018 it was revealed that Opus One, the iconic Napa winery jointly owned by Baron Philippe de Rothschild and Constellation Brands, was suing a cooper and barrel distributor for $471 356 in damages, resulting from alleged TCA contamination of 10 French oak barrels received in July 2016. At the time of writing the case was ongoing. The launch of this lawsuit received immediate publicity, including articles on numerous websites, including http://wine-searcher.com [91] and https://www.winebusiness.com [92].

Coopers, having become aware of the risks not only to wines but also their own reputations, have increased diligence and changed working practices to tackle any risk of haloanisole contamination. The implementation of quality management systems such as ISO 9000, together with HACCP and traceability of all materials are now commonplace in the industry. However, the risk of repercussions, bad publicity, lost contracts, and law suits will mean that preventive systems and actions to avoid haloanisole and related contaminations in new oak barrels will become even more important in the years to come. The Fédération Française de la Tonnellerie (FFT) has Suggested Maximum Acceptable Limits of haloanisoles and halophenols in cooperage oak. The figures are detailed in Table 3.3.

An interesting recent development in the battle against TCA and 2,4,6-TBA in cooperage, is the employment of sniffer dogs! TN Coopers, an international barrel-making company, has developed the 'Natinga Project' involving dogs for their highly sensitive noses (at least 10 000 times more powerful than humans) to seek out haloanisoles pre-cooperage and in finished barrels pre-despatch to clients. Michael Peters, who is the resident winemaker

Table 3.3 Suggested maximum acceptable limits of haloanisoles and halophenols in cooperage oak and oak wood alternatives.

Compound	FFT limit (ng/g)
2,4,6-Trichloroanisole	0.6
2,3,4,6-Tetrachloroanisole	2.8
Pentachloroanisole	No proposed limit
2,4,6-Tribromoanisole	0.6
2,4,6-Trichlorophenol	No proposed limit
2,3,4,6-Tetrachlorophenol	No proposed limit
Pentachlorophenol	18.8
2,4,6-Tribromophenol	No proposed limit

Source: Extracted from A *Guide de Bonnes Pratiques à Usage de la Tonnellerie*. FFT.

and sales manager in the Sonoma office of TN Coopers is quoted as saying, 'they're more accurate and effective than modern technology'. [93]. The dogs are trained to identify different concentrations of TCA, TeCA, PCA, and 2,4,6-TBA and also to avoid false positives. Their routine includes inspecting the wood yard, the supplies depot, barrel production areas, and the finished products. Any areas of contamination are pinpointed, and a clear signal given to the handler. The dogs have also been discovered to have the ability to identify PCP, TCP, and TeCP. Of course, the identification and elimination from the production process of wood containing halophenols removes the precursors to subsequent haloanisole formation. A photo of a Natinga dog at work is shown as Figure 3.3 and a video is available on YouTube [94].

As discussed earlier in this Chapter there are many instances of haloanisole and halophenol contamination of shipping containers, particularly the wooden floors. TN Coopers has an annual turnover of over 400 containers for both imports and exports, and the Natinga Project has now become a crucial procedure for contaminant control of these. The company has rejected some containers due to serious suspicion of contamination. Interestingly the Natinga sniffer dogs can have a further use in tracing locations of haloanisole contamination in wineries: when bentonite traps have indicated environmental contamination, there then begin detailed searches for the source(s). The dogs can become a valuable, efficient, and time-saving tool to this end.

3.12.3 Processing Aids and Additives as Sources of Contamination

There are many reported instances of contaminated additives and processing aids resulting in considerable loss to wine producers. In the years 2005 to 2006 batches of contaminated tartaric acid imported into Australia from China were contaminated with 2,6-dichlorophenol [95]. Tartaric acid is commonly used as an additive during the process of must adjustment, particularly in hot climates in order to produce balanced wines. According to AWRI [95], the cost to the industry of tainted wines produced with the addition of the contaminated acid amounted to $10 million (Australian dollars).

Figure 3.3   Natinga TCA sniffer dog. Source: Courtesy of TN Coopers.

In consequential legal action between a winery and the supplier, the court ruled (in 2011) in favour of the supplier. Although the Australian court agreed that the tartaric acid supplied was not of merchantable quality, and was not fit for use as an additive to wine, the contracting winery had, some five years before receiving the supplies in question, signed the supplier's terms and conditions that absolved the supplier of responsibility.

3.13   Laboratory Analysis for TCA and Other Haloanisoles in Corks and Wine

3.13.1   Haloanisoles in Wine and Releasable and Total TCA in Corks

The analysis of cork closures by the major cork producers and specialist laboratories has become highly sophisticated. Both releasable TCA and total TCA contained in corks, and the presence of these and other haloanisoles in wine can be analysed using chromatographic and bioanalytical techniques as discussed below. The International Organisation for Standardisation (ISO) has developed a standard for Determination of Releasable TCA from Cork Stoppers. This is ISO 20752:2014 (en) [96]. There is also a standard for a procedure of sensory analysis of corks, ISO 22308:2005 [97], which details methods to identify musty smells and other aromas that may possibly be present.

3.13.2 Laboratory Methods of Analysis

There are several laboratory methods for preparing samples, and sophisticated chromatographic methods for separating and detecting haloanisoles, and quantifying their concentration in wine and/or corks. The technology is now able to quantify the compounds to levels below 1 ng/l, which is below sensory threshold even for TCA (although TCA in sparkling wine may have a sensory threshold as low as 0.13 ng/l). The processes involve two-step procedures: sample preparation and analysis.

3.13.2.1 Sample Preparation

Samples may be prepared by several methods. The following are widely used:

- Solid-phase micro-extraction (SPME) and headspace solid-phase micro-extraction (HS-SPME). SPME is a solvent-free method of extracting analytes by partitioning them from a liquid or gaseous sample into an immobilised stationary phase. It is very simple to set up and the only instrumentation required is the conventional GC for the analysis, and accordingly is the most commonly used method. With SPME there is no need for pre-concentration, as the analytes are directly extracted into a polydimethylsiloxane (PDMS) coated fibre. This has been shown to be the most suitable fibre for the analysis of TCA in wine [98, 99]. The only drawbacks are that the PDMS coated fibre does have a limited life and is not inexpensive.
- Headspace liquid phase micro-extraction (HS-LPME). This is a recent and highly successful extraction method for chlorophenols [100]. This method is fast, simple, inexpensive, and environmentally friendly, requiring only a few millilitres of solvent.
- Dispersive liquid–liquid micro-extraction (DLLME), which was introduced as recently as 2006, has also become a popular method that can be used to extract many volatile compounds, including haloanisoles, geosmin, and volatile sulphur compounds. The method uses minimal organic solvents. Used with GC-MS it is a fast and efficient method for detecting halophenols and haloanisoles in a cellar environment [101].
- Stir bar sorptive extraction (SBSE) is another recent and highly sensitive method of extracting complex samples such as wine by scan mode, with excellent confirmation of identity of compounds such as TCA and other haloanisoles.
- Supercritical-fluid extraction (SFE) uses a medium such carbon dioxide in supercritical state (above its critical temperature and pressure) as an extractant as an alternative to organic solvents as an 'environmentally friendly extraction process'.

3.13.2.2 Separation and Quantification

Gas chromatography is now bar far the most popular method for separating the volatile components in a sample. The chromatograph is coupled with a system that analyses and quantifies the molecules eluded.

- *Gas chromatography with mass spectrometry (GC-MS)*: This is the most commonly used method. Mass spectral data is produced for each component of the chromatographic analysis enabling detected compounds to be identified by comparing to recorded spectra from commercially available libraries. Sensitivity can be to 0.5 ng/l or less for TCA, and 1 ng/l or less for other haloanisoles. Headspace solid-phase micro-extraction/gas chromatography-mass spectrometry (HS-SPME/GC-MS) is perhaps the most sensitive assay technique for haloanisoles: this system can detect the presence of TCA as low as 0.1 ng/l.
- *Gas chromatography with electron capture detection (GC-μECD)*: This method has the advantages of being the most selective detector with higher sensitivity for haloanisole containing analytes and at generally lower cost than GC-MS.
- *Tandem mass spectrometry (MS-MS)*: This system separates analytes by molecular weight in one mass spectrometer, and they exit as fragments. A second mass spectrometer then identifies the fragments.
- GC/MS/MS (QqQ) ('Triple Quadrupole') is very advanced technology used by some professional laboratories, which can quantify extremely low levels of haloanisoles. Analytes from a mass spectrometer react with an inert gas in a collision cell, resulting in further fragmentation, and are then identified in another mass spectrometer. The technique increases the selectivity and the sensitivity of the analytical methodology, avoiding ambiguous identification of compounds and matrix interferences remaining after the sample preparation step [100, 102].

There are other fast methods to determine haloanisole such as gas chromatography with atomic emission detection (GC-AE), which is an alternative to mass spectrometry. Atomic emission simultaneously determines the atomic emissions of many of the elements in analytes that are separated in a GC capillary column. Unfortunately, AE equipment is expensive. A 2015 study by Slabizki et al. showed good results in the quantification of several 'cork' off-flavour compounds in natural cork stoppers and wine by using HS-SPME and heart-cut multidimensional gas chromatography (MDGC) with tandem mass spectrometric detection [103].

It should be noted that whilst chloroanisoles and bromoanisoles may be analysed without prior preparation, chlorophenols must undergo derivatisation prior to analysis by GC or high performance liquid chromatography (HPLC) [99]. Special C18 columns with organic support may be used for this. C18 columns are HPLC columns that use a C18 substance as the stationary phase. C18 simply means that the molecules contain 18 carbon atoms.

3.13.2.3 Bioanalytical Methods

It is possible to analyse samples for haloanisoles using bioanalytical methods employing either antibodies, cells, or their combination as biorecognition elements or non-invasive optical-detection principles [104]. Biosensor-based approaches are still in the experimental stage, with limited testing on real samples, but in the future biosensors might introduce a revolutionary technology as alternatives to conventional analytical methods discussed above [105].

There is an International Organisation of Vine and Wine (OIV) Method for the Determination Of Releasable 2,4,6-TCA In Wine By Cork Stoppers (OIV-OENO 296-2009) [106] subsequently updated by Resolution OIV-OENO 623-2018 [107].

3.13.3   Inexpensive Alternatives

Unfortunately, the equipment needed for all the above methods of analysis is expensive, and beyond the means of many small wineries. Of course, the presence of haloanisoles in wines can be determined, if not quantified, by the nose of a sensitive and trained sensory assessor. This, of course, is not the case for corks. As Valeria Mazzoleni and Luana Maggi note, 'instrumental analysis for the systematic control of TCA in cork is seldom used by wineries' [108]. However, there is a simple method of screening batches of corks to determine levels of rapidly releasable TCA: group cork soaks. This method, which is also commonly used by laboratories, is discussed in Section 3.14.4.

Another interesting development for rapid, on site, and relatively inexpensive determination of TCA levels in wine is the zNose®. This is a small, portable, handheld instrument that utilises an ultra-fast (10 seconds) GC coupled with a single surface acoustic wave (SAW) solid-state vapour detector.

3.14   Prevention of Haloanisole Contamination of Wineries and Wines

3.14.1   Implementation of an HACCP System

The implementation of a strategy with processes, procedures and practices to mitigate the risk of haloanisole contamination of wines should be considered an essential part of a producer's quality and merchantability systems. Whatever the value of quality control systems such as ISO9001, they can instil an overconfidence that deviation from standards will not happen. However, the importance of the development and implementation of a thoroughly researched HACCP system, with the targeted hazards of halophenols and haloanisoles, cannot be overstated. At its simplest, the system should consider when and where in the production process haloanisoles might be produced, i.e. when one of group (a) below, and one or more of group (b) may be simultaneously present:

(a)   Filamentous fungi or xerophilic fungi;
(b)   Phenols: sources include grapes, must, wine, wood, synthetic polymers, halogens (chlorine, bromine, iodine) and water.

3.14.2   In the Vineyard

As discussed in Section 3.7.4, there are several species of fungi that can perform biomethylation from halophenols to haloanisoles, including species of the genera *Acremonium*, *Fusarium*, *Trichoderma*, and *Penicillium*. The limiting of the presence of these on grapes by biological or chemical treatments is desirable from a general winemaking perspective, as well as reducing the possibility of haloanisole formation.

3.14.3   In the Winery

As we have seen, there are numerous potential sources for haloanisole and halophenol contamination in wineries, and individual contaminated areas may a present real risk of an ambient presence. Constant vigilance is crucial. Particular attention should be paid to the topics detailed in Sections 3.14.3.1–3.14.3.10.

3.14.3.1 Cleaning and Sanitising

Regular cleaning and sanitation of the infrastructure (floors, walls, ceilings) to limit colonisation of microbes, including moulds and fungi, on surfaces should be a matter of routine, even if the surfaces show no sign of being dirty. The building should be ventilated after cleaning to reduce humidity and the surfaces should be dried.

3.14.3.2 Avoidance of Chlorine and Bromine

The use of cleansers and sanitisers containing active chlorine and bromine should be avoided. Common liquid bleach comprises sodium hypochlorite, formula NaOCl or NaClO. The compound is unstable in solution, and releases chlorine gas. The presence of 2,4,6-TCP in many winery facilities has been traced to historic use of chlorine bleaches that may have ceased years previously. Active chlorine-based cleaning agents are commonly used in the brewing industry; perhaps one of the reasons why TCA affected beer is far from uncommon. Wineries that also have an attached brewery (a small but growing number) should ensure that operatives are fully trained and supervised with regard to the dangers that the use of chlorine-based products present in a winery environment. The exclusion of such products should extend to the tasting room, restaurant, laboratory, and dry goods storage areas. Oxygen bleaches are a practical alternative, using sodium percarbonate which, when mixed with water, breaks down to hydrogen peroxide (water and oxygen) and sodium carbonate. Perhaps surprisingly, chlorine dioxide (ClO_2) is a safe sanitiser to use in wineries, as chlorine is not involved in any of the reactions. It is a cleaner and sanitiser that works by oxidation. It must be manufactured on site, but it is stable in a dilute solution in the absence of light. It is claimed that chlorine dioxide has 260% more disinfection power than chlorine-based products and does not allow any build-up of biofilms. It can kill most airborne and 99.9% of all waterborne microorganisms including bacteria, fungi, algae, and moulds [109]. However, it would appear that chlorine dioxide is ineffective as a barrel santiser: Maria de Lourdes et al. in 2015 found that it did not significantly reduce the number of wine spoilage organisms present in barrels [110].

3.14.3.3 Monitoring of Tap Water

Tap (municipal) water may have been treated with chlorine. Regular analysis should be undertaken for chlorine and also the presence of chlorophenols. Peat soils, for example, will impart phenols into a water system and, when chlorinated, chlorophenols will be formed. Further, with the widespread change from copper to high density polyethylene as a material for water pipe manufacture, chlorophenols may get in drinking water from the diffusion of organic additives in the material. Phenolic additives are used as antioxidants in the production of water pipes.

Tap water may be contaminated with haloanisoles. It may also be tainted with 2-isopropyl-3-methoxypyrazine (a wine taint that is more usually associated with ladybugs – see Chapter 13), and 2-isobutyl-3-methoxypyrazine (IBMP), also known as 2-methoxy-3-isobutyl pyrazine. Tap water should not be used for rehydrating yeast, or MLF bacteria. The winery water distribution system should also be periodically checked.

The use of carbon filters is recommended for water supplies: they will not remove TCA but may remove other musty tainting compounds such as 2-methylisoborneol (MIB) – see Section 3.17.5.

3.14.3.4 Air Renewal and Humidity Control

Good ventilation, particularly in confined spaces, will help avoid haloanisoles accumulating. Air should be renewed by either natural ventilation or mechanical systems. Temperature control to prevent the growth of moulds and fungi in key areas is also important. Humidity in production areas should be kept below 75% if possible to help limit the growth of microorganisms including filamentous fungi, although in the barrel store higher humidity, i.e. 80–85% is necessary for maturation and to avoid excessive evaporation losses.

3.14.3.5 Prohibition of PCP and 2,4,6-TBP

There must be a total prohibition of the use of PCP and 2,4,6-TBP treatment for wood or plastics of any kind and for any purpose, together with all other equipment introduced into the winery. A certificate of compliance should be obtained from all suppliers. Wooden pallets pose a high risk of bringing precursors or even haloanisoles into premises – the possibility of pallets, or indeed any equipment, having been washed or sanitised using chlorine-based bleaches should always be borne in mind. The use of plastic pallets has merits and demerits, and care should be taken that the material does not contain bromophenols. There has been a shift to their usage in the pharmaceutical industry, but the wine industry has been slower in take up. In the Napa Valley, Shafer Vineyards has implemented their usage throughout its facility as part of a TCA prevention programme [90].

3.14.3.6 Diligence with Regard to Oenological Products and Processing Aids

Diligence should be taken with all materials coming into the winery particularly, but not limited to, oenological products and processing aids. Products that might be contaminated include, bentonite, diatomaceous earth, polyvinylpolypyrrolidone (PVPP), ascorbic acid, citric acid, tartaric acid, yeast hulls, and wood chips and other oak products. These should be purchased first-hand from reliable and trusted sources and certification demanded from manufacturers that the products are haloanisole and halophenol-free. All oenological products should be screened for haloanisoles before use. Such screening might comprise a simple sensory assessment of 200 grammes of material per batch of each of the materials, with the exception of bentonite, where the screening should be of 1 kg [111]. Porous products should be stored away from production areas.

Bentonite has a high affinity for absorbing haloanisoles, and storage should always be in a scrupulously clean, dry with good air circulation. During the course of researching and writing this book, I visited a small producer who has a vineyard located on fairly heavy clay soil. I noticed bags of gypsum (calcium sulfate) in the dry goods store, close to bags of tartaric acid. In fact the 'dry' goods store was slightly damp. At one time gypsum was used in winemaking in hot regions (particularly in Jerez) in lieu of tartaric acid for 'plastering' grapes or acidifying grape must. It is no longer a permitted additive for wine, although it is still used in the brewing industry. The producer confirmed that the gypsum was to be used in the vineyard in order to improve soil structure. I considered that such storage represented a double risk – gypsum, like bentonite easily absorbs haloanisoles and in turn the tartaric acid could become contaminated. Many oenological products are susceptible to absorbing or adsorbing taint compounds; for example, there have been several instances of yeast hulls being tainted with 6-chloro-*O*-cresol, resulting in wine contamination.

3.14.3.7 Periodic Inspections and Testing of Production and Storage Areas

Wine can be contaminated by direct airborne transfer of related compounds. Periodic inspections, ideally semi-annually but at least annually, to determine if the production and storage areas are free of 2,4,6-TBP, 2,4,6-TBA, 2,4,6-TCP, TCA, 2,3,4,6-TeCP, and TeCA. The regular inspections should include testing the atmosphere (also known as aerial analysis), water supply, water storage, water pipes, drains, and all structural and inbuilt wooden materials, paint and wall coverings, plastics, vats, wooden barrel racks and chocks, pipes, and hoses. This list is far from exhaustive.

To sample the atmosphere of the winery and cellar, TCA traps which contain an absorbent (e.g. bentonite) can be placed at appropriate locations and held for a period of three weeks or so. Samples for aerial analysis can also be collected by the following methods:

- *Method 1:* The atmosphere can be captured with medium flow 2000–3000 ml/min pumps – air is drawn continuously through a thermal desorption cartridge filled with an appropriate adsorbent material. The absorbent can be removed and sent to a laboratory for analysis. If haloanisoles are found in the medium, then a more detailed site investigation is necessary to find the source(s) and locate precursors.
- *Method 2:* Solid-phase micro-extraction (SPME) fibres can be located wherever aerial contamination is possible.

3.14.3.8 Attention to the Barrel Room, Cellar, and Storage

Barrels, particularly pre-used barrels can be a source of chloroanisoles and precursors. In the case of the cellar contamination at Hanzel (see Section 3.4.3.1), 2,4,6-TCP was discovered on the exterior surfaces of some barrels. The winery made the decision to treat these by sandblasting. Of course, winemakers may wish to mature wines in second or third fill barrels in order to limit oak product uptake in the wine, reduce oxygenation (as the wood's pores become blocked with use), or for economic reasons. Ideally these should be from the winery's own stock, but if used barrels are purchased, they should be screened upon arrival and cleaned, sanitised, and dried prior to entering production and maturation areas. Attention should be paid to the system of barrel stacking. The use of concrete barrel chocks or metal racking is preferable to wooden staging or chocks, which may contain halophenols.

Cardboard, wooden wine cases, and wooden pallets should not be stored in or near production areas, including barrel cellars. All packaging materials should be stored at a humidity of less than 80% and in well-ventilated areas. Barrel cellars normally have a humidity in excess of 80% and sometimes as high as 90%, which can, of course, give ideal conditions for the development of moulds and filamentous fungi.

3.14.3.9 Regular Sensory Analysis of Wine Stock

Frequent sensory analysis of wines in the course of maturation and storage in vat, barrel, and bottle should take place in a tasting room that is situated well away from production and storage areas. The employment of external assessors is recommended.

3.14.3.10 Training Cellar Staff for Haloanisole Detection

The rapid detection of haloanisole-related odours by staff is crucial to rapid detection of contamination, and production staff should be trained accordingly, followed by refresher workshops every three or four years. Winemakers and cellar staff can become desensitised to low levels of haloanisole compounds in the winery atmosphere: this is also the case with sulfur dioxide.

3.14.4 Cork Screening and Group Cork Soaks

Corks should be screened before use. It is common for corks to be packed in plastic bags in an atmosphere containing SO_2 and it is advisable that the integrity of the packages is not broken until shortly before the corks are required.

- *Group cork soaks*: Of course, it is impractical for most wine producers to analyse every cork and problems may only be discovered after a wine has spent many years maturing in the bottle, much to the purchaser's chagrin and with little chance of any recompense. However, a very simple, low-cost, (but far from fool-proof) screening method is to soak a sample of corks from each package in a neutral, dry white wine for a period of 24–48 hours. The wine used for the soak is then subjected to sensory analysis, and/or laboratory analysis. Rapidly releasable TCA (RTCA) should be revealed during this process. An illustration of group cork soaks being undertaken is shown as Figure 3.4.

Figure 3.4 Group cork soaks.

The value of group cork soaks was shown in a 2006 study sponsored by the American Vineyard Foundation, and conducted by ETS Laboratories, which clearly confirmed a cork's RTCA as a good predictor of the TCA that will be transferred into bottled wine over a period of up to 20 months [39]. In this study bottled wines were segregated by cork bale and tested for TCA at intervals. The results after 20 months showed a distinct correlation between the RTCA that had been measured in the group cork soaks and the TCA found in the bottled wine. The predictive value of group cork soaks was well established with over 95% correlation between TCA levels in the cork soaks and the wines. The study also showed that tests of RTCA from group cork soaks, if applied in large-scale quality control programmes, can reduce dramatically the incidence of both 'corked' and 'muted' bottles [39].

However, it should be born in mind that TCA from contaminated corks may be absorbed by non-contaminated corks in the same soak, thus reducing the likelihood of organoleptic detection. For this reason, the number of corks in each soak should be restricted – perhaps 25 corks in 1.5 l of wine. From each package, several of these soaks should be undertaken to bring the total sample to 50–100 corks. Of course, all the surface area of the corks is soaked, but only a small part of this surface will be in contact with the wine. Accordingly, this could result in an over indication of absorption. Conversely it should also be borne in mind that these soakings will only reveal rapidly releasable TCA and further TCA may well migrate into wine over time – fine wines, of course, spend many years maturing in bottle. On balance, group soaks tend to underestimate the TCA that might eventually be transmitted into the wine. A final caveat is that members of the panel carrying out the sensory analysis should be routinely assessed for sensitivity – regular exposure to TCA, even at low levels, can lead to partial anosmia.

3.15 Treatment of Wines Contaminated with Haloanisoles

It is very often written that there is no effective treatment for wines that have been contaminated by TCA or other haloanisoles. In fact this is not the case, but they cannot be neutralised, or removed by standard filtration, reverse osmosis (RO) or basic fining, such as with isinglass, bentonite or albumen. However, fining with a material with a higher chemical affinity with TCA and other haloanisoles than is the case with wine can have some effect. Haloanisoles are highly non-polar. TCA has $\log P = 5.7$ and water solubility of 0.01 g/l and accordingly hydrophobic materials such as plastics can be effective at removing TCA (and other haloanisoles) from affected wines [34].

3.15.1 Fining with Half and Half

Fining with half and half (half cream) which comprises 50% whole milk and 50% light cream, with a fat content of 10–12% can be effective. TCA is adsorbed into the butterfat. The milk products are allowed to settle to the bottom of the tank, the wine is then racked off and filtered. Of course milk products are known allergens, although EU wine regulations do not require their addition to be stated on wine labels.

3.15.2 Filtration Using Zeolites Y-Faujasites

Special filter pads containing Zeolites Y-faujasites, Aluminosilicates (which are widely used as ion exchange beads in both domestic and commercial water purification systems), can reduce to sub-sensory threshold levels TCA and 2,4,6-TBA from a tainted wine, providing the concentration is not high. The filter pads are manufactured by a Swiss company, Filtrox, and marketed under the brand name Fibrafix® TX-R. The company claims that the aroma and flavour profiles of treated wines remains undamaged. The pads are available as plate filter or stacked modular canister filter formats [112]. The treatment must be carried out on clarified wines. The use of the pads for the purpose of reducing the concentration of haloanisoles to below their sensory perception thresholds is permitted by the International Organisation of Vine and Wine and the European Union Wine Regulations (EU 2019/934).

3.15.3 Absorption/Adsorption/Filtration Using Polymeric Media

3.15.3.1 Use of Polythene Resin Beads and Polymer 'Fining'
Haloanisoles absorb into and adsorb onto certain polymers, including polyvinylidene chloride (PVdC), and polyethylene.

In California, the company Winetech offer a haloanisole removal service, available at their Napa facility but also as a mobile service. Treatment can be undertaken for as little as one barrel. The system uses chromatographic methods to detect TCA and/or 2,4,6-TBA. Special polyethylene resin beads in a column are used to specifically absorb these compounds. The cost depends inter alia on the number of beads needed, which increases with the level of haloanisole contamination. Another Napa based company, VA Filtration, utilises a polymeric medium based on a material originally developed by Chevron to remove pesticide residues from wine [113]. For high levels of haloanisole taints there is a process that separates the wine into a phenol-free stream and a phenol-rich stream. Only the phenol-free stream is treated with a proprietary adsorbent. This process has a low impact on aroma and flavour profiles, as the phenols remain in the wine.

There are several polymers with an affinity for TCA sorption and which have potential for use as fining agents. Polyaniline emeraldine base (PANI-EB), polyaniline emeraldine salt (PANI-ES), and polyamidoamine (PAMAM-GE-CD) have been shown to retain greater than 75% of the TCA in a methanol reconstituted wine solution [114].

3.15.3.2 Filtration Using Adsorbing Polymers
TCA and 2,4,6-TBA polymers can also be incorporated into filter pads and can effectively remove haloanisoles at modest levels, perhaps up to 5 ng/l. Some phenols, including some desirable phenols, will also be removed in the filtration process.

3.15.3.3 Adsorption by Polyvinylidene Chloride (PVdC)
At a domestic level, if a consumer is faced with a tainted bottle, it is sometimes possible to partially clean up the wine by inserting some Saran wrap into a glass or bowl and swirling the wine around. The original Saran wrap formula created in 1933 by the Dow Chemical company was based upon PVdC, but the product is now made with polyethylene, which is a little less effective. A study by Capone et al. [80] showed that 40 cl of tainted wine kept in

contact with 800 cm^2 of 100 μm polyethylene for three days retained only 4% of the original TCA, 2% of TeCA and just 1% of PCA. Some desirable aromas and flavours are stripped from the treated wine. The use of adsorbing polymers on a commercial scale is time consuming and expensive.

3.15.4 Adsorbing by Yeast Hulls

Yeast hulls are highly adsorbent and are often used in winemaking to improve fermentation conditions. They adsorb many toxic compounds from must and wine, including fatty acids. They can also adsorb TCA, PCA, TeCA, and 2,4,6-TBA. The recommended dose is 20–40 g/hl.

3.15.5 Filtration Through Cork Dust

Of course, cork has a very high affinity to absorbing haloanisoles, and filtering affected wine through a bed of cork dust has shown some success in partial removal of the compounds. The treatment is not generally legally permitted.

3.15.6 Deodorising with Mustard Seeds

Mustard seeds can be used as a haloanisole deodorant, but treated wines suffer a severe aroma and flavour modification. The treatment is not generally legally permitted.

3.16 Chlorophenols and Bromophenols as Taints

Chlorophenols and bromophenols can taint wines. The odours imparted are very different to those of haloanisoles.

3.16.1 The Odours of Chlorophenols and Bromophenols

Although chlorophenols and bromophenols are often considered only as the precursors to related chloroanisole and bromoanisole taints, their presence in wine can give off-odours. Fortunately the perception thresholds are much higher than for the related haloanisoles, and thankfully their occurrence is fairly rare. Typically the nose and palate of affected wines will show notes of plastic or paint, sometimes with medicinal tones. Disinfectant and swimming-pool associated aromas may also be present. Plastic odours can also be due to indole (see Chapter 14).

The following phenols have been known to taint wine:

- 2,4-Dichlorophenol
- 2,6-Dichlorophenol
- 2,4,6-trichlorophenol
- 2-Chloro-6-methylphenol
- 2-Bromophenol
- 2,6-Dibromophenol
- 2,4,6-tribromophenol

3.16.2 2,4-Dichlorophenol

Molecular formula: $C_6H_4Cl_2O$

The perception threshold is 896 ng/l. Above the threshold the compound exudes odours of plastic or paint, usually with medicinal tones. There are recorded instances of wines tainted with 2,4-dichlorophenol in Australia, with a serious incidence back in in 2002. The source was discovered to be agro-chemicals [95]. The compound is widely used is some cleaning products.

3.16.3 2,6-Dichlorophenol

Molecular formula: $C_6H_4Cl_2O$

As with the other chlorophenols, this compound has a medicinal, paint, chlorine, or plastic-like smell. The odour perception threshold of 2,6-dichlorophenol in wine is 32 ng/l, which is by far the lowest of all the chlorophenols; the taste perception threshold is perhaps a little higher at 48 ng/l. Incidentally the threshold in wine is much lower than in water, which is 300 ng/l. The compound is present in many cleaning products and biocides. Investigations conducted by AWRI in 2017/2018 into a wine contaminated with 2,6-dichlorophenol concluded that the source was contaminated bentonite [115]. AWRI have noted concentrations in wine up to 236 ng/l [116].

3.16.4 2,4,6-Trichlorophenol

Molecular formula: $C_6H_3Cl_3O$

This compound, the precursor of TCA, has an odour of plastic and chemicals. The sensory perception threshold in wine is 2 µg/l.

3.16.5 6-Chloro-o-cresol

Molecular formula: $C_7H_5Br_3O$.

Although usually known in the wine industry as 6-chloro-o-cresol, the International Union of Pure and Applied Chemistry [IUPAC] name for the compound is 2-chloro-6-methylphenol. Odour characteristics are antiseptic, disinfectant, and plastic. Affected wine can also show a chlorine-like nose, sometimes with a burning finish on the palate. The sensory perception threshold (in white wine) was determined by AWRI to be 70 ng/l [116], but Geoff Cowey reported it to be as low as 16 ng/l [95]. The compound is known to taint many other drinks and foodstuffs, including soft drinks, chicken, and biscuits: a historic source of the compound contaminating biscuits was found to be aerial 'transport' of effluent from a factory producing herbicides some 8 km away.

This potent compound has been found as a contaminant in yeast hulls – the use of contaminated hulls resulted in a level of 500 ng/l in a wine analysed by AWRI [95]. It is also found is some proprietary cleaning products, including drain cleaners.

3.16.6 2-Bromophenol

Molecular formula: C_6H_5BrO

The compound can impart ink-like or 'old museum' odours. The sensory threshold of 2-bromophenol in wine is 1–5 µg/l.

3.16.7   2,6-Dibromophenol

Molecular formula: $C_{12}H_8Br_4O_2$
  The sensory perception threshold is 50 ng/l.

3.16.8   2,4,6-Tribromophenol

Molecular formula: $C_6H_3Br_3O$
  The precursor of 2,4,6-TBA, this compound is used as an intermediate in the manufacture of flame retardants and as a timber preservative, has been ubiquitous in the home and work environment [117], and is widely found in winery structures. The sensory perception threshold in wine is 890 ng/l.

3.17   'Musty' Taints Unrelated to Halophenols and Haloanisoles

3.17.1   Other Musty Taints in Wine

There are several compounds not related to halophenols or haloanisoles that can give musty taints to wines:

- MDMP
- 2-Isopropyl-3-methoxypyrazine
- 2-Methoxy-3-isopropylpyrazine
- 2-MIB
- 1-Octen-2-one
- 1-Octen-3-ol
- 1-Octen-3-one
- Various monoterpenes and sesquiterpenes
- Guaiacol
- *cis*-1,5-Octadien-3-one
- *cis*-1,5-Octadien-3-ol
- Geosmin

3.17.2   2-Methoxy-3,5-dimethylpyrazine (MDMP) or 'Fungal Must Taint'

Several methoxypyrazines may be present in wine, many of which may give positive aromas and flavours depending upon the individual methoxypyrazine, the concentration and the grape variety, and general wine matrix. Some are responsible for the grassy, herbaceous, green bell pepper and general green aromas and flavours of under-ripe Cabernet Sauvignon, Cabernet Franc, and the largely desirable gooseberry and asparagus aromas of Sauvignon Blanc and, to a lesser extent, Semillon. The most concentrated methoxypyrazine in these varieties is IBMP. The lowest detection threshold in white wine has been stated as 1 ng/l [118], although the figure of 2 ng/l is commonly quoted. In red wine the threshold is perhaps 15 ng/l [119].

However, there is a methoxypyrazine whose presence in wine gives rise to odours that render affected wine to be clearly faulty. This is MDMP, whose sensory characteristics are mushroom, mould, earth, potato, green hazelnut, dust, and the odour of an engineering workshop. MDMP is an extremely potent compound. The perception threshold in white wine is noted as 2.1 ng/l [120]. Waterhouse et al. [34] quote a general threshold of 2 ng/l.

MDMP can be found in cork and has been identified as the second most likely source of 'cork taint' after TCA in Australian wines. It was identified as the source of so-called 'fungal must taint' as recently as 2004 [120]. MDMP and TCA, when co-occurring in cork, can both absorbed by wine in bottle [120, 121]. In fact, the presence of MDMP in cork will, in practical circumstances, always co-occur with the presence of chloroanisoles [Dr. Palo Lopes, Amorim R&D, personal discussion]. MDMP can be biosynthesised from the amino acids l-alanine, l-leucine, and to a lesser extent l-phenylalanine and l-valine by the bacterium *Rhizobium excellensis*, the main organism responsible for the presence of the compound in cork. However, the rod bacterium *Serratia odorifera* and the fruiting body-forming myxobacterium *Chondromyces crocatus*, commonly present in soil are other bacteria that can synthesise MDMP, and the storage of oak or cork oak planks directly on the ground is a pathway to contamination. The molecules can be present in cork bark and also oak, including wood chips as shown in the study by Chatonnet et al. [122], and the punches used to cut corks from cork oak planks are another possible source of the contaminant. It extracts easily into wine. The Chatonnet study noted that a high percentage of corks show contamination: 51.2% of cork batches showed a level of 5.1–15.0 ng/l, indicating a medium risk of contaminating wine and batches with 16.3% of batches showed a level of over 15.1 ng/l giving a very high risk of wine contamination [122].

MDMP can also be synthesised from amino acids. Oak barrels are an unlikely source in wine as most of the compound is destroyed during the toasting process, where 93% of MDMP is destroyed at a temperature of 220° C. However, poorly processed and untoasted or lightly toasted oak chips, as often used for 'oaking' of inexpensive wines, do represent a possible source of the taint.

3.17.3   2-Isopropyl-3-methoxypyrazine

This compound is best known as being responsible for 'ladybeetle taint' and will be discussed in Chapter 13.

3.17.4   2-Methoxy-3-isopropylpyrazine

Organoleptic characteristics: earthy aroma.
    This compound can migrate in wine from mouldy barrels. It is also known to taint cocoa.

3.17.5   2-Methylisoborneol (MIB)

The compound is also known as 1,2,7,7-tetramethyl-*exo*-bicyclohepthane-2-ol. Organoleptic characteristics include mould, earth, mud, peat, dirty water, and camphor. The sensory perception threshold is perhaps 30 ng/l in a dry white wine, although Waterhouse et al. [34] states a general threshold of 55 ng/l.

The common source of the compound in wine is its presence in tap water, which is used for cleaning tanks, other equipment and, particularly barrels. However, along with geosmin, 1-octen-3-ol, and 1-octen-3-one, it is also present at high levels in grapes affected by *B. cinera* [34]. The compound is unstable and levels of these are considerably decreased during the fermentation process, and further diminished during maturation and storage. Geosmin and 1-octen-3-ol are more stable and therefore may well remain at above detection threshold.

3.17.6 1-Octen-3-ol

This compound typically imparts mushroom aromas and can have metallic notes on the palate, if the taster gets that far. Sensory perception threshold is stated by several sources as 40 µg/l [123], by AWRI as 20 µg/l [116] but some other authors note it as low as 0.5–7 µg/l.

3.17.7 1-Octen-3-one

Organoleptic characteristics: odour of mushrooms, metallic at high concentrations. Stated figures for the perception threshold range from a high of 70 ng/l [34], 20 ng/l (AWRI) [116] to a low of 7–15 ng/l. This compound can be present in grapes affected by powdery mildew (*Uncinula* necator) and also in rotten grapes. It is a mould metabolite, that is formed by the degradation of lipids. The compound possibly degrades in wine over time but is sometimes detected in wines made with Sauvignon and Pinot Noir grapes [123].

3.17.8 1-Octen-2-one

Perception threshold: 20 ng/l
   This is another mould metabolite.

3.17.9 *cis*-1,5-Octadien-3-one

The compound has an earthy odour. The sensory odour threshold is 20 ng/l.

3.17.10 *cis*-1,5-Octadien-3-ol

This compound can give a musty odour, sometimes with geranium or dried fig tones. Sensory threshold is 9 ng/l.

3.17.11 Geosmin

The organoleptic characteristics are: beetroot, earth, camphor, stale water.
   Geosmin is also known as *trans*-1,10-dimethyl-*trans*-9-decanol and *trans*-1,10-dimethyl-*trans*-9-decalol. The name geosmin means 'odour (osmo) of earth (geo)' and is derived from the Greek.
   The perception threshold in wine is 25 ng/l, although some researchers state higher thresholds – a review by Maria Carla Cravero notes a range of 30–45 ng/l [124]. However,

using a panel of six expert tasters, Georg Weingart et al. determined a group olfactory (odour) threshold level as low as 1.1 ng/l with individual tasters varying from 0.1 up to 22.6 ng/l [125]. Gustatory (taste) detection threshold was nearly identical. This research determined geosmin and TCA in white and red Austrian wines by headspace SPME-GC-MS and comparison with sensory analysis yielded some fascinating results. Geosmin was above the SPME-GC-MS limit of detection (0.5 ng/l) in 93% of the wines analysed, and over 2 ng/l in. 55% of the wines. TCA was detected only in 20% of all investigated wines and above 2 ng/l in 4% of all wines. The researchers state that their results show that geosmin is a much more prevalent cause of 'off-flavours' than TCA [125]. Geosmin is very unstable in wine and may degrade over time.

Although geosmin may be metabolised by some microorganisms in tainted cork bark, it can also stem from damaged or mouldy grapes, particularly those affected by *B. cinerea*, certain strains of which can produce 494 ng/l ($\pm$30 ng/l) [126]. The presence of the compound in wine can also come from ageing in affected barrels. Geosmin is also a known metabolite of soil bacteria and algae. It is really an earthy, not a musty taint (although inexperienced tasters may confuse these), and accordingly this topic is further discussed in detail in Chapter 14.

3.17.12 'Musty' Monoterpenes and Sesquiterpenes

Various sesquiterpenes, present in grapes, can make a major contribution to wine aromas. For example, Syrah (Shiraz) grapes can contain high levels of the sesquiterpene rotundone, which gives aromas of black pepper, rosemary and thyme, and is largely desirable in wines made from the variety, unless the levels are really excessive. However, there are some mono and sesquiterpenes that may be present in rotten grapes that can lead to musty taints: fenchol, fenchone, 2-heptanol, and 2-octen-1-ol.

3.17.13 Guaiacol

Molecular formula: $C_7H_8O_2$.

Guaiacol has a smoky or burnt odour, often reminiscent of smoked cheese or smoked fish, and is usually derived from guaiacum or wood creosote. The sensory perception threshold is 23 µg/l [127]. Its presence in wine may come from a number of alternative sources. These include contaminated corks, oak barrels and oak products such as powder, chips and staves, and smoke from wildfires or controlled burns. The topic of smoke taint, which has become of major concern in recent years, is covered in Chapter 12. However, guaiacol can act in synergy with chloroanisoles to produce the musty odours associated with so-called 'cork taint', even when the presence of both compounds is below their individual sensory DT [128].

3.17.14 Treatment of Musty Taints Unrelated to Halophenols and Haloanisoles

The addition of nitrogen compounds (e.g. the amino acid glycine or the tripeptide glutathione) to the altered wines can lead to lower concentrations of 1-octen-3-ol and 1-octen-3-one in wines and decrease the fresh mushroom odour intensity [129]. Fining with bentonite, PVPP or carbon are effective treatments for wines affected by 1-octen-3-ol, 1-octen-3-one, and geosmin. Such treatments may have a negative impact upon desirable odour compounds.

3.18   Final Reflections

It's October 2019, and I'm in the classic surroundings of Brown's Courtrooms in London's Covent Garden. The building is now a restaurant and bar and people are no longer judged here, except perhaps by service staff. In the upstairs function room, there is a small wine tasting taking place hosted by a highly regarded importer and merchant. I arrived early and the hosts have opened the wines, but not yet assessed them. I begin tasting the white wines; alongside me is Carolyn Bosworth-Davies, wine educator and competition judge. The very first wine is very clearly faulty: haloanisole impacted. The ninth wine is really musty – so-called 'corkiness' at it very worst. Wine number 11 is affected too, albeit to a lesser extent. We ask for another bottle of each of the faulty wines: these are fine, indicating that on this occasion cork was almost certainly the culprit. Later in the tasting one of the reds suffers the same fate: a total of four of the 53 wines shown have haloanisole issues. As I leave the building, I reflect that, notwithstanding all the measures taken by the cork industry and others to combat the problem, and mindful of all that I have written noting the reduction in the number of affected bottles, there are far too many haloanisole-affected wines yet to unleash their sadness; they are lying in wait in wine cellars, on the shelves of retailers, and in glasses of on dining tables.

References

1 Takeuchi, H., Kato, H., and Kurahashi, T. (2013). 2,4,6-trichloroanisole is a potent suppressor of olfactory signal transduction. *Proceedings of the National Academy of Sciences of the United States of America* 110 (40): 16235–16240. https://www.pnas.org/content/110/40/16235.long.

2 McCoy, M. (2010). J&J recalls tainted Tylenol. *Chemical and Engineering News* 88 (4): 18. https://cen.acs.org/content/cen/articles/88/i4/JJ-Recalls-Tainted-Tylenol.html.

3 Drug GMP Report FDA News (2011). https://www.fdanews.com/ext/resources/files/marketing_files/archives/d/DGR040511web.pdf (accessed 3 March 2020).

4 Hardisty, J. (2011). Good pallet practice – product recalls highlight need for clearer guidance. *International Pharmaceutical Industry* 3 (4): 126–129. http://ipimediaworld.com/wp-content/uploads/2012/03/Packaging-article-22.pdf.

5 Leblanc, R. (2011). New pharma recall leaves more bad smell, points to the value of effective pallet management, *Reusable Packaging News*, 22 June 2011. https://packagingrevolution.net/tba-pallets (accessed 4 April 2020).

6 Lambert, D.E., Shaw, K.J., and Whitfield, F.B. (1993). Lacquered aluminium cans as an indirect source of 2,4,6-trichloroanisole. *Chemistry & Industry* 6: 461–462.

7 Nystrom, A., Grimvall, A., Krantzrulcker, C. et al. (1992). Drinking-water off-flavor caused by 2,4,6-trichloroanisole. *Water Science & Technology* 25: 241–249. https://doi.org/10.2166/wst.1992.0058.

8 Piriou, P., Malleret, L., Bruchet, A., and Kiene, L. (2001). Trichloroanisole kinetics and musty tastes in drinking water distribution systems. *Water Science and Technology: Water Supply* 1 (4): 11–18. https://doi.org/10.2166/ws.2001.0061.

9 Zhang, N., Xu, B., Qi, F., and Kumirska, J. (2016). The occurrence of haloanisoles as an emerging odorant in municipal tap water of typical cities in China. *Water Research* 98: 242–249. https://doi.org/10.1016/j.watres.2016.04.023.

10 Taber, G.M. (2007). *To Cork or Not to Cork*. New York: Scribner, a division of Simon and Schuster Inc.

11 Moore, E., Pravda, M., and Guilbault, G.G. (2003). Development of a biosensor for the quantitative detection of 2,4,6-trichloroanisole using screen printed electrodes. *Analytica Chimica Acta* 484 (1): 15–24. https://doi.org/10.1016/S0003-2670(03)00311-8.

12 Fuller, P. (1995). Cork taint: closing in on an industry problem. *Australian and New Zealand Wine Industry Journal* 10: 58–60.

13 European Commission CORDIS (2016). Electronic Nose To Detect Haloanisoles In Cork Stoppers – Final Report Summary. https://cordis.europa.eu/project/rcn/111040/reporting/en (accessed 8 March 2020).

14 Bordeaux Wine Enthusiasts and the entry was made in 2012. http://www.bordeauxwineenthusiasts.com/viewtopic.php?f=4&t=4073 (accessed 8 March 2020).

15 Laube, J. (2007). Pillar Rock battles TCA-tainted Wine. https://www.winespectator.com/blogs/show/id/14600.

16 Penn, C. (2003). Hanzell tells the truth about TCA. *Wine Business Monthly,* December 2003. https://www.winebusiness.com/wbm/?go=getArticle&dataId=29793 (accessed 20 Feb 2019).

17 Laube, J. (2003). Gallo of Sonoma's battle with TCA. https://www.winespectator.com/webfeature/show/id/Gallo-of-Sonomas-Battle-With-TCA_21796 (accessed 9 May 2019).

18 Patterson, T. (2007). T'ain't necessarily corks – Prevention and eradication of winery TCA. *Vines&Wines*. June 2007. https://www.winesandvines.com/features/article/48435/Taint-Necessarily-Corks (accessed 14 April 2e019).

19 Laube, J. (2014). Chateau Montelena battles "cellar funk." Discovery of TCA in Napa winery and in its wines forces a renovation. *Wine Spectator*. https://www.winespectator.com/webfeature/show/id/Chateau-Montelena-Battles-Cellar-Funk_22109 (accessed 14 April 2019).

20 Conan, E. and Géné, J.-P. (1998). L'inavouable maladie du vin. *L'Express* 14 December 1998. https://www.lexpress.fr/informations/l-inavouable-maladie-du-vin_631572.html (accessed 21 January 2019).

21 Wine Berserkers forum (2018). Buyer Beware - Ducru Beaucaillou 1988. https://www.wineberserkers.com/forum/viewtopic.php?t=150377 (accessed 16 January 2019).

22 Lebeau, A. (2018). SpitBucket - Bordeaux futures 2017 – Gruaud-Larose, Lagrange, Ducru-Beaucaillou, La Croix Ducru-Beaucaillou – 29 October 2018. https://spitbucket.net/2018/10/29/bordeaux-futures-2017-gruaud-larose-lagrange-ducru-beaucaillou-la-croix-ducru-beaucaillou (accessed 16 January 2019).

23 Robinson, J. (2007). Battle of the bottle stoppers. https://www.jancisrobinson.com/articles/battle-of-the-bottle-stoppers (accessed 15 January 2019).

24 Molesworth, J. (2007). *Chile's Viña Errázuriz confronts contaminated cellar.* http://www.winespectator.com/webfeature/show/id/Chiles-Vina-Errazuriz-Confronts-Contaminated-Cellar_3409 (accessed 15 May 2019).

25 Tempere, S., Schaaper, M.H., Cuzange, E. et al. (2017). Masking of several olfactory notes by infra-threshold concentrations of 2,4,6-trichloroanisole. *Chemosensory Perception* 10 (3): 69–80. https://doi.org/10.1007/s12078-017-9227-5.

26 Teixeira, M.I.V., San Romão, M.V., Bronze, M.R., and Vilas Boas, L. (2006). 2,4,6-Trichloroanisole: a consumer panel evaluation. *Ciência Técnica Vitivinicola* 21 (2): 53–65.

27 Prescott, J., Norris, L., Kunst, M., and Kim, S. (2005). Estimating a "consumer rejection threshold" for cork taint in white wine. *Food Quality and Preference* 16 (4): 345–349. https://doi.org/10.1016/j.foodqual.2004.05.010.

28 Darriet, P., Pons, M., Henry, R. et al. (2002). Impact odorants contributing to the fungus type aroma from grape berries contaminated by powdery mildew (*Uncinula necator*); incidence of enzymatic activities of the yeast *Saccharomyces cerevisiae*. *Journal of Agricultural and Food Chemistry* 50: 3277–3282. https://doi.org/10.1021/jf011527d.

29 Mobley, E. (2019). The mysterious case of the cork-tainted carrots. *San Fransisco Chronicle* 12 February 2019. https://www.sfchronicle.com/wine/article/The-mysterious-case-of-the-cork-tainted-carrots-13595196.php (accessed 3 March 2019).

30 Pollnitz, A.P., Pardon, K.H., and Liacopoulos, D. (1996). The analysis of 2,4,6-trichloroanisole and other chloroanisoles in tainted wines and corks. *Australian Journal of Grape and Wine Research* 2: 184–190. https://doi.org/10.1111/j.1755-0238.1996.tb00107.x.

31 Sefton, M.A. and Simpson, R.F. (2005). Compounds causing cork taint and the factors affecting their transfer from natural cork closures to wine – a review. *Australian Journal of Grape and Wine Research* 11: 226–240. https://doi.org/10.1111/j.1755-0238.2005.tb00290.x.

32 Tanner, H., Zanier, C., and Buser, H.R. (1981). 2,4,6-Trichloranisol: Eine dominierende Komponente des Korkgeschmacks. *Schweizer Zeitschrift für Obst- und Weinbau* 117: 97–103.

33 Buser, H.R., Zanier, C., and Tanner, H. (1982). Identification of 2,4,6-trichloroanisole as a potent compound causing cork-taint in wine. *Journal of Agricultural and Food Chemistry* 30: 359–362.

34 Waterhouse, A.L., Sacks, G.L., and Jeffery, D.W. (2016). *Understanding Wine Chemistry.* Chichester: Wiley.

35 McKay, M., Bauer, F.F., Panzeri, V., and Buica, A. (2018). Testing the sensitivity of potential panelists for wine taint compounds using a simplified sensory strategy. *Foods* 7 (11) https://doi.org/10.3390/foods7110176.

36 Duerr, P. (1985). Wine quality evaluation. *Proceedings of the International Symposium on Cool Climate Viticulture and Enology.* Eugene, USA (25–28 June 1985).

37 Suprenant, A. and Butzke, C.E. (1997). Implications of odor threshold variations on sensory quality control of cork stoppers. *American Journal of Enology and Viticulture* 48 (2): 269.

38 Goode, J. (2002). Grain of truth. http://www.wineanorak.com/sabate_altectrial.htm (accessed 23 August 2020).

39 Cork Quality Council (2017). Screening for TCA with group cork soaks. https://www.portocork.com/wp-content/uploads/2017/01/CQC-UTILIZING-SPME-Screening-for-TCA-2017.pdf (accessed 2 January 2020).

40 Copete, M.L., Zalacain, A., Lorenzo, C. et al. (2009). Haloanisole and halophenol contamination in Spanish aged red wine. *Food Additives & Contaminants. Part A* 26: 32–38. https://doi.org/10.1080/02652030802273130.

41 Cravero, M.C., Bonello, F., Alvarez, P. et al. (2015). The sensory evaluation of 2,4,6-trichloroanisole in wines. *Journal of the Institute of Brewing* 121: 411–417. https://doi.org/10.1002/jib.230.

42 Chatonnet, P., Bonnet, S., Boutou, S., and Labardie, M.D. (2004). Identification and responsibility of 2,4,6-tribromoanisole in musty, corked odors in wine. *Journal of Agricultural and Food Chemistry* 52: 1255–1262. https://doi.org/10.1021/jf030632f.

43 Goode, J. (2018). *Flawless*. Berkeley: University of California Press.

44 Coque, J.J.R., Álvarez-Rodríguez, M.L., Goswami, M., and Martínez, R.F. (2006). *Causes and origins of wine contamination by haloanisoles (chloroanisoles and bromoanisoles)*. León: Institute of Biotechnology of León.

45 Czaplicka, M. (2004). Sources and transformations of chlorophenols in the natural environment. *Science of the Total Environment* 322 (1–3): 21–39. https://doi.org/10.1016/j.scitotenv.2003.09.015.

46 Michałowic, J. and Duda, W. (2007). Phenols - sources and toxicity. *Polish Journal of Environmental Studies* 16 (3): 347–362.

47 Olaniran, A.O. and Igbinosa, E.O. (2011). Chlorophenols and other related derivatives of environmental concern: properties, distribution and microbial degradation processes. *Chemosphere* 83: 1297–1306. https://doi.org/10.1016/j.chemosphere.2011.04.009.

48 Burttschell, R.H., Rosen, A.A., Middleton, F.M., and Ettinger, M.B. (1959). Chlorine derivatives of phenol causing taste and odor. *Journal of the American Water Works Association* 51: 205–214.

49 Tracy, R. and Skaalen, B. (2008). 2,4,6-TBA: next 2,4,6-TCA in US wine industry. *Practical Winery & Vineyard Journal*, 11–12 November/December 2008.

50 Coque, J.J.R., Álvarez-Rodríguez, M.L., and Larriba, G. (2003). Characterization of an inducible chlorophenol O-methyltransferase from *Trichoderma longibrachiatum* involved in the formation of chloroanisoles and determination of its role in cork taint of wines. *Applied and Environmental Microbiology* 69 (9): 5089–5095. https://doi.org/10.1128/AEM.69.9.5089-5095.2003.

51 Fischer, B. (1991). Pentachlorophenol: toxicology and environmental fate. *Journal of Pesticide Reform* 11: 2–5.

52 Cedar, F. (1984). Historical and current developments of antisapstain chemical use. In: *Symposium on Preservatives in the Wood Industry*. Vancouver, Canada: University of British Columbia.

53 ChemBK. (2015) https://www.chembk.com/en/chem/DOWICIDE%202S (accessed 2 March 2020).

54 Canosa, P., Morales, S., Rodríguez, I. et al. (2005). Aquatic degradation of triclosan and formation of toxic chlorophenols in presence of low concentrations of free chlorine.

Analytical and Bioanalytical Chemistry 383: 1119–1126. https://doi.org/10.1007/s00216-005-0116-4.

55 Environmental Protection Agency (2008). Pesticides: Reregistration. Triclosan facts. https://archive.epa.gov/pesticides/reregistration/web/html/triclosan_fs.html (accessed 2 August 2019).

56 United States Environmental Protection Agency. (2014). Priority Pollutant List. https://www.epa.gov/sites/production/files/2015-09/documents/priority-pollutant-list-epa.pdf (accessed 2 August 2019).

57 United States Environmental Protection Agency (ATSDR) (2017). Comprehensive Environmental Response, Compensation, and Liability Act (CERCLA). Priority List of Hazardous Substances. https://www.atsdr.cdc.gov/spl/index.html#2017spl (accessed 2 August 2019).

58 IUCLID (2003). Data Set for 2,4,6-Tribromophenol. Ispra, European Chemicals Bureau, International Uniform Chemical Information Database.

59 Koch, C. and Sures, B. (2018). Environmental concentrations and toxicology of 2,4,6-tribromophenol (TBP). *Environmental Pollution* 233: 706–713. https://doi.org/10.1016/j.envpol.2017.10.127.

60 Jung, R. and Schaefer, V. (2010). Reducing cork taint in wine. In: *Managing Wine Quality: Oenology and Wine Quality*, vol. 2 (ed. A.G. Reynolds). Cambridge: Woodhead Publishing.

61 91/173/EEC. (1991). Council Directive of 21 March 1991 amending for the ninth time Directive 76/769/EEC on the approximation of the laws, regulations and administrative provisions of the Member States relating to restrictions on the marketing and use of certain dangerous substances and preparations.

62 République Français Ministère de la Transition Écologique et Solidaire. (1994). Décret No. 94-647 du 27/07/94 relatif à la limitation de la mise sur le marché et de l'emploi du pentachlorophénol, du cadmium et de leurs composés.

63 Stockholm Convention on Persistent Organic Pollutants (POPs) (2009). Geneva, Secretariat of the Stockholm Convention on Persistent Organic Pollutants. https://www.wipo.int/edocs/lexdocs/treaties/en/unep-pop/trt_unep_pop_2.pdf (accessed 3 January 2020).

64 Saintsbury, G. (1920). *Notes from a Cellar Book*. London: Macmillan.

65 Balcke, G.U., Wegener, S., Kiesel, B. et al. (2008). Kinetics of chlorobenzene biodegradation under reduced oxygen levels. *Biodegradation* 19 (4): 507–518. https://doi.org/10.1007/s10532-007-9156-0.

66 Abhilash, P.C. and Singh, N. (2008). Distribution of hexachlorocyclohexane isomers in soil samples from a small scale industrial area of Lucknow, North India, associated with lindane production. *Chemosphere* 73 (6): 1011–1015. https://doi.org/10.1016/j.chemosphere.2008.07.037.

67 Alvarez-Rodríguez, M.L., López-Ocaña, L., López-Coronado, J.M. et al. (2002). Cork taint of wines: role of the filamentous fungi isolated from cork in the formation of 2,4,6-trichloroanisole by O methylation of 2,4,6-trichlorophenol. *Applied Environmental Microbiology* 68 (12): 5860–5869. https://doi.org/10.1128/aem.68.12.5860-5869.2002.

68 Maggi, L., Mazzoleni, V., Furni, M.D., and Salinas, M.R. (2008). Transformation ability of fungi isolated from cork and grape to produce 2,4,6-trichloroanisole from

2,4,6-trichlorophenol. *Food Additives & Contaminants Part A* 25 (3): 265–269. https://doi.org/10.1080/02652030701522991.

69 Simpson, R.F. and Sefton, M.A. (2007). Origin and fate of 2,4,6-trichloroanisole in cork bark and wine corks. *Australian Journal of Grape and Wine Research* 13: 106–116. https://doi.org/10.1111/j.1755-0238.2007.tb00241.x.

70 Confédération Européenne du Liège. (2019) Systecode: Systecode - CE Liège (celiege.eu). http://www.celiege.eu/systemcode (accessed 16 May 2019).

71 Simpson, R.F. and Lee, T.H. (1990). The microbiology and taints of cork and oak. In *Proceedings of 9th International Oenological Symposium*, Cascais, Portugal (24–26 May 1990). Breisach, Germany: International Association for Modern Winery Technology and Management, pp. 653–667.

72 Anderson, K., Nelgen, S., and Pinilla, V. (2017). *Global Wine Markets, 1860 to 2016: A Statistical Compendium*. Adelaide: University of Adelaide Press. License: CC-BY 4.0. https://doi.org/10.20851/global-wine-markets.

73 Liacopoulos, D., Barker, D., Howland, P.R. et al. (1999). Chloroanisole taint in wines. In: *Proceedings of the 10th Australian Wine Industry Technical Conference*, Sydney (2–5 August 1999) (eds. R.J. Blair, A.N. Sas, P.F. Hayes and P.B. Høj), 224–226.

74 Soleas, G.J., Yan, J., Seaver, T., and Goldberg, D.M. (2002). Method for the gas chromatographic assay with mass selective detection of trichloro compounds in corks and wines applied to elucidate the potential cause of cork taint. *Journal of Agricultural and Food Chemistry* 50: 1032–1039. https://doi.org/10.1021/jf011149c.

75 Macku, C., Gonzalez, L., Schleussner, C. et al. (2009). Sensory screening for large-format natural corks by "dry soak" testing and its correlation to headspace solid-phase microextraction (SPME) gas chromatography/mass spectrometry (GC/MS) releasable trichloroanisole (TCA). *Journal of Agricultural and Food Chemistry* 57 (17): 7962–7968. https://doi.org/10.1021/jf901135p.

76 Goode, J. and Harrop, S. (2011). *Authentic Wine toward Natural and Sustainable Winemaking*. Berkeley: University of California Press.

77 Easton, S. (2016). Are we reaching closure on the closure debate? http://www.winewisdom.com/articles/closures/are-we-reaching-closure-on-the-closure-debate (accessed 27 March 2020).

78 Shaw, L. (2017). Notion that 1 in 10 wines is corked is 'a myth'. *The Drinks Business* 21st September 2017. https://www.thedrinksbusiness.com/2017/09/notion-that-one-in-10-wines-is-corked-is-a-myth (accessed 8 January 2020).

79 Robinson, J. (2017). Top of the stoppers. *FT Magazine* 18 August 2017. https://www.ft.com/content/60ff7c02-8215-11e7-a4ce-15b2513cb3ff (accessed 6 January 2019).

80 Capone, D.L., Skouroumounis, G.K., Barker, D.A. et al. (1999). Absorption of chloroanisoles from wine by corks and by other materials. *Australian Journal of Grape and Wine Research* 5: 91–98. https://doi.org/10.1111/j.1755-0238.1999.tb00292.x.

81 Perrotti-Brown, L. (2017). Industry Insider: Corked wine reality check. *Michelin Guide Website* 29 November 2017. https://guide.michelin.com/us/new-york/features/industry-insider-corked-wine-reality-check/news (accessed 16 January 2019).

82 Sullivan, S.P. (2018). Yes, cork taint is still a problem. *Wine Enthusiast* 2 February 2018. https://www.winemag.com/2018/02/02/cork-taint-still-a-problem (accessed 16 January 2019).

83 Atkin, T. (2019). Finding closure in the cork-vs-cap conundrum. Harpers Wine & Spirit, February 15.

84 Amorim. https://www.amorim.com/en/how-we-do-it/r&d-innovation (accessed 4 August 2019.

85 Millar, R. (2016). Laroche adopts Amorim's NDtech. *The Drinks Business* 30 November 2016. https://www.thedrinksbusiness.com/2016/11/laroche-adopts-amorims-ndtech (accessed 4 January 2020).

86 Schmitt, P. (2018). Amorim to eradicate TCA by 2020. *The Drinks Business* 13 July 2018. https://www.thedrinksbusiness.com/2018/07/amorim-to-eradicate-tca-by-2020 (accessed 14 August 2019).

87 Cork Supply. http://www.corksupply.com/us/products/ds100-2 (accessed 15 July 2019).

88 Chatonnet, P., Fleury, A., and Boutou, S. (2010). Identification of a new source of contamination of *Quercus* sp. oak wood by 2,4,6-trichloroanisole and its impact on the contamination of barrel-aged wines. *Journal of Agricultural and Food Chemistry* 58 (19): 10528–10538. https://doi.org/10.1021/jf102571v.

89 Léauté, B. (2019). Haloanisoles and new barrels: highlighting the role played by molds. *International Viticulture & Oenology Society Technical Reviews,* 13 October 2019. https://ives-technicalreviews.eu/article/view/2585# (accessed 23 February 2020).

90 Shafer entitled 'Making TCA a relic of the past'. https://www.shafervineyards.com/news/pr_2016_beating_TCA.php (accessed 23 February 2020).

91 Zimmerman, L. (2018). Opus One sues over tainted barrels. https://www.wine-searcher.com/m/2018/12/opus-one-sues-over-tainted-barrels (accessed 16 January 2019).

92 Todorov, K. (2018). Opus One sues supplier alleging barrels were contaminated with TCA. *Wine Business.com* 13 December 2018. https://www.winebusiness.com/news/?go=getArticle&dataId=207240 (accessed 4 January 2020).

93 McCoy, E. (2019). In wine country, dogs are sniffing out threats to $325 Cabernet. Cocaine- and contraband-sniffing dogs train their noses on cork taint and vineyard pests. *Bloomberg News* 23 May. https://www.bnnbloomberg.ca/in-wine-country-dogs-are-sniffing-out-threats-to-325-cabernet-1.1263399 (accessed 26 May 2019).

94 TN Coopers. https://www.youtube.com/watch?v=MJ0tRQiR2QQ (accessed 4 January 2020).

95 Cowey, J. (2012). *The Avoidance of Taints and Contaminations during Winemaking.* Adelaide: The Australian Wine Research Institute.

96 ISO 20752:2014(E) (2014). *Cork stoppers – determination of releasable 2,4,6-trichloroanisole (TCA).* Geneva, Switzerland: International Organization for Standardization. https://www.iso.org/standard/62646.html (accessed 18 April 2020).

97 ISO 22308:2005 (2005). *Cork stoppers – sensory analysis.* https://www.iso.org/standard/34938.html (accessed 18 April 2020).

98 Riu, M., Mestres, M., Busto, O., and Guasch, J. (2002). Determination of 2,4,6-trichloroanisole in wines by headspace solid phase microextraction and gas chromatography-electron-capture detection. *Journal of Chromatography A* 977: 1–8. https://doi.org/10.1016/S0021-9673(02)01276-1.

99 Özhan, D., Anli, R.E., Vural, N., and Bayram, N. (2009). Determination of chloroanisoles and chlorophenols in cork and wine by using HS-SPME and GC-ECD detection. *Journal of the Institute of Brewing* 115 (1): 71–77. https://doi.org/10.1002/j .2050-0416.2009.tb00346.x.

100 Fontana, A.R. (2012). Analytical methods for determination of cork-taint compounds in wine. *Trends in Analytical Chemistry* 37: 135–147. https://doi.org/10.1016/j.trac.2012 .03.012.

101 Cacho, J.I., Nicolás, J., Viñas, P. et al. (2016). Control of halophenol and haolanisole concentration in wine cellar environments, wines, corks and woodstaves using gas chromatography and mass spectrometry. *Australian Journal of Grape and Wine Research* 22: 391–398. https://doi.org/10.1111/ajgw.12231.

102 Hjelmeland, A.K., Collins, T.S., Miles, J.L. et al. (2012). High-throughput, sub ng/L analysis of haloanisoles in wines using HS-SPME with GC-triple quadrupole MS. *American Journal of Enology and Viticulture* 63: 494–499. https://doi.org/10.5344/ajev .2012.12043.

103 Slabizki, P., Legrum, C., Wegmann-Herr, P. et al. (2015). Quantification of cork off-flavor compounds in natural cork stoppers and wine by multidimensional gas chromatography mass spectrometry. *European Food Research and Technology* 242 (6): 977–986. https://doi.org/10.1007/s00217-015-2604-x.

104 Tarasov, A., Rauhut, D., and Jung, R. (2017). "Cork taint" responsible compounds. Determination of haloanisoles and halophenols in cork matrix: A review. *Talanta* 175: 88–92. https://doi.org/10.1016/j.talanta.2017.07.029.

105 Mavrikou, S. and Kintzios, S. (2018). Biosensor-based approaches for detecting ochratoxin A and 2,4,6-trichloroanisole in beverages. *Beverages* 4 (1): 24. https://doi.org/10 .3390/beverages4010024.

106 International Organisation of Vine and Wine (OIV). (2009). Resolution OIV/OENO 296/2009. http://188.165.107.123/public/medias/1074/oiv-oeno-296-2009-en.pdf (accessed 2 April 2020).

107 OIV (2009). Determination of Releasable 2,4,6-Trichloroanisole in Wine by Cork Stoppers; Method OIV-MA-AS315-16; Compendium of International Analysis of Methods: Paris, France, Amending resolution. http://www.oiv.int/public/medias/6447/oiv-oeno-623-2018-en.pdf (accessed 2 April 2020).

108 Mazzoleni, V. and Maggi, L. (2007). Effect of wine style on the perception of 2,4,6-trichloroanisole, a compound related to cork taint in wine. *Food Research International* 40 (6): 694–699. https://doi.org/10.1016/j.foodres.2006.11.014.

109 Hawkins. (2018). GO2 Sanitizer: The Gold Standard For Brewery & Winery Disinfection. https://www.hawkinsinc.com/groups/water-treatment/go2-water-disinfection (accessed 24 February 2020).

110 de Lourdes, M., Solis, A.A., Gerling, C., and Worobo, R. (2015). *Sanitation of Wine Cooperage with Five Different Treatment Methods*. Practical Winery & Vineyard. https:// www.researchgate.net/publication/305654287_An_in_vivo_study_Sanitation_of_wine_ cooperage_with_five_different_treatment_methods.

111 AWRI (2017). Ask the AWRI – taints in wine. *Grapegrower & Winemaker* 638: 4. www .awri.com.au/wp-content/uploads/2018/03/s1903.pdf.

112 Jung, R., Schaefer, V., Christmann, M. et al. (2008). Removal of 2,4,6-trichloroanisole (TCA) and 2,4,6-tribromoanisole (TBA) from wine. *Mitteilungen Klosteneuberg* 58: 38–67.

113 Patterson, T. (2008). TCA testing and removal. *Wines and Vines.* https://
winesvinesanalytics.com/features/article/56627.

114 Valdés, O., Marican, A., Avila-Salas, F. et al. (2019). Simple approach for cleaning up
2,4,6-trichloroanisole from alcoholic-beverage-reconstituted solutions using polymeric
materials. *Australian Journal of Grape and Wine Research* 25 (3) https://doi.org/10
.1111/ajgw.12396.

115 AWRI (2020). Project 2.2.1. www.awri.com.au/research_and_development/2017-2025-
rde-plan-projects/project-2-2-1 (accessed 30 March 2020).

116 AWRI (2020). Wine Flavours, Faults and Taints. https://www.awri.com.au/industry_
support/winemaking_resources/sensory_assessment/recognition-of-wine-faults-and-
taints/wine_faults/ (accessed 30 March 2020).

117 Suziki, G., Takigami, H., Watanabe, M. et al. (2008). Identification of brominated and
chlorinated phenols as potential thyroid-disrupting compounds in indoor dusts. *Environmental Science & Technology* 42 (5): 1794–1800. https://doi.org/10.1021/es7021895.

118 Allen, M.S., Lacey, M.J., Harris, R.L.N., and Brown, W.V. (1988). Sauvignon Blanc varietal aroma. *Australian Grapegrower & Winemaker* 292: 51–56.

119 Roujou de Boubée, D., Van Leeuwen, C., and Dubourdieu, D. (2000). Organoleptic
impact of 2-methoxy-3-isobutylpyrazine on red Bordeaux and Loire wines. Effect of
environmental conditions on concentrations in grapes during ripening. *Journal of
Agricultural and Food Chemistry* 48: 4830–4834. https://doi.org/10.1021/jf000181o.

120 Simpson, R.F., Capone, D.L., and Sefton, M.A. (2004). Isolation and identification of
2-methoxy-3,5-dimethylpyrazine, a potent musty compound from wine corks. *Journal
of Agricultural and Food Chemistry* 52: 5425–5430. https://doi.org/10.1021/jf049484z.

121 Simpson, R.F., Capone, D.L., Duncan, B.C., and Sefton, M.A. (2005). Incidence and
nature of 'fungal must' taint in wine corks. *Australian and New Zealand Wine Industry
Journal* 20 (1): 26–31.

122 Chatonnet, P., Fleury, A., and Boutou, S. (2010). Origin and incidence of
2-methoxy-3,5-dimethylpyrazine, a compound with a "fungal" and "corky" aroma
found in cork stoppers and oak chips in contact with wines. *Journal of Agricultural
and Food Chemistry* 58 (23): 12481–12490. https://doi.org/10.1021/jf102874f.

123 La Guerche, S., Dauphin, B., Pons, M. et al. (2006). Characterization of some mushroom and earthy off-odours microbially induced by the development of rot on grapes.
Journal of Agricultural and Food Chemistry 54 (24): 9193–9200. https://doi.org/10.1021/
jf0615294.

124 Cravero, M.C. (2020). Musty and moldy taint in wines: a review. *Beverages* 6: 41.
https://doi.org/10.3390/beverages6020041.

125 Weingart, G., Schwartz, H., Eder, R., and Sontag, G. (2010). Determination of geosmin
and 2,4,6-trichloroanisole in white and red Austrian wines by headspace SPME-GC/MS
and comparison with sensory analysis. *European Food Research and Technology* 231:
771–779. https://doi.org/10.1007/s00217-010-1321-8.

126 La Guerche, S., Senneville, L., Blancard, D., and Darriet, P. (2007). Impact of the *Botrytis cinerea* strain and metabolism on (−)-geosmin production by *Penicillium expansum*
in grape juice. *Antonie Van Leuwenhoek* 92: 331–341. https://doi.org/10.1007/s10482-
007-9161-7.

127 Parker, M., Osidacz, P., Baldock, G.A. et al. (2012). Contribution of several volatile phenols and their glycoconjugates to smoke-related sensory properties of red wine. *Journal of Agricultural and Food Chemistry* 60: 2629–2637. https://doi.org/10.1021/jf2040548.

128 Prak, S., Gunata, Z., Guiraud, J., and Schorr-Galindo, S. (2007). Fungal strains isolated from cork stoppers and the formation of 2,4,6-trichloroanisole involved in the cork taint of wine. *Food Microbiology* 24: 271–280. https://doi.org/10.1016/j.fm.2006.05.002.

129 Pons, M., Dauphin, B., La Guerche, S. et al. (2011). Identification of impact odorants contributing to fresh mushroom off-flavor in wines: incidence of their reactivity with nitrogen compounds on the decrease of the olfactory defect. *Journal of Agricultural and Food Chemistry* 59 (7): 3264–3272. https://doi.org/10.1021/jf104215a.

4

Brettanomyces (Dekkera) and Ethyl Phenols

Brettanomyces strikes fear in the hearts and minds of most winemakers. The compounds metabolised by this 'rogue' yeast can impart off-odours, including those of BAND-AID®, stables, and vomit. The yeast may grow in wine during winemaking, maturation, and/or storage, and off-odours may appear years after the wine has been bottled. In this substantial chapter, *Brettanomyces* and its related volatile phenols and other metabolised compounds will be discussed. Volatile phenols related to 'smoke taint' will be covered in Chapter 12, and excessive levels of volatile acidity and ethyl acetate, which are often found in *Brettanomyces*-affected wines, will be discussed as a general topic in Chapter 7.

4.1 Introduction

Brettanomyces is a genus of yeast of the family *Saccharomycetaceae*. The presence in wine of the compounds metabolised by this yeast may be regarded as a serious fault, as a flaw, or occasionally even as a desirable attribute, depending upon the concentration of the compounds, the wine matrix, and the viewpoint of the taster. These compounds can give odours and flavours that, when pronounced, are objectionable to a great many wine drinkers. They include the odours of antiseptic, BAND-AID adhesive plasters used for small wounds, stables, farmyards, and even faeces. These smells and flavours also mask primary, secondary, and tertiary characteristics and can negate the personality, individuality, and 'sense of place' of affected wines. Even if they are below their individual sensory thresholds, the compounds may impact negatively, but also in some cases positively, upon the overall sensory characteristics of affected wines. High concentrations of these compounds mostly affect red wines. In fact, the yeast has been described as 'the main microbiological threat to red wine quality at present' [1]. On the one hand, the impact of compounds metabolised by *Brettanomyces* may affect the quality of wine in bottle, and on the other hand, even if they are not present, the steps taken to prevent or remove them, including sterile filtration, are deleterious to quality. Numerous authors have stated that the growth of *Brettanomyces* results in huge economic losses for the wine industry, although these have never been reliably globally quantified. However, preliminary data from a study conducted by University of California Davis (UCD) Robert Mondavi Institute Center for Wine Economics indicate that of 162 winery respondents to a survey, about 13% have had to recall a wine due to Brett in the bottled product [2].

Wine Faults and Flaws: A Practical Guide, First Edition. Keith Grainger.
© 2021 John Wiley & Sons Ltd. Published 2021 by John Wiley & Sons Ltd.

Contrary to popular belief, both white and sparkling wines can also be affected. Amongst still white wines, those made from Chardonnay and Gewürztraminer varieties are perhaps the most common to exhibit associated characteristics. The yeast does not usually impact upon fortified wines as it cannot survive at alcohol levels above approximately 15% alcohol by volume (abv). In some countries and wine regions, incidences of wines impacted by *Brettanomyces* have fallen in recent years, but in others they are increasing. This rise may be attributed to several factors that heighten conditions conducive to its development. Grapes have generally been harvested with a higher degree of ripeness in the last couple of decades, and riper grapes can contain higher levels of *Brettanomyces* precursors. The pH in musts and wines are generally higher, and high must weights result in high potential alcohol levels that can make *Saccharomyces cerevisiae* yeast activity sluggish in the later stages of alcoholic fermentation (AF). There has been a trend for increased residual sugars, particularly in so-called 'entry-level' red wines, and the use of oak barrels is common. Also in recent years, winemakers have strived to reduce levels of sulfur dioxide (SO_2), perhaps to be seen to reduce the additions of this regulated chemical, but particularly in order to achieve softer, rounder red wines that are attractive for early drinking. SO_2 retards the polymerisation of tannins [3]. *Brettanomyces* is colloquially known in the wine industry as 'Brett' (the term is often used un-capitalised), and wines exhibiting organoleptic characteristics resulting from its growth may be described as 'bretty'.

The three main chemical compounds metabolised by *Brettanomyces* and largely responsible for 'Brett' odours are the following:

- 4-Ethylphenol (4-EP);
- 4-Ethylguaiacol (4-EG);
- Isovaleric acid – also known as 3-methylbutanoic acid.

There are other compounds metabolised by the yeast that also have an aroma and flavour impact:

- 4-Ethylcatechol (4-EC);
- Decanoic acids;
- 2-Methylbutanoic acid;
- Ethyl-2-methylbutyrate;
- Ethyl decanoate;
- Tetrahydropyridines;
- Guaiacol;
- *trans*-2-Nonenal;
- 2-Phenylethanol;
- Isoamyl alcohol;
- Isobutyric acid.

Several of the compounds metabolised by *Brettanomyces* are invariably present together in affected wines, but the ratio of each will depend on a number of factors, including the grape variety. The individual strains of the yeast exhibit diverse metabolic and physiological characteristics and produce varying quantities of the individual odorous compounds. However, there are numerous fatty acids, esters, aldehydes, and ketones metabolised by *Brettanomyces*, and many of these have pleasant and desirable aromas that are valued or regarded as positive in certain wine styles or matrices [4]. Terpenes and norisoprenoids are glycosides with generally positive aromas and are naturally present in grapes, but in bound (locked-in) forms, and these can be 'unlocked' and released by *Brettanomyces*.

Brettanomyces is also a producer in wines of biogenic amines, notably cadaverine (which at high concentrations can have a smell of rotting flesh), putrescine, and spermidine. The presence of biogenic amines is undesirable from a health perspective, which is discussed in Chapter 14.

4.2 Background and History

The genus *Brettanomyces*, the name derived from the Greek for 'British fungus', was first identified by Niels Hjelte Claussen, laboratory director of the New Carlsberg Brewery in Copenhagen and the name introduced at a meeting of the Institute of Brewing in April 1904 [5]. The species he identified was to be named *Brettanomyces claussenii* (now usually called *Dekkera anomala/Brettanomyces anomala*), although Claussen initially classified it as a *Torula* species and the *Brettanomyces* genus name was not suggested until 1921. Four other species were identified over the next few decades: *Brettanomyces bruxellensis* (also called *Brettanomyces lambicus* and historically known as *Brettanomyces intermedia*) that was isolated from Belgian Lambic beer in 1921, *Brettanomyces custersianus*, *Brettanomyces naardenensis*, and *Brettanomyces nanus*. The first systematic study of *Brettanomyces* was undertaken as a thesis by M.T.J. Custers [6]. There have been many reclassifications and changes in names of the various species over the last 100 years, and some genera and species originally classified as *Brettanomyces* spp. have been reclassified as *Candida* spp. *B. bruxellensis* was historically regarded as commercially very important and was the first microorganism ever to be patented (UK patent number GB190328184) [7]. A further new species, proposed name *Brettanomyces acidodurans* sp. nov. (species novae) has recently been isolated from olive oil [8]. In wine, the genus subsequently to be re-classified as *Brettanomyces* was first identified in 1933 by Krumbholz and Tauschanoff, who named the isolate *Mycotorula intermedia* [9]. Their findings were confirmed in the 1950s by several researchers, including Emile Peynaud, now regarded by many as the grandfather of modern oenology, at Université de Bordeaux [10, 11]. The species usually responsible for Brett-associated compounds in wine is *B. bruxellensis*, although *B. anomala* can occasionally be involved. Other species, although not important in wine, have been found to populate carbonated drinks, olives, and beer. The *Dekkera* genus was only designated in 1964 with the first observation of spore formation – *Brettanomyces* is an anamorph, i.e. the non-spore-forming (asexual) form of the fungus. *Dekkera* is the rarely observed (sexual) spore-forming teleomorph of *Brettanomyces* and has two known species: *Dekkera bruxellensis* (previously known as *Dekkera intermedia*) and *Dekkera anomala*. Each species of *Brettanomyces* has several strains that differ in their genetic make-up, the quantity and type of volatile phenols produced and, crucially, their tolerance of sulfur dioxide. In the 1980s, the work of Tamila Heresztyn of the Australian Wine Research Institute (AWRI) in the study of the volatile phenolic compounds metabolised from hydroxycinnamic acids by *Brettanomyces* yeasts was groundbreaking [12]. Another important breakthrough from AWRI came in 2011 when a team led by Chris Curtin, then the Institute's Senior Research Scientist, succeeded in describing the entire genome sequencing, revealing the blueprint for one of the *Brettanomyces* strains (AWRI1499), regarded as an important spoilage strain in the Australian wine industry [13]. Unexpectedly this strain showed a triploid genome, having three copies of its chromosomes. Other triploid strains have since been discovered, but many strains have a diploid genome. The genome sequencing of several other important strains found in wine have now been detailed, including CBS2499 (of French origin), AWRI 1608 and 1613, and LAMAP 2480 (of Chilean origin).

Brettanomyces spp. like *Saccharomyces* spp. features a positive Crabtree effect, producing alcohol and high amounts of acetic acid under aerobic conditions (see Glossary), but the yeast also functions anaerobically. The wide genetic diversity amongst the strains results in a large variation in performance characteristics including resistance to sulfur dioxide (SO_2) and the amount and type of volatile phenols and other compounds produced.

It should be noted here that molecular DNA detection has uncovered no difference between the anamorph and teleomorph states of the yeast [14, 15]. Today, the botanical name of a genus is that of the sexually reproductive (teleomorph) state. From the beginning of 2013, the *International Code of Nomenclature for algae, fungi, and plants* (the Melbourne Code) permits each fungus to have only one name. Thus it is technically correct to refer to the genus only as *Dekkera*, even though the teleomorph state is very rarely observed. The wine (and also the brewing) industry almost invariably continue to call the genus 'Brettanomyces', and it is usually referred to as such in research papers, including those published since 2013. Accordingly, I will generally refer to the genus as *Brettanomyces* throughout this chapter, and elsewhere in this book, although this may be regarded as technically incorrect.

4.3 The Brett Controversy

4.3.1 *Brettanomyces* Paranoia?

Of all the topics discussed in this book, there is no other subject that raises such heated discussion as that of *Brettanomyces*. Today, there is perhaps a paranoia amongst many winemakers and oenologists, particularly in Australasia, should a hint of any Brett-associated aromas be present in a wine or if laboratory analyses shows any viable cell count or the presence of related volatile phenols other than at a few μg/l. In 2007, Vincent Renouf et al. stated that 'among all possible microbial alterations of wines, volatile phenols production by the yeast *B. bruxellensis* is one of most feared by the winemaker and probably one of the most undesired by consumers' [16]. In 2010, the Australian wine writer and critic Huon Hooke expounded that 'Winemakers call it The Pox. It is the dreaded social disease of wine. It is shared around unwittingly – passed by those who usually don't know they've got it, to those who are unaware they're running a risk of contracting it. If a winemaker buys or borrows someone else's used barrels, they may find their wine is infected...' [17]. In December 2017, the Final Report from AWRI to the Australian Grape And Wine Authority entitled 'Ensuring the continued efficacy of *Brettanomyces* control strategies for avoidance of spoilage' stated that the 'spoilage of wine by *Brettanomyces* yeasts remains one of the foremost microbiological problems faced by winemakers' [18]. In a thorough review in 2017, Juliana Milheiro et al. simply refer to *Brettanomyces* as a 'serious wine sensory defect' [19].

It is likely that Brett-associated compounds have been present in many red wines throughout the 8000 years or so of vinous history. It is believed that the lineages of *Saccharomyces* and *Brettanomyces* separated some 200 million years ago [20]. Even in recent times, wines were often lauded for exuding characteristics that we now know are consequential to the actions of the *Brettanomyces* and are generally considered to constitute a fault. Wine writers would use terms such as 'leathery', 'horsey', 'sweaty saddles', 'animal', and 'stables' when describing the nose and palate of affected wines, and many regarded these as positive characteristics. Such aromas, often attributed to growing conditions, were described as 'goût de terroir' and indeed could be found in wines of the finest pedigrees. In 1982, Master of Wine and Burgundy expert Anthony Hanson wrote in the first edition of his critically acclaimed book *Burgundy:* 'great Burgundy smells of shit' [21]. In a study in 2000, *Brettanomyces* was found to be present in about 25% of Burgundy wines when examined in bottle and 50% of the wines in barrel, post malolactic fermentation (MLF) [22]. Put simply, Brett characteristics were very often a major contributor to a wine sensory profile and were not regarded as objectionable.

In recent years, with the growth of varietal-led and fruit-driven wines and probably due to increased awareness, Brett-associated odours and flavours became more unacceptable. Perhaps the world is a cleaner place in the last 20 years or so – in my youth, rooms often smelled of people, town streets often smelled of dog excrement, and the country smelled of animals! The trip to the farm was an experience for all the senses, and the visual aspect was but a part of the picture. Today, a great many consumers simply want uncomplicated (and cheap) fresh and fruity wines that do not tax the palate or the brain. Winemakers, led by marketing departments who in turn may be influenced by focus groups, use recipes and formulae to consistently and inexpensively produce these wines and meet perceived consumer preferences. If we accept that it is the clean fruit- and variety-driven wines that satisfy these wants, they will, by definition, be totally Brett-free.

4.3.2 The Role of *Brettanomyces* in Craft and Lambic Beer Production

It is pertinent to note that although most brewers, including all the large 'factory' brewers, also consider *Brettanomyces* as a rogue contaminant yeast, it is key to the production of many Belgian beers (e.g. Orval and Rodenbach Grand Cru) and is sought after by many craft brewers, especially those producing Lambic beers and, increasingly, in specialist India Pale Ales (IPAs) and *coolship* ales. Indeed, some craft brewers only use *Brettanomyces* for both primary and secondary fermentations, whilst others use it together with strains of *S. cerevisiae* and *Saccharomyces pastorianus* (sometimes referred to as *Saccharomyces carlsbergensis,* although whether these species are identical is disputed). In his 1904 paper, Claussen stated that '...it is evident that the secondary fermentation effected by *Brettanomyces* is indispensable for the production of the real type of English beers' and 'the action of *Brettanomyces* is absolutely necessary to bring English stock beers into proper cask and bottle condition, and to impart to them that peculiar and remarkably fine flavour which in a great measure determines their value' [5]. In his detailed but controversial 2015 work *Terroir and Other Myths of Winegrowing*, Mark Matthews, professor of viticulture at the Robert Mondavi Institute at the University of California at Davis, has no hesitation in simplistically describing *Brettanomyces* as 'a spoilage yeast in beer and wines' [23]. In doing so, Matthews is following the precedents set by many research scientists who also use this descriptor. However, in the last couple of decades, the huge rise in popularity in real ales and craft beers, including Lambics and coolship ales, is testament to the fact that many beer lovers actively seek out particular Brett-associated aromas and flavours, including those that a great many winemakers, oenologists, and even wine writers and critics now seek to eliminate. Amongst the flagship artisan brewers who revel in the notes the *Brettanomyces* imparts to beers are Vinnie Cilurzo of Russian River Brewing in Sonoma County, California, Vince Marsaglia and Tomme Arthur of Lost Abbey Brewing Company of San Marcos, California, and Chad Yacobson of Crooked Stave in Denver, Colorado. Yacobson stated, '…we search out the complex, delicate characteristics that *Brettanomyces* can provide. We find that when great care is taken in the selection of strains, and when the right conditions are provided for the yeast to flourish, they lend anything from orange, lemon, or lime characteristics to stone fruit, apples, ripe mangoes, and pineapple. Brett will also impart dry, earthy, leathery notes, and we feel that these flavors and aromas are complementary to balance the tart and vinous characteristics while also adding a layer of complexity not found in most beers [24]'. Other US brewers using *Brettanomyces* for fermentations are Modern Times in San Diego, who make a 100% Brett IPA, and the Anchorage Brewing Company.

Chris White and Jamil Zainasheff noted in their book *Yeast: the Practical Guide to Beer Fermentation* that 'these flavors are like adding the spices of an Indian grocer to your local supermarket……..with the right skill and right balance these flavors create a beer that is both complex and delicious' [25]. With their willingness to experiment with yeasts deemed by most winemakers to be 'spoilage yeasts', brewers of craft and real ale beers have taken a much more adventurous approach. One might note that taken as a whole in the period 2006–2019, wine sales stalled or fell in most major European markets [26], but in the same period there was a general rise in sales of craft beers, Lambic beers, and real ales.

Although the industrial producers of cider consider *Brettanomyces* as a spoilage yeast, some craft cider makers are using it to impart wild, 'funky' notes to their products. Home brewers, particularly in the United States, are perhaps at the forefront of experimentation, and many commercial 'artisan' breweries have been founded by highly enthusiastic home brewers. A professor of viticulture at Davis or an Australasian winemaker might note, with some amusement, a request from the contributor 'Beerswimmer' on the Lambic and Wild Brewing forum of *homebrewtalk.com*: 'But it takes so long to get where I really enjoy the stronger funk of the Brett Brux. I'm wondering what I can do to get it there faster/stronger? I want a super horsey/…/melted plastic band-aids funky beer…'. [27]. It should be noted that in beers the concentration of 4-EP (medicinal aromas) is generally lower than that of 4-EG (smoky, spicy aromas), whilst in wine the situation is reversed [28].

Before we return to discussing *Brettanomyces* in wine, we might note that Brett characteristics are also sought in some other food and drinks. By way of example, they are, at low levels, known to contribute metabolites desirable in sour dough bread and kombucha tea. However, the yeast is always unwelcome is the dairy industry: the products of its role in any fermentation are distinctly unpleasant.

4.3.3   *Brettanomyces*: An Extension of Terroir?

Whilst few would deny that the technical standard of wines produced today is higher than at any time in the 8000 years of vinous history, the diversity of flavours once present in wines from individual regions has perhaps become less marked. A great many wines, and not just those at 'entry level', proudly announce themselves as fruit driven – fresh, lively, and abundant with the aromas of primary fruits. The appeal of such wines is self-evident. However, the absence of 'organic' tones in the vast majority of modern wines is mourned by some wine lovers. The word 'bouquet' was often historically used as a general descriptor for the nose of a wine, and the word meant much more than simple odours, but a whole melange of complementary aromas. A wine's bouquet might have included notes of truffle, the earthy aromas of kicking through woodland leaves and 'sous bois', and the heady perfumes of leather and sweat: any or all of these could turn wine tasting into a sensuous, even exciting experience perhaps sending the taster in an instant to another time and another place. The clean primary and secondary aromas exuded by technically correct, fruit- or variety-driven modern wines may appeal to the head and nose, but perhaps not to the soul. A great many modern wines may also be criticised for having become more standardised and globalised. This is not just down to formula-led winemaking: changes in viticultural practices in second half of the twentieth century have negated wine's sense of place and structure. As Clark Smith said, 'wine quality was eroded in the modern era by the replacement of organic principles of viticulture with the facile but destructive farming solutions of petrochemical agriculture……… these have robbed contemporary wines of flavour interest as well as longevity [29]'.

Some, mostly small, wine producers in the southern Rhône valley still describe the aroma and palate characteristics of bretty wines as 'goût de terroir', believing that the presence of *Brettanomyces* is in fact an extension of terroir and that its manifestations and the associated aromas and flavours are desirable. To a lesser extent, this view was also the case in Bordeaux, but the number of Bordeaux châteaux whose wines show Brett characteristics has diminished in the last couple of decades due in a large part to the pioneering work of Dr Pascal Chatonnet, whose Laboratoire Excell is based in Merignac on the outskirts of the city of Bordeaux. In 2005, Chatonnet noted, 'When I started out as a winemaker 20 years ago, I identified what I thought was a typical Graves character, a goût de terroir. But it wasn't the Graves region that was producing those aromas and flavours, it was Brett [30]'. Winemakers have often noted that grapes from certain vineyards have a tendency to produce wines with Brett characteristics. As discussed below, certain agronomic conditions do indeed lead to an increased likelihood of Brett, so it may be argued that there is substance in the 'terroir' association. However, critics often state that, with its own pronounced but perhaps simplistic aroma and palate profile, Brett does in fact detract from 'the sense of place', that it masks both varietal character and fruitiness, and that its products make wines produced in different regions taste more similar.

4.3.4 Artisans Choosing *Brettanomyces* for Wine Fermentations

In recent years, there are a growing number of small artisan winemakers who are seeking more diverse, broader, 'funky', and less refined aroma and flavour profiles. To this aim, many are letting the 'natural' yeasts present on the grape skins and in and about the winery carry out the AF. Others are experimenting with co-inoculations of *S. cerevisiae* and non-*Saccharomyces* species, such as *Metschnikowia pulcherrima*, *Lachancea kluyveri*, *Lachancea thermotolerans*, and *Torulaspora delbrueckii*. An interesting side note is that some of these non-*Saccharomyces* species, e.g. *M. pulcherrima*, when co-innoculated with *S. cerevisiae* or *Saccharomyces uvarum* produce less ethanol than pure *S. cerevisiae* inoculates [31, 32], consuming sugar by respiration. Many wine lovers berate the increase in alcoholic content of many or perhaps even most wines. This has been in the region of 2% abv over the last few decades. In fact, an increase of 2% abv represents an actual increase of about 15% or more in the alcohol content of wines. A typical red Bordeaux from a 'ripe' year in the 1980s had an abv in the region of 12–12.5%; in recent years an abv of 14% or 14.5% has been commonplace.

There are a few winemakers who are deliberately producing *Brettanomyces*-fermented wines. They include the artisan California winery Merisi, who produces a Pinot Gris called Manic White (fermented away from the main winery). Winemaker Mandy Heldt Donovan describes it as 'a wine fermented intentionally with *Brettanomyces*. Bottled fresh like Beaujolais Nouveau, slightly spritzy – like Vinho Verde, with exotic aromas like cider, it is a humble story of a yeast trying to make good [33]'.

Whilst most scientists look at the world of so-called 'natural' wine production and biodynamic wine production with both disbelief and disdain, there are those who are prepared to think outside of the box and consider possibilities that are dismissed by the scientific mainstream, including the employment of yeast species that mainstream science considers to be spoilage organisms. A study of *Brettanomyces* strains by the University of California at Davis found 18 strains that imparted positive characteristics and five that were distinctly negative [34]. Linda Bisson, professor of yeast microbiology and functional genetics, thinks the yeast may have potential for wine production. Whilst noting the major problems with *Brettanomyces*, including its hypermutable and hypervariable characteristics, Bisson said, 'I think that ultimately we will understand enough about this yeast to be able to use it with confidence and control its activity in the winery. I liken it to the malolactic fermentation, which was considered a spoilage fermentation 60 years ago' [34]. Bisson elucidated in a January 2013 seminar at UC Davis that when a trained panel of 14 assessors examined the sensory impact of 35 strains grown in a Cabernet Sauvignon, many strains were on the 'positive' side, and some of the *Brettanomyces*-infected wines were preferred over a control wine [35]. The concept of possible 'positive' Brett strains is further discussed in Section 4.3.7.

4.3.5   Some Wines with a Notable Brett Influence

As we have seen, in the not-too-distant past, many wines with very high levels of Brett-related compounds, often including high levels of volatile acidity, were highly regarded and sought after for their very distinctive characteristics. Indeed, it might be argued that some regions such as Toscana and Rioja and some producers made their wine styles and reputations upon Brett compounds. Examples of notable producers include Chateau Musar from Bekaa Valley, Lebanon, Crus Classés Château Bouscaut from Pessac-Léognan, Château Gruaud-Larose from Saint-Julien in Bordeaux, and Château de Beaucastel from Châteauneuf-du-Pape in the southern Rhône valley. The 2000 vintage of Château Gruaud Larose in Saint-Julien is described as 'a Brett monster' by Tom Higgins, owner and winemaker at Heart & Hands Wine Company on https://delectable.com [36]. It should be noted that the level of *Brettanomyces* characteristics can vary from bottle to bottle of the same wine, depending upon several factors, particularly including the storage temperature. The 1989 vintage of Château Margaux made under the auspices of the late Paul Pontalier, a brilliant and fastidious technical director, has been noted as having Brett tones. Dr Tom Ostler, reporting on a tasting held in 1994 at the Cornell University Agricultural Research Station, Geneva, noted that a 1989 Château Pichon Longueville Comtesse de Lalande (a Cru Classé property in Pauillac) had a 'classic round Bordeaux bouquet......cassis, with just a hint of burnt wet wood offering complexity ......a distinctive barnyard aroma.......desirable.......quite harmonious'. [3]. Ostler noted that tests showed this wine had (a staggering) 15 800 ng/ml (15 800 mg/l) of 4-ethyl-phenol and also that *The Wine Spectator* magazine had given it a score of 92 points, equating to 'outstanding: a wine of superior character and style' [37]. Somewhat tongue in cheek Ostler commented that 'the questions that rose uniformly across the lecture hall were: How much Brett do you need in order to be awarded a 92 by the Wine Spectator? How do we attain it?' [3].

The positive notes that Brett can give to mature Bordeaux wines were well illustrated in April 2019 by two of the Cru Classé wines available to accompany lunch at the Union des Grands Crus *En Primeurs* tasting held in Bordeaux city. The 1999 Château Rauzan Gassies from Margaux, which has had mixed reviews since its first release, showed game, tobacco, and leather aromas but with distinct animal notes of 4-EP and 4-EC. The 2004 Château Léoville-Poyferré from Saint-Julien, generally highly regarded since release, was unctuous with tones of bramble and cassis, all overlaid with an array of barnyard and medicinal odours. Both were delicious, particularly with hard cheeses and chocolate petit fours. In May 2019, at a Waitrose press tasting in London, the 2010 Château Gravier Figeac, Saint-Emilion, clearly had Brett influence and was exuding a broad palette of delightful savoury flavours – I found this to be a fascinating wine, an opinion endorsed by Tim Atkin MW who happened to be tasting alongside me [personal discussion].

4.3.6 An Old World/New World Divide?

As we have seen, and with the historic exception of South Africa, most New World have sought to eliminate Brett from their wineries and wines, possibly on account of not having had a history of living with related aromas and flavours. The Australian wine industry grew dramatically from the 1970s onwards, before which production concentrated on sweet and fortified wines (as was also the case in the United States). Brett does not grow at alcohol levels above 15% abv and so had not been an issue. In the 1980s, the total planted vineyard land in Australia was only 64 000 hectares [38]. The move to 'light' wine production and its subsequent rapid growth (by the first decade of this millennium vineyard land had increased to 161 000 ha) [38] brought new problems, including many Brett-dominated red wines. In the 1990s and early 2000s, faced with growing negative comments on Brett-associated aromas, the Australian wine industry, guided by the AWRI, instigated a strident approach to address the problem. Legendary wineries and winemakers had not been immune to falling victim. Henschke, whose flagship Eden Valley wine 'Hill of Grace' sells for stratospheric prices, became known for heavy Brett odours in the 1990s and early 2000s [39]. At Cape Mentelle, in Western Australia's Margaret River region, several red wines made from 'Bordeaux' varieties demonstrated overt Brett characteristics in vintages from the late 1990s and early 2000s, leading to a raft of fastidious measures to tackle the issue by John Durham, then chief winemaker. I can still remember seeing the look of horror on John's face when

I pronounced a Cabernet Merlot wine as having Brett tones when we were tasting together in the rammed earth winery in 2004. The AWRI approach certainly achieved the desired objective: their statistics show that in Australia's Cabernet Sauvignon-producing regions, the average levels of 4-EP in wines fell from 1000 ppb in vintage 2000 to less than 100 ppb by vintage 2005 [40]. Consumer panel studies conducted by the AWRI showed that (Australian) consumers invariably preferred wines free from Brett characteristics to those that had been spiked with related compounds. Of course, the results of research and feedback from focus groups do not worry those taking a more eclectic approach. The 2015 Express Winemakers Great Southern Syrah shows overt 4-EP and 4-EG characteristics, but a tasting note from Tim Wildman MW said it all: 'Wild yeast? You betcha. Touch of Brett? Get over it. 100% whole bunch, foot trodden, unfiltered drinky-thinky deliciousness. Darn tootin!' [41]. This is perhaps a wine that would satisfy the request by 'Oldbel' on *Reddit*: 'Recommend wines with Brett(anomyces) flavors please', the poster stated. 'I am heavily interested in those barnyard, saddle leather, horsey, manure, musty flavors. I was hoping people might recommend wines that I might find this in' [42]. I've no doubt that Oldbel's wants will send an unwanted shiver down the spine of certain UC Davis professors and AWRI researchers!

In some countries and regions of the Old World, a view that Brett is regarded as natural and desirable in wine is sometimes more than evident. Speaking to Tim Atkin MW in 2011, the late Serge Hochar of Chateau Musar enthused, 'I like Brett, and I believe that volatile acidity helps my wines to age' [43]. Back in 2004, Marc Perrin of Château de Beaucastel commented, 'there are certainly some *Brettanomyces* in every natural wine, because *Brettanomyces* is not a spoilage yeast (as many people think) but one of the yeasts that exist in winemaking. ...... Of course, you can kill all natural yeasts, then use industrial yeast to start the fermentation, saturate the wine with SO_2 and then strongly filtrate your wine. There will then be no remaining yeasts, but also no taste and no typicity. That is the difference between natural wine and industrial wine, between craftsmanship and mass-market product'. [44]. Grenache, the most widely planted variety in the southern Rhône valley, has high levels of the 4-ethylcatechol precursor caffeic acid, in the region of 270–460 mg/l, some four to eight times the average in *Vitis vinifera* grape varieties. Some producers in the southern Rhône valley also proudly stated that Mourvèdre, one of the varieties that may be used (usually in small quantities) for red Châteauneuf-du-Pape and other southern Rhône wines, as part of its varietal make-up, has high levels of the other volatile phenol precursors.

If not all New World winemakers are totally anti-Brett, whether or not they determine related compound levels that are acceptable will depend upon the grape variety and the matrix and context of the wine. Matt Thomson, consultant for several producers and winemaker at Rapaura Springs and his own Blank Canvas winery in New Zealand, said 'it all depends on the level, the grape variety and region. In aromatic varieties like Pinot Noir it's disastrous, and most unwelcome in Syrah from cool climates. But sometimes it can act as a bridge between fruit and oak, and also add a little spice' [personal discussion]. It is certainly the case that many critics applaud the clove and bacon notes often present on many warm climate New World Syrahs (Shirazes), including many examples from the Californian region of Paso Robles and the Barossa Valley in South Australia.

So, responses to Brett characteristics vary from taster to taster, region to region, country to country, and even continent to continent. Despite many tasters, both professional and amateur, claiming to be objective, in truth the objective taster does not exist. Whilst wine professionals may take steps to eliminate personal preference and bias, we are all determined by history, education, and culture. Commenting on Andrew Adams online article 'New Thinking in the Brett Debate' (2013), William Hipp noted, '…..I was chairman of a 120 person wine society and we did blind tastings on differing varietals each month. For the American born members, wines with even the slightest hint of Brett always came in last. European members, however, were much more favourable. The American palate just doesn't like Brett, which is a bit regrettable, but a fact of life' [45]. Indeed, many British wine lovers are perhaps traditionally more excited about Brett characteristics in some wines, leading a few French producers to jest about making wines for 'la Grande Bret(t)agne!' The level of wine knowledge and experience of the taster also influences their preferences as to whether some *Brettanomyces*-related compounds in wine are to their liking or distaste. In a 2019 study by Schumaker et al. [46], clusters of tasters assessed a Shiraz wine at three different levels of Brett compounds and also different levels of 'green character' and 'oak character'. The base (control) wine had minimal Brett metabolites, and two other examples of the same wine were spiked with ethyl phenols and isovaleric acid at medium and high levels. The base wine was most liked by the cluster with the greatest number of tasters with high wine knowledge, but another, generally less experienced, cluster had as their second preference the base wine spiked with high levels of Brett compounds (a staggering 1500 µg/l of 4-EP, 300 µg/l of 4-EG, and 2000 µg/l of isovaleric acid). A third cluster had the wine with high Brett compounds together with oak character as their second preference. In 2017, a previous study by Schumaker et al. [47], involving consumers in the United States and Portugal, had shown that cultural differences affect perceptions of *Brettanomyces* and also that consumers had a high ability to detect differences in red wines due to volatile compounds.

4.3.7 Good Brett/Bad Brett?

There is sometimes discussion amongst winemakers and critics as to the possibility of 'good' and 'bad' strains of *Brettanomyces* although such dialogue is largely academic as, unlike in beer, strain selection in wine, in practical terms, perhaps lies in the future. It has often been argued that some Brett strains can result in desirable aromas in wine, and in 2015, a study of 95 *Brettanomyces* strains was made by C.M. Lucy Joseph et al. to identify if some strains consistently give positive aroma characteristics. Results were obtained using solid-phase micro-extraction with gas chromatography–mass spectrometry (SPME GC-MS) coupled with olfactory analysis [48]. It was found that none of the strains yielded universally positive aromas for the team of evaluators. In all 22 compounds associated with *Brettanomyces* were found to have an impact on aromas, although not all the compounds identified are metabolised only by the organism [48]. Countering any propositions that there are good and bad strains, the oenologists, researchers, and critics who have a total aversion to these compounds, sometimes colloquially referred to as 'the Brett Gestapo', expound that accepting low levels of Brett is like accepting that a little disease is a good thing. To them it is an all or nothing situation and the answer has to be nothing. Chris Curtin of the AWRI, examining the 'myth' that 'Brett can be good, or at least not all bad', is unequivocal that, drawing on a wealth of AWRI research, including tastings of 'spiked' wines with consumers in Sydney, the 'myth' has been 'busted' [49]. Of course, it might be argued that data recording tasters' perceptions of wines that have been spiked with

Brett-related compounds have limited validity: the results might be different if the wines had naturally formed similar levels of the compounds and undergone extensive maturation. Also perhaps there are signs of a New World/New World divide in this area too, for the general dismissing of any 'positive' characteristics of Brett by AWRI researchers is not necessarily shared by those at UC Davis, as discussed in Section 4.3.4.

For the many wine lovers who argue that a little Brett 'character' adds complexity to a wine, there are paradoxical questions: how much is a little, how much is too much, and crucially how do you control the concentration? To add complexity to wine but not the unwanted Brett characteristics, should the amount of 4-EP and/or 4-EG be well below, just below, or even just over sensory perception thresholds? Back in 2003, Virgílio Loureiro and Manuel Malfeito-Ferreira commented that at a concentration of less than 400 µg/l 4-EP 'contributes favourably to the complexity of wine aroma by imparting aromatic notes of spices, leather, smoke or game, appreciated by most consumers' [50]. Depending upon the grape variety, wine style, and matrix, this figure is, roughly, around the sensory perception threshold – see Section 4.4.3. It might also be noted that a little 'Brett' character has been valued by some in the industry for giving young and simplistic red wines the sought after notes of complexity and the illusion of maturity.

4.4 Sensory Characteristics and Detection of *Brettanomyces*-Related Compounds in Wine

The first indication of the presence of *Brettanomyces* in a red wine may be a loss of colour due to the hydrolysis of anthocyanins. However, colour loss may result from other factors, including heat damage, as discussed in Chapter 14.

The range of aroma and taste characteristics imparted by the compounds metabolised by *Brettanomyces* is very broad. However, in a study that assessed wines spiked with Brett-associated compounds, using professional tasters involved in the Bordeaux wine industry, all of the odour descriptors given by the tasters were of a pharmaceutical or animal nature [51]. Whilst each of the compounds imparts its own distinctive odours and flavours, there is a wide variance in these when produced by individual strains of the yeast and also a variance in how these odours are perceived by individual tasters. Smelling an empty glass after draining a Brett-impacted wine often gives elevated aromas, particularly those of 4-EP.

4.4.1 The Odours Associated with *Brettanomyces*-Related Compounds

Below are descriptors given for the odours of compounds metabolised by *Brettanomyces* – it should be noted that this list is not exhaustive:

- *4-EP*: The nose and palate has numerous descriptors including medicinal, antiseptic, BAND-AID, Elastoplast, ink, animal, stable-like, leather-like, 'sweaty saddles', and faeces. On the back palate, there can be metallic notes.
- *4-EG*: Smoky and spicy aromas and flavours, including cloves, tobacco, cooked bacon, and truffles.

- *Isovaleric acid*: The nose can be rancid with odours of sweaty saddles, cheese, mouse, metal, and goat, and this compound can even emit an odour of vomit.
- *4-EC*: Odours of sweat, cheese, stables, horses, medicine, and coffee (the compound is present in arabica coffee).
- *Ethyl-2-methylbutyrate*: Pineapple, mango, strawberry, green apple, cider.
- *Ethyl decanoate*: Odour of plastic and oil.
- *Tetrahydropyridines*: At high concentrations, these impart 'mousy' aromas and flavours, the smell of urine, and sometimes, even at low concentrations, a nose of tortilla chips, Cheerios®, cream cracker biscuits, and popcorn. Tetrahydropyridines are often even more noticeable on the palate than the nose. However, it should be noted that the presence of tetrahydropyridines does not necessarily indicate *Brettanomyces* activity unless accompanied by 4-EP and 4-EG, as they may also be synthesised by heterofermentative lactobacilli.
- *Guaiacol*: Nose of smoke and smoky bacon crisps (chips). The presence of guaiacol has several sources other than *Brettanomyces*, including oak, cork, and smoke.
- *trans-2-Nonenal*: Nose of burning rubber.
- *2-Phenylethanol*: Aromas of almond, honey, rose, lilac, jasmine, and bread.
- *Isoamyl alcohol*: Nose of whisk(e)y, brandy, molasses, and banana.
- *Isobutyric acid*: Odour of sweat, cheese, and rancid.

Wines affected by *Brettanomyces* are also likely to have unacceptably high levels of volatile acidity (VA), which has aromas of vinegar, a marked piquancy on the palate, and an acrid, bitter finish. The presence of oxygen stimulates production of large amounts of acetic acid by *B. bruxellensis*. Under anaerobic conditions, the acetic acid production is very low.

Some of the odours and flavours associated with *Brettanomyces* are those of 4-vinylphenol, the precursor of 4-EP, and 4-vinylguaiacol, the precursor of 4-EG. These compounds, the presence of which in wine is usually at sub-sensory threshold levels, are metabolised from their individual precursors by *S. cerevisiae*, lactic acid bacteria, or other organisms. Vinyl phenols are discussed in Section 4.5.

4.4.2 Sensory Detection Thresholds

To give figures for the sensory detection thresholds of each of the Brett-metabolised compounds is far from straightforward: they cannot with any certainty be stated in isolation and require some discussion. As we have seen, the main compounds invariably co-occur albeit in differing ratios. The threshold at which any compound is detected is generally lower when accompanied by other Brett-related compounds. In particular, the threshold of 4-EP is considerably reduced when accompanied by 4-EG, even if the concentration of 4-EP is 10 times greater that, in wine, is very often the case. The ratio depends upon many factors including the grape variety. Also, the characteristics of individual grape varieties impact upon the sensorial perception of ethyl phenols. Brett characteristics may also masked by various other components of the wine matrix, the impact and balance of which will change with maturation. In a young wine, primary fruit characteristics can cover up the odours of Brett compounds, but as the wine ages the primary fruit will wane, allowing related characteristics to come to the fore. Overt *Brettanomyces* odours on mature wine may be due to the subsiding of this masking effect, as tertiary characteristics will outweigh the primary and secondary aromas, but may also be due to the growth of the yeast in the bottled wine and consequential production of ethyl phenols.

4.4.3   4-Ethylphenol (4-EP)

A simple, but far from ideal, way of stating the sensory detection threshold for Brett characteristics in red wine is just to consider that of the concentration of 4-EP that, depending upon the wine matrix, is usually detectable somewhere in the range of 300–600 μg/l, although figures ranging from 230 μg/l to over 650 μg/l have been stated by authors. Figures from the Curtin et al. of the AWRI [52], examining three full-bodied Australian Cabernet Sauvignon wines, showed varied perception levels for 4-EP: 368 μg/l in a neutral character wine, 425 μg/l in a 'green' wine possessing intense methoxypyrazine characteristics, and 569 μg/l in a wine with heavy oak influence. A 2018 study of 260 suspect wines by Csikor et al. [53] revealed that wines showing above 968 μg/l of 4-EP could be determined as having Brett character with a 5% probability of error and wines with a level below 245 μg/l could be determined as sound, again with a 5% probability of error. This is, of course, a spectacularly wide range. However, it is a sobering thought that red wines, at very worst, can contain up to 50 mg/l (50 000 μg/l) of 4-EP.

4.4.4   4-Ethylguaiacol (4-EG)

The figures quoted for the sensory threshold of 4-EG also vary considerably from 33 to 373 μg/l. The figures from Curtin et al of AWRI showed the perception levels in the full-bodied Australian Cabernet Sauvignon wines as 158 μg/l (neutral character wine), 209 μg/l (green wine), and 373 μg/l (heavily oaked wine) [52]. These figures are higher than those stated by some laboratories that suggest a sensory detection threshold of range from 'around 50 μg/l' to '135 μg/l' [personal discussions].

However, the co-presence of 4-EG invariably also lowers the perception threshold of 4-EP. Pascal Chatonnet et al. reported the limit sensory threshold of 4-EP as 605 μg/l and 4-EG as 140 μg/l [54]. With the two compounds combined in a ratio of 10 : 1, a common ratio in affected wines, the (combined) sensory threshold was found to be 426 μg/l [53]. The olfactory impact of 4-EG can also appear greater than 4-EP if the ratio is less than 8 : 1 4-EP/4-EG.

4.4.5   Isovaleric Acid and Isobutyric Acid

The sensory threshold of isovaleric acid (also known as 3-methylbutanoic or 3-methylbutyric acid) is in the range of 33–72 μg/l. The rancid, at times vomit-like odour of this compound, can be the dominating odour in a wine with heavy Brett contamination, outweighing the sensory impact of 4-EP and 4-EG. However, its role has perhaps been understated by many authors. Back in 1998, Licker et al. reported that isovaleric acid was determined to be one of the two predominant odour-active compounds responsible for Brett odour (the other was unidentified but was not 4-EP or 4-EG, both of which were identified by not regarded as predominant) [55]. Isobutyric acid, which has a sensory threshold of 2300 μg/l (2.3 mg/l), and isovaleric acid have a masking effect on the sensory detection of ethyl phenols. The detection threshold 4-EP and 4-EG can be three times higher when they are accompanied by these acids, as shown in a fascinating paper by Romano et al. [56]. In this study, a total of 51 Bordeaux wines of the 2005 vintage were analysed; 4-EP and 4-EG was present in all the wines to a greater or lesser concentration, with an average ratio of 10 : 1. The wines were nosed by expert tasters, trained in the recognition of 4-EP and 4-EG. Five were deemed to be heavily tainted, 10 mildly tainted, and 36 non-tainted. Accordingly, the wines were divided into three groups. However, the analysis by gas chromatography–mass spectrometry (GC-MS) revealed that the average 4-EP + 4-EG content of the wines was in fact highest in the group determined as non-tainted, with an average of 403 μg/l and one wine reaching a staggering 1370 μg/l. The study concluded that detection thresholds of 4-EP and 4-EG were three times higher when the ethyl phenols were accompanied by isobutyric acid and isovaleric acid due to a masking effect, even when the presence of the acids is below their own sensory detection thresholds.

4.4.6 4-Ethylcatechol

This compound is responsible for horse, leather, and farmyard odours. The lowest quoted figure for the sensory perception of 4-EC is 774 μg/l. The AWRI has noted an odour detection threshold of 1131 μg/l in 'green' red wine and 1528 μg/l in oaky red wine [52]. Much of the research work into *Brettanomyces*-metabolised volatile phenols has been focussed on 4-EP and 4-EG, and it is possible that the impact of 4-EC has been under reported.

4.4.7 Tetrahydropyridines Particularly 2-Acetyltetrahydropyridine (ATHP) and 2-Ethyltetrahydropyridine (ETHP)

Tetrahydropyridines are synthesised by *Brettanomyces* from lysine and ethanol. 2-Acetyltetrahydropyridine (ATHP) has an odour threshold of approximately 1.6 μg/l and a flavour threshold of approximately 5 μg/l, while 2-ethyltetrahydropyridine (ETHP), which is metabolised by *Brettanomyces* from ATHP, has the highest odour threshold of the tetrahydopyridines at approximately 150 μg/l and a flavour threshold of approximately just 3 μg/l. 2-Acetylpyrroline may also be synthesised by *Brettanomyces* and has an odour detection threshold of 0.1 μg/l.

4.4.8 Guaiacol

The odours of guaiacol are those of smoke, toast, or ash. The sensory threshold is generally accepted as 23 μg/l. Whilst guaiacol can be metabolised by *Brettanomyces*, it can also be present in wines as a result of oak ageing or contaminated corks and may be particularly dominant in those affected by smoke taint, which is discussed in detail in Chapter 12.

4.4.9 The Impact of the Wine Matrix

The sensory thresholds of Brett-related compounds are perhaps lower in lighter-flavoured and lighter-bodied wines than in full-bodied and concentrated wines – for this reason Brett characteristics are sometimes overt on red Burgundy wines that, with a few minor exceptions, are made exclusively from the somewhat delicate Pinot Noir variety. The aromas and flavours of oak products may also mask Brett compounds to a large extent. In short, the sensory thresholds of compounds metabolised by *Brettanomyces* vary considerably according to the myriad of factors that constitute each individual wine's make-up [57, 58].

4.5 The Origins of *Brettanomyces* and Formation of Related Compounds in Wines

4.5.1 Microflora on Grapes and Present in Wineries

Brettanomyces yeasts may originate in the vineyard, on the skins of grapes. This important fact had been long suspected (and sometimes denied) but only confirmed in 2007 [59, 60]. In 2019, Oro et al. [61] verified that the grape bunch is one of the main sources of *Brettanomyces* spp. contamination in wineries. It is perhaps worth reminding ourselves at this point that a grape berry naturally carries between 10 000 and 1 million microbial cells, depending on its size. Grape skins, or to be precise grape surfaces, contain a great diversity of microflora including filamentous fungi, bacteria, and up to 15 genera of yeasts. These include the following:

- *Hanseniaspora* spp. (teleomorph *Kloeckera* spp.) – apiculate (lemon) shape.
- *Saccharomyces* spp. including the species *S. cerevisiae* and *S. bayanus* – ellipsoid shape.
- *Brettanomyces* spp. including the species *B. bruxellensis* – ovoid shape.
- *Pichia* spp. including *Pichia guilliermondii* – spherical or ellipsoid shape.
- *Candida* spp. – spherical or ellipsoid shape.
- *Zygosaccharomyces* spp. – normally spherical shape.

By far the most dominant of these are *Hanseniaspora* spp./*Kloeckera* spp., which account for over 50% of the yeast populations on grape surfaces. In fact, *Saccharomyces* spp. represent a tiny percentage of the volume of yeasts on grapes: between 0.05% and 0.1%. However, they are generally by far the most prevalent community within the winery and will be the dominant yeast undertaking AF once the abv of the juice has surpassed 3% or 4%. *Brettanomyces* spp., if present, account for less than 3% of the yeast population of the grape cluster. Many yeast genera are regarded by winemakers as spoilage organisms, and steps may be taken to inhibit them, or at least limit their growth, during the early stages of winemaking. Such control mechanisms generally include pre-fermentation additions of potassium metabisulfite or another form of SO_2 and may sometimes include the additions of 'killer' strains of commercial yeasts. Of all the so-called spoilage yeasts, *Brettanomyces* is the only one that can survive in the alcohol levels found in fully-fermented wine.

Brettanomyces yeasts are generally ovoid or ogive shape but may also be found as ellipsoidal, cylindrical, or spherical forms. They are normally between 2 and 7 μm in size, 5×3 μm and 7×4 μm being typical. Reproduction is by budding, and the yeasts are facultative anaerobes. The two species most important in wine are *B. bruxellensis* and *B. anomala* [19]. There are thousands of strains, and amongst these is a great genetic diversity. The strains are structured according to ploidy level, substrate of isolation, and geographical origin [62]. Wine genotypes are highly disseminated across wine regions, but strains of the yeast do vary according to individual vineyard location [63].

The presence of *Brettanomyces* on grapes is most likely in moist conditions. Colonies are often found on grape berries that suffer from sour rot and also those affected by *Botrytis cinerea*. However, anti-*Botrytis* treatments containing procymidone are known to lessen the impact or *Brettanomyces* and thus limit the production of volatile phenols. Overripe grapes contain more volatile phenol precursors [64], and delaying the harvest may increase the risk of volatile phenol production – the quest for super-ripeness (with consequential high pH levels) is one of the reasons why Brett is increasing in some regions. The strains of *Brettanomyces* that are present on the grapes may remain present throughout the winemaking and maturation processes. Of course, once the yeast has entered the winery, whether from grapes, insects, infected wines, contaminated used barrels, or other source, it will colonise the equipment, including vats and barrel stock, and also the atmosphere. Indeed, winery contamination is one of the main sources of the yeast in wine, and some authors remain unconvinced that the presence of *Brettanomyces* on grapes is the primary source of infection. There are no known cases of *Brettanomyces* being found on new equipment prior to use.

4.5.2 The Formation Pathways of *Brettanomyces*-Related Compounds

4.5.2.1 Hydroxycinnamic Acids Precursors

Several hydroxycinnamic acids (sometimes referred to as phenolic acids), which are non-flavonoid phenolic compounds, are present in grapes in both free and bound forms. Those of particular importance to the production of Brett related compounds are *p*-coumaric, ferulic, and caffeic acids. It is only the free forms of these acids, which are usually extracted from the skins of red grapes during maceration, that are transformed by *Brettanomyces* into ethyl phenols as the second stage of a two-stage process. *p*-Coumaric acid is the precursor of 4-ethylphenol, ferulic acid the precursor of 4-ethylguaiacol, and caffeic acid the precursor of 4-ethylcatechol.

4.5.2.2 The Formation of Vinyl Phenols

The first stage of the formation of ethyl phenols is the enzymatic formation of vinyl phenols from hydroxycinnamic acids. The enzyme 4-hydroxycinnamate decarboxylase converts p-coumaric acid into 4-vinylphenol, ferulic acid into 4-vinylguaiacol, and caffeic acid into 4-vinylcatechol. There are a number of organisms that may be present in must or wine that are able to carry out the decarboxylation of hydroxycinnamic acids and produce vinyl phenols. These include:

- *Saccharomyces cerevisiae;*
- *Brettanomyces bruxellensis;*
- *Brettanomyces anomala* (sometimes referred to as *B. anomalus*);
- *Pichia guilliermondii;*
- *Candida versatilis*
- *Candida fermentati;*
- *Candida mannitofaciens;*
- *Lactobacillus plantarum;*
- *Torulaspora* spp.;
- *Zygosaccharomyces* spp.

4.5.2.3 The Formation of Ethyl Phenols

The second stage of the process is the reduction of vinyl phenols to ethyl phenols. *Brettanomyces* yeast contains the enzyme vinylphenol reductase (VPR) that reduces 4-vinylphenol into 4-ethylphenol, 4-vinylguaiacol into 4-ethylguaiacol, and 4-vinylcatechol into 4-ethylcatechol. Polyphenols drive these enzymatic reactions, and a high polyphenol content in the wine is likely to produce high levels of ethyl phenols [65]. In a 1995 study undertaken by Chatonnet et al. [66], using a model medium, *Brettanomyces* (*D. intermedia* MUCL 27 706) synthesised 2915 µg/l of 4-EP and 2947 µg/l of 4-EG. Other organisms included in the study were strains of three species of *Lactobacillus*, two of *Pediococcus*, three strains of *Leuconostoc oenos*, and a strain of *S. cerevisiae*. In this study, the only microorganism other than *Dekkera* to produce volatile phenols at a concentration above double figures was a strain of *L. plantarum* that produced 230 µg/l of 4-EP.

It is generally accepted that in wine, and in practical circumstances, *Brettanomyces* is the only major organism responsible for producing significant quantities of ethyl phenols and the presence of 4-EP is almost always on account of its action. *S. cerevisiae* does not contain the enzyme VPR needed for this biosynthesis. Some other organisms have been shown to carry out the degradation in laboratory conditions [67], but there is a general consensus amongst researchers that their action can only produce 4-EP (and other ethyl phenols) in very limited quantities in practical circumstances. Lactic acid may also produce significant amount of vinyl phenols, but, in practical circumstances in wine, it will only produce trace amounts of 4-EP. *Pichia guilliermondii* is often present in musts and is perhaps able to produce some 4-EP prior to the commencement of AF. Thus, the employment of the technique of pre-fermentation maceration (cold soaking) poses a danger if viable *P. guilliermondii* is present. The quantity of 4-EP and 4-EG produced is directly related to the quantity of *p*-coumaric acid and ferulic acid, respectively, and the greater the biomass of *Brettanomyces*, the greater the potential quantity of ethyl phenols. Synthesis can take place at all stages of winemaking, but it is post AF that large quantities may be produced.

The ratio of 4-ethylphenol to 4-ethylguaiacol in a 'bretty' wine can range from 3 : 1 to 40 : 1, the average being 10 : 1. This ratio will depend upon the ratio of the precursors, *p*-coumaric acid and ferulic acid. Individual grape varieties have different levels of the precursors: typical 4-EP/4-EG ratios in impacted wines are 10 : 1 for Cabernet Sauvignon, 9 : 1 for Syrah (Shiraz), 8 : 1 for Merlot, and 3.5 : 1 for Pinot Noir. The ratio of 4-EP to 4-EC is generally in the region of 3.7 : 1, and the ratio of 4-EG to 4-EC approximately 0.7 : 1 [68].

As shown in Section 4.2, *Brettanomyces* yeasts also produce other volatile compounds, several of which have significant sensory impact. Tetrahydropyridines are synthesised by the yeast from the amino acid lysine [12]. It should be noted that *Brettanomyces* yeasts contain esterases that, during winemaking, break down esters, which are compounds contributing to fruity secondary aromas. It is this process that may result in a reduction in these aromas in *Brettanomyces*-affected wines, even when the associated odours from metabolised compounds are below sensory detection thresholds.

Brettanomyces can grow in anaerobiosis that, of course, includes wines in totally filled or gas-blanketed vats. However, when oxygen is present, its development is facilitated. When growing in anaerobic conditions, *Brettanomyces* produces little alcohol – an example of the so-called Custer effect. Although it is often considered to be a slow-growing yeast, in a wine that has completed its AF, the population of *B. bruxellensis* can already have reached

10^4–10^6 colony-forming units (CFUs), that is, cells/ml [69]. In fact there is generally no or little growth during the early and middle stages of the AF process, but when it nears completion, the growth can be rapid. In positive fermentation conditions, which result in the complete exhaustion of sugars, there is usually no appearance of any non-*Saccharomyces* yeast when the fermentation ends [69]. However, Brett is generally more resistant to high levels of ethanol than *S. cerevisiae* and can continue to develop in wines with up to 15% abv and, unlike most other yeasts, in a low-sugar environment. It has meagre needs for nutrients, enabling it to survive in nutrient deficient wines, which presents a major challenge to its control [29]. It can use ethanol, which first converts into acetic acid, and unfermentable sugars, i.e. those that are not 'six carbons', as nutrients. The amino acid proline, which is not metabolised by *S. cerevisiae*, can be used by *Brettanomyces* as a source of carbon and nitrogen nutrients [70]. The yeast can also utilise arginine, another amino acid, and although *S. cerevisiae* does consume this acid, there is often a surplus, particularly if judicious quantities of yeast nutrients have been added.

4.6 The Danger Periods and Favourable Conditions for the Growth of *Brettanomyces*

4.6.1 The Danger During Winemaking and Cellar Maturation

The danger periods for the growth of *Brettanomyces* are as follows:

- Near the end of a slow or troubled AF.
- Between the end of AF and the onset of the MLF.
- During the MLF. The longer period of the MLF, the greater the biomass of *Brettanomyces* that may be produced.
- Immediately following MLF.
- During barrel ageing. There is no doubt that this is most likely period for rapid growth of *Brettanomyces*.
- In bottle.

Conditions and operations that favour Brett growth include the following:

- High pH. A pH above 3.8 is dangerous [71], and ideally a pH of 3.65 or below should be aimed for in red wines. Red wines with a pH below 3.5, usually from cool climates, have a much reduced risk of infection, but very low pH can present difficulties with the onset of the MLF. The lower pH values of white and sparkling wines is one of the reasons that *Brettanomyces* growth in them is much less common than in red wines. However, the yeast is resistant to carbon dioxide (CO_2) and can use autolysis compounds as a nutrient source, sometimes enabling development in sparkling wines that undergo a secondary fermentation in bottle. Many authors note that a high pH reduces the efficacy of SO_2, due to much less being available in molecular form as discussed in Section 4.8.7, but the growth of microbes is also generally reduced in high-acid environments.
- Delay between alcoholic and MLFs. This is a particularly high-risk period on account of, inter alia, the necessity of keeping SO_2 levels low in order not to inhibit the MLF. The

ideal level of SO_2 for the MLF process is close to zero, and the ideal temperature is 20 °C (68 °F) or just above. Each of these conditions is favourable for *Brettanomyces* growth.

- Warm temperatures. Any temperature above 15 °C (59 °F) may enable *Brettanomyces* growth, although the yeast may still develop at 10–15 °C (50–59 °F). Somewhat controversially Clark Smith stated that Brett cannot grow below 59 °F (15 °C) [29], contrary to the view of many researchers. The optimal temperature range for growth is 25–28 °C (77–82 °F).
- Maturation on lees. Populations of *Brettanomyces* are often concentrated in the lees, and any reincorporation of the lees into wine, including stirring in vat or bâtonnage in barrel, may pose a risk.
- Additions of oxygen. Barrel ageing, micro-oxygenation (MOX), cliqueage, and additions that take place during cellar operations such as racking can all give a boost to developing Brett colonies.
- Low levels of SO_2. Molecular SO_2 below 0.5 mg/l may be regarded as dangerous.
- Residual (unfermented) sugars. Higher residual sugar (RS) levels (over 4 g/l) pose the greatest risk, but Brett can grow in wines with as little as 275 mg/l of RS [66, 72], with some authors even quoting levels as low as 200 mg/l [73], or 210 mg/l of glucose, fructose, and trehalose combined [74]. Trehalose is not consumed by *S. cerevisiae*. A 4-EP content of 1000 µg/l (well over the perception threshold) can result from the consumption of 300 mg/l of sugars. There has been a trend in the last 20 years or so for the residual sugar levels in so-called 'commercial' or 'entry-level' red wines to increase. Australian red wines saw an average residual sugar increase from 0.5 g/l in 1998 to 2.1 g/l in 2010 [75].
- Poor cellar hygiene. This always used to be blamed for the presence of *Brettanomyces*. In recent years, there has been a trend for authors to claim that the occurrence of Brett-related faults is not a consequence of unhygienic cellars, but in reality there can be little doubt that a lack of cleanliness and sanitisation facilitates its growth.
- Contamination at the bottling. It is important that all parts of the bottling line, and particularly the corker, are effectively sanitised.

4.6.2 Growth in Bottle

As we have seen, *Brettanomyces* can grow in anaerobiosis and in wines with very low dissolved oxygen (DO), including wines in bottle. Indeed live *Brettanomyces* cells have been found in bottles of 100-year-old wines. Knowing how wine will develop in bottle is a skill that is essential to both winemakers and critics, but the presence of Brett can send bottle ageing down an unexpected and unwanted road. Robert Joseph, co-chair of the International Wine Challenge (IWC) that claims to be the world's largest wine tasting competition, said, 'I was once given an Italian wine that was so faecal and so harshly textured that I found it undrinkable. The bottle bore a sticker proclaiming that it had won an IWC trophy three years earlier when, in the company of a team of professionals, I had happily agreed to its award. None of us had noticed or objected to the Brett that must have lurked there, but since that tasting it had evidently developed into a monster' [76]. In fact, *Brettanomyces* can flourish in bottle: in the study by Joana Coulon et al. [72] in which Bordeaux wines with varying levels of Brett were stored in bottles at 20 °C (68 °F) for 130 days, the increase in ethyl phenol concentration was up to 104-fold during the period of storage.

4.7   Why Are *Brettanomyces*-Related Compounds Found Mostly in Red Wines?

There are numerous reasons why *Brettanomyces*-related compounds mostly affect red wines:

1. Red wines generally have a higher pH than white and rosé wines. On the fairly rare occasions that Brett character is found in white wines, it is often in those that are made with high pH varieties, e.g. Gewürztraminer.
2. Red wines generally have a higher phenolic content than whites and rosés, resulting in the presence of a much greater amount of the precursor compounds. Red wines, especially those designed for ageing, are generally made from grapes that are crushed, and the juice and solids are macerated during fermentation and sometimes pre- and/or post-fermentation. It is during maceration that extraction of hydroxycinnamic acids in their free form takes place. Due to the general absence of the maceration stage, newly fermented white wines contain a concentration of *p*-coumaric acid at a level of less than 25% compared with red wines. However, 'orange' wines, i.e. white wines that have been fermented on grape skins in a similar way to red wines, are prone to exhibiting Brett-related volatile phenols.
3. The maceration process can enable the colonisation of *Brettanomyces* even if a selected *S. cerevisiae* starter culture is utilised.
4. Red wines are often matured for long periods in barrels.
5. Additions and levels of SO_2 are very often less in red wines than in white or rosé wines.
6. Fine red wines may be aged for long periods – Brett can grow in bottle if the wine contains viable cells, particularly if not stored in cool conditions.
7. Red wines are less likely to be sterile-filtered than white or rosé wines.

4.8   Prevention: Formulation and Implementation a Brett Control Strategy

4.8.1   HACCP

Having glanced at the danger periods and favourable conditions for the growth of *Brettanomyces*, the formulation and implementation of a Brett control strategy is now considered. The instigation and deployment of a system using the principles of HACCP (see Chapter 18), in which the growth of *Brettanomyces* is an identified hazard, is valuable, but it must be kept in mind that there will be a large number of control points. The HACCP system should take account of the possible multiple origins of *Brettanomyces*, including grapes, insects, the winery and cellar, the atmosphere, and barrels. It is perhaps unrealistic to aim for a complete absence of *Brettanomyces* [16], but the aim should be to limit multiplication of the yeast to below a level at which significant quantities of volatile phenols might be formed. To this end, the utilisation of 'the hurdle concept' (commonplace in the food industry) by which *Brettanomyces* would have to surmount a series of 'control' hurdles to grow to an unacceptable concentration is effective.

4.8.2 The Vineyard

Prevention starts in the vineyard. Balanced nutrition of the vines is vital to obtain balanced musts and wines. An excess of potassium (K) can lead to musts and wines high in pH, and low in acids, which will require early must correction in order for sulfiting to be effective. A deficiency in nitrogen (N) or water stress can lead to slow and possibly incomplete fermentations that will favour Brett development. The yeast is often found on *Botrytis*-affected grapes, and the use of an anti-*Botrytis* treatment containing procymidone has been found to limit the presence of *B. bruxellensis* on berries [16].

4.8.3 Harvest

Delaying harvest in order to boost so-called phenolic ripeness increases the risk of later problems with Brett. The population of the organism on the clusters can increase, and a high pH in must and wine results in a more favourable growth medium for Brett development and reduces the efficacy of SO_2 additions. Whilst acidification can take place at the must adjustment stage, this is not without drawbacks. The harvesting of rotten, diseased, or damaged fruit should be avoided by utilising selective picking or, when machine harvesting is undertaken, removing such fruit by pre-picking and post-harvest sorting.

4.8.4 The Winery

Cleaning and sanitising the winery and equipment, with the with the aim of achieving a 'minimum Brett' winery, must be a matter of routine. Efficient cleaning will not only minimise the population of microorganisms but also remove the substrates on which Brett cells can live. However, as with other yeasts and bacteria, *B. bruxellensis* cells adhere to and colonise inert surfaces [77] and crucially form biofilms that will give them protection in harsh environments. Biofilms can resist sanitation products. Particular attention should be paid to cleaning sorting tables, crushers, presses, fixed pipes, flexible hoses, vats, and barrels. All equipment including tanks, pumps lines, hoses and filters should be sanitised before and between uses. Brett has been found in tank valves at an order of magnitude of 10^4 CFU/ml [61]. Avoiding cross-contamination is crucial – losing a barrel to Brett is bad, but losing several tanks is disastrous.

Brettanomyces (along with many other yeasts) have been found in the gastrointestinal tract of *Drosophila* (vinegar) flies and fruit flies, and every effort should be made to exclude these from fruit and the winery.

4.8.5 Winemaking Practices

4.8.5.1 Approaches to *Brettanomyces* Management

There are several, at times contrasting, winemaking approaches to minimising the risk of *Brettanomyces* growth, and there may be conflict as to the route taken and even precisely where in the destination region one wishes to be. Clark Smith, from a postmodern viewpoint, noted that modern oenological practices cause *Brettanomyces* and exacerbate its effects [29]. His concept of *Integrated Brettanomyces Management* (IBM), including microbial equilibrium in the cellar, will almost certainly jar with many researchers, who believe that Brett should be totally excommunicated from the winery and wines. Clark's three prin-ciples of IBM are as follows:

- Create a nutrient desert;
- Foster microbial balance;
- Achieve aromatic integration through good structure.

If you talk to 50 different winemakers, you will hear at least 100 different views as to what constitutes good practice. Accordingly some of the steps suggested below to minimise growth of *Brettanomyces* may raise eyebrows in some quarters.

4.8.5.2 Pre-fermentation
Sulfur dioxide in the form of potassium metabisulfite should be added to the picking bins containing harvested grapes or at the crusher. After analysis of the must, the free SO_2 should be adjusted to between 20 and 50 mg/l, depending upon the pH and sanitary condition of the fruit. An initial concentration of 50 mg/l may lessen the amount required in subsequent additions and fits well with the general rule that SO_2 is better added in a small number of large additions rather than a large number of small additions. Of course, such SO_2 additions also reduce the risk of other unwanted microbial activity and generally inhibit yeasts other than *S. cerevisiae*. This is one reason why producers wishing to undertake 'wild' ferments will limit or even avoid early SO_2 additions. In any event, excessive pre-fermentation SO_2 additions, i.e. in excess of 80 mg/l, are undesirable for several reasons, including the conse-quential possible delay in the onset of the MLF, which may per se facilitate the growth of *Brettanomyces*.

4.8.5.3 Cold Soaks and Hot Pre-fermentation Maceration
Cold soaks should not be considered if the must includes grapes that are rotten or dam-aged, which are the primary source of spoilage yeasts. With healthy grapes, cold soaking at a temperature below 10 °C (50 °F) will usually prevent the growth of *Brettanomyces* at this stage [78]. A low maceration temperature prevents stabilisation of the precursor hydrox-ycinnamic acids.

Following cold soaking, the must should be warmed rapidly to the initial fermentation temperature, and the desired culture of *S. cerevisiae* added as a starter or 'pied de cuve'. If the winemaker wishes to ferment using natural yeasts, cold soaking is particularly risky.

Hot pre-fermentation maceration, at a temperature above 65 °C (149 °F), will kill any *Brettanomyces* cells present (together with desirable yeasts and bacteria), so subsequent inoculations with *Saccharomyces* cultures and MLF bacteria will be essential.

4.8.5.4 Yeast Starters, Nutrients, and Enzymes
The decision by winemakers as to whether to use cultured yeasts or allow natural yeasts to carry out the AF will rest on numerous criteria. Of course, producers of so-called 'natural wines' generally believe that cultured yeasts are a factory product that negatively impact upon desired aroma and flavour profiles. Further, a great many producers of 'terroir-driven wines' prefer to use natural yeasts, regarding them as almost as much a part of terroir as top-soil, subsoil, aspect, and mesoclimate. The views of Jerome Fious of Burgundy's Domaine Faiveley are clear, 'if the vintage is balanced, the crop is healthy, and the pH as I would wish,

Table 4.1 Influence of yeast and sulfite additions on a Pinot Noir vintage contaminated with *Brettanomyces*.

AF = Alcoholic fermentation MLF = Malolactic Fermentation	'Lightly' sulfured harvest: (37 mg/l) with no cultured yeast	'Normally' sulfured harvest: (75 mg/l) with no cultured yeast	'Lightly' sulfured harvest: (37 mg/l) with cultured yeast
Brettanomyces at the end of AF: CFU/ml	40 000	3 000	700
Volatile phenols at the end of AF: μg/ml	94	12	17
Volatile phenols at the end of MLF: μg/ml	467	68	75

Source: Data from *Brettanomyces* et phénols volatils. Les Cahiers Itineraires d'itv France.

then natural yeasts will always be used. But in the event of difficult vintages, and those with a high pH, we will choose cultured yeasts to avoid potential difficulties' [personal discussion].

From a technical winemaker's point of view, a *Saccharomyces* starter culture helps to ensure an even and complete fermentation. It is preferable to add the yeast as a starter following the first grapes that go into the vat to deter the development of unwanted yeasts, including *Brettanomyces*. Having the strongest possible starter culture is recommended, especially if the musts are of high gravity [79]. The starter can decarboxylate the hydroxycinnamic acid precursors of 4-EP and 4-EG. Table 4.1, using data from ITV France [80], illustrates the influence of using a starter culture for a Pinot Noir harvest in restricting the growth of *Brettanomyces* by the end of the AF and the production of volatile phenols at the end of both the AF and MLF.

The addition of yeast nutrients should be limited, as any excess may also provide nutrients for *Brettanomyces*. To help avoid a sluggish or incomplete fermentation (and to reduce the risk of the formation of volatile sulfur compounds), it is important that sufficient yeast assimilable nitrogen (YAN) is present in the must. Approximately 110 mg/l of YAN are needed by yeasts for satisfactory red wine fermentations, and 165 mg/l for white wines. However, the use of diammonium phosphate (DAP) should be avoided unless the musts are very nitrogen deficient. Although DAP minimises yeast stress and helps a vigorous fermentation without the production of sulfides, such ferments may retain micronutrients that may subsequently be used by Brett as a food source. Further, any excess in fermentable nitrogen increases the substrates upon which Brett may grow. *Thiamine* (vitamin B_1), together with other B group vitamins, is sometimes added in the early stages of fermentation to help boost yeast populations and prolong their life. However, *Brettanomyces* needs thiamine to grow, so the additions need to be undertaken with caution. In a recent study, Hucker et al. [81] determined that the presence of 50 μg/l of thiamine and/or riboflavin stimulated the growth of *Brettanomyces* in artificial minimal media. The growth of *Brettanomyces* was classified as medium in the absence of either of these vitamins and heavy in their presence. Biotin (vitamin B3, also called vitamin H) is also a known Brett growth stimulant.

Enzyme preparations containing cinnamyl esterase (sometimes referred to as cinnamoyl esterase) should be avoided as this helps release hydroxycinnamic acids from esters [82], leaving them able to be converted into volatile phenols. Enzymes described as 'purified' do not usually contain cinnamyl esterase.

4.8.5.5   The Alcoholic Fermentation (AF), Post-Fermentation Period, and Pressing

The oxygen level in the fermentation should be boosted as necessary by aerated pump overs and/or rack and returns in order to maintain healthy *Saccharomyces* yeast colonies.

The 'modern' trend of leaving some residual sugar in red wines should be avoided. Brett loves carbon! Residual sugar should be kept below 1 g/l as an absolute maxim and ideally below 300 mg/l. This will limit the main nutrient for Brett growth.

Whole grape fermentations, including intracellular fermentations, present a danger as additional sugars may be released following the completion of the fermentation (including during the pressing process), which is a particularly dangerous time for *Brettanomyces* growth.

High turbidity during and particularly post fermentation can provide substrates upon which Brett can grow. Clarification procedures should be undertaken if necessary.

Temperature shock at, and immediately after, pressing must be avoided: the wine should be kept within 2 °C (4 °F) of the temperature of the withdrawal vat. This will help any remaining sugars to ferment out.

Dead yeast cells (gross lees) should be removed by racking the wine off these as soon as possible to reduce the compounds released during autolysis [83].

Post-fermentation warm macerations, i.e. at a temperature of 35 °C (95 °F) or higher, and MOX (at this time) should be avoided.

4.8.5.6   The Pre-MLF Lag Phase and the MLF

The periods near and at the end of AF and between AF and MLF are dangerous. Microbial competition for *Brettanomyces* is very low or non-existent at this time. The time lag between completion of the AF and the onset of the MLF should be minimised by, inter alia, ensuring the wine is held at a temperature of approximately 20 °C (68 °F), SO$_2$ levels are very low, and sufficient quantities of the desired MLF bacteria are present. To reduce solids, wine may be racked into a buffer tank and rested for three days or so after completion of the AF before racking again and inoculating, using a strong MLF culture if necessary. Such use of commercial MLF bacteria avoids the possible formation of volatile phenols by lactic acid bacteria [84] and reduces the risk of *Brettanomyces* cell growth. The MLF should be completed rapidly. Co-inoculations of a *Saccharomyces* yeast strain and an MLF starter bacteria culture, such as a strain of *Oenococcus oeni*, to induce simultaneous AF and MLF fermentations might be considered to ensure a rapid start of the MLF. This is regarded by some researchers as the best strategy in comparison with the sequential or late inoculation [85, 86], but many winemakers fear that this can lead to a loss of colour or a rise in volatile acidity as the malolactic bacteria can consume sugar and produce acetic acid as a by-product [87]. Should the AF stick there would be a major problem. Co-inoculation is not recommended if the more than 25 mg/l SO$_2$ has been added to the must – it must be kept in mind that most yeast produce SO$_2$ during the AF, and accordingly the MLF might be inhibited.

Identification of the completion of the MLF should be made as quickly as possible in order for the wine to be protected by the addition of SO$_2$. At the completion of MLF, the wine will have its lowest SO$_2$ content, and pH is likely to be at its highest, both conditions favouring

Brett growth. The molecular SO_2 should be adjusted to between 0.5 and 0.625 mg/l as soon as possible after completion of MLF. The addition of dimethyl dicarbonate (DMDC) (marketed under the trade name Velcorin®), which inhibits the enzymes involved in glycolysis, may also be considered.

4.8.6 Maturation, Barrel Ageing, and Barrel Care

4.8.6.1 Barrel Maturation

The worldwide increase in the use of oak barrels in the last 25 years has coincided with an increase of *Brettanomyces*-related compounds in wines. Barrels are difficult to effectively sanitise and may provide a nutrient source and substrate for growth. They also allow an ingress of oxygen into wine, significant in stimulating the growth of *Brettanomyces* [88]. Up to 30 mg/l of oxygen is dissolved into the wine per annum during barrel ageing; the actual figure will vary due to many factors, including the size of the barrel, the type and origin of the oak (or other wood), and the storage temperature. The oxygen uptake is highest in wine in new barrels, as the pores in the wood become partially blocked with tannins with repeated use. This favours a relatively high redox potential and results in greater DO. Consequently, wines maturing in new barrels lose active or molecular SO_2 faster than those in previously filled wood, a factor to be kept in mind when making the initial and subsequent SO_2 additions.

Cellobiose, a disaccharide compound formed in the toasting of oak barrels, can provide a nutrient for Brett (and other spoilage yeasts). The B-glucosidase enzyme of *Brettanomyces* produces glucose by cleaving cellobiose. The level of cellobiose is highest in new barrels and will be reduced as the barrels age and with subsequent fillings. It may be argued that some coopers, in their publicity and technical papers, understate the risk of Brett using cellobiose as a nutrient source. There are also winemakers who believe that Brett is present deep in new barrels from some coopers, although there would appear to be no validated evidence of this. However, the focus of most winemakers with regard to the possibility of growth of Brett in wine maturing in barrels is very much on previously filled wood, and many very much underestimate the risk of *Brettanomyces* growth in wine maturing in new barrels due to their being a source of cellobiose and relatively high levels of oxygen ingress. There can be no doubt that used barrels remain a major source of contamination, largely on account of the major challenge in effectively cleaning and sanitising. Maturation of wines in second, third, or fourth fill barrels needs constant vigilance: one or two infected barrels can contaminate many more, following racking, topping, or transfer operations. As Manuel Malfeito-Ferreira noted, 'wineries rarely monitor the contamination level of wine aged in oak barrels and, more rarely, the microbiological quality of wine put in the barrels' [89].

There is anecdotal evidence that barrels made of American oak (*Quercus alba*) are prone to higher levels of contamination, possibly due to greater porosity of the oak than the European *Quercus robur* and *Quercus petrea* and also due to the fact that American oak is sawn, not split, during the timber preparation and cooperage processes.

4.8.6.2 Barrel Cleaning and Sanitising

Thorough cleanliness and sanitisation of the winery's barrel stock is vital; if used barrels are purchased, these should be screened, and immediate cleaning and sanitisation undertaken upon arrival at the winery. However, when ageing of component wines in second fill, third fill, or older barrels is required, ideally these should preferably be from the winery's carefully maintained stocks that have had longitudinal monitoring. Any used barrels that show even the smallest sign of Brett characteristics should be destroyed away from the cellar.

Researchers have found Brett as deep as 8 mm in the wood of barrels [90, 91]. Much pioneering work into Brett in oak barrels (and also the use of Chitosan to prevent growth of the yeast) has been undertaken at the School of Food Science at Washington State University by Zachary Cartwright et al. [92], who confirmed deep Brett penetration in staves. The yeast can be particularly prevalent in areas where SO_2 contact is limited, for example, around the bunghole [93] in the grooves and on the sides of the staves [91]. Brett can form sanitation-resistant biofilm structures between the wood fibres.

4.8.6.3 Methods of Barrel Cleaning

There are several barrel cleaning methods, each with merits and demerits:

- *Cold water and hot water rinses followed by sulfurisation*: A thorough rinsing with cold water should be followed by three 70 °C (158 °F) water rinses and sanitising with SO_2 gas or filling with aqueous SO_2 at a concentration of 200 mg/l. An alternative method recommended by the AWRI is spraying with high-pressure (100–3000 psi) cold water for three minutes followed by hot water 60–82 °C rinsing for three to five minutes [93].
- *Steam and hot water, under pressure, followed by sulfurisation*: A thorough rinsing with cold water should be followed by rinsing with hot water at 70 °C (158 °F) and then a low-pressure steam treatment for 12 minutes. Unfortunately this is a fairly slow process. The 2018 study by Cartwright et al. [92] reported that steaming staves for 9–12 minutes eliminated yeast populations, with variances depending upon oak species and toasting level. After the nine-minute use of steam, Brett recovery is prevented in staves at a depth of 4 mm, and after 12 minutes, at a depth of 8 mm.
- *Filling with very hot water*: Barrels should be rinsed with cold water and then filled with hot water. The AWRI recommends heat as the most effective sanitisation method for barrels and specifies filling barrels with water at a temperature of at least 85 °C (185 °F) and keeping the water inside for at least 15 minutes [93]. It is easy to see that, keeping in mind that a standard barrique has a capacity of 225 l and ideally fresh water should be used for each filling, copious amounts of hot water are required! A study published in 2019 revealed that water heated to 80 °C eradicated *Brettanomyces* at depths of 5–9 mm after 20 minutes, although it should be noted that this study was undertaken using cubes of wood taken from barrels [94].
- *Ozone gas or aqueous ozone*: Ozone is a surface cleaning agent and does not penetrate into wood. However, it does degrade surface microbial biofilms. Care must be taken to avoid ozone contact with materials made from, or containing rubber, as they will be rapidly degraded. Prior to cleaning with ozone gas, barrels should be given a high-pressure cold water wash and blasted with a stream of hot water at 70 °C (158 °F) or above. They should then be rinsed to remove all debris and cooled. As an alternative, ozone may be dissolved in water, at a concentration of 2–2.5 mg/l, by using a bubble diffuser, venturi injector, or static mixer. Ideally, the water temperature at the time of the ozone addition should be less than 10 °C (50 °F) to maximise solubility, and the water should be filtered and

deionised prior to the dissolving of the gas. Ozone treatments achieve a 99.99% removal of Brett cells but as previously stated only on, or close to, the surface of the wood.

- *The use of microwaves*: Sanitation incorporating the use of microwaves has been used with some success in France. The procedure is a hot water pressure clean, rinsing with proprietary chemicals followed by the use of microwaves. A 2013 study found that populations of *Brettanomyces* were reduced by 35–67%, following a three-minute microwave treatment time [95].
- *The use of ultrasound*: Ultrasound treatment of barrels can remove over 90% of viable *Brettanomyces* cells to a depth of 2–4 mm in the barrel staves [96]. The procedure requires expensive equipment but can be undertaken by specialist contractors.
- *Sanitising with sulfur dioxide before use/reuse*: Whatever the method of barrel cleaning, sulfur dioxide (SO_2) in gaseous form should be used for sanitisation. The burning of sulfur candles is a traditional but effective way of achieving this and is recommended for new and used barrels. A study by Pascal Chatonnet et al. found that the burning of candles far more effective than adding SO_2 in solution [97]. Upon examining a red wine after six months of ageing in new one-year-old and two-year-old barrels, the ethyl phenols were considerably less when candles had been used, and the remaining free SO_2 was higher. In a later study by Solis et al., a three- and six-week treatment of burning sulfur candles in barrels that had previously been contaminated with *Brettanomyces* was found to decrease the yeast and other microbes to undetectable levels in most of the barrels [98]. To sanitise a 225 l barrel, at least 7 g of gaseous SO_2 is required [99].

4.8.6.4 Topping Up Barrels

Filled barrels should be regularly checked for ullage using a sanitised barrel thief and topped up as necessary. Even a tiny amount of oxygen encourages the growth of Brett and other spoilage organisms. However, the topping operations will result in a total oxygen ingress of between 0.2 and 1 mg/l per annum, depending upon the temperature, the method, and the frequency of the topping procedures. The wine used for topping must not have been stored on ullage and should be carefully nosed, checking for any trace of ethyl phenols or isovaleric acid. The risks of cross-contamination cannot be overstated – one barrique of 'rogue' wine used for topping could contaminate over 100 others! All transfer equipment should be sanitised prior to use. If there is any doubt about the topping wine being free from significant colonies of viable Brett cells, it should be sterile-filtered prior to use. The SO_2 level of each barrel should be adjusted following topping.

4.8.6.5 Temperature of the Barrel Store

The temperature or the barrel store should be controlled – ideally to maxima 12 °C (53 °F)–14 °C (57 °F). In producing regions where underground cellars are possible, such temperatures may be naturally maintained, but where above ground storage is the norm, as is the case in the chais of Bordeaux (the water table is too high for underground cellars), a temperature control system is necessary. A 2019 research paper by Alice Cibrario et al. [100] showed that (in the Bordeaux region) volatile phenols appear both earlier and faster in July, August, and September, and accordingly closer control of active yeast populations and closer monitoring of ageing wines is vital in the summer.

4.8.7 Maintaining Appropriate Molecular Sulfur Dioxide Levels

4.8.7.1 The Sulfur Dioxide Quandary

Individual strains of *Brettanomyces* vary considerably, perhaps threefold, in their tolerance to SO_2, the use of which remains the main chemical means of inhibiting growth. In Section 4.8.5.6, I suggested that immediately following completion of the MLF, the level of molecular SO_2 should be adjusted to between 0.5 and 0.625 mg/l. It should be noted that there is considerable dissention as to the level of molecular SO_2 required to limit or stop *Brettanomyces* growth, with various figures as low as 0.25 mg/l and as high as 1.25 mg/l having been shown in the studies [88]. The effectiveness of any concentration of molecular SO_2 is influenced by oxygen availability [101]. However, before discussing the molecular SO_2 concentrations that may be appropriate, perhaps we should consider another viewpoint. *Brettanomyces* is a survivalist that can thrive in hostile environments, including relatively high levels of ethanol, up to 15% abv, low levels of nutrients, anaerobic conditions, and levels of SO_2 that inhibit most 'natural' yeasts and bacteria. It is competitive, and the simultaneous presence of other yeasts can limit its development, although these are most likely to be present before and during the AF. As shown in Section 4.5.2, a wine that has completed its AF may already have a population of *B. bruxellensis* as high as 10^4–10^6 CFU/ml. As there are many strains that are tolerant to SO_2 at levels well above those that inhibit competition, adding high levels of SO_2 pre-fermentation may allow Brett to grow unhindered. Clark Smith states that 'Brett is like a hospital disease, fostered by the very sanitation measures designed to supress it' [29]. He also notes that 'sulfite-free producers report lower levels of *Brettanomyces* in their wines'. However, my experience has generally been rather different. Visits to some producers' tables at the 'Raw Wine Fair' and the 'Real Wine Fair' in London have often taken me back to those childhood farm experiences. Of course, desirable bacteria that can speed a wine into and through the MLF (a dangerous time for growth of *Brettanomyces*) are also inhibited by SO_2, and there is constant pressure to reduce total SO_2 content from a health perspective.

4.8.7.2 Appropriate Levels of Molecular Sulfur Dioxide

Maintaining a level of molecular SO_2 above 0.625 mg/l is recommended to block Brett activity, although 0.4 mg/l or 0.5 mg/l may be sufficient to limit its development. Many winemakers, considering only free SO_2, believe that a level above 30 mg/l will be adequate. However, it is only molecular SO_2 that has antimicrobial properties, and the amount of molecular SO_2 is just a fraction of the free SO_2, varying according to the wine's pH. Thus 30 mg/l free SO_2 might prove sufficient for wines with a pH of up to 3.5, but if it is as high as 3.8, then the amount of molecular SO_2 is effectively halved, and a much higher level of free SO_2 (60 mg/l) is required. The same amount of free sulfur dioxide is 10 times more effective against microbial activity at a pH of 3 than at pH 4. Conterno et al. [102] found that 17 of 35 *Brettanomyces* strains examined were tolerant to SO_2 at 30 mg/l in wine of pH 3.4, and there are numerous other reports of 30 mg/l being ineffective. Many winemakers underestimate the diminution of molecular SO_2 with escalating pH levels; for example, for a wine with 12.5% abv and a pH of 3.3, the percentage of free SO_2 that is molecular is 3%. This drops to 1% for a similar wine with a pH of 3.8. Accordingly, for a wine at a pH of 3.3 with a 25 mg/l concentration of free SO_2, the molecular SO_2 is 0.75 mg/l; for a wine with a pH of 3.8, it is just 0.25 mg/l, which is insufficient to prevent *Brettanomyces* growth. The role of sulfur dioxide in wine is further discussed in Chapter 6, and Appendix A shows the level of free SO_2 required to give molecular sulfur levels of 0.5 mg/l, 0.625 mg/l, and 0.8 mg/l at differing pH levels. It should be noted that the precise level required varies slightly according to the temperature of the wine.

The diminution of concentrations of molecular SO_2 during storage should also not be underestimated. Chatonnet et al. [97] showed that in a red wine of pH 3.65, initial doses of SO_2 of 15, 25, 30, and 35 mg/l had dwindled to 6, 11, 10, and 15 mg/l respectively, after four months of ageing in barrels. In a Joana Coulon et al. study [72], the levels of molecular SO_2 of each of the wines analysed dropped considerably after 80 days in bottle: in one example, a wine with an initial pH of 3.74 and molecular SO_2 of 0.52 mg/l fell to 0.19 mg/l after 80 days of bottle storage. In the 130 days from the beginning of the experiment, 1740 μg/l of ethyl phenols were produced in this wine, revealing the initial concentration of SO_2 to be totally ineffective against *Brettanomyces* growth.

Some authors sate that levels higher than 0.625 mg/l are required for safety, and it should be noted that the level of phenolics can affect SO_2 efficacy. Winery trials as part of the 2008 study by Barata et al. [73] showed that *D. bruxellensis* growth was only prevented in the presence of about 40 mg/l of free sulfur dioxide in a dry red wine, with 13.8% abv and a pH of 3.42, which was matured in oak barrels. This figure corresponds to about 1 mg/l of molecular SO_2. However, whilst individual strains of *Brettanomyces* vary considerably in their tolerance of SO_2, and there is much discussion about the existence of particularly resistant strains [18], a level of molecular SO_2 at a minimum of 0.625 mg/l may be regarded as adequate for a most strains. It should be noted that the addition of SO_2 can make *Brettanomyces* cells physically smaller [103, 104], possibly indicating that they have entered a viable but non-culturable (VBNC) state. In a VBNC state, the cells are viable but do not grow on culture medium, resulting from stress response to an environmental factor [7]. Only if the *Brettanomyces* cells are viable but culturable will there be risk of volatile phenols being produced. The VBNC state is reversible: the cells are able to resuscitate subsequent to the removal of the stress factor [7]. SO_2 additions, in appropriate quantities, have a rapid effect, and *Brettanomyces* cells can lose viability in less than six hours. The topic of VBNC state of *Brettanomyces* remains controversial, and the AWRI has long questioned its existence in wine, contrary to the results of work done elsewhere. In their 2017 report to the Australian Grape and Wine Authority Chris Curtin, Anthony Borneman and Paul Grbin of AWRI stated that, contrary to work done elsewhere, their studies did not find any evidence that *Brettanomyces* cells enter a VBNC state in response to sulfites [18].

The dichotomy between maintaining low SO_2 levels and minimising microbial growth is a constant concern. Speaking of the historic Brett problems that impacted upon Henschke Wines, including Hill of Grace, Stephen Henschke noted, 'We were trying to reduce our sulfur regime during the 1990s and we didn't understand at the time that this increased the risk of *Brettanomyces* … As an industry, we know much more about Brett now than we did a decade ago. We're now much more stringent with our barrel management and we sterile filter prior to bottling to eliminate the risk' [39]. However, a great many winemakers avoid sterile filtration, particularly for red wines, not only before bottling but also at any stage of the production process, believing that such treatment has numerous negative impacts, including reduction in fruitiness, stripping of complexity, modification of structure, and creation of a hard mouthfeel.

Of course, as discussed in Chapter 6, high SO_2 concentrations can result in a negative organoleptic impact and/or a risk of the wine exceeding legal limits. It is the total SO_2 content that is limited by law in all major markets, but much of the SO_2 in wine becomes bound and has no antimicrobial (or anti-oxidative) effect. It is generally better to add sulfur in a small number of relatively large doses, rather than a large number of small doses: this strategy leads to a higher ratio of free to total SO_2, increasing the inhibitive and preservative properties for the same amount of total sulfur.

A simple way to detect if Brett is growing in barrels is to regularly monitor the relationship between free and total SO_2. If the percentage of free SO_2 decreases continuously, this is an almost certain indication that Brett (or other microbes) are growing and converting the free SO_2 to the bound form.

4.8.8 Racking

From a microbial stabilisation perspective, rackings should take place regularly – perhaps every three months [105]. Until recently, it was commonplace in Bordeaux and many other regions to rack the red wines four times in the first year of barrel maturation and perhaps twice in the second. Many properties have reduced the number of rackings as 'minimal handling' has become popular and in order to reduce oxygen uptake. There are also cost savings in labour and the avoidance of wine losses during the operation. As some free SO_2 will be lost at each racking (approximately 5–12 mg/l), it is important to ensure the remaining levels are adequate, and additions should be made to the receiving barrel as required. The loss is partially due to the increase in DO – 1 mg of DO/l consumes about 2.5 mg/l of SO_2, and a standard racking, without deliberate aeration, will result in an increase in DO of 2–5 mg/l.

Racking presents a convenient opportunity to examine the wine visually and by nosing and tasting. It should be noted that the absence of any surface yeast film should not be taken as an indication that Brett is not present – the yeast grows in the body of wine, and surface growth is very rare.

4.8.9 Fining

Fining can help prevent the formation of Brett colonies and reduce the level already present in wine. It is perhaps a less controversial topic than filtration, but many winemakers feel that minimal or even no fining leads to greater aromatic complexity and avoids unwanted impact on texture and structure. Sebastian Beaumont, winemaker in the Bot River area in South Africa, said that fining certainly negates aromas and affects mouthfeel and added (tongue in cheek), 'I don't fine my wines – they are fine enough already' [personal discussion].

There are several groups of fining agents. These may be classified as protein, mineral, polysaccharide, and 'other'; protein agents are perhaps most appropriate. The use of fining proteins can reduce *Brettanomyces* populations by a factor of 40–2000 by flocculation [19, 64]. Protein fining agents include isinglass, egg albumin, gelatine, and casein (often in the form of potassium caseinate). Wine can be fined at any stage after MLF. Many producers fine wines before they are put into barrel, and the operation should be immediately undertaken if organoleptic assessment or laboratory analysis indicates any significant presence of *Brettanomyces* or its related compounds. Subsequent fining with egg whites (albumen) or gelatine might follow barrel ageing. The wine should be analysed for viable cells post fining – a level over 300 CFU/ml may give cause for concern, and further sulfurisation may be advisable. Filtration should be considered if the count remains very high – perhaps in excess of 1000 CFU/ml.

4.8.10   Filtration

Filtration is always a controversial topic, especially when discussing fine wines. Fine, and particularly sterile, filtration can impact upon organoleptic characteristics, including the reduction and/or modification of aromas together with a deterioration of the colloid structure, impacting upon the viscosity, body, and texture of a wine. When filtered through an absolute, sterile (0.45 μm) medium, the alterations can be pronounced. As Clark Smith noted, 'The benefits of good structure (profundity, aromatic integration and graceful longevity) appear to be lost by sterile filtration despite the fact that no tannin material may be retained by the filter'. [29]. Conversely, the AWRI's view is emphatic: 'While some winemakers seem hesitant to filter red wines, it is the AWRI's position that a well performed filtration is a much better option than taking the risk of post-bottling microbial spoilage' [106].

Filtration, as an 'insurance' measure to remove viable *Brettanomyces* cells, may be undertaken post MLF, before wines are put into barrels or tank for maturation, pre-blending, and pre-bottling. Interim filtrations may be generally regarded as unnecessary (and possibly damaging) unless the wines are showing a high cell count – 1000 CFU/ml or more. However, filtration immediately prior to barrelling as a general microbial control tool is gaining in popularity with some, mostly large, producers. It might particularly be considered a sensible precaution to filter wines used for topping barrels. Pre-bottling filtration to ensure viable Brett cells are not present in the finished product may be regarded as a sensible precaution; however, the question remains as to the desirable pore size. As with all cellar operations, care should be taken to avoid oxygen pick-up during filtration. It is recommended that lines and filters are purged with nitrogen or other inert gas.

Filtration as one of the stages of remedial treatments for *Brettanomyces* in wine is discussed in Section 4.10.1.

4.8.11   Bottling

Prior to bottling, analysis for viable cells should be undertaken as discussed in Section 4.9.1. Particular care should be taken with the cleaning and sanitation of bottling lines: there is anecdotal evidence of contamination taking place at bottling. In the case of small producers who employ the service of mobile bottlers, much is beyond their control, but assurances as to the sanitisation of the line should be sought from the contractor. Sulfur dioxide levels should be adjusted as discussed in Section 4.8.7. Tanks used for filling should be purged with inert gas and blanketed during filling. Levels of oxygen in the bottle headspace should be minimised by vacuum evacuation, and the DO in the wine analysed: 1.25 mg/l is the maximum recommended for red wines.

4.8.12 Storage

Bottled wines should always be stored in cool conditions. The ideal cellar temperature to help achieve an even maturation of wine and minimise the risk of defects due to inappropriate storage is generally regarded as 13–15 °C (55–59 °F), but, to further reduce the risk of the development of *Brettanomyces*, 12 °C (53 °F) or even a degree or two cooler is preferable [107]. This will not necessarily stop growth, but will slow the rate substantially. Storage below 10 °C (50 °F) may be regarded as an efficient measure to prevent the production of 4-EP [1]. In my experience, on most of the occasions that I have found pronounced ethyl phenols in relatively young red wines in restaurants and elsewhere, these have been stored at far above the ideal temperature, even often above 18 °C (64 °F). *Brettanomyces* can use ethanol as a carbon source, so even wines bottled without any residual sugar may be at risk of development of Brett characteristics if viable cells have entered the bottle. Interestingly in 2006, Conterno et al. [102] reported that 15 of 35 strains studied were capable of growth (in a synthetic wine medium) at temperatures below 10 °C.

4.9 Laboratory Analysis for *Brettanomyces* and Volatile Phenols

4.9.1 When Analysis Should Be Undertaken

Checking for signs of Brett-related compounds should be part the regular organoleptic analysis during a wine's production cycle. Laboratory analysis should also be carried out at least three stages in the production process:

- After AF;
- After MLF;
- Prior to bottling.

However, it is preferable to analyse at regular intervals during and following barrel maturation, and particularly before using wine for blending or topping-up purposes, although in practice this is rarely undertaken unless there are indications of a problem. For wine in barrel, analysis might be undertaken every three months, unless there is a history of Brett contamination. If the concentration of 4-EP exceeds 150 µg/l at any analysis, or the viable cell count exceeds 1000 CFU/ml, then immediate action must be taken, such as filtering the wine (1.2 µm or less) and sulfiting to at least 0.625 g/l molecular SO_2. When preparing wines for bottling, the criteria are perhaps more stringent, and if the cell counts exceeds 800 CFU/ml, a very fine or sterilising filtration is recommended. At lower counts, it is acceptable to control viable cells by the addition of SO_2 – 1 mg/l of molecular SO_2. In this case, bottling must be technically correct, and DO should be reduced to the lowest practical level. Otherwise, a sterile filtration is recommended, or, as an alternative, a soft thermal treatment of the wine to destroy viable cells can be used [89]. It should be kept in mind that if a cellular structured closure (such as a natural cork) is used, oxygen is contained within the closure.

4.9.2 Analysis for Viable Yeast Cells

As with all analysis, care should be taken regarding the locations from which the sample is extracted: the lees at the bottom of the barrel or tank (if unstirred) will contain by far the greatest concentration of microbes, including *Brettanomyces*. A sample taken using a barrel thief from the top or even the centre of the barrel leads to false analytical results – samples should always be taken from the lowest accessible position in the barrel or vat and not within a week or so of adding SO_2 as populations may be overstated, as non-viable cells may be included.

There are many proven methods of analysis for viable yeast cells, and in winery circumstances, the choice will depend upon the equipment available, having trained staff, and cost.

4.9.2.1 Plate Cultures

The growing of cell structures on nutritive gel plates (Petri dishes) followed by morphology under a microscope is the traditional method and is still the most widely used within the winery. It is cheap, requires no specialised equipment, and the consumables simply comprise a suitable medium with a specific antibiotic. However, the growth on Petri dishes is a relatively slow process, taking at least 5 days and often 7–10 days or sometimes even longer. When contamination is indicated, this time lapse may have permitted preventable growth and delayed the rapid implementation of remedial steps. Although it is generally a reliable method for identifying the cells by their morphology, and also the level of concentration, plating can result in both false positives due to confusion with other yeasts possessing similar growth characteristics, particularly *P. guilliermondii*, and false negatives, largely due to low cell viability subsequent to recent SO_2 additions. The most widely used sensitive and selective growth gel is *Dekkera/Brettanomyces* differential medium (DBDM) that uniquely amongst differential media contains *p*-coumaric acid, as well as cycloheximide, bromocresol green (a pH indicator), and ethanol. There are some Brett strains that do not grow well or at all on DBDM, but *P. guilliermondii* does grow on it [108]. There are several other available media: any chosen should contain cycloheximide, a naturally occurring fungicide that inhibits most yeasts except *Brettanomyces*. However, *Kloeckera* are resistant to cycloheximide, so it is necessary to take a first reading at three days, which will reveal non-*Brettanomyces* yeasts, followed by a second reading at seven to eight days, which will display *Brettanomyces*. Individual strains of *B. bruxellensis* cannot be identified from plate cultures, but at species level, the visual identification is reasonably straightforward. An illustration of *B. bruxellensis* is shown in Figure 4.1 and *S. cerevisiae* in Figure 4.2.

Small wineries without laboratory facilities (or even a microscope) may use a variation of plate culturing in the form of a simple test kit such as 'Sniff Brett'. This comprises a medium that contains all the nutrients for the growth of yeasts and cycloheximide (to inhibit non-*Brettanomyces* yeasts), together with a large amount of *p*-coumaric acid (the precursor of 4-EP). *Kloeckera* can grow in this environment (this is the only other species of yeast resistant to cycloheximide) but do not produce volatile phenols. The test is undertaken by inoculating the medium with 20 ml of wine, holding for several days

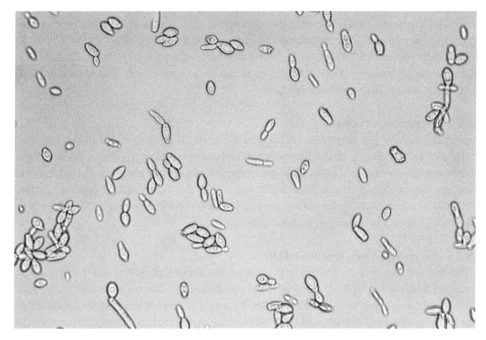

Figure 4.1   *Brettanomyces bruxellensis* x 400. Source: Courtesy of White Labs.

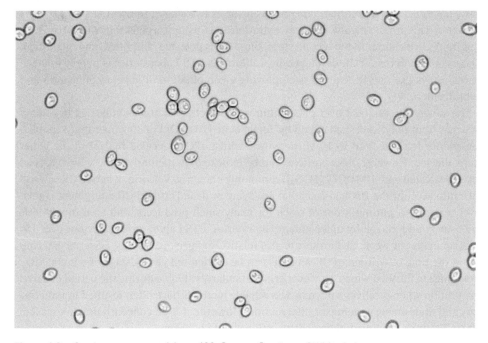

Figure 4.2   *Saccharomyces cerevisiae* x 400. Source: Courtesy of White Labs.

preferably at 30 °C (86 °F) and then checking every day for the smell of volatile phenols. There is a correspondence between the number of days before the appearance of the odour and the amount of *Brettanomyces* present in the wine: 2 days (at 30 °C) for 100 000–1 million cells per ml, 4 days for 10 000–50 000 cells m/l, 6 days for 100–1000 cells m/l, and 10 days for approximately 1 cell/ml. The kit is relatively inexpensive, and in case of high contamination, results are reasonably fast.

4.9.2.2 Fluorescence Microscopy

Simple microscopy for direct visualisation of the cells is unreliable due to the variable morphology and cell size of the various strains of *Brettanomyces*. However, fluorescence microscopy enables reliable immediate microscopic evaluation of samples. This technique uses fluorescent dyes to mark both viable and non-viable cells, and cell counts can be undertaken. Distinguishing between yeast and bacteria is relatively simple. The process is sometimes referred to as epifluorescence counts.

4.9.2.3 Polymerase Chain Reaction (qPCR)

A fast and reliable method for quantification of *Brettanomyces* cells is quantitative polymerase chain reaction (qPCR) also known as real-time PCR. This process uses molecular biology techniques that amplify and monitor the amplification of targeted ribosomal RNA and DNA, in this case the DNA of Brett. This is much faster than plate cultures, usually taking approximately 3–4 hours contrasted with the 7–10 days that typical plate cultures take. qPCR can detect *Brettanomyces* at a level as low as two cells/ml. In addition to *Brettanomyces*, the presence of the spoilage yeast *Zygosaccharomyces bailii* can be simultaneously identified and quantified. The process involves two stages. In the first stage, DNA is extracted. This process involves the concentration and washing of cells, followed by cell lysis and DNA purification. In the second stage, the yeast is detected, amplified, and quantified. The use of 'Scorpions' PCR developed by California's ETS Laboratories is hugely reliable. These assays use two PCR primers employing a combination of gene amplification and hybridization.

The superiority of qPCR over plate cultures in detecting Brett was indicated in a study of wines from Italy's Abruzzo region by Tofalo et al. [109]. Out of 30 wines analysed, plating counts detected Brett in 10 of the wines, whilst qPCR revealed that 22 of the wines were affected. However, false positives can be indicated, particularly if the wine has been recently treated with DMDC ($C_4H_6O_5$), commonly known as Velcorin, when qPCR can still take into account the DNA in cells that are dying or dead [110]. qPCR equipment is relatively expensive, putting it out of reach for many small producers, and accordingly tests are usually undertaken by independent laboratories. ETS Laboratories carry out over 150 PCR analyses per week. Such tests are also relatively expensive – about US $120 per sample at the time of writing. qPCR analysis can be carried out at any stage of winemaking, from must to finished wines. However, it is particularly valuable during the period of barrel maturation when positives shown by less reliable methods have often resulted in unnecessary and undesirable winemaker intervention. However, if Brett concentration in excess of 1000 CFU/ml is identified, then control action should be taken.

Whilst there are simple qPCR test kits available, such as 'VINEO™ *Brettanomy*test PCR Kit', these must be used in conjunction with a DNA extracting kit, and there are several necessary pieces of equipment including a centrifuge, real-time PCR detection system, and reaction block. Other simple systems suitable for use within the winery are 'vinoBRETT' and GEN-IAL® QuickGEN test kits. These also require a centrifuge and thermocycler.

4.9.2.4 Flow Cytometry and Fluorescence In Situ Hybridisation

Flow cytometry coupled with fluorescence in situ hybridisation (FCM-FISH) probes with markers for the 26S ribosomal RBA of Brett, or with antibody antigen type markers, gives fast (between 4 and 72 hours) and accurate results at concentrations above 100 CFU/ml. This is a culture-independent method that indicates both the quantity and quality of the cells. However, as with qPCR, flow cytometry requires expensive equipment and trained operators. For that reason, it is generally only used by specialist laboratories.

4.9.2.5 Raman Spectroscopy

It is possible to use Raman spectroscopy and chemometrics to identify *Brettanomyces* in a medium. Using these techniques, Rodriguez et al. [111] reported an accuracy of 93.3% at species level and a high accuracy at strain level.

4.9.2.6 Experimental Technologies for Detection of *Brettanomyces*

There are several experimental technologies for the determination and quantification of *Brettanomyces* cells. Biosensors, either impedance or amperometric based, have shown promising results. With impedance-based sensors, spoilage yeast adhere to and grow onto the surface of the biosensor that changes the measured impedance. These biosensors can detect the presence of spoilage yeasts (approximately 100 CFU/ml), with their sensitivity increasing by 10% with the use of immobilised anti-*Brettanomyces* antibodies [112].

4.9.3 Analysis for Volatile Phenols Using Gas Chromatography with Detection by Mass Spectroscopy (GC-MS)

The methods described in Section 4.9.2 will generally indicate the quantity and perhaps the quality of *Brettanomyces* cells, but further analysis is required to determine the produced level of the individual volatile phenols. Gas chromatography with detection by mass spectroscopy (GC-MS) is the most widely used method for determining volatile phenols. The compounds are generally extracted from the sample by solid-phase micro-extraction (SPME) (the most commonly used method), solid-phase extraction (SPE), or stir bar sorptive extraction (SBSE). Using GC-MS, 4-EP and/or 4-EG can be detected at a level of 4 µg/l or above, well below sensory detection thresholds. GC-MS will also detect the presence of 4-vinylphenol and 4-vinylguaiacol.

4.10 Treatment of Affected Wines

There are numerous treatments available for wines, pre-bottling, that have high numbers of *Brettanomyces* CFUs and/or unacceptably high concentration of Brett-related compounds. A complete treatment is a two-stage operation: the removal of the viable *Brettanomyces* cells and removal of the odorous synthesised compounds, including 4-EP and 4-EG. There are several methods of achieving these ends.

4.10.1 Reduction or Removal of Brettanomyces Cells

4.10.1.1 Racking

Racking can result in significant reduction of Brettanomyces population levels: the cells are invariably concentrated in the lees. Care must be taken to minimise oxygen pick-up during the operation. Jerome Fious of Domaine Faiveley noted that racking is normally enough to remedy modest viable Brettanomyces cell counts but said that if these exceed 1000 CFU/ml, treatment will be considered. This will definitely take place as a count above 1500 CFU/ml, usually by the addition of chitosan [personal discussion]. However, racking alone is unreliable, and unacceptably high population levels may remain in the wine.

4.10.1.2 Fining

The use of multiple fining agents may considerably reduce populations, and intense fining can remove the yeasts almost entirely [113].

4.10.1.3 Filtration

Filtration can effectively remove Brettanomyces yeasts from an impacted wine, but the winemaker is reminded that the process will not remove the volatile phenols responsible for the unwanted odours. Accordingly filtered wines must undergo further treatment to repair the organoleptic characteristics.

Although often considered as a pre-bottling operation, interim filtrations are advisable if Brett contaminations are discovered. Filtering of wine upon initial transfer to barrels (providing that the MLF has been completed) might be considered as a biological control tool but, as noted in Section 4.8.10, is only necessary if the Brett population is above 1000 CFU/ml. The process may need to be a multistage operation as the avoidance of clogging of dead-end sterile filtration material is essential. Prior to first filtration, the wine should be allowed to settle for at least two weeks, racked and fined if necessary, before being racked for the second time. It may then be filtered, initially through a depth (diatomaceous earth or sheet) filter or by cross-flow (tangential) filtration. Sterile membrane filtration may then be undertaken.

Brettanomyces cells are usually between 2 and 7 μm in size, but they can be smaller particularly depending upon the strain, and the cell size is likely to be reduced when the yeast is stressed. Accordingly, many strains of the yeast will be removed by filtration using 1.2 or 1 μm material, but others are only removed with 0.8 μm [114], 0.6 μm, or even 0.45 μm (sterile) material. When wine is adjusted to molecular sulfur levels of at least 0.5 mg/l and ideally 0.625 mg/l or above prior to filtration, 1.2 or 1 μm material should, in most cases, result in a stable wine with non-detectable Brettanomyces cells. Sterile filtration, using a medium with a pore size no greater than 0.45 μm, is almost always effective in removing Brettanomyces [113, 115], although it is possible that cells in a VBNC state may pass still through this size. This may possibly happen when the cells are stressed, as is the case following sulfur additions: the yeasts may become smaller or elongated. Complete assurance would be obtained with 0.22 μm material, but there is little doubt that this can result in serious organoleptic modification. Sterile filtration is most inadvisable at intermediate production stages.

However, the composition of the filter material is pertinent, as even with larger pore sizes some materials can be effective and there is less impact upon the colloidal structure of the wine. In 2017, in a study by Duarte et al. [116], polypropylene filters showed poor efficacy – *B. bruxellensis* colonies were too numerous to count following filtration. Using polyethersulfone (PES) filters PES 1.0 µm, PES 0.65 µm, and PES 0.45 µm (two lots and respective duplicates), no cells were detected in any of the filtered wines analysed. Incubations using DBDM confirmed the absence of growth. The PES filters showed high *B. bruxellensis* removal efficacy in all micron rating tested. Similar efficacy was achieved for X-grade borosilicate glass microfibre filter material. Polypropylene, V-grade borosilicate glass microfibre was also shown to offer less retention.

Filter presses are particularly effective and can eliminate of 99.5% of *Brettanomyces* CFUs. A reduction from 110 000 CFU/ml of *Brettanomyces* to just 440 CFU/ml (99.96%) in a wine following the use of a filter press has been noted by Bio-Rad Laboratories [117]. Of course, it should be remembered that any filtration is only a transitory action, and the process will not protect the wine against subsequent *Brettanomyces* contamination and development [72].

4.10.1.4   Heat Treatments

- *Heat treatment*: Heat treatment can eliminate 94% Brett CFU/m/l. Bio-Rad Laboratories have noted a reduction from 150 000 to 9300 CFU/ml (a reduction of 93.8%) following heat treatment of (basically) the same wine in the same winery referred to in the previous paragraph [117]. However, this number of remaining CFUs is still high, leaving the treated wine at risk of future volatile phenol development. Heat treatment should only be used immediately prior to bottling.
- *Flash pasteurisation*: Heating wine to between 75 °C (167 °F) and 80 °C (176 °F) and holding for 30 seconds will kill all *Brettanomyces* cells. Of course, other microbes are also inhibited.
- *Pasteurisation*: Pasteurisation, which involves heating the wine to 60 °C and holding for several minutes, will kill all *Brettanomyces* cells. Other microbes are also inhibited.

4.10.1.5   Chemical Treatments

- *Dimethyl dicarbonate (Velcorin)*: DMDC ($C_4H_6O_5$) inhibits the enzymes involved in glycolysis. The chemical is marketed under the trade name Velcorin. *B. bruxellensis* will not survive at DMDC levels above 150 mg/l (a study by Renouf et al. found this was the concentration required to inhibit 10 strains [118]), but the recommended dosage is 200 mg/l, which is the EU and International Organisation of Vine and Wine (OIV) limit, for wines. It should be noted that DMDC is quickly hydrolysed into methanol and carbon dioxide (CO_2) [19]. It will degrade in six to eight hours, and there will be no protection subsequent to treatment, so it is important to maintain appropriate SO_2 levels. It is permitted by OIV Regulations [119], and its use as a preservative (E242) is permitted for wine in the European Union (EU) under Regulation 2019/934 [120] (which amends previous regulations). It is approved as an additive for wine in the United States by FDA 21 Code of Federal Regulations, providing 'the viable microbial load has been reduced to 500 microorganisms per millilitre (500 CFU/ml) or less by current good manufacturing practices such as heat treatment, filtration, or other technologies prior to the use of dimethyl dicarbonate' [121]. It is also approved as a *processing aid* for use in wine in Australia and New Zealand under Food Standard Code 1.3.3. [122]. DMDC must not be present in the wine when sold. As DMDC must be completely homogenised into wine, and the dosing apparatus is expensive, its use 'in-house' is perhaps out of reach for many small wineries, although mobile systems are available in some major wine producing areas. Yeasts vary in their susceptibility to DMDC, and bacteria, including lactic and acetic acid bacteria, are more resistant than yeasts. Consequently, this preservative should not be regarded as a 'catch-all' or used alone [123]. DMDC will degrade in six to eight hours, and there will be no protection subsequent to treatment. Accordingly, DMDC should be used together with sulfite during wine storage or at bottling.

- *Chitosan*: Chitosan, a polysaccharide derived from chitin and isolated from *Aspergillus niger* mycelium, can prevent growth of *Brettanomyces* (together with that of some bacteria). Chitin is a naturally occurring polymer found in crab shells and the skeletons of some other crustaceans and insects, as well as in the cell walls of fungi. However, only the fungal form derived from *A. niger* using a process patented by the company *KitoZyme* is permitted by OIV (since July 2009) and EU (since December 2010) to be used in wine. The action of chitosan would appear to be both biological and physical, with an alteration of the yeast membrane, leading to cell death [124]. The product is marketed by Lallemand as 'No Brett Inside™'. An image of chitosan, magnified 600 times, is shown as Figure 4.3. The legal limit of addition, for this purpose, under EU (and OIV) regulations is 10 g/hl (0.1 g/l) (100 mg/l). The usual rate of application is 4 g/hl, and this is best added incrementally. Chitosan is generally effective at considerably reducing the number of viable cells, but some strains of *Brettanomyces* are less sensitive [18]. As the yeasts may grow again after treatment a racking [124], immediate addition of SO_2, followed by regular monitoring and maintaining an adequate molecular SO_2 level, is essential [125].
- *Ionised radiation*: The use of ionised radiation to reduce viable *Brettanomyces* cells has shown some success on an experimental basis, but the treatment would be expensive, requiring very specialised equipment.

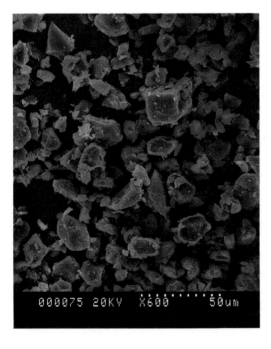

Figure 4.3 Chitosan x 600. Source: Courtesy of C.G. Edwards, Washington State University, Pullman, WA, USA.

4.10.2 Reduction in Volatile Phenols

There are several possible methods of reducing the level of volatile phenols, but each has a sensory impact of each upon desirable compounds to a greater or lesser extent:

- *Reverse osmosis*: Reverse osmosis (RO) is followed by processing the permeate through a column containing hydrophobic adsorptive resin. A reduction in 4-EP and 4-EG levels of up to 25% per pass can be achieved. In a study by Ugarte et al., the technique has been found to be successful in reducing to acceptable levels both 4-EP and 4-EG from a *Brettanomyces*-affected wine [126]. The three-hour process comprised a two-stage integrated process, a membrane with cross-flow filtration, and a hydrophobic absorbent resin (Amberlite XAD-16 HP). The reduction in the 4-EP and 4-EG ranged from 50% to 70%, but there was also a significant loss of desirable aroma compounds including ethyl and methyl vanillate and other esters. The process is not authorised under EU wine regulations or by the OIV International Code of Oenological Practices, although the oenological equipment manufacturer *Bucher Vaslin* was permitted to conduct trials (in association with *Inter Rhône*). It must also be stressed that RO and ion exchange systems will remove or reduce the levels of 4-EP and 4-EG, but will not remove *Brettanomyces* cells, so the wine must be sterile-filtered, or more volatile phenols may be metabolised by the yeast.
- *Yeast hulls*: It is possible to reduce the levels of volatile phenols by the addition of the inactivated hulls of *S. cerevisiae*. The hulls absorb phenols, but reduction in concentration is generally less than 50%. The organoleptic impact upon desirable compounds is relatively small.
- *Activated charcoal*: Treatment of wine containing volatile phenols with activated charcoal, also known as activated carbon, can to a degree be effective. Of eight fining materials tested by Milheiro et al. [19], activated carbon was by far the most effective. The challenge is getting the right medium and dose, as charcoal will remove wanted as well as unwanted flavours. Depending on the level of odours, the recommended doses are from 0.015 to 0.24 g/l of charcoal for slight off-odours and from 0.12 to 0.96 g/l for more intense off-odours. The differing characteristics of activated carbon, particularly pore size, makes a difference to the amount of colour and phenols removed, but at best a reduction of 73% 4-EP and 4-EG can be achieved [127]. Another study by Milheiro et al. showed that lower reductions were obtained when using egg albumin, about 19% decrease in the wine and a 30% decrease in the headspace concentration [128].
- *Polyvinylpolypyrrolidone (PVPP)*: Polyvinylpolypyrrolidone (PVPP) can also be used to reduce the level of ethyl phenols [128] and will also reduce any bitterness from polyphenols. The required dosage is 0.06–0.48 g/l. The compound should be mixed with a small quantity of distilled water to make a slurry before addition.

4.10.3 Future Possibilities for Physical Brett Inactivation

Research has been taking place into the application of high hydrostatic pressure (HHP), also referred to as high pressure processing (HPP), to inactivate *B. bruxellensis* [129–131]. This has proved to be successful, with complete elimination of the yeast (together with lactic and acetic acid bacteria). However, a very high pressure is required – 400 MPa for five seconds was required to inactivate the resistant strain AWRI 1499 [129, 130], and there can be a loss of phenol content. Research into the use of pulsed electric field (PEF) has produced conflicting results. In 2009, Puértolas et al. found that a PEF treatment of 186 kJ/kg at 29 kV/cm would inactivate the spoilage flora such as *Brettanomyces* and *Lactobacillus* in must or wine with a 99.9% reduction in populations [132]. However, in 2018, van Wyk et al. found the PEF process generally ineffective [131]. Conversely, Lucía González-Arenzana studying continuous PEF treatments' impact on the microbiota of red Tempranillo wines aged in oak barrels found that the treatment led to an immediate and significant microbial inactivation of most important spoilage microorganisms and a significant reduction of volatile phenols in wines with high *Brettanomyces* population [133]. However, the possible future wide-scale use of PEF as a wine treatment will perhaps depend upon the diminution of the very high costs of the PEF generator [134]:

- *Low electric current (LEC)*: The effective inactivation of *B. bruxellensis* using low electric current (LEC) has been demonstrated. As with HPP, no chemicals are involved, and accordingly the development of this technology could be of particular interest to organic and biodynamic wine producers [135]. As with HPP the utilisation of LEC could be a tool in the drive to reduce additions of SO_2.
- *Sorption by cork*: A 2015 study by Gallardo-Chacón and Karbowiak [136] claims that cork shows an active role in the sorption of volatile phenols from wine. However, as much as the cork industry might love to claim this as another reason for producers to choose natural cork as a closure, in practical terms the sorption achieved by such a small surface is inadequate, as anecdotally evidenced by the huge number of Brett-impacted wines where the closure has had little or no impact. However, the future possibility of utilising cork granules to absorb volatile phenols from bulk wines is perhaps worth exploring.

4.11 What the Future Might Hold for Microbiological Methods to Inhibit *Brettanomyces*?

Research into microbiological methods of combatting the growth of *Brettanomyces* in wine is proving interesting, and I will only touch on the future potential here. The use of 'killer' toxins, to which strains of *S. cerevisiae* are resistant, may be one way forward. The smut fungus *Ustilago maydis* is well known for its production of killer toxins, and one strain of this, CYC1410, has been shown to inhibit all of the tested strains of *B. bruxellensis* [137]. Several killer yeast species have also been shown to have toxins effective against *Brettanomyces*. For example, in a study published in 2014, two new killer toxins produced by the yeast *Candida pyralidae* were tested in wine [138]. These toxins have an antimicrobial effect against *B. bruxellensis*, and their activity was found not to be affected by sugar concentrations in must and the alcohol levels typically found in wine. *C. pyralidae* causes cell membrane and cell wall damage in *B. bruxellensis*, giving rise to the possibility of its use as a biocontrol agent [139]. In 2017, two killer toxins from *Pichia membranifaciens* (also called *Pichia membranaefaciens*) were studied by Belda et al. and found to be able to inhibit *B. bruxellensis*, without adverse effect on *S. cerevisiae* [140]. *M. pulcherrima* secretes pulcherriminic acid, which inhibits the growth of *B. bruxellensis* [141]. The use of non-Saccharomyces yeasts in red winemaking is attracting much interest – some have a role in increasing pigments, stabilising colour, or giving positive aromatic characteristics [142], as well as having potential for inhibiting *Brettanomyces* or other microbes. Another possible future method of biocontrol is the use of antimicrobial peptides (AMPs) that possess an antifungal action mechanism. The peptides are secreted by microorganisms to ensure their survival. A 2019 study by Rubén Peña et al. [143] identified that *Candida intermedia* LAMAP1790 secretes peptides with antimicrobial activity, which can control the growth of *B. bruxellensis* without affecting *S. cerevisiae*.

However, bearing in mind that the contamination and development of *Brettanomyces* in wine may occur late in the production process and during barrel or other maturation, the effectiveness in practical terms of biological methods of attack may prove to be limited. The use of starter cultures as a biocontrol strategy is already being practised, but further investigation regarding killer yeast activity might prove valuable [84].

4.12 Final Reflections

It is October 2019, and I'm in the Tate Modern Gallery in London at a tasting of wines from Rioja, along with several importers, merchants, and a few of the UK wine press. The wines are grouped in flights of 10, each flight representing a particular category, and have been judged by trade experts as the best of each category. Of the 10 red Rioja Reservas in the £10–14.99 category (which by law must spend at least 12 months in barrel), 5 showed levels of 4-EP markedly above sensory threshold. There were no adverse comments from the tasters, at least as far as Brett was concerned.

Fast forward to January 2020, I am on the 28th floor of Millbank Tower overlooking the River Thames in West London. Today's event is a portfolio tasting of wines represented by 'Maisons Marques et Domaines'. There are some wonderful wines on show, but, of course, most are relatively young. However, one mature wine stands out as truly outstanding, sending a shiver down my spine. It's from the Bordeaux commune of Pauillac: 1995 Château Pichon Longueville Comtesse de Lalande. The nose shows spice, black fruits, cedar, truffle, graphite, tobacco, and leather – my tasting note could fill a chapter. And there, amongst this melange of aromas, assertive but not brutal, is the unmistakable 'BAND-AID', medicinal, and antiseptic nose of 4-EP. The Brett notes are confirmed by Charles Metcalfe, co-founder and co-chair of 'The International Wine Challenge'. This is a stunning wine, overflowing with aromas that winemakers today strive to avoid. My mind goes back to the previous month: when asking Larry Cherubino of the eponymous winery and also winemaker for Robert Oatley Wines, whether he was prepared to accept any 4-EP or 4-EG in any of his wine, his answer was emphatic. Although he has not experienced an issue with Brett, he says, 'I am terrified of it – you don't spend so much time making expressive wines only to have them ruined'. But Larry admits to going his own way, including breaking many of the 'rules' for prevention detailed in Section 4.8. He ferments oxidatively, ages semi-reductively, does not rack wines in barrel at all, and does not filter at any stage. But the level of care and attention to detail is apparent in each and every glass of wines whose purity has provided the perfect platform upon which both purity and complexity beyond imagination has been built.

And now, I am in the barrel store of the Chateau St Hilarion winery in the village of Geçitköy in the northern 'unrecognised' part of Cyprus. This is a small operation, just 5 ha of vineyards, and I have been the winemaking consultant since its inception. Producing high quality wine here is challenging – summers are really hot, irrigation water is not always available, and there can be invasions of European hornets that devour rows of fruit when at optimum ripeness. However, in 15 years of production, there has never been a vat or barrel exhibiting a fault. Today, the owner Mustafa Seyfi and I are assessing the wines to determine their quality, progress, and destination. The 2018 Cabernet Sauvignon has now been in new French oak barrels for a year. The wine from each barrique is tasted, and we are really happy until we hit the unmistakeable odours of stables, farmyard, and sweat. Mustafa cannot tolerate a hint of Brett, and this one barrel has ethyl phenols in abundance. How? Why? There is no obvious answer: the winery and vineyard are quite isolated, and the nearest other commercial producer is 15 miles away. The affected barrel is isolated. Do we really want to sterile-filter the wine using hoses, pump, and filter that have never touched Brett and subsequently chemically treat the ethyl phenols? And what would we do with the treated wine? The decision is taken to destroy the wine and burn the barrel away from the winery. Just one infected barrel has just cost Mustafa €2000. The sleepless nights to come are unquantifiable.

References

1 Malfeito-Ferreira, M. (2018). Two decades of "horse sweat" taint and *Brettanomyces* yeasts in wine: where do we Stand now? *Beverages* 4 (32) https://doi.org/10.3390/beverages4020032.
2 Rieger, T. (2018). A *Brettanomyces* economics survey seeks more winery participants/feedback. *Wine Business Monthly* 18 September 2018. https://www.winebusiness.com/news/?go=getArticle&dataid=203428 (accessed 20 April 2020).
3 Ostler, T. (2009). Understanding *Brettanomyces*. https://www.burgundy-report.com/wp/wp-content/uploads/2009/10/Understanding-Brettanomyces.pdf (accessed 11 July 2019).
4 Joseph, C.M.L., Albino, E., and Bisson, L.F. (2017). Creation and use of a *Brettanomyces* aroma wheel. *Catalysts* 1: 12–20. https://doi.org/10.5344/catalyst.2016.16003.
5 Claussen, N.H. (1904). On a method for the application of Hansen's Pure Yeast System in the manufacturing of well-conditioned English stock beers. *Journal of the Institute of Brewing* 10 (4): 308–331.
6 Custers, M.T.J. (1940). Onderzoekingen over het Gistseglacht *Brettanomyces*; thesis. Delft, The Netherlands: Delft University.
7 Avramova, M. (2017). Population genetics and diversity of the species *Brettanomyces bruxellensis*: a focus on sulphite tolerance. Doctoral thesis Agricultural sciences. Université de Bordeaux. English. NNT: 2017BORD0911.
8 Gábor, P., Dlauchy, D., Tóbiás, A. et al. (2017). *Brettanomyces acidodurans sp. nov.*, a new acetic acid producing yeast species from olive oil. *Antonie van Leeuwenhoek, International Journal of General and Molecular Microbiology* 110 (5): 657–664. https://doi.org/10.1007/s10482-017-0832-8.
9 Krumbholz, G. and Tauschanoff, W. (1933). *Mycotorula intermedia* n. sp., ein beitrag zur Kenntnis der Gärungserreger in Wein. *Zentralblatt für Bakteriologie und Parasitenkunde Abteilung* 88 (2): 366–373.

10 Agostino, F. (1950). Polimorfismo esogene et endogenonei lieviti del genere *Brettanomyces*. *L'Agricoltura italiana* 50: 193–198.

11 Peynaud, E. and Domercq, S. (1956). Sur les Brettanomyces isolées de raisins et de vins. *Archiv für Microbiologie* 24 (3): 266–280.

12 Heresztyn, T. (1986). Metabolism of volatile phenolic compounds from hydroxycinnamic acids by Brettanomyces yeast. *Archives of Microbiology* 146: 96–98.

13 Curtin, C.D., Borneman, A.R., Chambers, P.J., and Pretorius, I.S. (2012). De novo assembly and analysis of the heterozygous triploid genome of the wine spoilage yeast *Dekkera bruxellensis AWRI 1499*. *PLoS One* 7 (3): e33840. https://doi.org/10.1371/journal.pone.0033840.

14 Loureiro, V. and Malfeito-Ferreira, M. (2006). Spoilage activities of *Dekkera/Brettanomyces* spp. In: *Food Spoilage Microorganisms* (ed. C.W. Blackburn), 354–398. Cambridge: Woodhead.

15 Oelofse, A., Pretorius, I.S., and du Toit, M. (2008). Significance of *Brettanomyces* and *Dekkera* during winemaking: a synoptic review. *South African Journal of Enology and Viticulture* 29 (2): 128–144. https://doi.org/10.21548/29-2-1445.

16 Renouf, V., Lonvaud-Funel, A., and Coulon, J. (2007). The origin of *Brettanomyces Bruxellensis* in wines: a review. *Journal International des Sciences de la Vigne et du Vin* 41 (3): 161–173. https://doi.org/10.20870/oeno-one.2007.41.3.846.

17 Hooke, H. (2010). The Dreaded Pox. *The Real Review*. https://www.therealreview.com/2010/04/01/the-dreaded-pox (accessed 11 July 2019).

18 The Australian Wine Research Institute (2017). Final report to Australian Grape and Wine Authority: ensuring the continued efficacy of *Brettanomyces* control strategies for avoidance of spoilage. https://www.wineaustralia.com/getmedia/72923111-4201-44ac-be1b-503712f4fff9/AWR-1304-Final-Report (accessed 6 April 2020).

19 Milheiro, J., Filipe-Ribeiro, L., Vilela, A. et al. (2017). 4-Ethylphenol, 4-ethylguaiacol and 4-ethylcatechol in red wines: microbial formation, prevention, remediation and overview of analytical approaches. *Critical Reviews in Food Science and Nutrition* 59 (9): 1367–1391. https://doi.org/10.1080/10408398.2017.1408563.

20 Steensels, J., Daenen, L., Malcorps, P. et al. (2015). *Brettanomyces* yeasts – from spoilage organisms to valuable contributors to industrial fermentations. *International Journal of Food Microbiology* 206: 24–38. https://doi.org/10.1016/j.ijfoodmicro.2015.04.005.

21 Hanson, A. (1982). *Burgundy*. London: Faber.

22 Gerbaux, V., Jeudy, S., and Monamy, C. (2000). A study of phenol volatiles in Pinot Noir wines in Burgundy. *Bulletin de l'OIV* 5 (73): 581–599.

23 Matthews, M.A. (2015). *Terroir and Other Myths of Winegrowing*. Oakland: University of California Press.

24 Crooked Stave (2020). *Brettanomyces* Project. http://www.crookedstave.com/brewery/brettanomyces-project.

25 White, C. and Zainasheff, J. (2010). *Yeast: The Practical Guide to Beer Fermentation*. Boulder, CO: Brewers Publications.

26 International Organisation of Vine and Wine (2020). *State of The World Vitivinicultural Sector in 2019*. Paris: OIV. http://www.oiv.int/public/medias/7298/oiv-state-of-the-vitivinicultural-sector-in-2019.pdf.

27 https://www.homebrewtalk.com/forum/threads/wanting-to-get-the-most-brett-character-funk-possible.473350 (accessed 8 October 2019).

28 Schifferdecker, A.J., Dashko, S., Ishchuk, O., and Piškur, J. (2014). The wine and beer yeast *Dekkera* bruxellensis. *Yeast* 31 (9): 323–332. https://doi.org/10.1002/yea.3023.

29 Smith, C. (2014). *Postmodern Winemaking*. Berkeley, University of California Press. Essay related to further references in this Chapter is also online: Integrated Brett Management. *Wines Vines Analytics*. https://winesvinesanalytics.com/columns/section/92/article/88228/Integrated-Brett-Management (accessed 23 July 2019).

30 Atkin, T. (2005). Losing by a nose. *The Guardian*, 1 May 2005. https://www.theguardian.com/lifeandstyle/2005/may/01/foodanddrink.shopping3 (accessed 25 August 2019).

31 Contreras, A., Hidalgo, C., Schmidt, S. et al. (2014). Evaluation of non-*Saccharomyces* yeasts for the reduction of alcohol content in wine. *Applied and Environmental Microbiology* 80: 1670–1678.

32 Contreras, A., Curtin, C., and Varela, C. (2015). Yeast population dynamics reveal a potential 'collaboration' between *Metschnikowia pulcherrima and Saccharomyces uvarum* for the production of reduced alcohol wines during Shiraz fermentation. *Applied Microbiology and Biotechnology* 99: 1885–1895. https://doi.org/10.1007/s00253-014-6193-6.

33 http://www.merisiwines.com/whites (accessed 4 March 2019).

34 Carpenter, K. (2014). Is a wine with Brett a bad thing? *Wine Spectator*. https://www.winespectator.com/articles/is-a-wine-with-brett-a-bad-thing-49761 (accessed 25 August 2019).

35 Bisson, L.F. (2013). Overview of the biology of *Brettanomyces:* A new look at an old problem. https://www.slideserve.com/euclid/overview-of-the-biology-of-brettanomyces-a-new-look-at-an-old-problem (accessed 11 April 2020).

36 https://delectable.com (accessed 20 January 2019).

37 https://www.winespectator.com/articles/scoring-scale (accessed 23 September 2019).

38 Anderson, K., Nelgen, S., and Pinilla, V. (2017). *Global Wine Markets, 1860 to 2016: A Statistical Compendium*. Adelaide: University of Adelaide Press. License: CC-BY 4.0. https://doi.org/10.20851/global-wine-markets.

39 Stelzer, T. (2010). Return to Grace – Henschke Release. https://www.tysonstelzer.com/return-to-grace-henschke-release (accessed 24 August 2019).

40 Curtin, C., Borneman, A., Zeppel, R. et al. (2014). Staying a step ahead of 'Brett'. *Wine & Viticulture Journal* 29 (5): 34–37.

41 Wine Australia (2017). *Inspired Tasting Brochure*. Wine Australia. https://www.wineaustralia.com/getmedia/85a73b20-2590-44bc-b9d0-2150f377932a/UK_InspiredTasting2017_Brochure_Final-web-version.pdf (accessed 15 May 2019).

42 Reddit https://www.reddit.com/r/wine/comments/68294v/recommend_wines_with_brettanomyces_flavors_please (accessed 17 August 2019).

43 Atkin, T. (2019). Lebanon. Cork Talk. https://timatkin.com/cork-talk/lebanon (accessed 25 August 2019).

44 Goode, J. (2003). *Brettanomyces*, 42–46. Harpers Wine and Spirit Weekly, 18 April 2003. http://www.wineanorak.com/brettanomyces.htm (accessed 16 August 2019).

45 Adams, A. (2013). New thinking in the Brett debate. https://www.winesandvines.com/news/article/114296/New-Thinking-in-the-Brett-Debate (Accessed 10 Aug 2017).

46 Schumaker, M.R., Diako, C., Castura, J.C. et al. (2019). Influence of wine composition on consumer perception and acceptance of *Brettanomyces* metabolites using temporal check-all-that-apply methodology. *Food Research International* 116: 963–872. https://doi.org/10.1016/j.foodres.2018.09.034.

47 Schumaker, M.R., Chandra, M., Malfeito-Ferreira, M., and Ross, C.F. (2017). Influence of *Brettanomyces* ethylphenols on red wine aroma evaluated by consumers in the United States and Portugal. *Food Research International* 100 (1): 161–167. https://doi.org/10.1016/j.foodres.2017.06.057.

48 Joseph, C.M.L., Albino, E.A., Ebeler, S.E., and Bisson, L.F. (2015). *Brettanomyces bruxellensis* aroma-active compounds determined by SPME GC-MS olfactory analysis. *The American Journal of Enology and Viticulture* 6: 379–387. https://doi.org/10.5344/ajev.2015.14073.

49 Curtin, C. (2013). Good Brett and other urban *Bretanomyces* myths. www.awri.com.au/wp-content/uploads/2013/09/curtin-w08-awitc15.pdf (accessed 16 August 2019).

50 Loureiro, V. and Malfeito-Ferreira, M. (2003). Spoilage yeasts in the wine industry. *International Journal of Food Microbiology* 86 (1–2): 23–50. https://doi.org/10.1016/S0168-1605(03)00246-0.

51 Tempère, S., Cuzange, E., Schaaper, M. et al. (2014). "Brett character" in wine: is there a consensus among professional assessors? A perceptual and conceptual approach. *Food Quality and Preference* 34: 29–36. https://doi.org/10.1016/j.foodqual.2013.12.007.

52 Curtin, C., Bramley, B., Cowey, G. et al. (2008). Sensory perceptions of 'Brett' and relationship to consumer preference. In: *Proceedings of the Thirteenth Australian Wine Industry Technical Conference*, Adelaide, Australia - 27July - 2 August 2007 (eds. R.J. Blair, P.J. Williams and I.S. Pretorius), 207–211. Australian Wine Industry Technical Conference Inc.

53 Csikor, Z., Pusztai, E., and Barátossy, G. (2018). Statistical evaluation of 4-ethylphenol and 4-ethylguaiacol concentrations to support sensory evaluation of "Brett Character" of wines: a proposed threshold. *Periodica Polytechnica Chemical Engineering* 62 (4): 450–456. https://doi.org/10.3311/PPch.12857.

54 Chatonnet, P., Dubourdieu, D., Boidron, J. et al. (1992). The origin of ethylphenols in wines. *Journal of the Science of Food and Agriculture* 60: 165–178. https://doi.org/10.1002/jsfa.2740600205.

55 Licker, J.L., Acree, T.E., and Henick-Kling, T. (1998). What is "Brett" (*Brettanomyces*) flavor? A preliminary investigation. In: *Chemistry of Wine Flavor, American Chemical Society Symposium Series*, vol. 714 (eds. A.L. Waterhouse and S.E. Ebeler), 96–115. American Chemical Society.

56 Romano, A., Perello, M.C., Lonvaud-Funel, A., and de Revel, G. (2009). Sensorial aspects of "Brett Character": re-evaluation of the olfactory perception threshold of volatile phenols in red wine. *Food Chemistry* 114 (1): 15–19.

57 Petrozziello, M., Asproudi, A., Guaita, M. et al. (2014). Influence of the matrix composition on the volatility and sensory perception of 4-ethyl phenol and 4-ethyl guaiacol in model wine solutions. *Food Chemistry* 149: 197–202. https://doi.org/10.1016/j.foodchem.2013.10.098.

58 Bramley, B., Curtin, C., Cowey, G. et al. (2008). Wine style alters the sensory impact of 'Brett' flavour compounds in red wines. In: *Proceedings of the 13th Australian Wine Industry Technical Conference*: 28 July – 2 August 2007; Urrbrae, South Australia (eds. R.J. Blair, P.J. Williams and I.S. Pretorius), 73–80. Adelaide: I.S. Australian Wine Industry Technical Conference Inc.

59 Agnolucci, M., Scarano, S., Rea, E. et al. (2007). Detection of *Dekkera/Brettanomyces bruxellensis* in pressed Sangiovese grapes by real time PCR. *Italian Journal of Food Science* 19: 153–164.

60 Renouf, V. and Lonvaud-Funel, A. (2007). Development of an enrichment medium to detect *Dekkera/Brettanomyces bruxellensis*, a spoilage wine yeast on the surface of grape berries. *Microbiological Research* 162 (2): 154–167. https://doi.org/10.1016/j .micres.2006.02.006.

61 Oro, L., Canonico, L., Marinelli, V. et al. (2019). Occurrence of *Brettanomyces bruxellensis* on grape berries and in related winemaking cellar. *Frontiers in Microbiology* 10: 415. https://doi.org/10.3389/fmicb.2019.00415.

62 Avramova, M., Cibrario, A., Peltier, E. et al. (2018). *Brettanomyces bruxellensis* population survey reveals a diploid-triploid complex structured according to substrate of isolation and geographical distribution. *Scientific Reports* 8: 4136. https://doi.org/10 .1038/s41598-018-22580-7.

63 Cibrario, A., Avramova, M., Dimopoulou, M. et al. (2019). *Brettanomyces bruxellensis* wine isolates show high geographical dispersal and long persistence in cellars. *PLoS One* 14 (12): e0222749. https://doi.org/10.1371/journal.pone.0222749.

64 International Organisation of Vine and Wine (OIV) (2014). Code of good vitivinicultural practices in order to avoid or limit contamination by *Brettanomyces*. http://www .oiv.int/public/medias/4831/code-brett-oiv-oeno-462-2014-en.pdf (accessed 17 January 2020).

65 Wedral, D., Shewfelt, R., and Frank, J. (2010). The challenge of *Brettanomyces* in wine. *LWT - Food Science and Technology* 43 (10): 1474–1479. https://doi.org/10.1016/j.lwt .2010.06.010.

66 Chatonnet, P., Dubourdieu, D., and Boidron, N. (1995). The influence of *Brettanomyces/Dekkera* sp. yeasts and lactic acid bacteria on the ethylphenol content of red wines. *American Journal of Enology and Viticulture* 46 (4): 463–468.

67 Conterno, L. and Henick-Kling, T. (2010). *Brettanomyces/Dekkera* off-flavours and other wine faults associated with microbial spoilage. In: *Managing Wine Quality, Volume 2: Oenology and Wine Quality* (ed. A.G. Reynolds), 346–387. Cambridge: Woodhead Publishing.

68 Larcher, R., Nicolini, G., Bertoldi, D., and Nardin, T. (2008). Determination of 4-ethylcatechol in wine by high-performance liquid chromatography–coulometric electrochemical array detection. *Analytica Chimica Acta* 609 (2): 235–240. https://doi .org/10.1016/j.aca.2007.12.038.

69 Ribéreau-Gayon, P., Dubourdieu, D., Donèche, B., and Lonvaud, A. (2006). *Handbook of Enology: The Microbiology of Wine and Vinifications*, vol. 1. Chichester: Wiley.

70 Smith, B.D. and Divol, B. (2016). *Brettanomyces bruxellensis*, a survivalist prepared for the wine apocalypse and other beverages. *Food Microbiology* 59: 161–175. https://doi .org/10.1016/j.fm.2016.06.008.

71 Benito, S., Palomero, F., Morata, A. et al. (2009). A method for estimating *Dekkera/Brettanomyces* populations in wines. *Journal of Applied Microbiology* 106: 1743–1751. https://doi.org/10.1111/j.1365-2672.2008.04137.x.

72 Coulon, J., Perello, M.C., Lonvaud-Funel, A. et al. (2010). *Brettanomyces bruxellensis* evolution and volatile phenols production in red wines during storage in bottles. *Journal of Applied Microbiology* 108 (4): 1450–1458. https://doi.org/10.1111/j.1365-2672 .2009.04561.x.

73 Barata, A., Cadeira, J., Botelheiro, R. et al. (2008). Survival patterns of *Dekkera bruxellensis* in wines and inhibitory effect of Sulphur dioxide. *International Journal of Food Microbiology* 121 (2): 201–207. https://doi.org/10.1016/j.ijfoodmicro.2007.11.020.

74 Sturm, M.E., Assof, M., Fanzone, M. et al. (2015). Relation between coumarate decarboxylase and vinylphenol reductase activity with regard to the production of volatile phenols by native *Dekkera bruxellensis* strains under 'wine-like' conditions. *International Journal of Food Microbiology* 206: 51–55. https://doi.org/10.1016/j.ijfoodmicro .2015.04.023.

75 Godden, P. and Muhlack, R. (2010). Trends in the composition of Australian wine, 1984–2008. *The Australian and New Zealand Grapegrower & Winemaker* 558: 47–61.

76 Joseph, R. (2015). The great Brettanomyces taste test. *Meininger's Wine Business International* 14 October 2015. https://www.meininger.de/en/wine-business-international/ great-brettanomyces-taste-test (accessed 4 February 2019).

77 Rubio, R., Garijo, P., Santamaría, P. et al. (2015). Influence of oak origin and ageing conditions on wine spoilage by *Brettanomyces* yeasts. *Food Control* 54: 176–180. https:// doi.org/10.1016/j.foodcont.2015.01.034.

78 Renouf, V. (2015). *Brettanomyces et Phénols Volatils*. Paris: Lavoisier Tec & Doc (in French).

79 The Australian Wine Research Institute (2015). Controlling *Brettanomyces* during winemaking. www.awri.com.au/wp-content/uploads/2014/05/Brett-fact-sheet.pdf (accessed 18 August 2019).

80 ITV France (2006). *Brettanomyces et phénols volatils. Les Cahiers Itineraires d'itv France*, vol. 12. Institut Français de la Vigne et u Vin (in French).

81 Hucker, B., Christophersen, M., and Vriesekoop, F. (2017). The influence of thiamine and riboflavin on various spoilage microorganisms commonly found in beer. *Journal of the Institute of Brewing* 123: 24–30. https://doi.org/10.1002/jib.385.

82 Gerbaux, V., Vincent, B., and Bertrand, A. (2002). Influence of maceration temperature and enzymes on the content of volatile phenols in Pinot noir wines. *American Journal of Enology and Viticulture* 53: 131–137.

83 Pinto, L., Baruzzi, F., Cocolin, L., and Malfeito-Ferreira, M. (2020). Emerging technologies to control *Brettanomyces* spp. in wine: recent advances and future trends. *Trends in Food Science & Technology* 99: 88–100. https://doi.org/10.1016/j.tifs.2020.02.013.

84 Berbegal, C., Spano, G., Fragasso, M. et al. (2018). Starter cultures as biocontrol strategy to prevent *Brettanomyces bruxellensis* proliferation in wine. *Applied Microbiology and Biotechnology* 102 (2): 569–576. https://doi.org/10.1007/s00253-017-8666-x.

85 Muñoz, V., Beccaria, B., and Abreo, E. (2014). Simultaneous and successive inoculations of yeasts and lactic acid bacteria on the fermentation of an unsulfited Tannat

grape must. *Brazilian Journal of Microbiology* 4: 59–66. https://doi.org/10.1590/S1517-83822014000100009.

86 Berbegal, C., Garofalo, C., Russo, P. et al. (2017). Use of autochthonous yeasts and bacteria in order to control *Brettanomyces bruxellensis* inwine. *Fermentation* 3 (4): 65. https://doi.org/10.3390/fermentation3040065.

87 Margalit, Y. (2012). *Concepts in Wine Technology: Small Winery Operations*, 3e. San Francisco, CA: The Wine Appreciation Guild.

88 Curtin, C., Varela, C., and Borneman, A. (2015). Harnessing improved understanding of *Brettanomyces bruxellensis* biology to mitigate the risk of wine spoilage. *Australian Journal of Grape and Wine Research* 21: 680–692. https://doi.org/10.1111/ajgw.12200.

89 Malfeito-Ferreira, M. (2019). Spoilage yeasts in red wines. In: *Red Wine Technology* (ed. A. Morata), 219–235. London: Academic Press.

90 Malfeito-Ferreira, M., Laureano, P., Barata, A. et al. (2004). Effect of different barrique sanitation procedures on yeasts isolated from the inner layers of wood. *American Journal of Enology and Viticulture* 55 (3): 304A.

91 Barata, A., Laureano, P., D'Antuono, I. et al. (2013). Enumeration and identification of 4-ethylphenol producing yeasts recovered from the wood of wine ageing barriques after different sanitation treatments. *Journal of Food Research* 2 (1): 140–149.

92 Cartwright, Z.M., Glawe, D., and Edwards, C.G. (2018). Reduction of *Brettanomyces bruxellensis* populations from oak barrel staves using steam. *American Journal of Enology and Viticulture* 69 (4): 400–409. https://doi.org/10.5344/ajev.2018.18024.

93 Cowey, G. (2019). What does the latest research say about barrel sanitation against Brett? *Grapegrower & Winemaker* 669: 76–77.

94 Edwards, C.G. and Cartwright, Z.M. (2019). Application of heated water to reduce populations of *Brettanomyces bruxellensis* present in oak barrel staves. *South African Journal of Enology and Viticulture* 40 (1) https://doi.org/10.21548/40-1-3008.

95 González-Arenzana, L., Santamaría, P., López, R. et al. (2013). Microwave technology as a new tool to improve microbiological control of oak barrels: a preliminary study. *Food Control* 30 (2): 536–539. https://doi.org/10.1016/j.foodcont.2012.08.008.

96 Breniaux, M., Renault, P., Meunier, F., and Ghidossi, R. (2019). Study of high power ultrasound for oak wood barrel regeneration: impact on wood properties and sanitation effect. *Beverages* 5 (10): 1–13. https://doi.org/10.3390/beverages5010010.

97 Chatonnet, P., Dubourdieu, D., Boidron, J.N., and Lavigne, V. (1993). Synthesis of volatile phenols by *Saccharomyces cerevisiae* in wines. *Journal of the Science of Food and Agriculture* 62 (2): 191–202. https://doi.org/10.1002/jsfa.274062021368.

98 Solis, A., de Lourdes Alejandra, M., Gadoury, D.M., and Worobo, R.W. (2017). Efficacy of sulfur dioxide and sulfur discs against wine-spoilage yeasts: in vivo and in vitro trials. *Food Safety Magazine* February/March 2017. http://www.foodsafetymagazine.com/magazine-archive1/februarymarch-2017/efficacy-of-sulfur-dioxide-and-sulfur-discs-against-wine-spoilage-yeasts-in-vivo-and-in-vitro-trials (accessed 3 April 2020).

99 Šućur, S., Čadež, N., and Košmerl, T. (2016). Volatile phenols in wine: control measures of *Brettanomyces/Dekkera* yeasts. *Acta agriculturae Slovenica* 107 (2): 453–472. https://doi.org/10.14720/aas.2016.107.2.17.

100 Cibrario, A., Sertier, C.M., Riquier, L. et al. (2019). Cellar temperature affects *Brettanomyces bruxellensis* population and volatile phenols production in aging Bordeaux wines. *American Journal of Enology and Viticulture* 71 (1): 1–9. https://doi.org/10.5344/ajev.2019.19029.

101 Kheir, J., Salameh, D., Strehaiano, P. et al. (2013). Impact of volatile phenols and their precursors on wine quality and control measures of *Brettanomyces/Dekkera* yeasts. *European Food Research and Technology* 237 (5): 655–671.

102 Conterno, L., Joseph, C.M.L., Arvik, T.J. et al. (2006). Genetic and physiological characterization of *Brettanomyces bruxellensis* strains isolated from wines. *American Journal of Enology and Viticulture* 57: 139–147.

103 Millet, V. and Lonvaud-Funel, A. (2000). The viable but non-culturable state of wine micro-organisms during storage. *Letters in Applied Microbiology* 30: 136–141. https://doi.org/10.1046/j.1472-765x.2000.00684.x.

104 Agnolucci, M., Rea, F., Sbrana, C. et al. (2010). Sulphur dioxide affects culturability and volatile phenol production by *Brettanomyces Dekkera/bruxellensis*. *International Journal of Food Microbiology* 143: 76–80. https://doi.org/10.1016/j.ijfoodmicro.2010.07.022.

105 Renouf, V. and Lonvaud-Funel, A. (2004). Racking are key stages for the microbial stabilization of wines (in French). *Journal International des Sciences de la Vigne et du Vin* 38 (4): 155–162.

106 The Australian Wine Research Institute (2015). Fact Sheet: Controlling *Brettanomyces* during winemaking. www.awri.com.au/wp-content/uploads/2014/05/Brett-fact-sheet.pdf (accessed 30 November 2020).

107 Oswald, T.A. and Edwards, C.G. (2017). Interactions between storage temperature and ethanol that affect growth of *Brettanomyces bruxellensis* in Merlot wine. *American Journal of Enology and Viticulture* 68 (2): 188–194. https://doi.org/10.5344/ajev.2017.16102.

108 Benito, S., Palomero, F., Morata, A. et al. (2006). Detección de *Brettanomyces/Dekkera* en vinos tintos mediante el uso de medios selectivo-diferenciales. *Tecnoligia del Vino* 32: 27–31.

109 Tofalo, R., Schirone, M., Corsetti, A., and Suzzi, G. (2012). Detection of *Brettanomyces* spp. in red wines using real-time PCR. *Journal of Food Science* 77 (9): M545–M549. https://doi.org/10.1111/j.1750-3841.2012.02871.x.

110 Gerland, C. (2010). Nouvelles connaissances scientifiques et pratiques pour la prévention des deviations phénolées par les *Brettanomyces*. *Revue internet de viticulture et oenologie* N.4/1. https://www.infowine.com/intranet/libretti/libretto7934-01-1.pdf. (accessed 19 May 2019).

111 Rodriguez, S.B., Thornton, M.A., and Thornton, R.J. (2013). Raman spectroscopy and Chemometrics for identification and strain discrimination of the wine spoilage yeasts *Saccharomyces cerevisiae*, *Zygosaccharomyces bailii*, and *Brettanomyces bruxellensis*. *Applied and Environmental Microbiology* 79 (20): 6264–6270. https://www.ncbi.nlm.nih.gov/pmc/articles/PMC3811183/pdf/zam6264.pdf.

112 Tubia, I., Prasad, K., Pérez-Lorenzo, E. et al. (2018). Beverage spoilage yeast detection methods and control technologies: a review of *Brettanomyces*. *International Journal of Food Microbiology* 283: 65–76. https://doi.org/10.1016/j.ijfoodmicro.2018.06.020.

113 Suárez, R., Suárez-Lepe, J.A., Morata, A., and Calderón, F. (2007). The production of ethylphenols in wine by yeasts of the genera *Brettanomyces* and *Dekkera*: a review. *Food Chemistry* 102 (1): 10–21. https://doi.org/10.1016/j.foodchem.2006.03.030.

114 Umiker, N.L., Descenzo, R.A., Lee, J., and Edwards, C.G. (2012). Removal of *Brettanomyces bruxellensis* from red wine using membrane filtration. *Journal of Food Processing and Preservation* 37: 799–805. https://doi.org/10.1111/j.1745-4549.2012.00702.x.

115 Calderón, F., Morata, A., Uthurry, C., and Suárez, J.A. (2004). Aplicaciones de la ultra-fitración en la industria enológica. Ultimos avances tecnológicos. *Tecnología del Vino* 16: 49–54.

116 Duarte, F.L., Coimbra, L., and Baleiras-Couto, M. (2017). Filter media comparison for the removal of *Brettanomyces bruxellensis* from wine. *American Journal of Enology and Viticulture* 68: 504–508. https://doi.org/10.5344/ajev.2017.17003.

117 Coupez, C. (2012). *Brettanomyces*, or management of a contaminant micro-organism. http://www.bio-rad.com/webroot/web/pdf/fsd/literature/PF_236.pdf (accessed 8 April 2020).

118 Renouf, V., Strehaiano, P., and Lonvaud-Funel, A. (2008). Effectiveness of dimethlydicarbonate to prevent *Brettanomyces bruxellensis* growth in wine. *Food Control* 19 (2): 208–216. https://doi.org/10.1016/j.foodcont.2007.03.012.

119 International Organisation of Vine and Wine {OIV} (2021). *International Code of Oenological Practices*. Paris: OIV. https://www.oiv.int/public/medias/7713/en-oiv-code-2021.pdf.

120 EUR-Lex. (2019). Commission Delegated Regulation (EU) 2019/934 of 12 March 2019 supplementing Regulation (EU) No 1308/2013 of the European Parliament and of the Council as regards wine-growing areas where the alcoholic strength may be increased, authorised oenological practices and restrictions applicable to the production and conservation of grapevine products, the minimum percentage of alcohol for by-products and their disposal, and publication of OIV files. https://eur-lex.europa.eu/legal-content/GA/TXT/?uri=CELEX:32019R0934 (accessed 25 may 2020).

121 e-CFR. (2020). Code of Federal Regulations, Title 27, Chapter I, Subchapter A, Part 24. https://www.ecfr.gov/cgi-bin/text-idx?c=ecfr&sid=506cf0c03546efff958847134c5527d3&rgn=div5&view=text&node=27:1.0.1.1.19&idno=27 (accessed 14 June 2020).

122 Food Standards Australia New Zealand (FSANZ). (2015). Australia and New Zealand Food Standard Code 1.3.3. www.foodstandards.gov.au/code/Documents/1.3.3%20Processing%20aids%20v157.pdf (accessed 8 April 2020).

123 Costa, A., Barata, A., Malfeito-Ferreira, M., and Loureiro, V. (2008). Evaluation of the inhibitory effect of dimethyl dicarbonate (DMDC) against microorganisms associated with wine. *Food Microbiology* 25: 422–427. https://doi.org/10.1016/j.fm.2007.10.003.

124 Taillandier, P., Joannis-Cassan, C., Jentzer, J.B. et al. (2018). Effect of a fungal chitosan preparation on *Brettanomyces bruxellensis*, a wine contaminant. *Journal of Applied Microbiology* 118: 123–131. https://doi.org/10.1111/jam.12682.

125 Petrova, B., Cartwright, Z.M., and Edwards, C.G. (2016). Effectiveness of chitosan preparations against *Brettanomyces bruxellensis* grown in culture media and red wines. *Journal International des sciences de la vigne et du vin* 50 (1) https://doi.org/10.20870/oeno-one.2016.50.1.54.

126 Ugarte, P., Agosin, E., Bordeu, E., and Villalobos, J.I. (2005). Reduction of 4-ethylphenol and 4-ethylguaiacol concentration in red wines using reverse osmosis and adsorption. *American Journal of Enology and Viticulture* 56: 30–36.

127 Felipe-Ribeiro, L., Milheiro, J., Matos, C.C. et al. (2017). Reduction of 4-ethylphenol and 4-ethylguaiacol in red wine by activated carbons with different physicochemical characteristics: impact on wine quality. *Food Chemistry* 229: 242–251. https://doi.org/10.1016/j.foodchem.2017.02.066.

128 Milheiro, J., Filipe-Ribeiro, L., Cosme, F., and Nunes, F.M. (2017). A simple, cheap and reliable method for control of 4-ethylphenol and 4-ethylguaiacol in red wines. Screening of fining agents for reducing volatile phenols levels in red wines. *Journal of Chromatography B. Analytical Technologies in the Biomedical and Life Sciences* 1041–1042: 183–190. https://doi.org/10.1016/j.jchromb.2016.10.036.

129 van Wyk, S. and Silva, F.V.M. (2017). High pressure processing inactivation of *Brettanomyces bruxellensis* in seven different table wines. *Food Control* 81: 1–8. https://doi.org/10.1016/j.foodcont.2017.05.028.

130 van Wyk, S. and Silva, F.V.M. (2017). High pressure inactivation of *Brettanomyces bruxellensis* in red wine. *Food Microbiology* 63: 199–204. https://doi.org/10.1016/j.fm.2016.11.020.

131 van Wyk, S., Farid, M.M., and Silva, F.V.M. (2018). SO_2, high pressure processing and pulsed electric field treatments of red wine: effect on sensory, *Brettanomyces* inactivation and other quality parameters during one year storage. *Innovative Food Science & Emerging Technologies* 48: 204–211. https://doi.org/10.1016/j.ifset.2018.06.016.

132 Puértolas, E., López, N., Condón, S. et al. (2009). Pulsed electric fields inactivation of wine spoilage yeast and bacteria. *International Journal of Food Microbiology* 130 (1): 49–55. https://doi.org/10.1016/j.ijfoodmicro.2008.12.035.

133 González-Arenzana, L., Lopéz-Alfaro, I., Garde-Cerdán, T. et al. (2018). Microbial inactivation and MLF performances of Tempranillo Rioja wines treated with PEF after alcoholic fermentation. *International Journal of Food Microbiology* 269: 19–26. https://doi.org/10.1016/j.ijfoodmicro.2018.01.008.

134 Puértolas, E., López, N., Condón, S. et al. (2010). Potential applications of PEF to improve red wine quality. *Trends in Food Science & Technology* 21 (5): 247–255. https://doi.org/10.1016/j.tifs.2010.02.002.

135 Lustrato, G., Vigentini, I., de Leonardis, S. et al. (2010). Inactivation of wine spoilage yeasts *Dekkera bruxellensis* using low electric current treatment (LEC). *Journal of Applied Microbiology* 109: 594–604. https://doi.org/10.1111/j.1365-2672.2010.04686.x.

136 Gallardo-Chacón, J.-J. and Karbowiak, T. (2015). Sorption of 4-ethylphenol and 4-ethylguaiacol by suberin from cork. *Food Chemistry* 181: 222–226. https://doi.org/10.1016/j.foodchem.2015.02.102.

137 Santos, A., Navascués, E., Bravo, E., and Marquina, D. (2011). *Ustilago maydis* killer toxin as a new tool for the biocontrol of the wine spoilage yeast *Brettanomyces bruxellensis*. *International Journal of Food Microbiology* 145 (1): 147–154. https://doi.org/10.1016/j.ijfoodmicro.2010.12.005.

138 Mehlomakulu, N.N., Setati, M.E., and Divol, B. (2014). Characterization of novel killer toxins secreted by wine-related non-*Saccharomyces* yeasts and their action on *Brettanomyces* spp. *International Journal of Food Microbiology*. 188: 83–91. https://doi.org/10.1016/j.ijfoodmicro.2014.07.015.

139 Mehlomakulu, N.N., Setati, M.E., Prior, K.J., and Divol, B. (2017). *Candida pyralidae* killer toxin disrupts the cell wall of *Brettanomyces bruxellensis* in red grape juice. *Journal of Applied Microbiology* 122 (3): 747–758. https://doi.org/10.1111/jam.13383.

140 Belda, I., Ruiz, J., Alonso, A. et al. (2017). The biology of *Pichia membranifaciens* killer toxins. *Toxins* 9 (4): 112. https://doi.org/10.3390/toxins9040112.

141 Oro, L., Ciani, M., and Comitini, F. (2014). Antimicrobial activity of *Metschnikowia pulcherrima* on wine yeasts. *Journal of Applied Microbiology* 116 (5): 1209–1217. https://doi.org/10.1111/jam.12446.

142 Ciani, M. and Comitini, F. (2019). Use of non-*Saccharomyces* yeasts in red winemaking. In: *Red Wine Technology* (ed. A. Morata), 51–68. London: Academic Press.

143 Rubén Peña, R., Chávez, R., Rodríguez, A., and Ganga, M.A. (2019). A control alternative for the hidden enemy in the wine cellar. *Fermentation* 5 (1): 25. https://doi.org/10.3390/fermentation5010025.

5

Oxidation, Premox, and Excessive Acetaldehyde

Oxidation is one of the most common wine faults. Severely oxidised wines are distinctly unpleasant for several reasons: they lack brightness, suffer from browning, smell 'burnt' or toffee-like, and the palate is bitter and dried out. Less heavily impacted examples may be regarded as flawed, with a loss of fruit, complexity, structure, and balance. In this chapter, the production in wine of acetaldehyde that can be formed by yeast activity or chemical oxidation and, in excess, gives negative aroma characteristics is addressed. I detail the types, causes and effects of oxidation, including premature oxidation (premox) of wines before they reach maturity, which has only become apparent as a problem since the early years of the twenty-first century. Furthermore, aspects of oxygen management in winemaking are discussed in this chapter and also in Chapter 6. However, oxidised wines often have a high level of volatile acidity, which will be covered as a general topic in Chapter 7.

5.1   Introduction

5.1.1   Oxygen: The Enemy and the Friend

Oxygen is both the enemy and friend of wine and the winemaker. In 1873, Louis Pasteur stated that 'oxygen is the greatest enemy of wine' but also 'oxygen makes the wine, which ages under its influence' (translated from French) [1]. Pasteur had been commissioned in 1863 by Emperor Napoleon III to investigate why so much wine became spoiled. In fact the adverse effects of wine being exposed to the atmosphere have been known for millennia; the Romans would protect amphorae of wine from oxidation by covering the surface with olive oil. Up until the very late eighteenth and early nineteenth century, it was generally the youngest, freshest wines that were preferred. The concept of laying down bottled wine to mature only became possible with the change from 'squat' to cylindrical glass bottles that began in the mid-eighteenth century and was largely complete by the early 1800s. However, the labelling of oxygen as an enemy really began in the years post-World War II with the drive to produce fresh clean white wines driven by primary fruit aromas and flavours. Indeed, the late Professor Émile Peynaud, regarded by many today as the grandfather of modern oenology, famously declared that 'oxygen is the enemy of wine' in his book *Le Goût du Vin*, first published in 1980 [2]. As so often happens when new discoveries are made and new concepts developed, old beliefs and methods were rapidly discarded, and producers

Wine Faults and Flaws: A Practical Guide, First Edition. Keith Grainger.
© 2021 John Wiley & Sons Ltd. Published 2021 by John Wiley & Sons Ltd.

in many parts of the world began to practice anaerobic fermentations and to exclude oxygen as far as possible at every stage of white winemaking. By way of example, many of the white wines from Central and Northern Italy, which in the 1950s and even 1960s had their fruit characteristics irreparably damaged by oxidation, largely due to careless winemaking, began to be made in a clean, fresh, largely reductive style. Certainly, the dull, drying, bitter tones of oxidation disappeared. Unfortunately so did most of the aromas and flavours. Moreover, tasters would be hard-pressed to tell the difference between a Soave, Frascati, and Orvieto – describing such wines as of 'neutral' character was probably being polite. Many red wines too began to be made in a reductive manner, often resulting in aggressive tannins, metallic tones, or notes of garlic, and a roller coaster of a maturation ride. As Clark Smith says, 'it was to take half a century before people once again recognized oxygen's power to elaborate and refine structure' [3]. Today, few would deny that for full-bodied red wines, oxygen plays a vital role in developing tertiary aromas, softening tannins, and reducing astringency and bitterness.

Oxygen uptake begins as soon as the integrity of the grape is compromised in the atmosphere. This may be by damage or disease whilst on the vine, during harvesting, certainly during the crushing process and, unless presses employing inert gas are utilised, during pressing. Numerous subsequent cellar activities may be sources of oxygen pickup. However, the careful use and management of oxygen in winemaking and, depending upon the wine type, in maturation is a key component and tool to give a broad palette of aromas and flavours.

5.1.2 The Importance of Oxygen in Winemaking

Oxygen is required at various stages of winemaking, in varying amounts, and at various times depending, inter alia, on the type and style of wine being produced. Its role in the development of high-quality red wines is crucial. Ingress in the wrong amount or for the wrong duration, at the wrong time or in the wrong wine, will compromise quality. The uptake of high volumes may at least result in an unacceptable level of acetaldehyde or at worst the serious fault of heavy oxidation. Whilst the contribution that oxygen makes in the maturation of many red wines and some high-quality white wines is well known, as Professor Paul Kilmartin of the University of Auckland notes, 'predicting which wines will benefit from oxygen exposure, and the correct dosage to employ, is difficult, and the underlying processes are still not well understood' [4].

The role of oxygen in winemaking and oxygen management are briefly discussed in Section 5.11 and Chapter 6.

5.1.3 Oxygenation Is Not Oxidation

It is important at this point to distinguish between oxygenation and oxidation as the terms are often misused in the popular media, in specialist wine publications and, bearing in mind that the terms have evolved, sometimes even in research papers. Oxygenation is the controlled, and usually beneficial, addition or uptake of oxygen during the winemaking and/or maturation processes, whereas oxidation is the irreversible damage to must or wine due to unwanted and largely preventable oxygen ingress. Of course, oxidation can result from excessive oxygenation.

5.1.4   Good and Bad Oxidation?

Leaving aside, for the moment, the distinctive speciality and fortified wines deliberately exposed to high levels of oxygen, which is essential to their style and quality, there are producers of 'light' wines who desire and allow significant oxygen uptake during the production and/or maturation processes. Their white wines may have tones of nuts, beeswax, or honey, and the red wines aromas of soaked or dried fruits, and umami, together with a thousand and one tertiary flavours. To wine lovers who appreciate such oxidative styles, the wines can be superb, but to many oenologists, winemakers, and critics, particularly in New World countries, these wines are oxidised and accordingly perceived to be faulty. However, generally speaking, for white and rosé wine production and storage, any exposure to oxygen, apart from oxidative juice handling (if practised), and barrel ageing for some high-quality whites, may be considered negative [5]. The loss of fruit- and fermentation-derived aromas and flavours (the primary and secondary aromas and flavours) and the development of oxidised characteristics are considered to be detrimental and, at worst, destructive.

5.2   Oxidation in Must and Wine

5.2.1   Oxidation Defined

In chemistry, the definition of oxidation has changed in recent years to mean 'an atom, or molecule losing one or more electrons in a chemical reaction'. Conversely, reduction is 'an atom or molecule or gaining one or more electrons'. The word 'oxygen' is no longer in these definitions. However, oxidation in wine involves elements combining with oxygen, which relates to previous definitions of oxidation, the oxidising processes, and their results. The mechanisms of oxidation are very complicated, and these are only briefly discussed in this chapter. To quote from a 2011 paper by Carla Maria Oliveira et al., *Oxidation mechanisms occurring in wines*, '…the complexity of the mechanisms implicated in wine oxidation is not fully understood and the identification of all mediators' reactions and its characterization needs to be done' [6]. Readers seeking further scientific detail of the processes involved are particularly referred to this paper and also to *Mechanisms of oxidative browning in wine* by Hua Li et al. [7].

5.2.2   Types of Oxidation

There are three types of oxidation that may affect must or wines:

- Enzymatic – this only really affects must.
- Chemical – this mostly affects wine and is the most usual form of wine oxidation.
- Microbial – this can affect both must and wine.

These will be discussed in Sections 5.6–5.8.

Wine (as with any other liquid) is saturated with oxygen when no more oxygen can dissolve in it. The saturation level in wine at 25 °C (77 °F) is approximately 7 mg/l and increases to 10 mg/l at 5 °C (41 °F). The dissolved oxygen (DO) will slowly be removed from wine by reaction with certain molecules, forming oxygen-containing compounds. White wines may be considered completely oxidised with a DO content of 60 mg/l (6–10 saturations depending upon temperature). Full-bodied red wines will improve with over 60 mg/l DO and can take up to 150 mg/l (15–23 saturations) before oxidation becomes apparent. During production, DO can, to a large extent, be removed by the winemaker by sparging with inert gas, namely nitrogen (N), carbon dioxide (CO_2), or argon (Ar).

5.3 Sensory Characteristics and Detection of Excess Acetaldehyde and Oxidation in Wine

5.3.1 Excess Acetaldehyde

5.3.1.1 Background
Acetaldehyde is also known as ethanal, but to avoid any possible confusion with ethanol, the compound will be referred to as acetaldehyde. All wines contain some acetaldehyde, both in free and bound forms – it is the most important volatile wine carbonyl and may be produced by microbial (the actions of yeasts) and chemical (the oxidation of ethanol) pathways. Only when present at excessive concentrations may acetaldehyde be considered a flaw or fault – such level will depend upon the wine matrix. However, it should be noted that the human consumption of acetaldehyde in conjunction with alcohol is harmful to health. Although acetaldehyde is an 'oxidised' compound, many wines can show an excess, without exhibiting characteristics of other compounds of oxidation as shown in Table 5.1.

5.3.1.2 Detection
Excess acetaldehyde may be detected on the appearance, nose, and palate of a wine.

5.3.1.3 Appearance
In white wines there is a deeper than expected colour with pronounced yellow or gold tones. In red wines there are brick-red tones, veering towards garnet.

5.3.1.4 Nose
The wine has cut, bruised, or green apple aromas, often with notes of leaf, grass, hay, cherry, metal, the nutty aromas of Fino Sherry, and sometimes odours of solvent.

Table 5.1 Compounds responsible for wine oxidation.

Compound	Formula	Olfactory characteristics	Sensory detection threshold (as defect)
Acetaldehyde	C_2H_4O	Bruised apple, Fino Sherry, nuts	100–150 mg/l
Phenylacetaldehyde	$C_6H_5CH_2CHO$	Honey, rose	25 µg/l
Benzaldehyde	C_6H_5CHO	Amontillado Sherry, nuts	3 mg/l
Sotolon	$C_6H_8O_3$	Walnuts, rancid, curry	8–15 µg/l
Methional	$CH_3SCH_2CH_2CHO$	Cooked potato	0.5 µg/l

Acetaldehyde at high levels has also been shown to have important suppressive interactions with some aroma compounds, such as 3-mercaptohexan-1-ol and 3-isobutyl-2-methoxypyrazine [8], both important in the aromatic profile of Sauvignon Blanc wines. In other words, the compound mutes many desirable aromas.

5.3.1.5 Palate
The wine has reduced varietal characteristics, the taste of nuts, Sherry-like flavours, and a sour taste and finish.

5.3.1.6 Sensory Detection Threshold
A sensory detection threshold of 100–125 mg/l is widely stated, but figures as low as 40 mg/l have been noted. At a level above 125 mg/l, the nutty aromas are pronounced, and at higher levels, the Sherry-like aromas overwhelm other volatile characteristics.

Acceptable levels in white wines are generally in the region of 20 mg/l, and in red wines 40 mg/l or so. The impact upon aromas and flavours perhaps begins as low as 0.5 mg/l. The presence of the compound in wine at a level above 50 mg/l, although well below sensory detection threshold, generally indicates that the wine has been oxidised [4].

5.3.2 Oxidation

High levels of acetaldehyde are key to oxidised characteristics in wine, but several other compounds also contribute to oxidation. These are detailed in Table 5.1.

5.3.2.1 Appearance
White wines look dull flat and lack brightness, with darkening of colour. There will be straw, copper-gold, or brown tones, sometimes accompanied by cloudiness. Rosé wines may appear dull pink or orange. Red wines have a paling of colour showing orange and brick-red, brown-red, or brown-orange notes; they lack brightness, and appear flat.

White wines may suffer from pinking, which has no organoleptic impact other than the unwanted pink colour.

5.3.2.2 Nose
There is a distinct loss of freshness, varietal characteristics, and fruit. The odours are sometimes dank and a little musty, but they should not be confused with haloanisole taint (see Chapter 3). There may also be the vinegar-like odour of high volatile acidity. White wines can have notes of cider, hay, and straw, and red wines smell of cooked, or even stewed, fruits.

5.3.2.3 Palate
The wine exudes bitter, dried out, burnt, flat, nutty, and Sherry-like flavours, sometimes with increased and aggressive acidity. It shows green apple or curry notes. Dulling of expected primary fruit, loss of freshness, loss of varietal characteristics and loss of secondary and tertiary aromas. Cardboard, wax (particularly beeswax), honey, wet dog, wet wool, fleshy 'bruised' apple, quince, dried fruits, nuts, Oloroso Sherry, toffee, caramel, rancid. NB: Some white grape varieties, particularly Chenin Blanc, can exude 'wet wool' aromas as part of their varietal characteristics in sound, unoxidised, wines. Oxidised wines

generally have a loss of structure, balance, and complexity and, in the case of red wines, tannin. There may be stewed or baked fruit tones and prune flavours, especially those of cooked prunes. Fig and prune aromas may also be beneficially present in certain wine styles – particularly in some fortified wines, including Ports.

5.4 Deliberately Oxidised and Highly Oxygenated Wines

5.4.1 Fortified and Speciality Wines

As can be seen in Section 5.1.4, there are many wines where a high level of oxygenation is part of the production process, and some of the characteristics of very high levels of acetaldehyde or of oxidation are key to the sensory profile. The wines mentioned below are a few of the more important examples.

5.4.1.1 Speciality Unfortified Wines
- Vin Jaune, from France's eastern region of Jura.
- Château-Chalon, an appellation totalling 50 hectares, exclusively for Vin Jaune – made with the Savagnin grape variety. The wines must be aged for 60 months under a veil (voile) of 'flor-like' yeasts, which leads to high levels of acetaldehyde but controlled oxidation. The wines are expensive starting at €75/£65.
- Tokaji, 'old'-style wines.

5.4.1.2 Fortified Wines
- *Vin Doux Naturels*: Rivesaltes Rancio, Banyuls and Banyuls Rancio, and Maury. Each of these has a very distinctive and individual style. Some are matured in an oxidative environment, but others are now made in 'fresher' styles.
- *Sherries*: Fino and Manzanilla styles, aged under a film of 'flor' yeast, may have strong aromas of acetaldehyde, which contribute to their individual aroma profile. Amontillado, Palo Cortado and Oloroso styles, which undergo oxidative ageing in butts that are deliberately left on ullage, have the nutty and cooked dried fruit aromas of oxidation.
- *Ports*: Reserve Tawny, including 10-year-old, 20-year-old, 30-year-old, 40-year-old, and Colheita Ports. These have been aged in wooden 'pipes' of 534/550l capacity, and the nutty notes are regarded as an essential part of the aroma profile.
- *Madeira*: Particularly the Bual and Malmsey styles.
- Australia's Rutherglen Muscats in the Classic, Grand, and Rare styles and Classic Topaques.

5.4.2 Deliberate Oxidation in High-Quality 'Table' Wines

As mentioned in Section 5.1, there are many producers of high-quality wines who deliberately allow, and usually control, oxidation in some of their wines.

In the region of Rioja, in the north of Spain, wines in very diverse styles are produced, but today most are much fresher than the oxidative wines made up until the 1970s. However, there remain a few producers of Rioja Blanco crafting showcase oxidative wines with the broadest spectrum of aromas and flavours. A classic example is Lopéz de Heredia's Viña Tondonia Reserva Blanco.

In California, there are some artisan producers who are thinking outside of the box and working with oxygen, or more correctly the atmosphere, defying all the rules of winemaking and producing some exhilarating wines. Abe Schoener of the Scholium Project, a former professor of philosophy at St John's College, Maryland, is a brilliant example. He crafts in tiny quantities some totally crazy and stunning wines, and each wine is made in a very individual manner. Explaining his 2011 Babylon, made from 100% Petite Syrah, Abe states, 'we matured the wine for 36 months in two neutral 220 l barrels, without topping or SO_2 (sulfur dioxide) for the first two years, and only one topping and SO_2 addition in its whole life. We think that the slow constant exposure to oxygen, and simultaneous undisturbed resting on the lees, are essential to the development of the wine. The wine is bottled with zero free SO_2 and about 30 mg/l total (SO_2)' [9]. Most of the Scholium Project whites are made in an oxidative way. According to *Imbibe*, Schoener said, 'The wines are made in 60-gallon barrels with about five gallons of headspace. It's a crazy amount of breathing space, but it makes the wine so intense' [10].

5.4.3 The Strange Case of Orange Wines

'Orange Wines' is the newest and oldest category of wines on the market. Put simply, these are white wines made using the methods of red winemaking. Grapes are crushed but not pressed and allowed to ferment in contact with the skins and other grape solids, picking up colour and phenolic aromas and tastes from the skins in the winemaking process. Whilst many writers and critics describe the wines, or at least some of them, as having the characteristics of oxidation, Simon J. Woolf, who is a specialist writer on the topic, is at pains to point out that this is not generally the case. He says, 'as Orange Wines are skin fermented, there is a natural anti-oxidant effect from the polyphenols in the grape skins – so Orange Wines made with low intervention (as most are) are actually less likely to suffer from issues with oxidation than correspondingly made white wines, which don't have this natural built-in protection [personal correspondence].

5.4.4 Grape Varieties Susceptible to Oxidation

It is generally accepted that some varieties are more prone to oxidation than others. Of the red varieties, Grenache is known to oxidise easily, and of white varieties, Grenache Blanc, Malvasia, Palomino, Picpoul, Pinot Blanc, Riesling, and the Muscat family are sensitive. Wines from these varieties often show oxidative characteristics, and in many cases this can add complexity to what otherwise could be simple wines. Conversely, red wines made from Syrah may show aromas of reduction if inadequately oxygenated during the production process, or inappropriately bottled, and Sauvignon Blanc wines can have reductive characteristics as a trademark. Oxygen management in wine production is a skill: there is no one size fits all, and although winemakers may have basic recipes, breaking the rules can sometimes lead to the most exhilarating wines. Unfortunately, it can also lead to ruination.

5.5 Metal Ions and Substrates for Oxidation

5.5.1 The Role of Metal Ions

Molecular oxygen is in a triplet ground state containing two unpaired electrons and accordingly cannot form bonds by accepting electron pairs. However, the addition of a single electron, originating from a transitional metal ion, which is consequentially reduced, can overcome this limitation [7]. The bonding of the single electron leaves an unpaired electron in the resulting negatively charged superoxide radical, and a second transfer of an electron results in a peroxide anion. Iron and copper are two transition metals present in must and wine – without these metals chemical oxidation does not take place. Enzymatic oxidation in must takes place due to copper-containing enzymes. Hydrogen peroxide (H_2O_2) is formed during the oxidation of phenols, and in association with ferrous ions, hydrogen peroxide forms hydroxyl radicals, which are highly reactive [6] and can oxidise many wine compounds.

5.5.2 Substrates for Oxidation

Phenolic compounds are the primary substrates for oxidation, but there are many other substrate compounds in must or wine, including some metal ions, SO_2, sugars, lipids, organic acids including amino acids, and ethanol. Polyphenols, especially *o*-diphenols (flavanols), are considered to be the major and most readily oxidised compounds [6, 7], the process being catalysed by iron. White and rosé wines are more prone to oxidation than red wines, which have higher levels of polyphenols that paradoxically give some protection. Although polyphenols are major substrates for oxidation, they are also antioxidants, scavenging free radicals and chelating metal ions. The concentration of polyphenols in red wine is much greater than in white wine – a typical red wine might contain 1–5 g/l (solidly structured wines usually contain the higher levels), contrasted with 0.2–0.5 g/l in white wines. With higher levels of phenolics, wines are able to consume larger amounts of oxygen before there is a significant deterioration in appearance, aroma, and/or palate.

5.5.3 Classes of Phenolics

Phenolics are largely located in grape vacuoles and will be released upon damage, including crushing and pressing. There are two basic classes of phenolics that are derived from grapes: flavonoids and non-flavonoids.

5.5.3.1 Flavonoids
There are four main classes of flavonoids in wine: anthocyanins, catechins, flavonols, and condensed tannins. Anthocyanins are located in the skins of red varieties and are primarily responsible for the colour of red wines but have no aroma or flavour impact. Catechins, also known as flavan-3-ols, are located primarily in the pips and have a bitter taste. Flavonols are located in grape skins and are strong antioxidants. Lastly, condensed tannins, also known as proanthocyanidins, are located in grape skins and seeds and to a small extent the flesh. They are the most abundant class of phenolics found in grapes (and red wines) and give astringency and 'grip' on the palate.

5.5.3.2 Non-flavonoids

Non-flavonoid phenolics are primarily located in the pulp of grapes. The two most well-known classes of non-flavonoids present are hydroxycinnamates and stilbenes. Hydroxycinnamates, of which caftaric acid is the most abundant, are present in the pulp of the grape. Hydroxycinnamates are odourless and tasteless. In white wines, they are the principal phenolics. Of the stilbene class, resveratrol is perhaps the best known.

Hydrolysable tannins are also non-flavonoids. In wine, these are not derived from grapes, but from oak barrels and oak products. Ellagitannins are important hydrolysable tannins noted for their strong antioxidant properties. Gallic acid, another non-flavonoid, is also a powerful antioxidant. Gallic acid is not found in grapes, but is generated in wine by the hydrolysis of the gallate esters found in condensed and hydrolysable tannins [11].

5.6 Enzymatic Oxidation

5.6.1 Substrates

Enzymatic oxidation will be briefly discussed first, as it almost always takes place before fermentation, resulting in the browning of must, the reduction in and/or destruction of desirable grape flavour compounds, and the synthesis of undesirable flavour compounds. In addition to giving off-aromas and off-flavours, these can mask desirable aromas. Enzymatic oxidation particularly impacts white grape must and can happen very quickly. This can easily be understood if one considers how quickly an apple goes brown after cutting or biting. However, enzymatic oxidation also effects red musts, resulting in a loss of polyphenols [7]. When must is first exposed to air, oxygen uptake is rapid and can range from 0.5–5 mg/l per minute (contrasted with 1.95–6.83 mg/l per day in wine) [12] but decreases as the caftaric acid substrate is depleted [13]. Non-flavonoid hydroxycinnamic acids, particularly caftaric (also known as caffeoyltartaric acid) and coutaric acid (also known as *p*-coumaroyltartaric acid), and their esters provide the main phenolic substrates for enzymatic oxidation. Accordingly, enzymatic oxidation can be limited to some extent by reducing the phenolic substrates, especially the non-flavonoids. This can be helped by soft pressing and the avoidance of skin contact.

5.6.2 Polyphenol Oxidases, Tyrosinase, and Laccase

In order to catalyse the enzymatic oxidation process, it is necessary for one of two copper-containing enzymes to be present: a polyphenol oxidase (PPO) or laccase. PPOs are a group of metalloproteins that, in the presence of the oxygen (including the atmosphere), have catechol oxidase ability, oxidising certain phenols (*o*-diphenols) into *o*-quinones. Some PPOs have tyrosinase activity, which hydroxylises monophenols to *o*-diphenols [14]. The oxidised quinones can be reduced back to the original catechol through a coupled oxidation of ascorbic acid or other phenolic compounds, including flavonols or sulfur dioxide (SO_2) [15].

PPOs are also responsible for the enzymatic browning of many fruits other than grapes (especially apples and pears) and vegetables and even of crustaceans. It might be noted that many authors consider PPO to be an individual enzyme (not part of a group), that tyrosinase is also called PPO, and that laccase is a PPO.

Tyrosinase is located in the chloroplasts of the grape berry and hence forms part of the grape berry's cell membrane system. It has the ability to catalyse the oxidation of monophenols, o-diphenols, and catechols. Accordingly, caftaric acid and p-coumaric acid are oxidised by PPO, producing o-quinones that are powerful oxidants and able to oxidise other compounds in wine. o-Quinones can almost oxidise any substrate with lower potentials such as other phenols, ascorbic acid, and SO_2 [7].

5.6.3 Glutathione and Grape Reaction Product (GRP)

Glutathione (GSH) is present in grapes at varying concentrations, with a chemical formula of $C_{10}H_{17}N_3O_6S$. It is a tripeptide of glutamic acid, cysteine, and glycine and is the most abundant non-protein intracellular thiol [16]. It is a strong antioxidant and as such will slow the process of must oxidation through reacting with the o-quinones produced from caftaric acid by tyrosinase, to produce 2-S-glutathionyl caftaric acid, commonly known as grape reaction product (GRP). GRP is not a substrate for any further oxidation by tyrosinase, and it will limit the oxidative browning. Accordingly, if there is enough GSH in grape must, it can trap the quinones produced during enzymatic oxidation and decrease the amount of browning pigments [7]. When the GSH is used up, quinones will react with the phenolic compounds. However, in a 2018 published study of Chardonnay wines, Jordi Ballester et al. concluded that pre-fermentation actions, such as must protection, which reduce the production of GRP and maintain high levels of flavanols, could lead to lower oxidation notes in finished wines [17].

5.6.4 Prevention of Tyrosinase-Catalysed Enzymatic Oxidation

In juice, tyrosinase is attached to grape particles. It is not particularly soluble, and levels can be reduced by must settling or juice clarification. It functions best at higher pH levels and is inactive in wine. It is also sensitive to SO_2 and so can be rendered largely inactive by adding SO_2 or potassium metabisulfite to the picking bins/grape crush/must. The concentration of SO_2 required at the must stage to inactivate tyrosinase usually ranges from 25 to 50 mg/l. Sulfur dioxide denatures the PPO enzyme that can no longer function as its shape is changed in this process. Must may be clarified by settling (débourbage) with or without chilling, centrifugation (now often regarded as a harsh process), or fining. The addition to must of adsorbent copolymers of polyvinylimidazole/polyvinylpyrrolidone (PVI/PVP) or activated chitosan can reduce PPO activity by removing copper. These products are approved by the European Union (EU) [18] and the Organisation Internationale de la Vigne et du Vin (OIV) [19]. Accordingly, enzymatic oxidation consequential to the action of tyrosinase may be regarded as entirely preventable.

5.6.5 Laccase

Laccase is produced by moulds contained in and on *Botrytis-infected* grapes. It has the ability to catalyse the oxidation of a wide range of substrates. Laccase will readily oxidise GRP [6].

It is soluble, and settling operations have limited effect of reducing concentration. Further, laccase is much more resistant to SO_2 than is tyrosinase.

5.6.6 Hyperoxygenation

In the case of must for white wines, there is an alternative juice handling technique that is diametrically opposed to protecting juice from oxidation. White musts that are protected from oxidation result in wines that are much more likely to be oxidised [13]. Hyperoxygenation (also known as hyperoxidation) is the deliberate browning of juice prior to fermentation by means of a high oxygen addition in order to remove potential browning components from the juice [20]. It is a technique that may only be used for white must, but it is not appropriate if a fresh, reductive style of wine is required. Phenolics are the major oxygen-consuming substrate in wine. Hyperoxygenation of must oxidises these, and they settle. The must is then removed from the precipitate by racking.

Hyperoxygenation decreases the browning potential of wine by the following:

- The disappearance of tyrosinase and the depletion of phenolic substrates that could be subsequently oxidised [7, 15].
- The formation of GRP that is resistant to further browning [7].

Accordingly, wines made from hyperoxygenated musts will have low polyphenol and high GRP contents and accordingly are more resistant to oxidation than those made from non-oxidised juice, which contain high polyphenols retaining a high browning potential [6, 7, 13]. As it is important that tyrosinase activity proceeds unhindered, little or no SO_2 should be added to the harvested grapes, at crushing, or to the crushed must. A maximum SO_2 level of 35 mg/l (3.5 g/hl) will reduce the microbial risk whilst maintaining the tyrosinase activity [21]. At a SO_2 concentration of 50 mg/l, tyrosinase activity is reduced by 75–90%.

There are several methods of oxygenating the must, including adding with a diffuser, pumping in whilst the juice is circulating in tank or injected in line when the wine is pumped from tank to tank [22]. The usual oxygen addition ranges from 8 to 30 mg/l over two or three hours. It is of the utmost importance that the oxygenated must is clarified before fermentation. Suspended solids should be reduced, and the majority of the phenolic precipitates eliminated. The effect of hyperoxygenation is cancelled when unclarified or poorly clarified juice is fermented [23], but it must be noted that excessive clarification can lead to fermentation problems [24]. Juice settling (débourbage) overnight at a low temperature to prevent the onset of fermentation, preferably in horizontal tanks to reduce the settling distance, may be sufficient. If flotation is used for clarification in conjunction with a fining agent, oxygen can be used as the flotation gas instead of nitrogen [22], thereby hyperoxygenating and clarifying in a single process.

Sulfur dioxide may be added following must clarification. However, hyperoxygenation is a valuable tool for producers who produce wines with no added sulfites, as it reduces the phenolic content early in the vinification process and helps stabilise the wine produced against future oxidation, one of the purposes for which SO_2 is conventionally used.

5.7 Chemical Oxidation

Chemical oxidation, by redox reactions, is the most usual form of wine oxidation. It is often referred to as non-enzymatic oxidation that can take place in both must and wine but usually takes place post fermentation. There are several pathways to chemical oxidation. The enzyme PPO does not exist in wine, due to the presence of alcohol and sulfur dioxide. Chemical oxidation is a much slower process than enzymatic oxidation. The oxidation of alcohol increases as wines age, and oxidation in bottle is commonly encountered.

5.7.1 Substrates

As in must, phenolics are the main substrates of oxidation in wines. Oxidation of polyphenols occurs due to the presence of catechol compounds such as catechin, epicatechin, catechin gallate, gallic acid, caffeic acid, and anthocyanins [25]. Ascorbic acid (normally considered as an antioxidant) and tartaric acid are lesser substrates. Particularly as wines age, ethanol becomes an important substrate as it can react with hydrogen peroxide to produce acetaldehyde.

5.7.2 Chemical Oxidation Pathways

Phenolic compounds are the main chemical oxidation substrates, and it is reaction between phenolics and their oxidation products that cause wine browning [7]. The process begins by the oxidation of polyphenols containing a catechol or galloyl group (anthocyanins and tannins) into semiquinone free radicals and quinones [6]. Phenols react indirectly with oxygen in the presence of transition metal ions, particularly iron but also copper, to produce quinones, and oxygen is reduced to hydrogen peroxide (H_2O_2) [22]. The transition metal ions are essential for the reaction. A significant amount of iron in must is lost during the fermentation process as it is consumed by the yeast, which may be from 25% to 80% of the original content. The amount of iron remaining in wine is usually between 0.5 and 5.5 mg/l. The concentration of copper is 0.2–0.3 mg/l.

Using the Fenton reaction, hydrogen peroxide can be reduced by iron to form a hydroxyl radical, a strong oxidising agent. The radical will oxidise ethanol to the 1-hydroxyethyl radical, which will in turn be oxidised to acetaldehyde [26]. Hydrogen peroxide will also oxidise organic acids, including tartaric acid, and in fact is a stronger oxidising agent than PPO, which is the major catalyst for enzymatic must oxidation.

5.8 Microbial Oxidation

Acetaldehyde may be derived from yeast metabolism during the alcoholic fermentation (AF) or produced from alcohol (together with acetic acid) by acetic acid bacteria in the presence of oxygen. This is usually the main source of acetaldehyde production. Paradoxically, although yeast typically reduces acetaldehyde to ethanol, when there is high oxygen (either surface or dissolved), yeast can also convert ethanol back to acetaldehyde. Protection from oxygen uptake by keeping containers full or gas-blanketed and if necessary sparging the wine of dissolved oxygen is key to limiting microbial oxidation.

5.9 Acetaldehyde

Acetaldehyde is also known as ethanal with a chemical formula of C_2H_4O. In addition to its sensory impact upon wine, it is known to be a contributor to hangovers. All wines contain some acetaldehyde, both in free and bound forms. It is the most important volatile wine carbonyl, and its formation can be by microbial and/or chemical pathways. As we have seen, the sensory threshold for acetaldehyde is 120–125 mg/l. In 2016, of wines submitted for analysis to Vintessential Laboratories in Australia, about 13% had a level in excess of 125 mg/l [27].

5.9.1 Biological Production

Acetaldehyde is produced as a fermentation product by yeasts as the penultimate step before the formation of alcohol (ethanol) [28]. As such it is particularly manufactured at the start of the alcoholic fermentation process. Sugar is first converted to pyruvate by yeast via the glycolytic pathway. The pyruvate is then decarboxylated to acetaldehyde with the associated release of carbon dioxide. As fermentation approaches completion, acetaldehyde intermediate is transported back into yeast cells and then reduced to ethanol. There is a risk of excessive acetaldehyde being produced during this process if:

- A commercial yeast strain that produces high levels of acetaldehyde has been used. There are many strains whose metabolic characteristics include low production of acetaldehyde.
- SO_2 is added during fermentation.
- There is an increase in pH.
- There is an increase in the temperature during fermentation. Of course temperature increases during early tumultuous fermentation are common and may be desired to bring the vat up to the required core fermentation temperature.

It might be noted that in the United States, acetaldehyde is permitted for use in grape must as a processing aid to stabilise colour.

In a well-made wine, most of the acetaldehyde actually stems from yeast activity, but it can also originate from the activity of other microbes, including lactic acid bacteria and acetic acid bacteria [11]. The production of excessive acetaldehyde can lead to stuck fermentation, as although it is produced by *Saccharomyces cerevisiae* paradoxically it is toxic to it. In the latter stages of the alcoholic fermentation, viable yeasts will degrade acetaldehyde, and malolactic bacteria will also significantly degrade the compound during the malolactic fermentation (MLF). Acetic acid bacteria contribute to the production of acetaldehyde, as an intermediate stage in the production of acetic acid – see Chapter 7.

Several surface film-forming bacteria and yeasts including *Pichia membranaefaciens* (also called *Pichia membranifaciens*) promote acetaldehyde production, particularly during bulk storage. Although normally regarded as spoilage yeasts, there are some valuable film-forming yeasts, including *Saccharomyces beticus*, the main yeast responsible for production of the surface 'flor' that forms on the surface of Fino and Manzanilla Sherries being matured in butts. These butts are deliberately left on ullage to enable its growth, and levels of acetaldehyde in these wines can be very high. The compound can also be produced post fermentation as the enzyme ethanol dehydrogenase catalyses the oxidation of ethanol to produce acetaldehyde.

5.9.2 Chemical Production

Acetaldehyde is also produced by the chemical autoxidation of ethanol in the presence of phenolic compounds. First, oxygen interacts with phenolic compounds in the wine. Hydrogen peroxide (H_2O_2) is produced as a by-product and is a strong oxidising agent. Second, hydrogen peroxide then interacts with alcohol (CH_3CH_2OH) and produces acetaldehyde (CH_3CHO) and water (H_2O). The alcohol and hydrogen peroxide reaction can be shown as

$$CH_3CH_2OH + H_2O_2 \rightarrow CH_3CHO + 2H_2O$$

Ethanol + Hydrogen peroxide → Acetaldehyde + Water

In an oxidative environment, levels can easily become unacceptable. In these conditions, acetic acid and ethyl acetate can also be formed. However, a study by Bueno et al. suggested that the ability of a red wine to accumulate acetaldehyde is negatively related to its content of aldehyde reactive polyphenols (ARPs), comprising anthocyanins and small tannins [29]. In this study, bottled red wines under two years old hardly accumulated any acetaldehyde, no matter how much oxygen they had consumed.

Methional and phenylacetaldehyde ($C_6H_5CH_2CHO$) are two aldehyde compounds associated with the oxidised aroma of white wine [28]. Methional contributes cooked or even rotten potato odours (sensory detection threshold of 0.5 µg/l), and phenylacetaldehyde has honey aromas, sometimes with notes of old roses (sensory detection threshold of 25 µg/l) [30]. The compounds can be formed from the amino acids methionine and phenylalanine by the Strecker degradation or alternatively from the respective peroxidation of methionol and phenylethanol, which are fermentation products. It is suggested that the alcoholic fermentation can be a major origin for methional, phenylacetaldehyde, and other Strecker aldehydes [31].

5.9.3 Acceptable and Unacceptable Levels of Acetaldehyde

An acceptable acetaldehyde level in a well-made light red wine is in the region of 20–40 mg/l, with an average of 25 mg/l [32] or 30 mg/l [33]. In fact, the level of acetaldehyde in red wine can vary from 4 to 212 mg/l [33]. White wines have an average of 40 mg/l [32], although certain styles commonly exhibit 60–80 mg/l. The range in dry white wines is 11–493 mg/l and 188–248 mg/l in sweet white wines [33]. Vins Jaunes from France's eastern region of Jura and Fino Sherries, which have undergone a lengthy microbiological ageing under the film of 'flor' yeast, often contain as much as 300 mg/l. Sherries generally can have a concentration that varies from 90 to 500 mg/l. As we have seen the elevated acetaldehyde content of these wines is part of their very character, but in most other wines the high levels noted would be regarded as a serious fault.

5.9.4 Binding of Sulfur Dioxide

Acetaldehyde is the main compound that binds SO_2, thereby reducing its effectiveness. In a dry wine, about 45% or so of the total SO_2 content will be bound to acetaldehyde. The reaction between acetaldehyde and bisulfite produces hydroxysulfonate that is odourless. This reaction is rapid, and, at a pH of 3.3, 98% of the acetaldehyde will be combined with the sulfite within 90 minutes [11]. A high ratio of bound to free SO_2 often indicates that a large proportion of the added SO_2 has being remedying an oxidation problem by combining with acetaldehyde and other carbonyl compounds. Acetaldehyde also binds to other wine compounds, particularly pyruvic acid and 2-oxoglutaric acid (ketonic acids) [13].

5.10 Sotolon

Sotolon (3-hydroxy-4,5-dimethylfuran-2(5H)-one) is commonly found in certain fortified and 'flor'-aged wines with a chemical formula of $C_6H_8O_3$. Sotolon has two enantiomers, R and S, each of which has individual olfactory characteristics, and the ratio between these two varies from one wine to another. The S form has the most prominent aromas that include curry and walnuts. The R form smells largely of rancid walnuts. Other sensory characteristics associated with sotolon include honey and maple syrup (low concentration of the compound), caramel, toffee, spicy, and 'burnt' aromas. The sensory detection threshold of sotolon generally has been reported as 8 µg/l [28]–15 µg/l [34]. In 'flor'-aged Sherries, it has been found at levels of 22–72 µg/l where it contributes to nutty aromas, and in Vins Jaunes at 75–143 µg/l [34]. Sotolon has been determined as a key compound giving rise to the typical aromas of oxidatively aged (Tawny) Ports [35, 36].

Sotolon's 'off-flavour' characteristic is often found in oxidised white wines overlaying any fruity or flowery notes. The compound can be formed by several pathways. It may be produced by an oxidative mechanism, based on the peroxidation of acetaldehyde [37], and its high level in some fortified wine, including many Madeiras, may be due to this pathway. The oxidative degradation of ascorbic acid in the presence of ethanol is also an important mechanism, and related aromas often develop in bottle, particularly when ascorbic acid has been added as an antioxidant. In dry white wines, it can also be formed by a reaction between 2-ketobutyric acid and acetaldehyde [38]. Sotolon is sometimes found in mature and over-mature red wines and is regarded as a marker of oxidation. When studying the influence of oxidation in the aromatic composition and sensory profile of Rioja red aged wines, Mislata et al. were surprised to find that sotolon was only detected as active compounds in the oldest wines with low prices [39].

5.11 Oxygen Management in Winemaking

The topic of oxygen management in winemaking is broad and includes desired oxygen additions, unwanted oxygen uptake, scavenging of oxygen by yeast lees, and (indirect) reactions with sulfur dioxide. Aspects of oxygen management, including the utilisation of micro-oxygenation, will be further discussed in Chapter 6.

5.11.1 The Role of Oxygen During Fermentation

Although *S. cerevisiae* can ferment in anaerobic conditions, the presence of oxygen is important in boosting yeast colonies to levels desirable for a trouble-free alcoholic fermentation. Lack of sufficient oxygen can lead to stuck fermentations and the production of hydrogen sulfide (H_2S). Depending upon handling, and unless protected by inert gas (or dry ice) or hyperoxygenated, grape musts will usually contain in the region of 5 mg/l of dissolved oxygen. During fermentation, oxygen is required to create, boost, and maintain healthy yeast colonies: yeasts scavenge oxygen that strengthens their cell walls and helps them resist stress due to high alcohol levels. Yeasts need oxygen to produce sterols and unsaturated fatty acids that play a key role in the fluidity and activity of membrane-associated enzymes, which influence ethanol tolerance, fermentative capability, and viability of yeast [22]. Oxygen will help regulate the redox potential helping to prevent the production of hydrogen sulfide and reductive aromas, as discussed in Chapter 6. At the start and up to the end of the yeast growth phase, which is approximately for the first third of the fermentation process, about 4–6 mg/l of oxygen are required to ensure a successful finish to the alcoholic fermentation for white wines, with perhaps 6–10 mg/l being ideal for red wines. From the point of view of a smooth and complete alcoholic fermentation, the optimum time for oxygen additions is generally 36–48 hours into the fermentation or up to 72 hours after yeast inoculation. Additions will, inter alia, beneficially oxidise some volatile sulfur compounds to less odorous forms [40]. Almost all oxygen added at this stage will be consumed by yeast or dissipated with the rising carbon dioxide.

The benefits of adding significant amounts of oxygen to fermenting red wines include the following:

- Less need to add nitrogen supplements, e.g. diammonium phosphate (DAP).
- Prevention of the formation of low levels of volatile sulfide off-odours.
- Softening of tannins during fermentation. This may reduce the time required for cellar maturation prior to bottling and make the wine available for market several months earlier [41].

Contrary to the perceptions of many winemakers, there are many benefits to the managed addition of oxygen during white wine fermentation, with the main impact on the kinetics of fermentation [42].

5.11.2 Post-fermentation Oxygenation

Post-fermentation oxygenation may take place in three distinct phases:

- Pre-MLF to stabilise colour and build structure.
- Post-MLF to integrate pyrazines, soften tannins, reduce astringency, lessen reductive tendencies and for general structural refinement.
- Post-barrel ageing to harmonise oak tannins and negate pre-bottling reduction [43].

During maturation, a controlled amount of oxygen uptake can help polymerise long-chain tannins, improve texture, reduce astringency, and, by reacting with anthocyanins, improve, brighten, and fix colour.

5.11.3   Barrel Maturation

The process of barrel maturation is generally the winemaking procedure that results in the largest uptake of oxygen. The total annual uptake can perhaps range between 12 mg/l per year for a 225 l barrel [44] and 20–45 mg/l [45]. Barrel maturation, particularly in new barrels, also involves the absorption of oak products, including vanillin and tannins. The antioxidant role of ellagitannins, which are hydrolysable tannins absorbed from oak wood, is discussed in Section 5.16.17.

5.11.4   Lees Ageing

Yeast lees have a tremendous capacity to scavenge and consume oxygen. Ageing white wine on either gross lees or fine lees in tank or barrel can help maintain freshness and give creamy textural characteristics. The dead yeast cells absorb dissolved oxygen, and GSH and cysteine contained in lees can assist in the protection from oxidation. The implications of the process of lees stirring or bâtonnage are discussed in Section 5.15.3.7.

5.12   Oxygen Uptake During Cellar Operations

Having briefly considered the desirable and controlled oxygen additions during winemaking and barrel maturation, there are numerous occasions when there will be additional pickup during processing. Every cellar operation is a possible source of oxygen uptake, and those taking place after the end of the alcoholic fermentation can lead to an increase in acetaldehyde concentrations [32]. Some activities take place only once, but others, such as racking and topping, may be required on multiple occasions. The highest uptake is normally at the beginning and end of these operations; this can be considerably reduced if equipment is purged with inert gas.

Below is a list of indicative figures of oxygen uptake for each operation or stage. The figures given are averages or ranges using data from over 30 sources, including research papers, oenology books, and my own and colleagues' measurements in real winemaking situations. Some of the ranges stated are wide. One of the more recent research papers covering some of the operations, using a limited range of equipment, is by Calderón et al. [46]:

- Crushing grapes, 5–10 mg/l.
- Topping of barrel, 0.25 mg/l.
- Pumping, 0.1–2 mg/l.
- Pump-over, up to 8 mg/l depending upon method:
  Closed-circuit pump-over: less than 0.5 mg/l
  Splash-rack pump-over: 1.5 mg/l
  Pump-over with an in-line venturi: 2.5 mg/l
  Pump-over with diffusion stone with air: 4 mg/l.
- Racking, 0.5–3.5 mg/l.
- Transfer tank to tank, up to 4 mg/l (more if there is accidental oxygen ingress through a loose hose union). If the tanks are blanketed with nitrogen or argon, and hoses and pump flushed, the uptake may be 0.2 mg/l.

- Fining, 1.3–1.6 mg/l.
- Filtration, rotary vacuum, 2–4 mg/l:
  Diatomaceous earth: 0.14–0.7 mg/l – more in the early segment of the process due to the air-filled interstices in the filter medium.
  Sheet: 0.22–0.43 mg/l.
  Cross-flow: 0.18–0.51 mg/l.
  Membrane cartridge: 0.12–0.51 mg/l.
- Centrifuging, 0.35–2.5 mg/l, very careless up to 8 mg/l [22].
- Barrel ageing, 20–45 mg/l per annum.
- Cold stabilisation, 1.22–3.5 mg/l.
- Bottling, 0.2–4 mg/l. This is the last step and can ruin all previous good work. Oxygen uptake during the operation can result in a rapid loss of protective SO_2 in bottle.
- Transport of bulk wine, 0.1–2 mg/l.
- Uptake in bottle ageing, 0.005–5 mg/l per annum.

Methods to limit oxygen uptake in these operations/stages will be discussed in Section 5.16.

5.13 Containers and Closures

The role of oxygen in maturation and ageing of wines in bottle, or other container, is complex. It is often stated that the uptake of oxygen does not improve white wines in bottle, but interestingly it would appear that even a wine made from Sauvignon Blanc, a variety often loved for its fresh and sometimes slightly reductive characteristics, can actually be preferred when allowed a little oxygen during the ageing process [47]. The influence of the type of container and closure upon packaged wine and in particular its oxygen uptake and dissolved oxygen content has also been the subject of considerable research in the last 20 years or so, much of it funded by packaging manufacturers, especially the makers of closures. Glass bottles and cans are impermeable to oxygen, but wine packed in 'bag-in-box' or polyethylene terephthalate (PET) bottles, usually only used for inexpensive product, will suffer some uptake. There is a considerable variation in oxygen ingress into wine depending upon the type of closure. Closures with high oxygen transmission rates (OTRs) can lead to the development of significant oxidised characteristics, and even moderate OTRs can lead to the development of cooked and stewed fruit characteristics [48]. The topics of containers, closures, and storage are discussed further in Sections 5.16.21 and 5.16.22 and Chapter 17.

5.14 Pinking

Pinking is a phenomenon that may affect white wines that have been made under reductive conditions. Although many varieties are susceptible, there is a relatively high incidence with Sauvignon Blanc, Albariño, Loureiro, Malvasia, and Verdejo. Affected wines may take on a salmon pink or blush colour. Whilst the issue may occur at any time after (or occasionally before) the alcoholic fermentation, it is most likely to appear after bottling. It usually occurs when wines have a sudden exposure to oxygen, as may take place during bottling operations. The origin is anthocyanins contained in grapes, mainly malvidin-3-O-glucoside [49]. There are alternative hypotheses of the mechanisms of the pinking phenomenon.

It is possible that early picking and de-stemming of grapes may help reduce the phenomenon [50]. The pinking potential of a fermented wine may be tested with hydrogen peroxide – 1.25 ml of 0.3% hydrogen peroxide is added to 100 ml of wine, and the colour changes are observed after 12 hours. Wines suffering from pinking may be corrected by fining with polyvinylpolypyrrolidone (PVPP), PVPP plus bentonite, and PVPP plus ascorbic acid. These may also be used as a precaution to prevent pinking. Fining with potassium caseinate is also effective. The amount of addition required should be determined by laboratory trials. Other possible methods to help correct pinking tint are the exposure to UV light [51] or the use of ultrafiltration – see Chapter 16.

5.15 Premature Oxidation (Premox)

5.15.1 Premature Oxidation Rears Its Head

Premature oxidation is generally referred to as 'premox' and, keeping in mind the losses suffered by many wine collectors, is perhaps unsurprisingly sometimes further abbreviated to 'POX'. A bottled wine may be described as premoxed when it shows the characteristics of oxidation at a point in its life cycle when, in normal circumstances, it would be immature or just mature. The problem has been particularly evident with white Burgundies but also impacts other white and red wines. Impacted wines show many of the classic oxidation characteristics. White wines exhibit shades of orange and brown and have a flat appearance. The nose can have loss of fruit, tones of honey, beeswax, and nuts. Sotolon, formed from acetaldehyde, is regarded as a universal marker of premoxed wines, and in severe cases, the wines have the appearance and nose of Sherries that have had oxidative ageing and aromas of curry and, at worst, are undrinkable. Red wines take on garnet and brown colour, with strong aromas of prunes, figs, and dried or soaked fruits.

Premox became a particularly pertinent topic in 2002, with the first discussion in the press, relating to the problem with some white Burgundies, appearing that year in Steve Tanzer's bimonthly magazine, *International Wine Cellar* [52]. Interestingly, premox is sometimes linked by researchers with 'atypical ageing' that had been first noted in some German wines in 1988 (see Chapter 8). It was initially considered to be a random fault affecting some white Burgundies and impacting just a small percentage of bottles. However, in 2005, many wine critics, journalists, and collectors of 'fine' white Burgundy were regarding it as a major problem. In fact, the vintages badly affected had begun with the 1995 and 1996 harvests, and many wines produced in the next 15 years show the fault. Although incidences of the problem have declined in recent years, there is still much dispute as to the reasons for the particularly high number of affected bottles from many years starting in the late 1990s. Although at first the issue was believed to be a problem that only affected white Burgundies, but in vintages from the early years of this century, the fault was found in bottles of dry white wines from Bordeaux and subsequently in other whites. Writing in *The World of Fine Wine* magazine in 2014, Burgundy expert Jasper

Morris MW noted that he has experienced the phenomenon 'with California Chardonnay, white Rhône wines, white Bordeaux including Sauternes, and Loire and Alsace wines, to offer a selection that comes immediately to mind but is not exhaustive' [52]. In recent years, red wines, including some crus classés from Bordeaux and fine red wines from other regions, have also been implicated. The late professor Denis Dubourdieu whose team conducted research into the 'prune aromas' of prematurely aged red wines said, 'I believe there is a similar scandal with red wine, and that in 10 years' time it will be just as explosive as the one affecting white Burgundy has been. And it's not limited to one region... [53]'. However, Burgundy was the region that really sent shockwaves to wine lovers and induced a degree of panic amongst collectors of its 'fine' wines. In his book *My Favorite Burgundies* Clive Coates MW (2013) [54] noted gloomily '...reports began to circulate of more and more oxidized bottles, and from just about every single estate, even the very prestigious – Lafon, Bonneau du Martray, Roulot, Ramonet. No one seemed to be spared. It was rather like a very contagious plague'. According to Coates, the worst years for Burgundies exhibiting premox were 1996, 1997, and 1998. He and other authors and critics noted that (anecdotal) evidence indicated that some domaines seem to have had a greater propensity for the problem, which included Sauzet, Colin-Déleger, and Domaine Leflaive. Interestingly, Domaine Leflaive seemed to have escaped the first 'bout' of affected wines, i.e. those from vintages in the late 1990s. However, it is well documented that the Domaine suffered badly from premox in the vintages from 2006 [55, 56]. Coates and other observers noted that there appeared to be less incidence at Domaine Coche-Dury and the flagships Domaine Leroy and Domaine de la Romanée-Conti [54]. Whilst incidences of premox have declined in recent years, the problem and perceptions of it have certainly not gone away. Writing in the influential magazine *Decanter* in August 2017, William Kelley noted, 'What's more, in the era of premature oxidation, cellaring of many of the Côte de Beaune's whites for more than a decade has become a dubious proposition...' [57]. On 12 November 2017, there was a series of forum posts on 'WINE Beserkers' that describes itself as 'the world's largest and most active online wine community', entitled 'Premox is still with us big time' [58].

5.15.2 Reversible Oxidation?

Although oxidation is an irreversible process, there is anecdotal evidence that, following the passage of time, wines that had been considered to be suffering from premox became 'unoxidised' again. Jasper Morris MW and Michel Bettane (the highly regarded French wine critic) are amongst those who have noted the apparent reversal of apparent oxidation [52], raising the question as to whether the symptoms of premox really relate to oxidation at all! However, as premox has often been found to affect bottles at random, one can never know whether those in question were, or were not, suffering from the condition. Perhaps more interesting are impressions that some oxidised wines have become unoxidised after some time in the glass [personal discussion with some Burgundy amateurs]. It might be noted that consumer perceptions have changed, and it may be argued that even some lovers of fine wines have become conditioned to expect aged wines to exude primary fruits and perhaps do not tolerate the oxidative characteristics that can be part and parcel of advanced tertiary development.

5.15.3 Causes of Premature Oxidation

5.15.3.1 Possible Causes

There has been much discussion amongst researchers, producers, consultants, critics, and knowledgeable consumers as to the possible causes of premox. These include the following:

- Clonal selection of grape varieties.
- The harvesting of fruit with greater phenolic ripeness and higher pH than was historically the case.
- Lack of GSH in the fruit and resulting wines.
- Less oxidative pressing than in the past particularly with the advent and widespread use of membrane presses and tank presses from which oxygen may be excluded.
- The use of winemaking enzymes.
- The increased use of new oak barrels.
- The role of bâtonnage.
- Insufficient sulfur dioxide additions relative to the pH of the wine.
- Interventionist operations undertaken in order to speed the wine to market.
- Ingress of oxygen when bottling.
- Low-quality cork closures and corks washed in peroxide.

Each of these possible reasons warrants consideration.

5.15.3.2 Clonal Selection

It is generally accepted that low levels of phenolics promote oxidation. Grégory Viennois, winemaker at Chablis Domaine Laroche, believes that clonal selection bears much responsibility [59]. 'If we compare a grape from a clone with one from a massale selection, then you find that when you taste the grapes, the one from the massale selection has a lot of tannins, it has thick skins, and with the clonal selection, you don't have that', he said. Continuing, he said that 'as a measure to ensure Laroche wines will age and develop slowly, the Domaine has now decided to protect its old vine stock, and only replant by taking massale selections from such vineyards' [59].

5.15.3.3 The Phenolic Ripeness and High pH of Fruit

There is no doubt that growers are harvesting fruit with greater ripeness of both sugars and phenols than was the case up until the mid-1990s. The topic of climate change is as much discussed in the wine industry as elsewhere, and meteorological figures show that, on average, summers have been warmer in the Burgundy region in the last couple of decades. It can be argued that producers in the 1990s lacked experience in handling such ripe fruit, and consequently disastrous mistakes were made. Further, many growers have perhaps become a hostage to the concept of phenolic ripeness, delaying harvesting until the grape skins are translucent and the seeds totally brown. Of course with the increase in ripeness, the pH of the must is higher, which presents a more favourable climate for the growth of bacteria. As grapes ripen, acidity levels decrease, sometimes necessitating the addition of tartaric acid to grape must, where such additions are legal. Tartaric acid and other grape acids are antioxidants and essential components of the structure of a wine to achieve satisfactory ageing. Although a high alcohol level is regarded by many as key to the longevity of both red and white wines, the level of acidity is every bit as, or even more, important. I still have in my wine store several old Mosel wines from great estates. Those from less 'ripe' years, whose acidity was perhaps regarded as swingeing in youth, still taste wonderful. The 1987 and 1997 vintages of 'Wehlener Sonnenuhr Riesling Kabinett from S.A. Prüm' have an alcohol level of just 8%.

Of course with an increase in grape ripeness, the pH of the must is higher, which also presents a more favourable climate for the growth of bacteria. Dr Valérie Lavigne from the Faculty of Oenology at the Bordeaux Institute of Vineyard and Wine Sciences said that 'as a result of leaving bunches longer on the vine, more grapes are overripe, and we think that this means that the wines are losing their ageing ability' [60]. She noted that this is 'a particular problem in Bordeaux for the Merlot'.

5.15.3.4 Low Levels of Glutathione in Grapes

Hydric stress and low levels of soil nitrogen result in low vine vigour. The late Denis Dubourdieu of the University of Bordeaux 2 found that this results in one of the main causes of premox. He stated, 'Vines that are too weak, and with a poor nitrogen intake, produce grapes low in GSH. Summer drought conditions and/or competition from grass left to grow between the vine rows also worsen this deficit' [61]. It is certainly my experience that many white wines from the hot and drought-impacted 2003 vintage in Bordeaux showed all the characteristics of premox. It was already well known that hydric stress of the vines and high yields invariably lead to low levels of nitrogen in the must, resulting in many winemakers making DAP additions in order to avoid delayed, sluggish, or incomplete fermentations. GSH is permitted to be added to musts or at the start of fermentation at a maximum dose of 20 mg/l by the OIV [19] and EU [18]. It is important to ensure that at the start of and during alcoholic fermentation, the level of yeast assimilable nitrogen (YAN) is sufficient to avoid the metabolism of GSH by the yeast.

5.15.3.5 The Use of Pneumatic Presses

Traditionally, pressing was always a somewhat oxidative process. The basket press, introduced by the monks of the middle ages, continued in use for centuries and only began to be widely substituted by horizontal plate presses after World War II, when a version of the 'Vaslin' press, which had been patented in the mid-1920s, began mass production. It is interesting to note that basket presses have undergone a renaissance in the last decade or so, being used not only for white wines but also for some red wines, and are particularly favoured for pressing Shiraz/Syrah. Horizontal plate presses also allow considerable oxygen uptake, particularly as the plates are released, the press cylinder is inversely rotated and the loosening grape cake is crumbled. Both of these press types yield juice with a high content of grape solids. Although horizontal pneumatic membrane presses had been invented in the 1950s, the advent of their use in many Burgundy vineyards was not until the mid-1990s. With this type of press, the uptake of oxygen in the pressing process was considerably reduced, and the introduction of the fully enclosed horizontal tank press, flushed with inert gas, enabled oxygen pickup to be minimised. This may not be beneficial, for when juice is exposed to oxygen at an early stage, the phenolics are protected from subsequent oxidation. The process of hyperoxygenation of must is often used for this effect. Further and importantly, pneumatic presses release juice with lower levels of solids. The solids contain yeast nutrients, and deficiency in these can delay the onset of fermentation, allowing the must to oxidise. Dr Valérie Lavigne argued that 'when the pneumatic press arrived in 1995–1996, it was very good, but the winemakers didn't adapt the way they used it – and if they didn't, the must was too clean… and you have less ability for ageing' [60]. Fermentation can be slower in low turbidity juice, and accordingly reductive qualities lost.

5.15.3.6 The Increased Use of New Oak Barrels

New oak barrels are more porous than those that have been previously used, and the amount of oxygenation of the content wine is higher. With repeated use the pores of oak become filled with compounds, particularly tartrates. Some producers whose practice was to rack their whites off the lees prior to ageing the wines in new oak would appear to have had greater issues with premox. In 2013, speaking at an Institute of Masters of Wine (MW) seminar entitled 'How to make the best Burgundy', Etienne de Montille of Domaine de Montille said, 'I definitely keep more lees now, because the phenolics protect the wine against oxidation. I'm convinced this is for the benefit of the wine in the future' [62].

5.15.3.7 Bâtonnage

During the widespread press coverage that followed the realisation that premox was widespread amongst white Burgundies, the finger of blame was often pointed at the practice of bâtonnage, i.e. the stirring of fine lees in cask. Much discussion was based upon the notion that this was a relatively recent introduction by many producers, in order to increase the richness and 'fatness' of their wines. The method normally used is removing the barrel bung and inserting a baton into the bunghole and stirring rigorously, rousing the lees from the bottom of the barrel, and thoroughly mixing them with the wine. The actual process of stirring is by nature oxidative, as the amount of oxygen dissolved in the wine increases and both SO_2 and CO_2 are released from the wine during the process. However, lees, particularly when in suspension, scavenge oxygen, and the process of bâtonnage can result in less dissolved oxygen in the wine. So if undertaken gently and carefully, the process does, in fact, offer protection whilst the wine is undergoing barrel maturation.

Speaking at a 2013 MW seminar, Benjamin Leroux of Domaine Comte Armand said, 'We have a long alcoholic fermentation which gives a natural lees stirring but we hardly do any-bâtonnage after 2–3 months because it can dissolve oxygen into the wine' [62]. Contrasting this view, Dr Valérie Lavigne noted that bâtonnage helps create a reductive atmosphere. 'In Burgundy they don't stir the lees so the wine has no protection at a stage when it is very sensitive' [63]. Without doubt, lees stirring has been undertaken more sparingly in the last decade by many Burgundy producers, and alternative methods of lees rousing are now undertaken. One alternative to bâtonnage is rolling of barrels that are stored on racks specifically designed for the purpose, such as the OXOline system. It should be noted that lees-aged wines have less pyruvic acid [28] and pyruvic acid binds SO_2 negating some of its antioxidant properties and decreasing its ability to reduce quinones. Lees also break down aldehydes and ketones.

5.15.3.8 Insufficient Free Sulfur Dioxide Levels

In the mid-1990s, with the aim of producing elegant, fresher wines and perhaps reacting to seemingly constant pressures to reduce levels of SO_2, winemaking decisions regarding protective SO_2 additions changed from the 'insurance' addition previously undertaken. Historically, when tasting young wines, a hint or even a whiff of sulfur was apparent, but the wines were made to last, and SO_2 levels fall during the ageing process. During the 1990s, free SO_2 levels were, at some domaines, adjusted at bottling to just 15–25 mg/l. As we have seen, this was at a time that harvested fruit was at ripeness levels not previously achieved, with higher sugars but lower acidity and higher pHs. It is the free sulfur dioxide component that possesses antioxidant properties, but the levels of free SO_2 require careful and constant monitoring. Shortly following bottling, some of the free SO_2 will be consumed, the amount depending largely upon the oxygen uptake in the bottling process. One milligramme per litre of oxygen will generally consume 4 mg/l of sulfur, so a wine that contains 4 mg/l of DO subsequent to bottling can have 16 mg/l of SO_2 rapidly depleted, leaving little left available as an antioxidant. Also, if there is microbial activity such as the growth of *Brettanomyces* or due to closure issues the wine is gaining dissolved oxygen, the level of free SO_2 can fall dramatically during bottle storage. In a world where 'natural' wines have gained considerable press (although far from all this is positive) and given the continuing public concerns regarding SO_2, the relentless pressure on producers to minimise sulfur dioxide additions continues.

5.15.3.9 Interventionist Operations Undertaken to Speed the Wine to Market

It can be argued that some producers are employing techniques that enable wine to be brought to market early, but do not allow them to age in the manner or for as long as was previously the case. Bringing to market early, and designing the wines for short-term drinking, is not simply a matter of satisfying the impatient consumer: early positive assessments by reviewers and critics help sell wine. Also, of course, time spent in the cellar is money, spent and not received (unless the wines have been sold *en primeur*). Microbiologist Terry Leighton of Kalin Cellars in California's Livermore Valley makes superb wines that are renowned for their longevity. The wines are marketed (if that is the right word for no real marketing activity is required) when mature. For example, the 1995 Cuvee CH Sonoma Chardonnay was released for sale in January 2017. He accepts that the reasons for premox are multifactorial. However, he believes that the need to get products to the market as quickly as possible has to take some of the blame and notes that consumers have little intention of cellaring wine today and expect white wines, even at investment prices, to be ready to drink upon release [64]. Kalin's white wines receive the minimum of fining and are bottled without filtration. He noted, 'Add nothing and take nothing away – a lot of people have abandoned this traditional winemaking because it takes six to nine years to release the wine' [64].

5.15.3.10 Ingress of Oxygen When Bottling

Careless bottling resulting in high levels of total package oxygen (TPO), coupled with low levels of free SO_2 as discussed above, are a recipe for disaster. Many producers are now bottling wines with a level of 800–850 mg/l of dissolved carbon dioxide (CO_2), in order to help protect from oxidation. However, historically wines were bottled in a high oxygen environment, although it might be noted that this seemed to present no problems prior to 1995.

5.15.3.11   Low-Quality Cork Closures and Corks Washed with Peroxide

Most fine Burgundies, with the exception of Chablis from a few producers, are sealed with a natural cork or Diam closure. In 2015, Domaine Laroche in Chablis announced, having bottled under screw cap for the previous decade, it had decided to return to natural cork [65]. The incidence of individual bottles of wines suffering from premox is very often random: sometimes some bottles in a case are, and others are not. Accordingly, many have blamed cork producers, citing the variable quality and OTRs of natural corks. A historic (1996) Australian study noted that oxygen permeation is a significant defect in a cork, and, for example, in the case of a wine stoppered with a 38 mm peroxide-treated cork and held for three years, about 36 mg/l of total SO_2 may be consumed by the process [66]. The study found that some bottled wines sealed with corks treated with peroxide, and stored from 6 to 18 months, had high levels of browning, typical sensory characteristics of oxidation, and a loss of SO_2. By the mid-1990s, bleaching of corks with chlorine had been largely replaced with peroxide, on account of chlorine's implication in the production of 2,4,6-trichloroanisole (TCA), as discussed in Chapter 3. Domaine Leflaive decided to switch to a Diam closure for wines from the 2014 vintage [55], and although at the time of writing it is perhaps early days, it would appear that incidences of premox have waned. In November 2017, when I spoke to Philip Tuck MW of Hatch Mansfield, agent for the Burgundy producer Louis Jadot, he stated emphatically, 'I have no doubt that it (premox) is purely a cork issue. Since we changed to Diam we have had no problem'. Philip is agreeing with the views of Chablis producers Alain Marcuello of William Fèvre and Benoît Droin of the eponymous family producer [67]. Both state that since they introduced the use of Diam closures for all of their production, they were happy with the results, noting no more 'corked' bottles or premox problems. However, cork manufacturers have raised the question that were the problem due to the corks, many wines other than Burgundies would have been affected, which is generally, with the exceptions noted above, not the case. The cork industry believes that the problem clearly is in the winemaking. Dr Valérie Lavigne stated that the type of closure is 'just a little part of the problem' [60]. Perhaps a potentially significant contributor to oxidation during the bottle ageing (and the apparent random nature of this) could be the transfer of oxygen at the interface between the cork stopper and the bottleneck. Thomas Karbowiak et al. found this to be unambiguously the case, leading to a notable modification of a wine's chemical makeup [68].

5.15.4   Prune Aroma

The aroma of prunes is sometimes found in red wines that have prematurely aged, including those from the Bordeaux region. There are several compounds that have been found to contribute to the aroma: 3-methyl-2,4-nonanedione (MND) (overt prune aromas), γ-nonalactone (overripe peach aromas), and β-damascenone (dried fruit and apple sauce aromas) [69]. In red wine, the perception threshold for the MND ketone in wine is 62 ng/l. However, concentrations up to 340 ng/l have been found in oxidised red wines, and 293.8 ng/l in oxidised botrytised white wines [70]. The presence above threshold of MND may be regarded as an oxidation marker.

5.16 Prevention of Excess Acetaldehyde and Oxidation

5.16.1 Basic Principles

The prevention of oxidation may be regarded as a full-process concern. Issues giving rise to oxidation may happen at any time during winemaking and subsequently during storage. However, the production steps and procedures to be undertaken to prevent oxidation and the production of excess acetaldehyde are largely those undertaken as a matter of course, utilising sound winemaking practices. Such preventative actions will also help limit the occurrence of some other faults including excess volatile acidity, ethyl acetate and *Brettanomyces*-related issues. There is now a vast array of highly technical equipment available utilising inert gas to limit oxygen exposure, including receiving hoppers and presses, but all this comes at high cost, often beyond the means of small wineries.

Measuring the amount of oxygen that enters wine during processing operations is easily achieved by calculating the decrease in free SO_2 after each operation. Measuring the free SO_2 values before individual cellar operations, and then measuring again four to five days after the operation, will give a good indication of the extent of oxygen pickup. Dissolved oxygen levels can be regularly monitored by the use of a DO meter.

There are basic principles to be considered:

- Limiting oxygen exposure.
- Removing phenols from must for white wines.
- Removing PPO and laccase.
- Removing iron/copper.
- Increasing GSH content.
- Scavenging radicals.
- Stabilising redox potential.
- Removing oxidation substrates – fining.
- Reducing temperature (when protected from oxygen) – this reduces speed of reactions.
- Keeping pH low – reactions happen slower.
- Inhibiting oxidase activity.
- Maintaining appropriate free sulfur dioxide.

5.16.2 Picking at Appropriate Ripeness

Oxidation can begin in the vineyard, particularly when fruit is overripe (or damaged). There can be little doubt that the number of wines showing oxidised characteristics has increased in tandem with the modern trend of leaving grapes on the vine to maximise so-called phenolic ripeness. Clark Smith explains field oxidation as taking place in the late stages of ripening, resulting in permanent structures that are oxidatively inactive and a loss of reductive strength in the wine [3, 71]. Anthocyanins are located only in grape skins, so they are not optimally incorporated into these structures. Accordingly these are prone to becoming over-polymerised and will precipitate after ageing for a few years. The palate will be dry and gritty, lacking texture and with the feel of coffee grounds. Field oxidation robs the wine of both anthocyanins and active tannins. Smith noted that a typical Napa Cabernet with three weeks of excessive ripeness will be able to consume only 30–40 ml of oxygen for three to

Figure 5.1   Grape clusters ready for sorting.

five days, having lost approximately 90% of its reductive strength. 'Three weeks on the vine in fact steal a decade of cellaring potential' [71].

5.16.3   Harvesting and Transporting Fruit

Hand-picked fruit should be put into small containers – e.g. 12–16 kg lug boxes, which will help prevent crushing and damage during transport to the winery. This is important for red grapes but should be considered absolutely essential for white grapes. Figure 5.1 shows grape clusters ready for sorting and crushing during the difficult 2018 vintage at Château Pontet-Canet in Pauillac, Bordeaux.

Enzymatic oxidation can commence as soon as the grape berry is split or damaged. As we have seen, tyrosinase is located in the chloroplasts of the grape berry, and its enzymatic oxidation actions will begin with the crushing of the cell walls. Loose-loaded trucks and trailers, although often used to transport grapes for inexpensive wines, are the worst options. Figure 5.2 shows white grapes, and juice from those already damaged, being discharged from a loose-loaded truck at an Italian winery.

There should be minimal delay between picking and processing. The blanketing of picking bins and receiving hoppers with dry ice can substantially reduce oxygen uptake. Fruit should be transported as cool as possible and in temperature-controlled containers if the journey will take many hours, as is often the case in Australia where multiregional blends are commonplace.

Figure 5.2 Loose-loaded grapes being discharged.

5.16.4 Sorting

A sorting of fruit is necessary to ensure that only clean, undamaged and disease-free fruit is used. Fruit affected by *Botrytis cinerea* contains the enzyme laccase, which catalyses enzymatic oxidation.

5.16.5 Crushing

Oxygen uptake at crushing can range from 5 to 10 mg/l, depending, inter alia, upon the type of crusher, temperature, and speed of the operation. Crushers using minimal agitation limit the uptake [72].

5.16.6 Pressing (for White Wines)

In a traditional horizontal press, most of the GSH is rapidly consumed. This issue can be addressed by the use of tank presses that can be purged with inert gas. The pressing and processing of grapes in this largely anaerobic manner is sometimes referred to as 'hyper-reduction'. Although presses employing inert gas are particularly valuable for the production of fresh white or rosé wines, more and more wineries are using them for post-fermentation pressing for red wines too. There are now many models available that have a recycling of gas system, such as Bucher Inertys® and Siprem vacuum system (VS) presses. The vacuum system enables a completely inert environment as pockets of air are completely removed. It is now possible to use carbon dioxide from fermentations to provide gas for the press.

High phenolic fractions from press should be separated – these are likely to come from later pressings and will generally have high pH. A stronger fining may be used for these.

5.16.7 Juice Clarification (White Must)

Enzymes are contained in solid matter, and the use of clarification enzymes may help reduce oxidised phenolics. Fining with potassium caseinate or casein will also help precipitate oxidised phenolics, or those susceptible to oxidation. Both treatments are permitted by the EU and OIV [18, 19]. Fining at juice stage can actually result in improved aromatics. Moreover, fining the must with PVPP can also be beneficial if the fruit has been infected with *botrytis*.

The addition of tannins to must can be beneficial as these can limit the enzymatic activity of laccase and also absorb oxygen. Proanthocyanidins, also known as condensed or catechin tannins, are oligomers or polymers of flavan-3-ols. These tannins may be added to grapes for red wines at a rate of 180–250 g/tonne or to must at a rate of 10–30 g/hl (healthy grapes) or 30–60 g/hl (botrytised grapes [as determined by a test for laccase activity]). Proanthocyanidins have strong antioxidasic action on musts and are radical scavengers. They inhibit laccase activity, due to grey rot (*B. cinerea*) and act on tyrosinase. Ascorbic acid is sometimes added to white musts in addition to sulfur dioxide for complimentary antioxidant action, but this remains controversial as it decreases the redox potential and can act as an oxidising agent in wine. Also, for white musts, after settling or early in the fermentation, inactivated yeasts that contain cysteine and GSH may be added at a maximum dose of 30 g/hl.

5.16.8 Inoculate with Commercial Strains of *Saccharomyces*

Vats should be inoculated with desired yeast cultures, unless the winemaker has decided upon 'wild' (natural) yeast fermentation. If the fruit shows *Botrytis* or mildews, the use of commercial strains is essential.

5.16.9 Fermentation

Although fermenting musts should be supplied with necessary oxygen, additions should cease in the later stages of the process – usually when the alcohol level has reached approximately 10% abv.

5.16.10 Post-fermentation Maceration

During post-fermentation maceration of red wines, the cap is exposed to air, and so is at risk of oxidation. Wetting the cap by a very brief pump-over as necessary reduces the risk. Alternatively, keeping the cap submerged under the top of the juice, using a metal grid or other means that holds the cap structure below the liquid surface, is a method of reducing oxygen uptake. Of course, a specific tank design with a metallic grid strong enough to resist the CO_2 pressure on the cap is necessary [73].

5.16.11   Malolactic Fermentation (MLF)

The level of acetaldehyde is generally reduced during MLF. The period between the end of alcoholic fermentation (AF) and start of MLF, if undertaken, should be kept as short as possible, as the wine will not be protected by SO_2 at this time. Co-inoculation of a *Saccharomyces* yeast strain and an MLF starter bacteria culture, such as a strain of *Oenococcus oeni*, to induce simultaneous AF and MLF fermentations might be considered – the total time for completion of the two fermentations may be reduced by 50%. Co-inoculation is not recommended if more than 25 mg/l of SO_2 has been added to the must. It must be noted that most yeasts produce SO_2 during the AF, and accordingly the MLF might be inhibited.

Aeration should be avoided during the MLF as there is a risk of bacterial production of acetic acid or other microbial spoilage. The identification of the completion of the MLF should be made as quickly as possible in order for the wine to be protected by the addition of SO_2.

5.16.12   Avoiding Ullage in Tanks and Barrels

Ullage in tanks and barrels of wine undergoing maturation or in storage must be avoided. The floating lids of tanks should be within 3 mm of the surface of the wine – cellar hands often place these several centimetres above the wine as installation is easier! Tank lids (paying particular attention to the inflatable tyres on floating lids), hatches, valves, air locks, sight tubes, and all seals (e.g. thermostat seals) should be regularly checked. Any fall in level, after the initial post-fermentation cooling, should be regarded with suspicion as evaporating wine means there is an ingress of oxygen.

Any tanks that are not completely full should be gas blanketed with CO_2, nitrogen, or argon, as appropriate.

5.16.13   Reducing Oxygen Uptake in Cellar Operations Including Rackings and Transfers

When transferring from tank to tank, the wine should be taken from the bottom of one tank and pumped into the bottom of the receiving tank, provided all fermentation activity has ceased. All hoses should be flushed with inert gas to displace oxygen. The transfer tank should be gas blanketed, and the receiving tank or vessel purged with inert gas [45]. Ideally, an inert gas should also be used to evacuate hoses and pumps. With very careful pumping utilising nitrogen/argon flushing of lines and hoses, the oxygen uptake can be as low as 0.2 mg/l, but without attention this can be 4 mg/l or more. The use of a high-quality pump helps to reduce uptake. Lobular pumps have low oxygen dissolution and can be used from crusher to the bottling line, and peristaltic pumps are particularly effective at gently moving high volumes. Hose unions should be carefully tightened. This is especially important on the inlet side of the pump; otherwise there will be an unwanted 'cracking the fitting', drawing air into the line. All pump seals should be checked.

It should be kept in mind that the amount of uptake during rackings and other operations is, inter alia, temperature dependent. A study published in 2004 found that racking at 10 °C (50 °F) resulted in three times the oxygen uptake than at 15–20 °C (59–68 °F) [74].

Figure 5.3 Rotary vacuum filter.

Rotary vacuum filtration can be a major source of uptake – 2–4 mg/l. This operation is often undertaken for basic filtration of large quantities of must for white or sparkling wines, but alternative methods of clarification may be more appropriate. A photo of a rotary vacuum filter is shown in Figure 5.3.

Dissolved oxygen should be checked before and after cellar operations (and routinely in maturation and storage) – see Section 5.18.2.

5.16.14 Maintaining Appropriate Levels of Free SO_2

The sulfiting of harvested grapes, although commonly practised, is not recommended (as far as avoiding oxidation is concerned) as this promotes the extraction of phenols [13]. However, the addition of SO_2 to must pre-fermentation is desirable for both its antimicrobial and antioxidant properties. An addition of 25–50 mg/l will inhibit tyrosinase. For every 10 mg/l of SO_2 added to the must, bound SO_2 levels in the final wine will increase by 3–7 mg/l [75]. The molecular weight of oxygen is 16 and of SO_2 is 64. Every 1 mg/l of DO will consume 4 mg/l of SO_2 in both must and wine, of which approximately 3 mg will be free SO_2. In fact, SO_2 does not react directly with oxygen, but with hydrogen peroxide. Sulfur dioxide reduces quinones back to phenols and converts hydrogen peroxide to water. Free SO_2 only offers limited protection against oxidation. At a level of 30 mg/l free SO_2, only half the oxygen content oxidises the SO_2, with the other half oxidising the wine's remaining, most oxidisable components [45].

Additions of SO_2 during fermentation are to be avoided as these can lead to an increase in acetaldehyde production (and may also lead to a stuck fermentation and inhibition of the MLF). Post-fermentation additions should be completed until the MLF (if desired) is completed. Thereafter adequate levels of free SO_2 should be maintained throughout the maturation and storage processes. From a prevention of oxidation perspective, this range may be between 25 and 45 mg/l, but if the wine is ageing on lees, lower levels are required.

5.16.15 Reducing Heavy Metals

Wines should be analysed for the level of heavy metals. Copper is the most frequently occurring, and its presence may be due to vineyard pesticides or brass fittings on tanks, hoses, etc. The presence of copper may also be a consequence of the use of copper sulfate for the treatment of reductive off-odours and off-flavours [76, 77]. At high levels in white wines (copper above 1 mg/l and iron above 10 mg/l), levels of free SO_2 should be maintained at a minimum of 45 mg/l [76]. NB: 1 mg/l is the maximum permitted level of copper in wine according to OIV and EU regulations [18, 19].

5.16.16 Controlling Cellar Temperature and Humidity

Attention should be paid to temperature control during maturation and storage. Wine evaporation from barrels is greater at higher temperatures, and the reduction in liquid is replaced by oxygen. The ideal temperature is generally regarded as 13–15 °C (55–59 °F) or perhaps a degree or two cooler if the growth of *Brettanomyces* is feared. However, at low temperatures, if wine has access to oxygen, the uptake will be rapid [74].

5.16.17 Barrel Ageing and/or the Addition of Ellagitannins

Oak wood releases into wine ellagitannins from ellagic acid and gallotannins from gallic acid. Ellagitannins, which are not present in grapes, are soluble in a hydroalcoholic solution and are gradually extracted into wine during ageing in oak barrels. In the mid-1990s, ellagitannins were revealed to be impressive oxidation regulators, quickly absorbing the dissolved oxygen and facilitating the hydroperoxidation of wine constituents [78].

The level of ellagitannins in barrel oak depends on many factors including its botanical and geographical origin, size of grain, seasoning methods, toasting degree, and, of course, the number of previous uses of the barrel [79]. *Quercus robur* (Limousin oak) has the highest levels of ellagitannins, *Quercus petraea* (Allier oak [90% of the oaks in the forest are Q. *petraea*]) has an intermediate level of ellagitannins, and *Quercus alba* (American oak) provides the lowest levels, less than half of Q. *robur* [79]. Although present in raw wood, how the wood is processed in cooperage, including the type and length of drying and toasting, also has an effect [80].

Ellagitannins help to protect anthocyanins from oxidation. Oxygen picked up by the wine will react with ellagitannins by the process of peroxidation and reduce the direct oxidation of other wine constituents [81]. In fact, ellagitannins are potent and effective antioxidants and help protect wine from oxidation [82]. During the first three months of barrel maturation, there is a substantial increase in the concentration in wine of ellagitannins

[83]. The time in barrel necessary for a wine to reach maximum ellagitannin concentration varies according to its matrix, including alcoholic degree and pH [84]. The level eventually decreases, perhaps due to their high reactivity in the presence of oxygen. The oxidation of ellagitannins leads to the formation of quinones that can undergo an attack from ethanol to form hemiacetal derivatives that then undergo another attack from ethanol to form acetal derivatives [85]. In any event, the level of ellagitannins imparted to a wine decreases considerably after one year of barrel use.

Ellagitannins are a permitted additive by the OIV and EU. An interesting study published in 2020 evaluating the oxygen consumption of commercial oenological tannins was undertaken by Jelena Jeremic et al. [86]. They found that the addition of ellagitannins to red wine increased the rate of oxygen consumption and provided an effective tool in winemaking operations with high levels of oxygen uptake, when it is important to have immediate protection against oxidation. However, they also found that the effect on oxygen consumption tended to drop with time. In contrast, although grape skin tannins showed more consistent reactivity and lower oxygen consumption rates in each saturation when added to red wine, grape seed tannins are more suitable than ellagitannins for guaranteeing fast oxygen consumption.

5.16.18 Lees Ageing

As previously noted, lees have a tremendous capacity to scavenge and consume oxygen. Lees yeast cells absorb dissolved oxygen, and GSH and cysteine contained in lees have an antioxidant effect. They can help absorb oxygen picked up during cellar operations and storage, but are not able to consume all of this oxygen. Whilst maintaining adequate levels of free SO_2 is generally considered important to avoid oxidation, it has recently been shown that sulfur dioxide strongly inhibits the reactions involved in oxygen consumption by yeast lees [87]. Accordingly, the antioxidant effects of both yeast lees and sulfur dioxide cannot be used in a complementary manner, and the protection against oxidation by yeast lees in wines with free SO_2 levels higher than 20 mg/l is less beneficial than in wines with low SO_2 concentrations [87].

5.16.19 Cold Stabilisation

This process, commonly used to precipitate tartrate crystals in white wines, can be the largest source of oxygen uptake. Oxygen dissolves into wine easier and faster at cold temperatures. It is vital that there is no headspace in the tank during this process. Cold stabilisation is inefficient and uses copious amounts of energy. Alternative methods to precipitate tartrates, if really needed, should be considered. These include electrodialysis, mannoproteins, and gum arabic, as discussed in Chapter 15.

5.16.20 Ultrafiltration Before Bottling

Ultrafiltration is an effective method for removing oxidised phenolics that would otherwise cause premature browning. For red wines, the precursors of browning can be selectively removed based upon molecular weight. In the case of white wine, the ultrafiltration process is capable of taking out tint and pinking compounds.

5.16.21 Bottling

5.16.21.1 A Major Source of Oxygen Uptake

The bottling of wine is one of the largest potential sources of oxygen ingress [88], and oxygen has the potential to dissolve into the wine at every stage of the process [89]. The winemaker may not be aware of the level of uptake at this time, which can vary considerably depending upon the point in the bottling run. During the bottling operation, white wines should not pick up more than 0.2 mg/l of oxygen, and red wines not more than 0.5 mg/l. The amount of oxygen introduced into the wine at bottling, together with the OTR of the closure, will impact upon the style, quality, development, and longevity of the bottled wine. Without any precautions, the uptake at bottling could be 4 mg/l or more, which would consume at least 16 mg/l of free SO_2. This might leave the bottled wine at risk of oxidation and/or bacterial spoilage.

5.16.21.2 Total Package Oxygen (TPO)

TPO comprises both dissolved oxygen in the wine and headspace oxygen (HSO) (which is the oxygen percentage in the headspace and headspace volume). Once sealed, the bottle will also contain oxygen from the closure application and, particularly if a cellular structured closure (such as a natural cork) is used, any oxygen contained within the closure. TPO levels after bottling should be less than 1–1.25 mg/l for red wines and less than 0.5–0.6 mg/l for white wines. However, Maurizio Ugliano of the closure manufacturer Nomacorc states that TPO can vary over a range of approximately 1–9 mg/l [90].

5.16.21.3 Dissolved Oxygen in the Wine

Dissolved oxygen in the wine should be measured before, during, and after bottling – prior to the commencement of the operation, it should be 0.5 mg/l or less. If the level is higher, the wine should be sparged with nitrogen. An in-line sparger is an effective tool for this. It is extremely important that the supply tank should be purged and gas blanketed.

5.16.21.4 Oxygen in the Empty Bottle Pre-filling

An empty bottle prior to preparation for filling contains 750 ml of air, which of course includes an oxygen content of over 200 mg. Left unpurged, this could increase the DO in the wine by 0.3–0.7 mg/l, depending on the filling technology adopted (carousel or orbital) [91]. If the bottle is purged with one or two volumes of inert gas in the form of nitrogen, carbon dioxide, or argon, this may reduce the potential contribution of oxygen previously in the bottle to 1–3 mg/l.

5.16.21.5 Headspace Oxygen

The headspace may include up to 2.8 mg/l of oxygen. Knowing that 4 mg/l of SO_2 reacts with 1 mg/l, this HSO can decrease free SO_2 by 11 mg/l. HSO can represent 50% of the TPO and can be reduced by gas flushing or vacuum application. Even if a vacuum system is used in corking, the HSO can still comprise 8–10% oxygen [92]. Wines sealed with screw caps should have a double inert gas injection both inside the screw cap and bottleneck. For cork-closed bottles, including technical corks and synthetic closures such as Nomacorc, the headspace atmosphere is compressed as the cork is inserted into the bottle, and, if no vacuum is created, this compression could lead to a fourfold increase in oxygen uptake into the wine. Non-destructive analytical instruments are now available, e.g. 'PreSens' and 'NomaSense', which allow TPO to be accurately measured. Also, there are luminescence-based technologies that allow for non-invasive measurement of dissolved oxygen and HSO through the sides of the glass bottle.

5.16.21.6 Oxygen Content of the Closure
Natural cork contains between 80% and 85% of air in its cellular structure, and a typical (long) cork with dimensions of 44 mm × 24 mm contains about 4.8 mg of oxygen. This can slowly diffuse into the headspace or wine during bottle storage [93]. During the 1980s, there was an increase in frequency of oxidised bottles of wine. This has been put down, in part, to the advent of high-speed bottling lines. For technical reasons, corks were specified to have a silicone coating, rather than wax that had been previously used. This made them easier to insert by the corking module but perhaps also allowed great oxygen ingress down the walls of the closure [94].

5.16.21.7 Addition of Glutathione Prior to Bottling
The addition of GSH at bottling time may lessen the risk of premature ageing [95]. A study by Pons et al. revealed that wines spiked with GSH (together with ascorbic acid) at bottling were preferred by panellists to control wines without GSH. Analytical results showed that the spiked wines also had the lowest sotolon content [96].

5.16.22 Storage

5.16.22.1 Oxygenation During Storage
It is generally accepted that, in the case of red wines, a low rate of controlled oxygenation during storage in bottle is important to the maturation process and has positive sensory benefits. It is also accepted that this is not the case for white wines, as such oxygenation can lead to the browning of colour, loss of fruit, and development of the undesirable characteristics of oxidation [97, 98]. The amount of oxygen ingress via the various bottle closure types is highly variable, which is discussed in Chapter 17. It is, however, important to keep in mind that cork closures have an oxygen content per se and this migrates into wine, albeit at a low transmission rate [99]. Corks can also contain quinones that have oxidative properties. The screening of cork closures for oxidants (and haloanisoles) before they are used is highly desirable.

Even if there is no or a negligible intake of oxygen, such as may be the case if a saran–tin screw cap closure is used, oxidation in bottle may take place as it not only is linked to the presence of oxygen but also relates to the action of oxidising–reducing pairs of metal ions and polyphenols, substances that catalyse the effect of oxygen [7].

5.16.22.2 Oxidation Due to Poor Storage
Poor storage is one of the most common reasons for oxidation and perhaps the one that is most likely encountered by the consumer. Fortunately, most inexpensive wine is usually consumed within months, or at most a year or two, after packaging, and problems may not have time to develop. If wine is to be stored for any length of time, it is vital

that good conditions are maintained, including keeping the wine cool and in the dark. It used to be generally accepted that bottles sealed with a natural or technical cork should be stored horizontally with the belief that this would prevent the cork from drying out, shrinking and allowing a passage for oxygen ingress between the bottle and closure. However, recent research has challenged this long-held dogma [100]. According to *The Drinks Business*, Miguel Cabral, Director of Research and Development at *Amorim*, the world's largest cork producer, states, 'The cork will never dry out with almost 100% humidity in the headspace, so it is a myth that you need to store a bottle on its side' [101]. Anecdotally, my experience is very different. On occasions, I have been able to compare old bottles of a wine that have been vertically stored, with those that have been stored horizontally. Almost without exception, the vertically stored bottles have had dried out corks that have crumbled upon removal, whilst for the horizontally stored wines, the closures (and wine) have been sound.

Maintaining an appropriate bottle storage temperature is very important for all wines, with white wines suffering dramatically if stored at an elevated temperature. For a white wine, the rate of browning and the rate of decline in total SO_2 are, respectively, 2.9 and 1.2 times faster at 20 °C (68 °F) compared with storage at 10 °C (50 °F) [102].

Without a preservation system, the wine in a part-empty opened bottle becomes oxidised within a couple of days. Bars and restaurants offering wines 'by the glass' should use a preservation system such as *Coravin®* or *Winesave®*, unless an opened bottle is completely used within a few hours. These use argon to fill the space left by dispensed wine. Argon does not dissolve in wine, unlike nitrogen that is slightly soluble. An illustration of a Coravin is shown in Figure 5.4.

An alternative is an *Enomatic®* dispensing cabinet, which holds several bottles and uses a patented argon/nitrogen gas mix for wine protection.

5.17 Additions of Ascorbic Acid: Antioxidant or Oxidising Agent?

The possible addition of ascorbic acid as an antioxidant to white musts and particularly to white wines prior to bottling remains highly controversial. Whilst in the short term it may be effective as an antioxidant, decreasing the redox potential of wines, with time the effect is reversed, and browning increases, whether or not SO_2 is present in the wine [103]. In 1998, Peng et al. concluded that the presence of the appropriate level of molecular SO_2 alone (rather than ascorbic acid alone or both SO_2 and ascorbic acid) at bottling is the most effective method of protecting white wine from oxidation [103]. Conversely, a more recent study by Morozova et al., looking at a Riesling, showed that 250 mg/l of ascorbic acid added to a wine with 45 mg/l of free SO_2 resulted in increased perceptions of fruity aromas and a lower intensity of oxidised aromas and a less oxidised and fresher palate when examined six months after bottling compared with a wine with SO_2 alone (68 mg/l of free SO_2) [104]. However, Hua Li et al. noted that more and more extensive studies have shown that ascorbic acid may have a pro-oxidant role rather than antioxidant depending on its levels in wine under some conditions [7]. It must be kept in mind that if ascorbic acid is used in combination with SO_2, then it can cause accelerated consumption of SO_2 and accelerate the production of yellow and browning pigments [7, 105, 106].

Figure 5.4 Coravin Wine Preservation System.

The addition of ascorbic acid to prevent pinking of white wines, including wines in bottle, has been shown to be effective. A recent study by Cojocaru and Antoce [107] found that for Sauvignon Blanc, additions of 20 or 30 mg/l of ascorbic acid were effective, but additions to Chardonnay were not sufficient to guarantee protection from pinking.

The legal limit (EU and OIV) for the addition of ascorbic acid to wine is 250 mg/l [18, 19].

5.18 Laboratory Analysis

5.18.1 Acetaldehyde

In this chapter, the discussion on acetaldehyde has been limited to its production and organoleptic impact, particularly when present in excess quantities. However, we should be reminded that acetaldehyde, when in association with alcohol consumption, is classified as 'a human carcinogen' (Group 1) by the International Agency for Research on Cancer

(IARC) [108]. Accordingly, from a health perspective, steps should be made to limit its production, and analysis of its concentration in wine pre-bottling should be considered essential.

Acetaldehyde can be analysed by gas chromatography, particularly gas chromatography with flame ionisation detector (GC-FID). Within the small winery, multi-analysis test equipment utilising spectrophotometer technology is useful for measuring acetaldehyde concentrations. Two popular models are CDR WineLab® (that has a measuring range of 18–300 mg/l) [109] and Megazyme MegaQuant (that has a lower limit of detection of 0.18 mg/l) [110].

5.18.2 Dissolved Oxygen

Dissolved oxygen in must and wine, prior to bottling, may be easily measured using a dissolved oxygen meter. Handheld portable devices are relatively inexpensive, and using Bluetooth technology, the data can be fed into a remote computer.

Systems are now available for measuring oxygen in bottled wine. OxySense® manufactures non-invasive optical oxygen measurement systems. These can measure oxygen in the headspace, as well as in the wine. The technique is based upon the fluorescence quenching of a metal organic fluorescent dye immobilised in a gas-permeable hydrophobic polymer. The dye absorbs light in the blue region and fluoresces within the red region of the spectrum. The presence of oxygen quenches the fluorescent light from the dye as well as its lifetime [111]. The systems comply with International Standard ASTM F2714 - 08 [112]. It is also possible to measure dissolved oxygen at various stages of the production process using this technology.

DO in wine in tank or barrel can be measured using a dipping probe that can also check the effectiveness of inert gas blanketing. During transfers, DO can be measured using sight glasses containing the immobilised fluorescent dye. Oxygen uptake during individual operations can be quantified by placing a sight glass before and after the equipment utilised, e.g. pump or filter.

There are alternative luminescence-based systems for measuring DO in bottled wine (and in production).

5.18.3 Oxidation

The analysis of wine oxidation is normally undertaken using organoleptic means. In principle, determining the state of oxidation by measuring a wine's redox potential can be achieved using platinum electrodes [113]. However, this does not reveal a true redox potential as there is no equilibrium present [28].

5.19 Treatments

5.19.1 Acetaldehyde

The addition of SO_2 can reduce the acetaldehyde content. The acetaldehyde will bind the SO_2, so repeated additions should be made until the level of free SO_2 begins to rise, indicating that the acetaldehyde has absorbed as much sulfur dioxide as it can take. Legal limits as to total SO_2 must be observed, keeping in mind that a further adjustment of free sulfur dioxide relative to the pH of the wine will be necessary at bottling time. The winemaker should also be cautious of the sensory implications. When SO_2 is added to wine containing free acetaldehyde, hydroxysulfonate is formed, which can result in a smell of sulfur, which influences the aromatic composition of a wine [11].

Fining with potassium caseinate or casein as discussed below may also prove effective in reducing high levels of acetaldehyde.

5.19.2   Oxidation

There are very limited possibilities of correction when a winemaker is face with oxidised wine. If the problem is relatively minor, fining together with the addition of SO_2 may reduce the characteristics to an acceptable level.

It may be possible to re-ferment the wine with a large quantity of another fermenting wine, but this procedure presents risks of producing an even larger quantity of unacceptable wine.

Fining with PVPP, which is a white powder that binds with hydroxyl phenolics, may be effective, but the process does have organoleptic implications. Such fining may strip body from the wine and have a negative effect on mouthfeel. Fining with casein, a milk product containing protein, can also be effective, and this can be undertaken subsequent to PVPP fining. There are also proprietary blends of casein and PVPP that can be used for one-stage fining, which perhaps have less of an organoleptic impact than fining with PVPP alone. The legal limit for PVPP is 80 g/hl (OIV and EU). Although casein is a milk-based product, it is not necessary, at the time of writing, to list this on the label as an allergen in wines marketed in the EU.

Fining with inactivated yeast, or inactivated yeast with guaranteed GSH levels, may have positive results. The process is permitted by the EU and OIV [18, 19].

5.20   Final Reflections

It is May 2019, and I have again been invited to be a member of the Jury for the 'Coupe des Crus Bourgeois du Médoc'. Together with the other professional tasters, I'm in the pristine Stade Matmut Atlantique close by the Bordeaux artificial lake. The wines we are tasting have already passed all the stringent criteria, including blind tasting by experts, which have resulted in them being chosen for inclusion in the 2016 vintage official selection of the French government-approved quality assurance classification: Crus Bourgeois du Médoc. They are already on the market. Today we are choosing 12 laureates from 166 candidate wines. Sommeliers are bringing the masked bottles in flights to the tables of panels of tasters, and we are tasting the wines blind. The process today consists of a series of eliminations of individual wines, which is not always easy due to the abundance of high-quality wines. Then in one particular flight, there is a wine that has intense Sherry-like aromas, with baked fruit and prune tones and a bitter palate with aggressive acidity. The other members of my panel and I all agree that the wine is badly oxidised and call for a second bottle from the sommeliers. Could the problem be the result of a failed cork? The second bottle is identical, and we swiftly mark the wine as eliminated. The question immediately enters my head: how many more of these faulty bottles, each bearing an authenticated Crus Bourgeois sticker, are on the market or already purchased, and just what will be their fate?

References

1 Pasteur, L. (1873). *Etudes sur le vin: Ses maladies, causes qui les provoquent, procédés nouveaux pour les conserver et pour les vieillir*. Paris: Imprimerie Royale.

2 Peynaud, E. (1980). *Le Goût du Vin*, 1e. Paris: Bordas.

3 Smith, C. (2014). *Postmodern Winemaking*. Berkeley: University of California Press.

4 Kilmartin, M.A. (2010). Understanding and controlling nonenzymatic wine oxidation. In: *Managing Wine Quality*, vol. 2 (ed. A.G. Reynolds), 432–458. Great Abington, Cambridge: Woodhead Publishing.

5 Lopes, P., Silva, M.A., Pons, A. et al. (2009). Impact of oxygen dissolved at bottling and transmitted through closures on the composition and sensory properties of a Sauvignon Blanc wine during bottle storage. *Journal of Agricultural and Food Chemistry* 57: 10261–10270. https://doi.org/10.1021/jf9023257.

6 Oliveira, C.M., Ferreira, A.C.S., de Freitas, V., and Silva, A.M.S. (2011). Oxidation mechanisms occurring in wines. *Food Research International* 44: 1115–1126. https://doi.org/10.1016/j.foodres.2011.03.050.

7 Li, H., Guo, A., and Wang, H. (2008). Mechanisms of oxidative browning of wine. *Food Chemistry* 108: 1–13. https://doi.org/10.1016/j.foodchem.2007.10.065.

8 Waterhouse, A.L. and Nikolantonaki, M. (2015). Quinone reactions in wine oxidation. In: *ACS Symposium Series*, vol. 1203, 291–302. American Chemical Society https://doi.org/10.1021/bk-2015-1203.ch018.

9 Scholium Wines (2014). 2011 Babylon Fact Sheet. https://www.scholiumwines.com/pdfs/2011-babylon-fact-sheet.pdf (accessed 29 April 2020).

10 Govinda, P. (2009). Oxidized wines. *Imbibe*, 14 August. https://imbibemagazine.com/oxidized-wines (accessed 29 April 2020).

11 Coetzee, C., Buica, A., and du Toit, W.J. (2018). Research note: the use of SO_2 to bind acetaldehyde in wine: sensory implications. *South African Journal of Enology and Viticulture* 39 (2) https://doi.org/10.21548/39-2-3156.

12 Carrascón, V., Vallverdú-Queralt, A., Meudec, E. et al. (2018). The kinetics of oxygen and SO_2 consumption by red wines. What do they tell about oxidation mechanisms and about changes in wine composition? *Food Chemistry* 241: 206–214. https://doi.org/10.1016/j.foodchem.2017.08.090.

13 Ribéreau-Gayon, P., Dubourdieu, D., Donèche, B., and Lonvaud, A. (2006). *Handbook of Enology, The Microbiology of Wine and Vinifications*, vol. 1, 418–419. Chichester: Wiley.

14 Sullivan, M.L. (2015). Beyond brown: polyphenol oxidases as enzymes of plant specialized metabolism. *Frontiers in Plant Science* 5 (783) https://doi.org/10.3389/fpls.2014.00783.

15 Lisanti, M.T., Blaiotta, G., Nioi, C., and Moio, L. (2019). Alternative methods to SO_2 for microbiological stabilization of wine. *Comprehensive Reviews in Food Science and Food Safety* 18 (2): 455–479. https://doi.org/10.1111/1541-4337.12422.

16 Kritzinger-Stadler, E.C., Bauer, F.F., and du Toit, W.J. (2013). Role of glutathione in winemaking: a review. *Journal of Agricultural and Food Chemistry* 61: 269–277. https://doi.org/10.1021/jf303665z.

17 Ballester, J., Magne, M., Julien, P. et al. (2018). Sensory impact of polyphenolic composition on the oxidative notes of Chardonnay wines. *Beverages* 4 (1): 19. https://doi.org/10.3390/beverages4010019.

18 EUR-Lex. (2019). Commission Delegated Regulation (EU) 2019/934 of 12 March 2019 supplementing Regulation (EU) No 1308/2013 of the European Parliament and of the Council as regards wine-growing areas where the alcoholic strength may be increased, authorised oenological practices and restrictions applicable to the production and conservation of grapevine products, the minimum percentage of alcohol for by-products and their disposal, and publication of OIV files. https://eur-lex.europa.eu/legal-content/GA/TXT/?uri=CELEX:32019R0934 (accessed 25 May 2020).

19 International Organisation of Vine and Wine (OIV) (2020). *International Code of Oenological Practices*. Paris: OIV. http://www.oiv.int/public/medias/7213/oiv-international-code-of-oenological-practices-2020-en.pdf.

20 Theron, C. (2017). Oxygen management during winemaking. *WineLand Media*. www.wineland.co.za/oxygen-management-during-winemaking (accessed 22 May 2020).

21 Lagarde-Pascal, C. and Fargeton, L. (2014). White grape must oxygenation: set up and sensory effect. *Internet Journal of Enology and Viticulture* N7/2. https://www.infowine.com/intranet/libretti/libretto11785-01-1.pdf.

22 du Toit, W.J., Marais, A., Pretorius, I.S., and du Toit, M. (2006). Oxygen in must and wine: a review. *South African Journal of Enology and Viticulture* 27 (1): 76–94. https://doi.org/10.21548/27-1-1610.

23 Schneider, V. (1998). Must hyperoxidation: a review. *American Journal of Enology and Viticulture* 49 (1): 65–73.

24 Theron, C. (2014). The clarification of white juice prior to alcoholic fermentation. *WineLand Media*. www.wineland.co.za/the-clarification-of-white-juice-prior-to-alcoholic-fermentation (accessed 22 May 2020).

25 Tarko, T., Duda-Chodak, A., Sroka, P., and Siuta, M. (2020). The impact of oxygen at various stages of vinification on the chemical composition and the antioxidant and sensory properties of white and red wines. *Hindawi International Journal of Food Science*: 7902974. https://doi.org/10.1155/2020/7902974.

26 Elias, R.J., Andersen, M.L., Skibsted, L.H., and Waterhouse, A. (2009). Identification of free radical intermediates in oxidized wine using electron paramagnetic resonance spin trapping. *Journal of Agricultural and Food Chemistry* 57 (10): 4359–4365. https://doi.org/10.1021/jf8035484.

27 Byrne, S. and Howell, G. (2017). *Acetaldehyde – How to Limit its Formation During Fermentation*. Vintessential. www.vintessential.com.au/acetaldehyde-how-to-limit-its-formation-during-fermentation (accessed 31 May 2020).

28 Waterhouse, A.L., Sacks, G.L., and Jeffery, D.W. (2016). *Understanding Wine Chemistry*. Chichester: Wiley.

29 Bueno, M., Marrufo-Curtido, A., Carrascón, V. et al. (2018). Formation and accumulation of acetaldehyde and strecker aldehydes during red wine oxidation. *Frontiers in Chemistry* 20 https://doi.org/10.3389/fchem.2018.00020.

30 Ferreira, A.C., Hogg, T., and de Pinho, P.G. (2003). Identification of key odorants related to the typical aroma of oxidation-spoiled white wines. *Journal of Agricultural and Food Chemistry* 51 (5): 1377–1381. https://doi.org/10.1021/jf025847o.

31 Bueno, M., Carrascon, V., and Ferreira, V. (2016). Release and formation of oxidation-related aldehydes during wine oxidation. *Journal of Agricultural and Food Chemistry* 64 (3): 608–617. https://doi.org/10.1021/acs.jafc.5b04634.

32 Jackowetz, J.N. and Mira de Orduña, R. (2013). Survey of SO_2 binding carbonyls in 237 red and white table wines. *Food Control* 32 (2): 687–692. https://doi.org/10.1016/j.foodcont.2013.02.001.

33 Liu, S.Q. and Pilone, G.J. (2000). An overview of formation and roles of acetaldehyde in winemaking with emphasis on microbiological implications. *International Journal of Food Science and Technology* 35: 49–61. https://doi.org/10.1046/j.1365-2621.2000.00341.x.

34 Martin, B., Etievant, P.X., and Henry, R.N. (1990). The chemistry of sotolon: a key parameter for the study of a key component of *flor* Sherry wines. In: *Flavour Science and Technology* (eds. Y. Bessiere and A.F. Thomas). Chichester: Wiley.

35 Ferreira, A.C.S., Barbe, J.C., and Bertrand, A.T. (2003). 3-Hydroxy-4,5-dimethyl-2(5*H*)-furanone: a key odorant of the typical aroma of oxidative aged port wine. *Journal of Agricultural and Food Chemistry* 51: 4356–4363. https://doi.org/10.1021/jf0342932.

36 Bakker, J. and Clarke, R.J. (2012). *Wine Flavour Chemistry*, 2e. Oxford: Wiley-Blackwell Publishing.

37 Pisarnitsky, A.K., Bezzubov, A.A., and Egorov, I.A. (1987). Nonenzymatic formation of 4,5-dimethyl-3-hydroxy-2(5*H*)-furanone in foodstuffs. *Prikladnaia Biokhimiia i Mikrobiologiia* 23 (5): 642–646.

38 Pons, A., Lavigne, V., Landais, Y. et al. (2010). Identification of a sotolon pathway in dry white wines. *Journal of Agricultural and Food Chemistry* 58: 7273–7279. https://doi.org/10.1021/jf100150q.

39 Mislata, A.M., Puxeu, M., Tomás, E. et al. (2020). Influence of the oxidation in the aromatic composition and sensory profile of Rioja red aged wines. *European Food Research and Technology* https://doi.org/10.1007/s00217-020-03473-4.

40 Specht, G. (2010). Yeast fermentation management for improved wine quality. In: *Managing Wine Quality*, vol. 2 (ed. A.G. Reynolds), 3–33. Cambridge: Woodhead Publishing.

41 Gómez-Plaza, E. and Bautista-Ortín, A.B. (2019). Emerging technologies for aging wines: use of chips and micro-oxygenation. In: *Red Wine Technology* (ed. A. Morata), 149–162. London: Academic Press.

42 Australian Wine Research Institute (2017). Project 3.3.2 – Influencing wine style through management of oxygen during winemaking. Final report to Australian Grape and Wine Authority . Principal investigator: Dr Paul Smith. https://www.wineaustralia.com/research/projects/influencing-wine-style-through-managemen (accessed 28 May 2020).

43 Smith, C. (2017). Issues and innovations in oxygen management. *Wine Business Monthly*, February. https://www.winebusiness.com/wbm/index.cfm?go=getArticle& dataid=179533 (accessed 30 April 2020).

44 del Alamo-Sanza, M., Cárcel, L.M., and Nevares, I. (2017). Characterization of the oxygen transmission rate of oak wood species used in cooperage. *Journal of Agricultural and Food Chemistry* 65: 648–655. https://doi.org/10.1021/acs.jafc.6b05188.

45 Ribéreau-Gayon, P., Glories, Y., Maujean, A., and Dubourdieu, D. (2006). *Handbook of Enology: The Chemistry of Wine: Stabilization and Treatment*, 2e, vol. 2. Chichester: Wiley.

46 Calderón, J.F., del Alamo-Sanza, M., Nevares, I., and Laurie, V.F. (2014). The influence of selected winemaking equipment and operations on the concentration of dissolved oxygen in wines. *Ciencia e investigación agraria* 41 (2): 273–280. https://doi.org/10 .4067/S0718-16202014000200014.

47 Coetzee, C., van Wyngaard, E., Šuklje, K. et al. (2016). Chemical and sensory study on the evolution of aromatic and nonaromatic compounds during the progressive oxidative storage of a Sauvignon Blanc wine. *Journal of Agricultural and Food Chemistry* 64 (42): 7979–7993. https://doi.org/10.1021/acs.jafc.6b02174.

48 AWRI (2016). *Fact Sheet: Non-Destructive Analysis of Oxygen Transmission Rate of Wine Packaging*. AWRI. www.awri.com.au/wp-content/uploads/otr_analysis_fact_sheet.pdf (accessed 30 April 2020).

49 Cosme, F., Andrea-Silva, J., Filipe-Ribeiro, L. et al. (2019). The origin of pinking phenomena in white wines: an update. *BIO Web of Conferences* 12: 02013. https://doi.org/ 10.1051/bioconf/20191202013.

50 Doronae. (2018). Guidelines to reduce pinking potential in white wines. Wineland. www.wineland.co.za/guidelines-reduce-pinking-potential-white-wines (accessed 25 April 2020).

51 AWRI (2020). *Fact Sheet: Pinking*. AWRI. www.awri.com.au/industry_support/ winemaking_resources/frequently_asked_questions/pinking (accessed 25 April 2020).

52 Morris, J. (2014). White Burgundy out of the Woods? *The World of Fine Wine* 43. http://www.worldoffinewine.com/news/white-burgundy-out-of-the-woods-4717208 (accessed 11 March 2020).

53 Anson, J. (2014). Premox: has the crisis moved to red wines? *Decanter*, 11 November. http://www.decanter.com/features/premox-has-the-crisis-moved-to-red-wine-245696 (accessed 10 March 2020).

54 Coates, C. (2013). *My Favorite Burgundies*. Berkeley: University of California Press.

55 Goode, J. (2016). *Domaine Leflaive's Switch to Diam*. Jamie Goode's Wine Blog. https:// www.wineanorak.com/wineblog/wine-science/domaine-leflaives-switch-to-diam (accessed 23 April 2020).

56 Cornwell, D. (2020). Oxidized Burgundies. Wiki. Guillaume Deschamps. http://www .gdeschamps.net/wiki/doku.php?id=domaine_leflaive (accessed 23 April 2020).

57 Kelley, W. (2017). Mâcon made good. *Decanter* August, pp. 34–40.

58 WINE Beserkers Forum. Premox is still with us big time. https://www.wineberserkers .com/forum/viewtopic.php?t=145802 (accessed 31 May 2020).

59 Schmitt, P. (2018). Clonal selection to blame for premox, says Laroche winemaker. *The Drinks Business* 3 September. https://www.thedrinksbusiness.com/2018/09/clonal-selection-to-blame-for-premox-says-laroche-winemaker (accessed 20 May 2020).

60 Schmitt, P. (2015). Premox study shifts to red wine. *The Drinks Business* 1 September. https://www.thedrinksbusiness.com/2015/09/premox-study-shifts-to-red-wine/2 (accessed 23 April 2020).

61 Anson, J. (2013). The premature oxidative ageing of wine, Dr Valérie Lavigne and Prof Denis Dubourdieu, Faculty of Oenology, Bordeaux Institute of Vineyard and Wine Sciences. New Bordeaux. https://www.newbordeaux.com/dubourdieu-premox-in-whites (accessed 30 May 2020).

62 Schmitt, P. (2013). Burgundy seminar report part 1: Reducing the risk of premox. 10 January https://www.thedrinksbusiness.com/2013/01/burgundy-seminar-report-part-1-reducing-the-risk-of-premox (accessed 30 May 2020).

63 Savage, G. (2013). Fermentation key to premox battle. *The Drinks Business* 26 September. https://www.thedrinksbusiness.com/2013/09/fermentation-key-to-premox-battle (accessed 23 April 2020).

64 Colman, T. (2013). The puzzling problem of premox. https://www.wine-searcher.com/m/2013/09/the-puzzling-problem-of-premox (accessed 23 April 2020).

65 Cork Quality Council (2016). *Domaine Laroche Changes from Screwcap to Cork.* CQC https://www.corkqc.com/blogs/cork-news/domaine-laroche-changes-from-screwcap-to-cork (accessed 31 July 2020).

66 Waters, E.J., Peng, Z., Pocock, K.F., and Williams, P.J. (1996). The role of corks in oxidative spoilage of white wines. *Australian Journal of Grape and Wine Research*: 191–197. https://doi.org/10.1111/j.1755-0238.1996.tb00108.x.

67 Jefford, A. (2016). Jefford on Monday: Debating Diam. *Decanter* 22 August. http://www.decanter.com/wine-news/opinion/jefford-on-monday/jefford-on-monday-debating-diam-8087 (accessed 20 February 2020).

68 Karbowiak, T., Crouvisier-Urion, K., Lagorce, A. et al. (2019). Wine aging: a bottleneck story. *npj Science of Food* 3: 14. https://doi.org/10.1038/s41538-019-0045-9.

69 Pons, A., Lavigne, V., Frérot, E. et al. (2008). Identification of volatile compounds responsible for prune aroma in prematurely aged red wines. *Journal of Agricultural and Food Chemistry* 56 (13): 5285–5290. https://doi.org/10.1021/jf073513z.

70 Pons, A., Lavigne, V., Darriet, P., and Dubourdieu, D. (2013). Role of 3-methyl-2,4-nonanedione in the flavor of aged red wines. *Journal of Agricultural and Food Chemistry* 61 (30): 7373–7380. https://doi.org/10.1021/jf400348h.

71 Smith, C. (2010). Field Oxidation. Postmodern winemaking. http://postmodernwinemaking.com/field-oxidation (accessed 31 May 2020).

72 Jackson, R.S. (2020). *Wine Science: Principles and Applications*, 5e. San Diego, CA: Academic Press.

73 Morata, A., González, C., Tesfaye, W. et al. (2019). Maceration and fermentation: new technologies to increase extraction. In: *Red Wine Technology* (ed. A. Morata), 35–49. London: Academic Press.

74 Castellari, M., Simonato, B., Tornielli, G.B. et al. (2004). Effects of different enological treatments on dissolved oxygen in wines. *Italian Journal of Food Science* 16: 387–396.

75 Jackowetz, N., Li, E., and Mira de Orduñ, R. (2011). Sulphur dioxide content of wines: The role of winemaking and carbonyl compounds. *Cornell Viticulture and Oenology Research Focus* 3. https://grapesandwine.cals.cornell.edu/sites/grapesandwine.cals .cornell.edu/files/shared/documents/Research-Focus-2011-3.pdf.

76 Clark, A.C., Wilkes, E.N., and Scollary, G.R. (2015). Chemistry of copper in white wine: a review. *Australian Journal of Grape and Wine Research* 21: 339–350. https://doi .org/10.1111/ajgw.12159.

77 Claus, C. (2020). How to deal with uninvited guests in wine: copper and copper-containing oxidases. *Fermentation* 6 (1): 38. https://doi.org/10.3390/ fermentation6010038.

78 Vivas, N. and Glories, Y. (1996). Role of wood ellagitannins in the oxidation process of red wines during aging. *American Journal of Enology and Viticulture* 47 (1): 103–107.

79 Zamora, F. (2019). Barrel aging; types of wood. In: *Red Wine Technology* (ed. A. Morata), 125–147. London: Academic Press.

80 Michel, J., Jourdes, M., Silva, M.A. et al. (2011). Impact of concentration of ellagitannins in oak wood on their levels and organoleptic influence in red wine. *Journal of Agricultural and Food Chemistry* 59 (10): 5677–5683. https://doi.org/10.1021/jf200275w.

81 Theron, C. (2019). The role of ellagitannins in the oxidative stability of wine. *Wineland* 1 July. www.wineland.co.za/the-role-of-ellagitannins-in-the-oxidative-stability-of-wine (accessed 30 April 2019).

82 Navarro, M., Kontoudakis, N., Giordanengo, T. et al. (2016). Oxygen consumption by oak chips in a model solution: influence of the botanical origin, toast level and ellagitannin content. *Food Chemistry* 199: 822–827. https://doi.org/10.1016/j.foodchem.2015 .12.081.

83 Watrelot, A.A., Badet-Murat, M.-L., and Waterhouse, A.L. (2018). Oak barrel tannin and toasting temperature: effects on red wine condensed tannin chemistry. *Food Science and Technology* 91: 330–338. https://doi.org/10.1016/j.lwt.2018.01.065.

84 Nikolantonaki, M., Coelho, C., Diaz-Rubio, M.-E. et al. (2019). Oak tannin selection and barrel toasting. *Practical Winery & Vineyard* March, pp. 56–59.

85 Nikolantonaki, M., Coelho, C., Diaz-Rubio, M.-E. et al. (2018). Impact of oak tannin and barrel toasting on the oxidative stability of dry white wine. *Australian and New Zealand Grapegrower and Winemaker*, August 2018, pp. 63–68.

86 Jeremic, J., Vongluanngam, I., Ricci, A. et al. (2020). The oxygen consumption kinetics of commercial oenological tannins in model wine solution and Chianti red wine. *Molecules* 25 https://doi.org/10.3390/molecules25051215.

87 Schneider, V., Müller, J., and Schmidt, D. (2016). Oxygen consumption by postfermentation wine yeast lees: factors affecting its rate and extent under oenological conditions. *Food Technology and Biotechnology* 54 (4): 395–402. https://doi.org/10 .17113/ftb.54.04.16.4651.

88 O'Brien, V., Colby, C., and Nygaard, M. (2009). Managing oxygen ingress at bottling. *Wine Industry Journal* January/February: 24–29. www.awri.com.au/wp-content/ uploads/wij_24_1_24-29.pdf.

89 Steiner, T.E. (2013). Strategies to manage dissolved oxygen. *Wines & Vines* August. https://winesvinesanalytics.com/features/article/119752/Strategies-to-Manage-Dissolved-Oxygen (accessed 25 April 2020).

90 Ugliano, M. (2013). Oxygen contribution to wine aroma evolution during bottle aging. *Journal of Agricultural and Food Chemistry* 61 (26): 6125–6136. https://doi.org/10.1021/jf400810v.

91 Letaief, H. (2016). Key points of the bottling process. *Wines & Vines* May. https://winesvinesanalytics.com/features/article/168227/Key-Points-of-the-Bottling-Process (accessed 30 April 2020).

92 Koralewski, M.-E. (2017). Best practices for gas management at bottling. *Wines & Vines* October. https://winesvinesanalytics.com/features/article/190162/Best-Practices-for-Gas-Management-at-Bottling (accessed 30 April 2020).

93 Karbowiak, T., Gougeon, R.D., Alinc, J.B. et al. (2010). Wine oxidation and the role of cork. *Critical Reviews in Food Science and Nutrition* 50: 20–52. https://doi.org/10.1080/10408398.2010.526854.

94 Taber, G.M. (2007). *To Cork or Not to Cork*. New York: Scribner, a division of Simon and Schuster Inc.

95 El Hosry, L., Auezova, L., Sakr, A., and Hajj-Moussa, E. (2009). Browning susceptibility of white wine and antioxidant effect of glutathione. *International Journal of Food Science and Technology* 44: 2459–2463. https://doi.org/10.1111/j.1365-2621.2009.02036.x.

96 Pons, A., Nikolantonaki, M., Lavigne, V. et al. (2015). New insights into intrinsic and extrinsic factors triggering premature aging in white wines. In: *Advances in Wine Research*, vol. 1203, 229–251. Washington, DC: American Chemical Society. https://doi.org/10.1021/bk-2015-1203.ch015.

97 Escudero, A., Asensio, E., Cacho, J., and Ferreira, V. (2002). Sensory and chemical changes of young white wines stored under oxygen. An assessment of the role played by aldehydes and some other important odorants. *Food Chemistry* 77: 325–331. https://doi.org/10.1016/S0308-8146(01)00355-7.

98 Oliveira, A.C., Silva Ferreira, A.C., Guedes de Pinho, P., and Hogg, T.A. (2002). Development of a potentiometric method to measure the resistance to oxidation of white wines and the antioxidant power of their constituents. *Journal of Agricultural and Food Chemistry* 50: 2121–2124. https://doi.org/10.1021/jf0112015.

99 Lopes, P., Saucier, C., Teissedre, P.L., and Glories, Y. (2007). Main routes of oxygen ingress through different closures into wine bottles. *Journal of Agricultural and Food Chemistry* 5: 5167–5170. https://doi.org/10.1021/jf0706023.

100 Lopes, P., Saucier, C., Teissedre, P.L., and Glories, Y. (2006). Impact of storage position on oxygen ingress through different closures into wine bottles. *Journal of Agricultural and Food Chemistry* 54: 6741–6746. https://doi.org/10.1021/jf0614239.

101 Schmitt, P. (2018). Storing wine on its side is nonsense, says scientist. *The Drinks Business* 11 June. https://www.thedrinksbusiness.com/2018/06/storing-wine-on-its-side-is-bullsht-says-scientist (accessed 31 May 2020).

102 Scrimgeour, N., Nordestgaard, S., Lloyd, N.D.R., and Wilkes, E.N. (2015). Exploring the effect of elevated storage temperature on wine composition. *Australian Journal of Grape and Wine Research* 21: 713–722. https://doi.org/10.1111/ajgw.12196.

103 Peng, Z., Duncan, B., Pocock, K.F., and Sefton, M.A. (1998). The effect of ascorbic acid on oxidative browning of white wines and model wines. *Australian Journal of Grape and Wine Research*: 127–135. https://doi.org/10.1111/j.1755-0238.1998.tb00141.x.

104 Morozova, K., Schmidt, O., and Schwack, W. (2015). Effect of headspace volume, ascorbic acid and sulphur dioxide on oxidative status and sensory profile of Riesling wine. *European Food Research and Technology* 240: 205–221. https://doi.org/10.1007/s00217-014-2321-x.

105 Goode, J. (2018). *Flawless*. Berkeley: University of California Press.

106 Barril, C., Rutledge, D.N., Scollary, G.R., and Clark, A.C. (2016). Ascorbic acid and white wine production: a review of beneficial versus detrimental impacts. *Australian Journal of Grape and Wine Research* 22: 169–181. https://doi.org/10.1111/ajgw.12207.

107 Cojocaru, G.A. and Antoce, A.O. (2019). Effect of certain treatments to prevent or partially reverse the pinking phenomenon in susceptible white wines. *42nd World Congress of Vine and Wine. BIO Web of Conferences 15*, p. 02003. https://www.bio-conferences.org/articles/bioconf/full_html/2019/04/bioconf-oiv2019_02003/bioconf-oiv2019_02003.html#S7 (accessed 16 June 2020).

108 International Agency for Research on Cancer [IARC], World Health Organisation (2019). List of classification. https://monographs.iarc.fr/list-of-classifications (accessed 1 June 2020).

109 https://www.cdrfoodlab.com/foods-beverages-analysis/acetaldehyde-analysis-wine (accessed 8 June 2020).

110 https://www.megazyme.com/acetaldehyde-assay-kit (accessed 8 June 2020).

111 https://www.oxysense.com/how-oxysense-works.html (accessed 31 May 2020).

112 ASTM F2714 - 08 (2013) *Standard test method for oxygen headspace analysis of packages* using Ffluorescent decay. West Conshohocken ASTM International.

113 Danilewicz, J.C. (2012). Review of oxidative processes in wine and value of reduction potentials in enology. *American Journal of Enology and Viticulture* 63: 1–10. https://doi.org/10.5344/ajev.2011.11046.

6

Excessive Sulfur Dioxide, Volatile Sulfur Compounds, and Reduced Aromas

It is convenient to discuss the topics of sulfur, sulfur dioxide (SO_2) including its excess in wine, sulfides, and volatile sulfur compounds (VSCs) in a single chapter. The relationship between elemental sulfur, sulfites, sulfates, and VSCs is clear to see, but some of the issues that arise are not related in ways that might be expected. Wines exuding the odours of unwanted low-molecular-weight VSCs are commonly described as 'reduced' or suffering from reduction – some purists consider the term 'reduction' is misused in this regard. Often, the 'stinky' sulfur compounds only appear, or sometimes reappear, after the wines have been bottled for some time (post-bottling reduction) or when bottles are exposed to light (lightstrike). Malodorous VSCs are now amongst the most common wine faults, and incidences certainly increased in the early years of this century. The odours can be more than unpleasant: hydrogen sulfide has the sickening smells of bad eggs and drains, and some other low-molecular-weight VSCs (mercaptans) reek of skunk or burnt rubber.

6.1   Introduction

6.1.1   Volatile Sulfur Compounds

VSCs are major contributors to wine aromas and flavours, and over 100 have been identified. Whether or not a VSC is desirable depends upon the compound in question, its concentration, and the matrix and style of the wine, including the grape variety. Many low-molecular-weight VSCs are unwanted in wine, and a dozen or so of these render the wine clearly faulty. Thankfully, just four of these compounds are regularly found at concentrations above sensory thresholds:

- Hydrogen sulfide (H_2S),
- Methanethiol (MeSH) also known as methyl mercaptans,
- Ethanethiol (EtSH) also known as ethyl mercaptans,
- Dimethyl sulfide (DMS).

The other undesirable VSCs that may be present are usually at concentrations below individual sensory thresholds [1], but they can still have negative aroma impacts, particularly when co-occurring with other unwanted compounds. Low-molecular-weight mercaptans have a skunk-like odour – indeed, they are the main odorants of the defensive spray of skunks.

Wine Faults and Flaws: A Practical Guide, First Edition. Keith Grainger.
© 2021 John Wiley & Sons Ltd. Published 2021 by John Wiley & Sons Ltd.

6.1.2 Thiols

Thiols are organosulfur compounds, i.e. containing a sulfyhydryl (–SH) group. They are significant contributors to the aromas and taste of wines. The presence of individual thiols may or may not be desirable depending upon the thiol in question, the concentration of each thiol, the relationship between them, the grape variety, the style of wine, and the region of origin. However, volatile thiols are generally associated with positive aromas. The coffee tones exhibited by some warm climate Pinot Noir wines, including many from California's Sonoma County, and the chocolate and liquorice notes of Malbecs are both on account of the high levels of individual thiols. In a Sauvignon Blanc wine, varietal thiols hugely influence the trademark aromas and flavours. By way of examples, grapefruit and passion fruit aromas stem from 3-mercaptohexan-1-ol (3-MH); boxwood, broom, and 'cat's pee' from 4-mercapto-4-methylpentan-2-one; and cut grass from 1-hexanol.

The concentration of volatile thiols in wine decreases with wine ageing, particularly in the presence of oxygen. Glutathione (GSH), SO_2, and anthocyanins have a protective effect and reduce degradation [2]. In the presence of oxygen and iron, certain volatile thiols can be oxidised and form pungent disulfides [3].

The terms 'thiol' and 'mercaptan' are in general usage synonymous. However, in the wine world, thiol is used for the desirable compounds containing a sulfyhydryl group, such as 3-mercaptohexan-1-ol (3-MH). Mercaptan is usually reserved for the distinctly malodorous compounds containing a sulfyhydryl group, which, if present in the finished wine, render it faulty.

The presence of VSCs in wine can result from enzymatic or chemical mechanisms – the latter includes photochemical or thermal reactions [2]. They may be formed in a reductive environment, i.e. in wine with a low redox potential. Redox potential is the tendency of molecules or ions to gain electrons. A compound becomes reduced when it loses oxygen atoms and gains hydrogen atoms and also gains electrons. Redox potential is affected by pH level – higher pH favours lower redox potentials, and lower pH favours higher redox potentials. Only wines with low redox potential contain VSCs. Hydrogen sulfide, if not removed, can form other VSCs, particularly mercaptans and disulfides.

6.2 The Presence and Role of Sulfur, Sulfur Dioxide, Sulfite, and Sulfate in Wine Production

The presence of sulfur in grape must and wine comes from four sources:

- Sulfur is a natural component of grapes.
- Sulfur-containing fungicides applied in the vineyard.
- Sulfur dioxide is generated by yeasts during fermentation.
- Sulfur dioxide is usually added during winemaking processes.

6.2.1 Sulfur Contained in Grapes

Sulfur (S) is a macroelement necessary for the growth of grapevines and, as such, must be obtained by the vine from the soil. It is a catalyst for chlorophyll production and is a structural component of protein and peptides. Only the topsoil, with its content of organic matter, contains sulfur. Elemental sulfur in the soil is not available to the vine as such, but must be converted into sulfates by soil microbes. If required, the amount of sulfates in the ground may be boosted by the addition of fertilisers containing sulfate or elemental sulfur. Accordingly, inorganic sulfate is the main form of sulfur that is present in grapes, together with some in the form of vitamins thiamine and biotin, and sulfur-containing amino acids including peptides of which the tripeptide GSH is particularly important [4]. The amino acids cysteine (which is also one of the GSH peptides) and methionine both of which contain sulfur are also naturally present in grapes. During the fermentation process, these can be converted into hydrogen sulfide (H_2S), sometimes at unacceptably high levels, and also into other VSCs both during fermentation and post-fermentation.

The level of sulfate contained in grapes juice, according to a recent review, is perhaps between 160 and 700 mg/l [5]. According to Smith et al. [6], drawing from Leske et al. [7], wine grapes in Australia contain an average of 260 mg/l (±121 mg/l) of sulfate.

6.2.2   Sulfur-Containing Fungicides

Elemental sulfur, other sulfur-containing pesticides, and/or copper sulfate may be used as a contact treatment on grapevines. Sulfur, in either powder form or solution, is used to combat powdery mildew, commonly known as oïdium, taxonomic name *Erysiphe necator* (also known as *Uncinula necator*). Four to six applications may be applied in a typical growing season, and such contact sprays work both as a preventative and an eradicant. It is a relatively inexpensive treatment, and there is a low risk of vines developing resistance. It is interesting to note that although many producers, including some who farm organically and biodynamically, espouse that sulfur is a 'natural material', almost all the elemental sulfur used in the wine industry is not mined, but in fact is a product from the petroleum industry. Copper sulfate, usually mixed with lime and water and known as 'Bordeaux mixture', may be used to counter downy mildew, commonly known as Peronospora, taxonomic name *Plasmopara viticola*. In the last decades of the twentieth century, the use of systemic treatments became commonplace, but several strains of downy mildew developed resistance to some of these resulting in a recent resurgence of the use of copper sulfate treatments. From the beginning of 2019, the amount of copper that may be used in the European Union (EU) has been reduced from 6 kg/ha per annum to 4 kg/ha, averaged over seven years. Sulfur and copper sulfate are permitted for use as fungicides by Demeter and Biodyvin, the two main certification agencies for biodynamic wine production. Indeed, these are the two main antifungal agents employed by organic and biodynamic producers, although alternative organic products such as potassium bicarbonate may sometimes be used in countries where these are permitted.

All sulfur-based treatments should cease several weeks before the harvest – there is much dispute as to how long this withholding period should be, with figures ranging from two to eight weeks being stated. However, there is no doubt that substantial sulfur residues on harvested crop often reach the winery. About 50% of any sulfur residue is transferred to the must during crushing, but in the case of white wines, the majority of this is usually eliminated by settling or must clarification. The maximum residue of elemental sulfur in the must should not exceed 10 mg/l; otherwise, H_2S can be produced by the yeast from the elemental sulfur content, in addition to other mechanisms. It has also been shown [8, 9] that elemental sulfur spray residues on grapes not only produce hydrogen sulfide during fermentation but also can form precursors capable of generating additional H_2S after storage in bottle.

6.2.3 Sulfur Dioxide Generated by Yeast During Fermentation

Sulfur dioxide is produced by yeasts during the fermentation process. The amount generated will vary considerably according to the yeast strain(s), the condition of the must including nutrient content, and the fermentation temperature. The range of production in currently available commercial strains is from less than 1 mg/l to over 90 mg/l, but historically, numerous authors state levels well in excess of 100 mg/l.

6.2.3.1 Sulfur-Containing Amino Acid

There are 20 or so amino acids, of which the most prevalent in grape must are usually proline (the most abundant), arginine, and glutamine and glutamic acid. *Saccharomyces cerevisiae* can release reductive compounds through the degradation of sulfur-containing amino acids, particularly cysteine and methionine, and also from elemental sulfur, sulfites, or sulfates. However, several amino acids, including proline and lysine, are unable to be used by yeast as a nitrogen source.

6.2.4 Additions of Sulfur Dioxide During the Winemaking Process

6.2.4.1 The Use of Sulfur Dioxide in Winemaking

Of course, SO_2 is used extensively during winemaking processes by the vast majority of winemakers for its universal antimicrobial, antioxidant, and antienzymatic properties. This is nothing new – its benefits as a preservative were recognised as far back as Roman times, when sulfur candles were burned in barrels and amphorae. Additions during winemaking processes may be in solid, gaseous, or liquid forms. The compound is often added in the form of potassium metabisulfite (molecular formula $K_2S_2O_5$) to grapes post-harvesting or to grape must to inhibit bacteria and unwanted (non-*Saccharomyces*) yeasts and also to avoid pre-fermentation oxidation. Large additions, generally in excess of 50 mg/l (as SO_2), are to be avoided at this stage, unless the crop is suffering from *Botrytis cinerea* or other sanitary issues. High levels of SO_2 may inhibit or slow down the alcoholic fermentation (AF), may be used by yeasts to produce large amounts of diacetyl, and may prevent the onset of the malolactic fermentation (MLF). Although additions should never be made during the alcoholic fermentation (as this may lead to the production of excessive acetaldehyde), SO_2 is usually added at several subsequent stages of the winemaking processes, which include immediately post-MLF, before and during tank or barrel maturation, when wines are racked, and pre-bottling.

The total SO_2 content in packaged wine is regulated in all major markets. The compound is a known allergen and may cause severe reactions in asthmatics.

6.2.4.2   Total, Bound, Free, and Molecular Sulfur Dioxide

A large portion of SO_2 added to wine immediately becomes 'bound' to various compounds, including phenols, aldehydes (particularly acetaldehyde), and sugars, and thus has no antimicrobial or antioxidant effect. The remaining 'unbound' portion is known as free SO_2 and although it is widely believed that this entire portion has antimicrobial properties, this is not in fact the case. A small part of the free SO_2 is present in the form of sulfite $(SO_3{}^{2-})$, a larger part as bisulfite (HSO_3), and the remainder as molecular SO_2. Bisulfite has antioxidant capacities, but only molecular SO_2 has antimicrobial properties. The portion of free SO_2 that is available as molecular SO_2 varies, particularly with the pH of the wine and, to a lesser extent, the temperature. The amount of molecular SO_2 required to offer effective protection against unwanted microbial growth will vary with many factors: a typical range is between 0.5 and 0.8 mg/l. Appendix A details the levels of free SO_2 required for 0.5, 0.625, and 0.8 mg/l of molecular SO_2 for differing pH values.

6.2.4.3   Additions of Sulfur Dioxide to Prevent Enzymatic Oxidation

Sulfur dioxide prevents enzymatic oxidation by denaturing the enzyme polyphenol oxidase (PPO), which has tyrosinase activity. The enzyme can no longer function as its shape is changed in this process. However, the powerful enzyme laccase, which is present on grapes affected by *B. cinerea*, whether in its highly damaging 'grey rot' or beneficial 'noble rot' form is largely tolerant to SO_2. Noble rot is only possible on white grapes and only in certain climatic and weather conditions but when present can impart the most exciting aromas to luscious sweet wines such as Sauternes, Bonnezeaux, and Trockenbeerenauslese.

6.2.4.4   The Use of Sulfur Dioxide to Prevent Chemical and Microbial Oxidation

As we have seen, the bisulfite component of the free SO_2 has antioxidant properties. It should be remembered that SO_2 does not react directly with oxygen, but with the reduced form hydrogen peroxide. Sulfur dioxide also reduces quinones back to their phenolic form and prevents their depletion [10, 11]. It is generally accepted that at least 10 mg/l of free SO_2 is needed to be effective as an antioxidant.

The role of sulfur dioxide in wine is discussed further in Chapter 18.

6.3   Excessive Sulfur Dioxide

6.3.1   High Levels of SO_2: Flaw or Fault?

A concentration of SO_2 that is apparent on the nose of a wine may be considered to be a flaw or fault, depending upon the sensory impact relative to the matrix of the wine. Historically, this was a relatively common occurrence, particularly in white wines and sweet wines, when winemakers were heavy handed with the compound in attempting to avoid the perils of oxidation and microbial degradation. Whilst incidences of apparent excessive SO_2 have declined considerably in the last 20 or so years, there are still too many white wines, both dry and sweet, where overt SO_2 characteristics take the wine into unacceptable territory.

6.3.2 Sensory Detection of Excess Sulfur Dioxide

Excessive sulfur dioxide is detectable upon the nose of wine. The odour is that of a struck match or the burning of coke (fuel). There may also be a nasal burning sensation and a tendency to cough. The detection threshold in white wine is approximately 60 mg/l, although some authors suggest a lower threshold – Maynard Amerine and Edward Roessler state that most people will begin to detect a burned match odour at about 15–40 mg/l [12]. However, the lower the pH of the wine, the higher the perceived intensity of the sulfur smell, which can perhaps be explained by the higher percentage of molecular SO_2 at a lower pH levels [13].

6.3.3 Possible Reasons for High SO_2 Levels

With the level of knowledge today held by winemakers and oenologists and the fast and convenient methods of analysis available, one might consider the presence of excessive SO_2 in wine to be a consequence of inexcusable lack of control. However, particular challenges may arise when the harvest includes damaged fruit, including that infected by mildews or *Botrytis*, when there have been delays in processing, when microbial activity in the must or wine is detected, or when wine has suffered an excessive exposure to oxygen. In these circumstances, maintaining adequate molecular SO_2 levels to achieve microbiological protection and adequate free SO_2 to have effective antioxidant capabilities may lead to much higher levels than might be considered to be desirable. Indeed, repeated additions may result in an excessive level, and the winemaker must remain constantly vigilant so as to stay within organoleptic thresholds and legal limits. Making a small number of (relatively) large additions is always preferable rather than a large number of small doses.

6.3.4 Removal of Excessive Sulfur Dioxide from Wine

A winemaker might consider several courses of action if faced with an excessively high level of SO_2 in a wine. The choice will depend, inter alia, upon the stage in the winemaking/maturation process, organoleptic impact, and legalities. The possibilities include the following:

- *Do nothing*: If the wine is still subject to cellar operations, there will usually be some oxygen uptake at each of these that will result in free SO_2 levels falling – 1 mg/l of oxygen will generally consume 4 mg/l of sulfur dioxide. Levels of SO_2 will fall during the barrel ageing process as a consequence of evaporation and the absorption of oxygen.
- Blend with a similar wine with a lower SO_2 level.
- *Aeration*: A simple method of aeration is by racking the wine, taking from a valve near the bottom of the tank, and splashing it into the top of the receiving tank (rack and splash). The method is less effective when the pH of the wine is high.
- *Oxygenation*: Macro-oxygenation or cliqueage – adding 8 mg/l will drive off some SO_2. Care must be taken to ensure that dissolved oxygen (DO) does not reach an unacceptable level.
- *Nitrogen bubbling*: The bubbling of nitrogen (N) through wine in tank can be effective in driving off SO_2 without the risk of unwanted oxygen uptake.

- The addition of hydrogen peroxide (H_2O_2). This additive is NOT permitted in the EU or by the International Organisation of Vine and Wine (OIV). Hydrogen peroxide binds with (oxidises) the free SO_2 as per the following equation:

$$SO_2 + H_2O_2 \rightarrow SO_4 + 2H$$

Sulfur dioxide + Hydrogen peroxide → Sulfate + Hydrogen

A 1% solution of hydrogen peroxide should be used, at a rate of 50 ml/hl, to achieve a SO_2 reduction of 10 ppm. This is the maximum addition that should be made in a single step – repeats of the treatment are possible, but great care must be taken as there is a risk of the wine becoming oxidised – since H_2O_2 is a strong oxidising agent.

6.4 Oxygen Management in Winemaking

Before discussing the formation of hydrogen sulfide and low-molecular-weight VSCs that impart reduced aromas to wine, it is appropriate to briefly return to the topic of oxygen management, introduced in Chapter 5.

6.4.1 The Role and Management of Oxygen During Fermentation

During the early stages of the alcoholic fermentation (AF), the presence of sufficient oxygen is important for the growth and maintenance of healthy colonies of yeasts. Insufficient oxygen can lead to a stuck fermentation or the production of hydrogen sulfide (H_2S). Yeasts require oxygen to strengthen their cell walls and help them resist stress due to high alcohol levels. They need oxygen to produce sterols and unsaturated fatty acids that play a key role in the fluidity and activity of membrane-associated enzymes, which influence ethanol tolerance, fermentative capability, and yeast viability [14]. Oxygen will help regulate the redox potential helping to prevent the production of hydrogen sulfide and volatile sulfide compounds. At the start and up to the end of the yeast growth phase, which is approximately for the first third of the fermentation process, 4–6 mg/l of oxygen is required to ensure a successful finish to the alcoholic fermentation for white wines, with perhaps 6–10 mg/l being ideal for red musts. Additions will, inter alia, beneficially oxidise some VSCs to less odorous forms [15]. Almost all oxygen added at this stage will be consumed by yeast or dissipated with the rising carbon dioxide.

6.4.2 Oxygenating During Fermentation

The necessary supply of oxygen for fermenting musts may be undertaken by several methods. Historically, oxygen, often in large quantities, would be naturally introduced during the pressing of grapes for white wines and the crushing or treading (pigeage) of grapes for red wines. If undertaken, the process of manual de-stemming of red grapes using an égrappage grille would introduce considerable amounts of oxygen. In comparison, modern de-stemmers and crushers are quick and relatively gentle. Depending upon handling and unless protected, grape musts will contain approximately 5 mg/l of dissolved oxygen.

During the production of red wines, particularly those produced from 'robust' grape varieties such as Cabernet Sauvignon or Syrah, aerated pump-overs are commonly practised. A valve near the bottom of a tank fermenting wine is opened, and the juice is allowed to splash into a large tub, where it will be foaming with oxygen. It is then pumped back to the top of the tank and run over a 'Chinese hat' or other device to splash onto the wine's surface, picking up a substantial (but unregulated) amount of oxygen in the process. Alternatively, vats may have a built-in pump-over system allowing the fermenting juice to be withdrawn from the bottom, pumped up a side-pipe to the top, and sprayed over the 'Chinese hat' to soak the cap of grape solids, still oxygenating the fermenting wine, but to a lesser extent. Of course, the pump-over process also facilitates colour, aroma, and flavour extraction.

The process of 'rack and return', also known as 'délestage', is often carried out once or twice during fermentation. A vat of fermenting wine is pumped into the top of a receiving vat and then back into the top of the discharging vat. This operation may, or may not, involve a high level of oxygenation, depending largely upon the level of splashing undertaken – running the juice down the side of the vat as opposed to splashing onto the surface reduces the uptake. The traditional method of 'pigeage', by which the cap of grape skins is punched below the surface, will also oxygenate the fermenting must.

Air may also be introduced in line during pump-overs. A popular way of doing this is by 'cracking the fitting', which is simply loosening a joint in the hose feeding the inlet side of the pump. Whilst this is usually a cost-free option, care must be taken that the pump impeller remains well-charged with juice, or impeller/rotor damage may result. An inexpensive but effective alternative is the use of a 'venturi'. This is an air injector, which comprises a tube section that narrows to restrict the flow of wine and creates a partial vacuum. At the restriction point, there is an air inlet, with a valve. The air volume may be controlled by computer software. A VinPilot® venturi is shown in Figure 6.1.

6.4.3 Micro-oxygenation During Fermentation

Micro-oxygenation (MOX) may be undertaken during fermentation of red wines. During the yeast growth phase, 6–10 mg/l of oxygen is needed. By the MOX process, a diffusion in the region of 2 mg/l per day may be undertaken early in the fermentation cycle, but additions should not be made near the end of the alcoholic fermentation as yeasts no longer require oxygen [16]. Macro-oxygenation is another commonly practised method – this involves rapid introduction of 8 mg/l of oxygen in early to mid-fermentation. It should be noted that the measurement amount of oxygen in mg is not the same as in ml – 1 mg = approximately 1.5 ml, but there is a variance in the ratio of ml and mg of oxygen depending upon the temperature. At 15 °C (59 °F), 1 mg of oxygen is equal to 1.47 ml, and at 20 °C (68 °F), 1 mg of oxygen is equal to 1.5 ml.

Contrary to the perceptions of many winemakers, there are benefits to the addition of oxygen during white wine fermentation, with the main impact on the kinetics of fermentation [16].

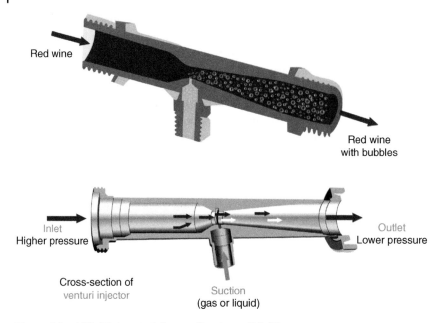

Red wine

Red wine
with bubbles

Inlet
Higher pressure

Outlet
Lower pressure

Cross-section of
venturi injector

Suction
(gas or liquid)

Figure 6.1 A VinPilot venturi. Source: Courtesy of VinPilot.

6.4.4 Micro-oxygenation Post-fermentation

Post-fermentation MOX may take place in three distinct phases:

- Pre-MLF to stabilise colour and build structure.
- Post-MLF to integrate pyrazines, soften tannins, reduce astringency, lessen reductive tendencies, and facilitate general structural refinement.
- Post-barrel ageing to harmonise oak tannins and negate pre-bottling reduction.

When I speak to other winemakers, the main benefits they expound are improvements in colour and mouthfeel together with decreased astringency. The dosage must be carefully controlled and will vary according to the type of wine and stage of winemaking. During maturation, a controlled amount of oxygen uptake can help polymerise long-chain tannins, improving texture and reducing astringency, and improve, brighten, and fix colour. Of course, well-managed MOX can speed cellar maturation and is used in association with oak products, replicating barrel ageing at much lower cost. However, the process needs care, attention, and experience: particularly, if a wine is 'over' micro-oxygenated, there may be a build-up of acetaldehyde or the aromas and flavours of full-blown oxidation and a risk of microbial spoilage.

The bubbling takes place at a temperature of approximately 15 °C (59 °F) or a little higher. Increases in acetaldehyde have been noted in the later stages of MOX. Oxygen may be applied to maturing wine to reduce green or vegetal character – MOX minimises the aroma perception of methoxypyrazines by masking. Professor Andrew Waterhouse of UC Davis noted that if you analyse a wine before and after MOX, the chemical evidence of pyrazines will be unchanged, but the aroma and flavour will be gone [17]. One of the important principles of MOX is that it must not lead to an increase of dissolved oxygen (DO) – the rate of application, at whatever stage in winemaking, should be just below that of consumption [18, 19]. Accordingly, the monitoring of a MOX is best done organoleptically rather than instrumentally. If, at any time, aldehyde becomes apparent, the MOX dosage should be reduced; if sulfide characteristics are noticeable, it should be increased [20, 21]. One of the advantages of MOX is that it provides direct control and more predictability than leaving to natural circumstances.

6.4.5 Barrel Maturation

The process of barrel maturation is generally the winemaking procedure resulting in the largest uptake of oxygen. The total amount of annual uptake can vary between 12 mg/l/year for a 225 l barrel [22] and 20–45 mg/l [23]. Barrel maturation, particularly in new barrels, also involves the absorption of oak products, including vanillin and tannins.

6.4.6 Lees Ageing

Lees have a tremendous capacity to scavenge and consume dissolved oxygen. Ageing white wine on either gross or fine lees in tank or barrel can help maintain freshness as well as give creamy textural characteristics. Lees contain amino compounds including GSH and cysteine that help protect the wine from oxidation. The bound sulfurous compounds are released into wine as lees decompose.

6.5 Reduction in Wine: Positive and Negative

6.5.1 Reducing Agents

Reducing agents in wine are those that easily release electrons. The main reducing agents are ascorbic acid (which acts very fast), sulfur dioxide (which acts somewhat slower), and various phenols (which are slow acting) [24]. Levels of GSH, another powerful reducing agent that is very active against free radicals and other oxygen reactive compounds, vary considerably, depending upon grape content and if additions are made. GSH may be added to wine, including pre-bottling, for its antioxidant properties.

6.5.2 Positive Reduction

6.5.2.1 Reduced Notes as a Hallmark?
A few winemakers believe that reduction must be avoided at all costs, as associated odours mask the very fruit character that they have strived to bring to the fore. However, wines from some regions and certain grape varieties have a particular propensity to produce reductive wines in certain agronomic conditions, and these can contribute to positive wine characteristics. These include Sauvignon Blanc, Riesling, Chardonnay, and especially Shiraz (Syrah). Syrah often has low levels of yeast assimilable nitrogen (YAN), which, unless compensated for by must additions, may lead to the formation of VSCs. Historically, and even

in some wines today, the Sauvignon Blanc wines from the Central Vineyards of the Loire Valley displayed reduced notes. 'Gunflint' is regarded as a trademark aroma characteristic of Pouilly Fumé and Menetou-Salon. Indeed, the term 'gunflint' is an aroma descriptor for the VSC benzenemethanethiol (also known as benzyl mercaptan) often found in Sauvignon Blanc wines produced from grapes grown on 'poor' soils such as those of Pouilly Fumé and Menetou-Salon. Benzenemethanethiol, which is also found in boxwood (*Buxus sempervirens*), can also give smoky notes to some Sémillons and Chardonnays [25].

6.5.2.2 Desirable Aromas of Some VSCs

Some VSCs produced in wine under reductive conditions can give desirable aromas and flavours. This 'positive reduction' is generally more applicable to white wines, but it is also key to the sensory characteristics of some reds. The passion fruit, tropical, and 'sweaty armpit' aromas of a Marlborough Sauvignon Blanc are on account of 3-mercaptohexan-1-ol (3-MH) and 3-mercaptohexylacetate (3-MHA). The savoury, mocha, roasted coffee and chocolate tones found in some Syrahs can also be on account of VSCs. The metallic notes often exhibited by a young red wine that has just been opened may be on account of some VSCs at low concentrations. As Matt Thomson of New Zealand's Blank Canvas and Rapaura Springs wineries says (when discussing sulfur compounds in Syrah), 'they can give a meatiness. I like some bacon with my pepper' [personal discussion].

6.5.2.3 Minerality

The topic of 'minerality' in white wines is somewhat controversial. Wine writers, critics, wine lovers, and winemakers often expound the pronounced minerality of certain wines on the nose and palate, and such reports increased in the first decade of this century. There are also reports by wine professionals that minerality is perceived not only as aromas and flavours but also a tactile, trigeminal sensation [26]. Of course, wines contain several minerals, which may originate from not only soil but also processing aids (e.g. bentonite) and cellar equipment (e.g. concrete tanks). Potassium is the most abundant, with a concentration of 125–3060 mg/l [27]. Other minerals found in wine include calcium, magnesium, phosphorous, and sodium – this is the most likely to have an impact upon taste. However, a large number of observers and researchers attribute 'minerality' on the nose and palate to sulfur and/or reduction. This view is hotly challenged by those who hold the concept of terroir dearly in their hearts and minds. There are countless references in wine magazines and books to wines made from grapes grown on chalk soils having aromas and flavours of minerality.

In his 2015 book *Terroir and Other Myths of Winegrowing*, Mark Matthews, professor of viticulture at the Robert Mondavi Institute for Wine and Food Science at the University of California, Davis, dismisses any notion of mineral flavours as being derived from the soil [28]. Wendy Parr et al. note that 'the popular notion that we are simply tasting inorganic minerals in the wine transmitted from the vineyard ground is not scientifically plausible' [29]. Writing in January 2020, Master Sommelier John Szabo notes 'wines high in sulphur compounds such as benzenemethanethiol, methanethiol, or hydrogen sulphide, often described as "flinty", "oyster shell", or "reductive", are often also described as "mineral".... They can result from struggling yeasts in nutrient-poor musts, a feature of mineral (nutrient)-poor terroir and thus a feature of place ... but even more frequently these sulphur compounds – for which hydrogen sulphide is used as a building block – result simply from late season sulphur sprays too close to harvest, which allows sulphur to find its way into must/wine and cause that flinty-reductive smell, quite unrelated to vineyard geology or soil' [30]. Sulfur compounds noted as having a positive correlation with sensory perceptions of minerality include free and bound SO_2, benzenemethanethiol, methanethiol, and disulfane [29].

The soils of the district of Chablis are largely Kimmeridgian and Portlandian limestones, and being largely unimpacted by oak characters, the best wines are noted for exuding 'mineral' characteristics. A 2017 published study by Heber Rodrigues et al., examining the minerality of Chablis Premier Crus, an appellation covering 780 ha on both sides of the River Serein, produced some fascinating results [31]. Four wines were chosen from vineyards on the left bank of the river (Cote de Léchet, Montmains, Vaillons, and Beauroy) and four from vineyards on the right bank (Montée de Tonnerre, Mont de Milieu, Fourchaume, and Vaucoupin). The wines were of the same (2013) vintage, made by the same producer, and in the same manner fermented and aged in stainless steel. The wines from the left bank were perceived by the tasting panel as having more minerality, and laboratory analysis showed concentrations of the VSC methanethiol, which has shellfish aromas and masks fruity and floral aromas, some 68% higher than the right bank wines [30].

6.5.3 Negative Reduction

Simply, to use the term 'reduction' in relation to wine is normally perceived as a negative descriptor. There is no doubt that the odours of some compounds that are products of reduction will render a wine faulty, if present above sensory detection threshold. But, of course, reduced aromas can appear, disappear, and reappear in bottle; so, sometimes, making a final pronunciation may not be wise. Obviously, all winemakers aim to avoid H_2S being apparent on the nose of a wine. The bad egg or drain smell of the compound, which can also develop in bottle, or the rotten cabbage odour of methyl mercaptan will render a wine undrinkable.

6.6 Hydrogen Sulfide

6.6.1 Hydrogen Sulfide: A Serious Fault When Present Above Sensory Threshold

The presence of hydrogen sulfide (molecular formula H_2S) in wine that has finished its production processes can be a very serious fault. If left untreated, or if present at levels that are too high for effective treatment, the wine may be totally spoiled and require destruction. Although hydrogen sulfide is regarded as an unwanted compound in wine, in fact, during the alcoholic fermentation stage, its production by yeasts is necessary as it detoxifies them and extends their lifespan. It also performs an important signalling function, in arresting the yeast respiration phase and signalling the onset of fermentation.

Hydrogen sulfide is highly toxic and very reactive. It can react with various compounds in wine, including ethanol, to produce other VSCs, such as dimethyl sulfide, and ethanethiol (also known as ethyl mercaptan). Reactions are dependent upon the redox state of the wine and also the pH and are also influenced by environmental factors, particularly ultraviolet (UV) light that can lead to the production of dimethyl sulfide and other mercaptans in bottled wine resulting in 'lightstrike'. It should be noted at this point that the formation of dimethyl disulfide (DMS) is not related to the presence of H_2S.

During initial wine ageing, VSCs including hydrogen sulfide may be produced below their sensory detection threshold, but during the reductive ageing conditions that may occur with extended ageing on lees, or particularly in bottle, different chemical forms of these VSCs may be generated that are above sensory detection thresholds.

6.6.2 Sensory Characteristics and Detection of Hydrogen Sulfide in Wine

Hydrogen sulfide has odours of bad egg, drains, hot sulfur springs, or volcanic gas.

The sensory threshold in wine is 1–3 µg/l [32, 33], although the threshold can sometimes be higher, depending upon the wine matrix. In white wine, the threshold was determined by Siebert as 1.6 µg/l [32], and the Australian Wine Research Institute (AWRI) quote a general threshold of 1.6 µg/l [34]. The range in wine can vary from 0 to 370 µg/l [34]. At levels below sensory threshold, H_2S can mask positive fruit aromas.

6.6.3 The Formation of Hydrogen Sulfide in Wine

There are several pathways by which hydrogen sulfide may be formed.

6.6.3.1 Formation by the Actions of Yeasts During the Alcoholic Fermentation

The majority of hydrogen sulfide in wine is produced by yeasts during the alcoholic fermentation [5]. As such, it may be regarded as a normal product of fermentation, and in fact, production of the compound is necessary for extending yeast life. Hydrogen sulfide and other sulfur volatile compounds can be metabolised by yeasts from organic sources, particularly by the degradation of amino acids that contain sulfur, or any sulfur containing source of nitrogen. Methionine and cysteine degrade as they are used as a necessary source of nitrogen by yeasts, particularly in fermenting musts that are low in nitrogen. As these acids degrade, H_2S, ammonia, pyruvate, and thiols are released. Diacetyl, which is produced by yeasts during the alcoholic fermentation and also by certain lactic acid bacteria, is highly reactive and can degrade methionine into H_2S and cysteine to dimethyl sulfide (DMS) and other compounds. Yeast stress due to a deficiency of nitrogen or vitamins is the major cause of the manufacture of high levels of H_2S, and the presence of metal ions can stimulate production. Paradoxically, metal ions are also necessary for oxidation.

Sulfides can be formed by reduction from sulfates, sulfites, elemental sulfur, and organic sources of sulfur by yeasts via the sulfate reduction sequence (SRS), sometimes called the sulfate assimilation pathway (SAP). Sulfates are mainly present in must due to their source in grapes, which, as we have seen, may contain between 160 and 700 mg/l [5]. The presence of sulfites results from additions of SO_2 to grapes post-harvesting and must, and elemental sulfur from its use as a fungicide in vineyards [35]. In the SRS process, *S. cerevisiae* enzymes pick up and activate sulfate, which is then reduced first to sulfite and later to sulfide [6, 24]. If the amount of YAN is insufficient, H_2S produced by yeast is converted to the sulfur-containing amino acids cysteine and methionine [36]. Sulfite present in the fermenting must (due to sulfur dioxide or potassium metabisulfite additions) may be subject to unregulated uptake by the yeast, and sulfite reduction is uncontrolled [6].

Yeast catabolism of the amino acids cysteine and methionine may also serve as a source of MeSH as well as H_2S [37]. Elemental sulfur from fungicide residues can also be easily converted to H_2S during fermentation through non-enzymatic reaction with yeast-derived GSH and also by enzymatic pathways [38]. The non-enzymatic pathways have only recently been recognised. It is interesting to note that fruit produced from vines grafted onto particular rootstocks can contain higher levels of sulfur-containing amino acids.

Yeasts require an adequate supply of nutrients to successfully complete fermentation without the formation of undesirable sulfur compounds. It is important that sufficient YAN is present in the must. Approximately 110 mg/l of YAN is needed by yeasts for satisfactory red wine fermentations and 165 mg/l for whites. As well as nitrogen deficiency, inadequate levels of vitamins (especially B-complex vitamins including pantothene and pyridoxine), insufficient oxygen, or temperature stress in the fermenting must may contribute to the formation of H_2S and other VSCs. *Diammonium phosphate* $((NH_4)_2HPO_4)$, commonly referred to as DAP, is often added to the must, usually at a rate of 200 mg/l, to help ensure that all the sugars are totally fermented and to help prevent the formation of H_2S. Its use is particularly prevalent in New World countries, even when musts are not nitrogen deficient. However, the use of DAP has drawbacks, including providing a possible nutrient source for *Brettanomyces* (see Chapter 4). If adding vitamins to the must is necessary, they should be included in the yeast starter culture and never added once fermentation is underway.

The production of H_2S from amino acids in the latter stages of fermentation varies considerably according to the yeast strain [39, 40]. Following considerable research into the creation of low H_2S-producing commercial yeasts, these are now marketed by several yeast suppliers. Unfortunately, as a result of the relationship between sulfite metabolism and H_2S, strains that produce low quantities of H_2S can, as noted by Jamie Goode, produce 'quite a bit of SO_2' [41]. This may delay the onset of the MLF or even result in excessive levels in the wine. Strains are available, such as Renaissance Yeast's Vivace and Maestoso that have been developed to have normal levels of SO_2 production, or even slightly reduce SO_2 levels relative to the concentration in the must before fermentation [42].

Most of the hydrogen sulfide produced during fermentation is blown off with the carbon dioxide gas, but as the fermentation slows in the latter stages, some may become trapped. Late production of the compound from amino acids is of particular concern, particularly as aeration of the ferment is not advisable at this time.

6.6.3.2 Production by Reduction of Elemental Sulfur or Inorganic Sulfur

Hydrogen sulfide can be produced by reduction of elemental sulfur and/or the reduction of inorganic sulfur. The chemical equation for the formation is simple:

$$S + 2H \rightarrow H_2S$$

$$\text{Sulfur} + \text{Hydrogen} \rightarrow \text{Hydrogen Sulfide}$$

As we have seen, grapes and soil contain elemental sulfur. A major source of elemental sulfur on grapes is vineyard spray residues. This can account for high levels of H_2S production early in the fermentation, particularly in red wines that are fermented on the skins. Inorganic sulfur compounds, particularly copper sulfate, may be used in the vineyards as sprays against downy mildew.

6.7 Prevention of Hydrogen Sulfide Formation

The following might be considered to help prevent the formation of H_2S and related off-odours:

- Sulfur-containing vineyard sprays should be withheld from four to six weeks before the harvest.
- Nitrogen deficiency in the vineyard can result in nutrient-deficient musts, which can lead to yeast stress and H_2S production. Of course, this can be addressed by appropriate fertilisation. Overcropping of fruit and vine virus can also lead to low YAN.
- Other deficiencies in grape nutrients can result in yeast stress and promote formation.
- In the making of white wines, consideration should be given to whole cluster pressing and/or the avoidance or reduction of skin contact, which will reduce vineyard spray residues;
- Must clarification by settling (débourbage) or other method should be undertaken for white wines, which will result in 90% or more of grape sulfur residues settling to the bottom of the tank.
- Low levels of must turbidity, between 50 and 150 nephelometric turbidity units (NTUs), should be aimed for before white fermentations, which will lessen the formation of reductive aroma compounds, and also facilitate a problem-free alcoholic fermentation. The use of a turbidimeter will help control the level of must clarification [43]. However, over-clarification is to be avoided as yeasts do need some solids to adhere to.
- Pre-fermentation fining with bentonite may be beneficial, but the fining will remove proteins, and, yeast nutrients may need to be added to the must.
- The maximum level of total SO_2 in the must should not exceed 80 mg/l and will ideally be well below this figure. Depending upon the sanitary condition of the fruit and the pH, free SO_2 levels should be adjusted between 20 and 50 mg/l.
- Ascorbic acid added to must at a maximum rate of 250 mg/l may help prevent the formation of H_2S and other volatile thiols (including those that may be desired!).
- Pre-fermentation cold soaks should be limited, as necessary yeast nutrients can be consumed by microbiota.
- If necessary, the must should be adjusted to ensure that YAN levels are sufficient. Fermentable nitrogen should be monitored in the must and early stages of the fermentation. Yeasts require ammonia nitrogen (NH_4) and free amino nitrogen (FAN) – alpha-amino acids. Both of these comprise YAN and will be assimilated by the yeast. However, FAN is less easy for the yeast to assimilate and utilise than the ammonia nitrogen. Approximately 110 mg/l of YAN is needed by yeasts for satisfactory red wine fermentations and 165 mg/l for whites [44]. In the case of very high must weights, more YAN will be required.

Excessive levels of YAN, particularly above 300 mg/l, are not recommended as nutrients will be available for unwanted yeasts including *Brettanomyces bruxellensis* and bacteria. The amount of ammonium required for fermentations is 50 mg/l for red wines and 140 mg/l for white wines [44]. Fermentable nitrogen may be measured by formal titration. DAP $(NH_4)_2 HPO_4$ may be added. One grammme of DAP per litre will increase YAN by 212 mg/l. However, the addition of inorganic nitrogen from ammonia salts such as DAP should be avoided if organic nitrogen is limited; otherwise, more hydrogen sulfide will be produced by the yeast as soon as the ammonium is absorbed [15].

- A low sulfide production yeast should be considered, particularly when musts are nutrient deficient.
- Although yeasts produce greater amounts of SO_2 during low-temperature fermentations (for example, the strain J10 can produce 15 mg/l of SO_2 at 28 °C but 40 mg/l at 16 °C), high fermentation temperatures can be favourable to sulfide production.
- Fermenting musts should be supplied with oxygen until approximately 10% alcohol has been produced. It used to be believed that oxygenation would drive off H_2S, but a 2013 study by Viviers et al. revealed that aerative winemaking practices commonly used to remove reductive aromas did not result in a physical displacement of VSCs, but that the O_2 was involved in a reactive manner for H_2S, as well as other VSCs [45].
- Wine should be racked off the gross lees, i.e. those formed within 48 hours or so of the alcoholic fermentation, as soon as possible.
- Lees ageing should be carefully monitored as sulfides can be released from sulfur-containing compounds with the yeast cells, and lees are oxygen scavengers.
- Care should be taken when burning sulfur candles to sanitise barrels. The candle should not touch the staves, and unburned sulfur must not be allowed to fall on them. Any residues will be converted to H_2S.
- Careful consideration should be given as to the choice of bottle closure and bottling adjustments, as H_2S can be formed in bottle. For example, a wine with undetectable levels of H_2S at bottling time later showed a high level (21.1 ng/ml) when sealed with a saran-tin screw caps, compared with 3.5 ng/ml when closed with Nomacorc [46–48].
- Winemakers use several 'unofficial' methods to avoid the formation of hydrogen sulfide. These include undertaking a 'late' copper sulfate spray in the vineyard, adding copper sulfate to the fermenting wine (legal in Australia), using some bronze or brass fittings (e.g. tank valves) in the winery, and even using sections of copper pipe.

6.8 Treatment for Hydrogen Sulfide in Wine

Several possible treatments are available for wines containing high concentrations of H_2S. However, all post-fermentation treatments also have a negative organoleptic impact, reducing the levels of desirable volatile aroma compounds including thiols, esters, and terpenes.

Some of the methods discussed in this section are effective at removing other VSCs. However, it is vital that analysis and bench trials are undertaken to establish the individual VSCs present, as using the wrong treatment, e.g. aeration for ethyl or methyl mercaptans, can make the problem much worse by oxidising sulfides to disulfides, which are much harder to remove.

6.8.1 Physical Treatments

6.8.1.1 Splash Racking
Splash racking, putting plenty of air into the wine, may drive off hydrogen sulfide. This should be undertaken not long after fermentation is completed [49], and no oxygen addition should be made if the wine contains mercaptans, as they could be converted to disulfides.

6.8.1.2 Aeration or Inert Gas Bubbling
Bubbling oxygen through the wine can be effective at driving off both SO_2 and H_2S particularly during the alcoholic fermentation. The equation is simple:

$$2H_2S + O_2 \quad \rightarrow \quad 2S + 2H_2O$$

$$\text{Hydrogen sulfide + Oxygen} \quad \rightarrow \quad \text{Sulfur + Water}$$

The addition of oxygen results in a loss of hydrogen sulfide due to the formation of quinone-sulfhydryl adducts, which are believed to be stable during wine storage [50]. If undertaken post-fermentation, the operation is perhaps less effective and should be undertaken with great care as it may have a detrimental effect on finished wine, with risk of oxidation and promoting an environment favourable to the growth of spoilage yeasts and bacteria. A study by Bekker et al. using a Shiraz wine showed oxygenation during fermentation to be effective at reducing VSCs [51]. This study did not find any significant effect consequential to post-fermentation aeration (splash-racking) treatment.

Inert gas bubbling: The bubbling of an inert gas, nitrogen, or CO_2 is regarded as preferable by some winemakers, but in the case of red wine, care should be taken to ensure the wine is fully degassed prior to bottling, or a 'prickly' taste sensation may ensue. The study by Bekker et al. [51] did not find nitrogen bubbling to be particularly effective – significantly, less H_2S remained in the wines treated with air or oxygen, than those treated with nitrogen. Neither of the above bubbling treatments should be used if the wine contains mercaptans.

6.8.2 Chemical Additions as Treatments

All post-fermentation chemical treatments are non-specific in the removal of volatile compounds and accordingly have a negative impact upon aromas and flavours. As well as the unwanted compounds, many desirable esters, terpenes, and thiols are also removed [52].

6.8.2.1 Sulfiting
The addition of sulfite can convert H_2S into sulfur and water. The equation is as follows:

$$2H_2S + H_2SO_3 \quad \rightarrow \quad 3S + 3H_2O$$

$$\text{Hydrogen sulfide + Sulfite} \quad \rightarrow \quad \text{Sulfur + Water}$$

A combination of aeration, preferably by bubbling, and sulfiting can be particularly effective, and the sulfite will also help protect the wine from oxidation promoted by the aeration.

6.8.2.2 Use of an Oxidising Agent

There are three oxidising agents that are used to eliminate hydrogen sulfide (and certain other VSCs):

- copper sulfate pentahydrate
- copper citrate
- silver chloride

Copper Sulfate Pentahydrate The use of copper sulfate pentahydrate ($CuSO_4 \cdot 5H_2O$ or $CuH_{10}O_9S$) to precipitate copper sulfide, which is insoluble, is relatively commonplace. The compound is toxic, and the operation should always be undertaken by qualified personnel. The rate of addition is 0.05–0.5 mg/l of copper. Bench trials should be undertaken to ascertain the amount to be added. Excessive additions may cause copper haze (see Chapter 10), but the AWRI has noted copper haze in wines with less than 0.5 mg/l copper. Copper haze can be removed by fining with so-called blue fining, i.e. potassium ferrocyanide, which should be undertaken under the supervision of a qualified personnel. The Hubach Test must be undertaken subsequent to the fining operation to ensure that there is no cyanide (which is, of course, poisonous) remaining in the wine.

The treatment with copper sulfate of wines containing VSCs increased during the first decade of this century, particularly in Australasia. The increase in the use of saran-tin lined screw-cap closures led to an increase in so-called post-bottling reduction, as discussed below. The copper fining would usually take place shortly before bottling, and crazily, some producers would even fine wines that contained no VSCs, leaving the copper to react with other wine components. If not precipitated, copper haze could be formed, and the copper act as a catalyst for oxidation. The ideal time for copper fining, if it must be undertaken, is during the latter stages of the alcoholic fermentation or shortly after, which allows time for the wine to self-stabilise and precipitate the metal [53].

Until recently, it was accepted that the addition of copper salts would result in the elimination or reduction in levels of hydrogen sulfide (and methanethiol) by the formation of copper-sulfhydryl complexes that could be removed by racking, centrifugation, or filtration. However, several recent studies have shown that only a negligible amount of added copper is in fact lost in the process [54, 55]. The sulfhydryl binding properties of copper could serve as latent precursors of H_2S and MeSH or other sulfhydryls [56, 57]. Nikolaus Müller and Doris Rauhut recently noted that the method to remove reductive off-odours by treatment with copper(II) salts (sulfate or citrate) 'is more and more questioned: the effectiveness is doubted, and after prolonged bottle storage, they reappear quite often' [37]. Furthermore, copper(II) addition may result in the formation of asymmetric disulfides, polysulfanes, and/or (di) organopolysulfanes from hydrogen sulfide and methanethiol, which could also serve as latent forms of sulfhydryls during bottle storage [50]. Copper remains in solution following addition to wine. Over 80% remains in a wine with excess of H_2S even after separation steps including filtration [54].

Copper sulfate also has an impact upon organoleptic qualities, reducing aromatics, muting varietal characters, and reducing complexity. The negative impact upon the sensory profile applies to both red and white wines. Recent research has sometimes produced conflicting results regarding effectiveness. The 2016 study by Bekker et al. [51] noted that with copper treatment, early treatment was the most effective in decreasing VSC concentrations, but the sensory profiles of the (Shiraz) wines were diminished, compared with wines that had aeration during fermentation. After extended bottle storage, the wines not having had oxygenated treatments showed increased levels of hydrogen sulfide, methanethiol, ethanethiol, methyl thioacetate, and ethyl thioacetate. As might be expected, dimethyl sulfide, dimethyl disulfide, diethyl disulfide, and carbon disulfide were not significantly affected by the aeration or copper treatments [51]. Interestingly, a study by Zhang et al., looking at Chardonnay and Shiraz wines, found that treating grape must with copper and sulfur dioxide reduced the level of hydrogen sulfide and methanethiol, but not other measured low-molecular-weight VSCs [58]. In fact, the practice of adding copper sulfate to wine just before bottling can actually lead to increased H_2S and methyl mercaptans [45, 47, 48]. As summarised by Mark Smith et al. in 2015, 'It is now clear that addition of Cu to treat VSCs is a double-edged sword affecting both unwanted and desirable sulfur compounds' [6].

Copper Citrate An alternative to copper sulfate is fining with copper citrate. This is an organic copper salt, which when dissolved in wine, the copper reacts very quickly with the VSCs in the wine to form an insoluble deposit of black copper sulfide [59]. It is usually added at 2% concentration as a coating to bentonite. The granules are then mixed in water, in a similar way as bentonite, before adding to the wine and thoroughly mixed. There is an immediate reaction with the sulfur compounds that flocculate and can be removed as a sediment after two days [59]. The maximum legal dose is 50 g/hl, which equals 1 g copper citrate/hl. The compound was legalised for use in Australia and New Zealand in 2004 and in the EU since August 2009.

If residual copper remains in a wine, it may be or become unstable, and copper haze (see Chapter 10) or oxidation may be the undesirable consequence. The concentration that can give rise to these situations ranges from 0.2 to 0.6 mg/l.

Legal Restrictions on the Use of Copper

The OIV limit on copper remaining in wine is 1 mg/l [60]. Treatment of wine with Cu(II), as an addition of either sulfate or citrate, is allowed to a maximum addition of 1 g/hl, which equates to 10 mg/l. The EU limits are the same, since the *International Code of Oenological Practices* was incorporated into EU wine regulations in 2019 [61]. In the United States, the Code of Federal Regulations permits copper sulfate to be added at up to 6 mg/l Cu(II), although the residual level in wine cannot be over 0.5 mg/l [62]. Amazingly, in Australia, there is no legal limit. Of course, whatever the legal limits in the country of production, wines must comply with the regulations applicable in the country of sale.

Incidentally, the consumer faced with a glass of wine with reduced odours might find that the 'penny treatment' removes some of the obnoxious smells. A couple of clean 'copper' coins may be dropped into the glass, and the wine may improve in a matter of minutes. However, today's coins in both the United Kingdom and the United States of America are not as effective as 'old' ones, i.e. pre-1992 in the case of the United Kingdom. Today's 1 and

2 pence coins (the United Kingdom) are copper-plated steel; before 1992, they were 97.5% copper, but the high price of the metal helped put an end to their manufacture.

Silver Chloride Silver chloride will react with sulfides to form silver sulfide. The silver chloride must be adsorbed onto an inert carrier material such as diatomaceous earth (kieselguhr), bentonite, or kaolin; the silver sulfide formed during the treatment will remain adsorbed by the inert carrier material, which can then be removed by filtration. The use of silver chloride was banned in the EU for many years but has again been permitted since August 2015, and its use is incorporated in the 2019 International Organisation of Vine and Wine (OIV) *International Code of Oenological Practices* [60] and EU wine regulations [61]. The maximum addition is 1 g/hl (10 mg/l), and the maximum residue in wine is limited to 0.1 mg/l. The operation must be undertaken under the supervision of a qualified wine technologist.

Silicon Dioxide Silicon dioxide, marketed as *Kieselsol* and *Bocksin*, is permitted by OIV, EU, and US Code of Federal Regulations for use as a wine processing aid, specifically as a clarification agent, to achieve flocculation of gelatin and to soften tannins. There is no reference to its use to remove hydrogen sulfide. However, there is anecdotal evidence of its effective use. It bonds to hydrogen sulfide and proteins, and settles out. The compound should be mixed in the wine at a rate of 100–150 ml per hectolitre, stirred several times during the first 24 hours, and the wine should be allowed to settle. Racking off the undisturbed sediment should take place after a period of seven days. The wine may need subsequent fining to remove resultant cloudiness. Filtration to remove all traces of silicon dioxide is essential. There will be a small loss of aromatics.

6.9   Mercaptans, Sulfides, Disulfides, Trisulfides, and Thioesters

6.9.1   Low- and High-Boiling-Point VSCs

VSCs may largely be divided into those with a low boiling point (below 90 °C), those with a high boiling point, i.e. above 90 °C (194 °F), and thiol-acetic acid esters. High-boiling-point VSCs are only produced by yeast metabolism during the fermentation process (they are not produced post-fermentation), and they remain stable in wine [63]. Accordingly, they will not be removed by aeration and do not react with copper. Although copper is often used as a treatment for reduced wines and can be effective with mercaptans and hydrogen sulfide, its use has organoleptic consequences and can sometimes make matters worse. When VSCs are present at levels a little below the sensory threshold, although the aromas of the compound are not detected, there will be a muting effect on the fruit and varietal characteristics of the wine. Affected wines may be regarded as flawed.

6.9.2   Mercaptans

As we have seen, the wine industry, critics, and writers often make a distinction between thiols and mercaptans that is not adopted by the general scientific community. Technically, mercaptans are thiols, but in the wine world, the word mercaptans is reserved for the foul smelling and tasting forms of the compounds.

All VSCs are very reactive, and changes may take place during the various stages of wine production, maturation, and storage. If hydrogen sulfide remains in wine, it can react with ethanol or acetaldehyde to form ethanethiol. Mercaptans can react with each other to form sulfides. If ethanethiol is not removed, then it can further react to form diethyl disulfide, which, in practical circumstances, is almost impossible to remove from wine.

Although mercaptans can be treated by the use of copper, this does reduce desirable aromas and flavours, particularly those of volatile thiols, as the copper binds with the sulfhydryl (–SH) group, precipitating copper sulfate. Treatment is usually with copper sulfate pentahydrate, which may be added at a rate of 0.05–0.5 ppm of copper. This is effective for mercaptans, but not for disulfides. Copper can also cause thiols to oxidise, forming disulfides, and lead to oxidative characteristics in the wine. Whilst oxygen can be used to remove H_2S from wine, adding oxygen to a wine containing mercaptans could make matters worse: the mercaptans may be converted to di-mercaptans (and/or other VSCs).

6.9.2.1 Methanethiol, Also Known as Methyl Mercaptan (MeSH)

The molecular formula of methanethiol is CH_3SH, and the boiling point is 6 °C (43 °F).

All wines contain some methanethiol, which is usually produced by yeast autolysis near the end of fermentation, or during maturation. As with other reductive compounds, high levels can be a consequence of nitrogen-deficient musts, including inadequate quantities of amino acids, resulting in yeast stress during fermentation. The compound is the most common low-molecular-weight problem VSC found in wine.

The sensory characteristics are at low levels shellfish and at high levels decaying rubbish, pungent, putrid, stagnant water, faeces (it is one of the gasses released from animal faeces), sewage, rotten cabbage, garlic, and burnt rubber.

The sensory threshold for methanethiol has been stated as low as 200 ng/l – 300 ng/l, but the figure of between 1.8 and 3.1 µg/l stated by Solomon et al. [33] and AWRI [34] is widely accepted. A typical range in wines is 0–11 µg/l. [34]

The laboratory detection is using gas chromatography + specific detector.

6.9.2.2 Ethanethiol, Also Known as Ethyl Mercaptans (EtSH)

The molecular formula of ethanethiol is CH_3CH_2S, and the boiling point is 35 °C (95 °F).

Ethanethiol is converted by yeast acting upon sulfur during fermentation, or post-fermentations by a reaction between alcohol and H_2S.

The sensory characteristics are burnt match, sulfur, earthy, garlic, raw onion, skunk, town (natural) gas, and phosphorus. Ethanethiol is sometimes referred to as the smelliest substance on the planet.

The sensory threshold is 1.1–1.8 µg/l. This compound is occasionally found in finished wine at a level above sensory threshold. A typical range in wines is from 0 to 50 µg/l [34].

6.9.3 Sulfides Other than H_2S

6.9.3.1 Dimethyl Sulfide (DMS)

The molecular formula is CH_3SCH_3, and the boiling point is 37 °C (98 °F).

The sensory characteristics of DMS are indicated by a nose of tomato ketchup, canned sweetcorn (maize), asparagus, cooked cabbage, asparagus, blackcurrant, olives, molasses, and sometimes clams. At low concentrations, DMS can have positive organoleptic attributes. In white wine, it can contribute to body and mouthfeel. In red wines, it can add truffle tones, notes of blackcurrant, and hints of cooked quince (raw quince is inedible) and contribute to the perception of fruitiness. Low levels of DMS may also add to the blackcurrant character of Cabernet Sauvignon and complexity of Shiraz-based wines.

The sensory detection threshold is 10–25 µg/l. The sensory threshold for off-flavour is 25–30 µg/l for white wines and over 50 µg/l for reds. A bottled wine usually contains less than 10 µg/l of DMS, but levels up to 980 µg/l are possible.

Dimethyl sulfide can be produced from amino acids during fermentation, but most of this is blown off by CO_2. The primary formation pathway appears to be through non-enzymatic hydrolysis of grape-derived *S*-methylmethionine [64]. However, *S*-methylmethionine, present in varying concentrations in grapes and must, is a precursor to the release of DMS during post-bottling ageing. [65]. Levels of DMS may increase the longer a wine ages, and the higher the storage temperature [2]. DMS cannot be treated with copper.

6.9.3.2 Ethyl Sulfide

The molecular formula for ethyl sulfide is $C_4H_{10}S$.

The sensory characteristics are nose of burnt rubber and palate of soap.

The sensory threshold is 1 µg/l.

Ethyl sulfide is very rarely found in wine above sensory threshold

6.9.3.3 Diethyl Sulfide

The molecular formula for diethyl sulfide is $CH_3CH_2SCH_2CH_3$, and the boiling point is 92° (198 °F).

The sensory characteristics are rubber, garlic, onion, and ether.

The sensory threshold is 920 ng/l–1.3 µg/l. A typical range in wines is 0–10 µg/l [34].

6.9.4 Disulfides

Disulfides are usually formed post-fermentation. The formation pathway involves the oxidation of the relevant precursors, mercaptans, or sulfides. They easily convert to mercaptans – they can be reduced back into mercaptans by ascorbic acid or sulfites. Disulfides cannot be treated with copper, and as with mercaptans, oxygenation may make matters worse. The 'penny treatment' as described in Section 6.8.2.2 does not reduce the sensory characteristics of disulfides.

The most common disulfides that may be present in wine are carbon disulfide, diethyl disulfide, and dimethyl disulfide. Their presence is usually below sensory detection threshold.

6.9.4.1 Carbon Disulfide

The molecular formula of carbon disulfide is CS_2, and the boiling point is 46 °C (115 °F).

The sensory characteristics are rubber, sulfide, and ether.

Sensory threshold is often quoted as over 38 µg/l, although Paulo Lopes note it as 5 µg/l [66]. A typical range in wines is 0–140 µg/l [34].

6.9.4.2 Diethyl Disulfide

The molecular formula of diethyl disulfide is $C_4H_{10}S_2$, and the boiling point is 110 °C (230 °F).

The sensory characteristics are a nose of rotten onion, garlic, cheese, and burnt rubber.

The sensory threshold is 4.3 µg/l. Typical range in wine is 0–85 µg/l [34]. It is rarely present in wine at supra-threshold levels.

The compound can be produced by the oxidation of ethanethiol.

6.9.4.3 Dimethyl Disulfide (DMDS)

The molecular formula of DMDS is CH_3SSCH_3, and the boiling point is 110 °C (230 °F).

The sensory characteristics are Brussel sprouts, boiled cabbage, rotten onions, asparagus, molasses, and quince.

The sensory threshold is 9.8–29 µg/l, the latter being the figure quoted by AWRI [34].

DMDS is produced as a result of the oxidation of methanethiol, often produced as a consequence of yeast stress during fermentation. It may also be produced from vineyard fungicides containing elemental sulfur. DMDS cannot be treated with copper. Exposure of wine to light is also a cause of the formation of dimethyl disulfide and other VSC off-odours, particularly in white and sparkling wines – the topic of lightstrike is discussed in Section 6.11.

6.9.5 Trisulfides: Dimethyl Trisulfide (DMTS)

The molecular formula is $C_2H_6S_3$. Boiling point is 170 °C (338 °F). Dimethyl trisulfide (DMTS) is the only trisulfide that may generally be found in wine.

Sensory characteristics: The compound has aromas of cabbage, onion, pickled radish, and sometimes fish or meat.

The sensory threshold is perhaps 200 ng/l. Concentrations have been found in white and red wines up to 0.9 µg/l (900 ng/l) [67]. DMTS may develop during wine storage. The findings of a recent study by Nahoko Nishibori suggest that certain compounds that originate from grapes are involved in DMTS formation in wine and that DMTS formation is accelerated when the must is exposed to oxygen [68].

6.9.6 Thioesters

There is a third group of VSCs, thioesters, which generally do not contribute to odours, unless they break down into mercaptans. They may originally be derived from sulfur-containing amino acids, or from thiamine.

6.9.6.1 Methyl Thioacetate

The molecular formula for methyl thioacetate is C_3H_6OS.

The sensory characteristics are cheese, rotten vegetables, egg, and sulfur.

Sensory threshold: There is some doubt as to the sensory perception threshold in wine – in beer, the range is 50–300 µg/l. Its presence in wine may be from 0 to 110 µg/l [34].

Methyl thioacetate is formed from acetic acid and methanethiol. If aged in anoxic conditions in bottle, the compound may be converted back to methanethiol.

6.9.6.2 Ethyl Thioacetate

The molecular formula of ethyl thioacetate is C_4H_8OS. The compound is derived from ethanethiol.

The sensory characteristics are cheese, onion, coffee, and sulfur.

The sensory detection threshold is 10–40 µg/l and may be found at levels of up to 180 µg/l [34].

6.10 Post-bottling Reduction

6.10.1 A New Wine Fault

Unwanted reduction in wine, in the forms of hydrogen sulfide and mercaptans, has probably been around for centuries. With the advent of reductive winemaking techniques, issues involving low-molecular-weight VSCs increased from the early 1980s. However, 'post-bottling reduction' was largely unheard of until the early years of this century, and the problem has mostly been associated with wines sealed with saran-tin lined screw-cap closures. There were a few reports of sulfurous off-aromas developing during bottle storage prior to 2004, and the paper by Alan Limmer in 2005 advanced understanding [69]. The commencement in the use of screw-cap closures, particularly in Australasia, was regarded as a panacea to the high incidence of corks contaminated with 2,4,6-trichloroanisole (TCA) and other issues encountered with cork. Having had several 'false starts' in the 1970s and 1980s, the rebirth of bottling under screw cap began in July 2000 with a Jeffrey Grosset Riesling, followed shortly after by other Riesling producers in the Clare Valley, South Australia [70]. Over the coming five years, bottling under screw cap became commonplace for Australasian wines and was soon adopted by some producers in many other countries, particularly for white and rosé wines, together with inexpensive reds. Some European countries, particularly France and Portugal were, and to a large degree remain, resistant to their use. However, more and more reports began to emerge of wines, both white and red, bottled under screw cap exuding some very stinky vegetal odours. The wines were suffering from reduction. Wineries began to adjust their bottling regimes in an attempt to address the problem. One of the first steps taken by many wineries in an attempt to address the post-bottling reduction issue was to reduce the levels of free SO_2 for wines sealed under screw cap. However, it seemed the panacea to the problems with cork had, to some extent, replaced one issue with another. Writing in 'Winefront Monthly' in March/April 2006, Campbell Mattinson said, 'This is the article I never wanted to write about screw caps. It is a major concern. I have never seen widespread reduction issues in a clutch of wines

like I did in the NZ wines. Do we need to introduce a Screwcap License system?' (www
.winefront.com.au).

6.10.2 The Causes of Post-bottling Reduction

The topic of post-bottling reduction remains controversial, with many winemakers and
researchers stating that reductive characteristics that appear after some time in bottle are
a 'winemaking issue' and others believing that the utilisation of certain bottle closures,
particularly saran-tin lined screw caps, can result in reduction in the bottle. It is now well
established that low oxygen concentrations during alcoholic fermentation and also during wine storage are associated with increased hydrogen sulfide concentrations in wines
post-bottling [71].

There are several hypotheses for the production of H_2S in bottle. These include the following:

- The reduction of sulfur residues from vineyard sprays.
- The reduction of disulfides.
- The reduction of SO_2 or bisulfite to H_2S.
- The degradation of the sulfur-containing amino acid cysteine.
- The reduction of copper sulfides.

However, hypotheses on the mechanisms of VSC formation in bottle are still not clearly
understood, and there is dissent amongst researchers as to valid pathways. Vincente Ferreira and Ernesto Franco-Luesma note that bonded forms of H_2S and MeSH present in
wine at bottling will be slowly released when the wine is stored in anoxic conditions as
copper ions are reduced [72]. They suggest that the tools to avoid the problem are 'extended
micro-oxygenation and avoiding the use of completely tight closures' [72]. The use of copper in an attempt to combat reduction was discussed earlier in this chapter in Section
6.8.2.2. Several studies have shown that H_2S and MeSH (and reduced aromas generally)
can increase in wines in bottle under low oxygen or anoxic conditions, e.g. when sealed
with a screw-cap with a saran-tin liner, particularly in the presence of transition metals.
The 2018 study by Kreitman et al. found that H_2S and MeSH can increase by over 40 µg/l
during anoxic storage, but some positive VSCs such as 3-sulfanyhexan-1-ol (3SH) that contribute pleasant grapefruit, passionfruit, and blackcurrant aromas also increased [50]. Marlize Viviers et al. note that 'investigations of interactions between metals and volatile sulfur
compounds are now revealing that metals can also promote the formation and release of
these unwanted aroma compounds, particularly in low oxygen storage environments' [73].
They also state that metal treatments initially decreased concentrations of VSCs but the
effects were reversed after four months of anaerobic storage [73]. In a 2018 study, Marlize
Bekker et al. found that copper, acting on unknown precursors, was associated with large
increases in hydrogen sulfide in Shiraz wines [74]. Also, in 2018 Vincente Ferreira et al. note
that total amounts of H_sS and MeSH can increase 'due to the metal-catalyzed desulfhydra-
tion of cysteine and methionine' [75]. It is possible that some sulfide-bound copper can
be removed from white wine pre-bottling by membrane filtration, but this is by adsorption rather than by particle size discrimination, and polysaccharides and proteins inhibit
adsorption [76].

Bottling at an appropriate redox potential is perhaps key to avoiding the formation of VSCs in bottle. Brian Croser of Tapanappa Wines in South Australia's Piccadilly Valley notes that 'Wines are being put into bottle with the potential, under very low redox conditions – a very reduced state where oxygen is excluded – for the sulphide to reappear under screwcap… this can be avoided by putting wine into bottle at the right redox potential. So it is possible to bottle under the most anaerobic screwcap (SaranTin liner) without sulphides appearing' [77].

6.10.3 The Oxygen Transmission Rates (OTRs) of Closures

From a practical point of view, the amount of permitted oxygen uptake at bottling and particularly the choice of bottle closure are key to avoiding the development of VSCs in bottle. There is considerable variance in the Oxygen Transmission Rates (OTRs) of the different closure types and, in the case of screw caps, the different type of cap liners. Studying two Shiraz wines, Maurizio Ugliano et al. note that for the two wines studied, a consumption of oxygen of 5 mg/l over 12 months was the most effective oxygen exposure regimen to decrease accumulation of methanethiol and hydrogen sulfide during bottle ageing [78]. This is far higher than a saran-tin lined screw cap permits.

The topic of the OTRs of various closure types is examined in Chapter 17.

6.11 Lightstrike

Lightstrike is an all too frequently encountered fault, caused by the formation of undesirable VSCs consequential to exposure of wine to light, particularly UV light, but also visible light at the blue end of the spectrum.

6.11.1 Wines Likely to Be Affected by Lightstrike

Lightstrike affects wines bottled in clear glass (flint) bottles. The fault, which can have serious organoleptic impacts, is most likely to affect sparkling wines and white and rosé wines – red wines have a higher degree of protection due to their higher levels of phenols. Lightstrike also affects milk, and beers – thankfully most ales or bottled in brown glass – and brewers have to adjust hopping regimes if the marketing departments insist that they are bottled in clear bottles.

Whilst there can be little doubt that white and rosé wines can look tempting when bottled in clear glass, there is no doubt whatsoever that this can be hugely damaging to the product, as discussed in Section 6.11.3. Figure 6.2, a photo taken in a supermarket in France, shows two bottles of Sauternes from the same property and same vintage. The bottle on the left has been taken from the shelf and is the victim of lightstrike, the bottle on the right is newly removed from the carton.

Louis Roederer Cristal, regarded by many as a particularly fine 'de-luxe cuvée' Champagne, is bottled in clear glass, but this is wrapped in orange coloured cellophane, which filters up to 98% of UV light [personal discussion with Alexis Deligny of Louis Roederer]. Interestingly, Louis Roederer was the first major Champagne house to switch to brown

Figure 6.2 Lightstrike.

(amber) bottles (for wines other than Cristal) in 2010; since then, several other houses have followed suit including Piper-Heidsieck and Drapier, so it would seem that at last the Champagne industry is beginning to take the problem seriously.

6.11.2 Sensory Detection

On the appearance, there is a deepening of colour, sometimes considerably so. The nose may exude odours of sulfur, skunk, garlic, or cooked cabbage. There can also be aromas of 'marmite', wet wool, or wet cardboard. However, wet cardboard aromas can also result from poor filtration, and wet wool aromas are sometimes associated with the Chenin Blanc grape variety.

6.11.3 Causes and Incidences of the Fault

6.11.3.1 The Transmission of UV Light Through Clear Glass Bottles

Lightstrike is found in wines that have been exposed to UV light and visible light at low wavelengths. The severity of the fault may depend upon the following:

- The spectrum of the light source
- The intensity of the light
- The irradiation duration
- The optical properties of the bottle
- The light absorbing species and composition of the wine

With regard to the spectrum of the light source, the most damaging wavelengths to wine are below 520 nm [79], especially in the range of 325–450 nm. UV has wavelengths below 400 nm and is invisible to most humans; visible light wavelengths are from 400 to 760 nm. Exposure of wine bottles in clear glass to light at wavelengths around 370 or 442 nm, particularly induce the lightstrike defect [80]. Critical wavelengths are 340, 380, and 440 nm – 440 nm is in the visible region [81]. UV light-resistant bottles can prevent the transmission at these wavelengths. It is generally accepted that clear glass filters just 10% of UV light, green between 50% and 90%, and brown from 90% up to 99%, at wavelengths below 520 nm. There are some variances, and particularly with clear glass, 'thicker' bottles filter out more UV light. There has been a growing movement to bottle wines in lighter-weight glass – this has been primarily led by environmental concerns, particularly with regard to the impact of transporting additional weight.

A recent study by Panagiotis Arapitsas considered the impact of light on four white wines packaged in flint glass and dark green glass and stored in 'supermarket-like' conditions [82]. The room was air-conditioned at a constant 20 °C, and four tube lamps were on for 12 hours. Only a few flint glass bottled wines developed the lightstrike fault after one to two days of storage, but all developed the fault after 20–40 days shelf life, and all white wines were susceptible, although some wines were more resistant than others. The wines packaged in green glass had a high degree of protection for 50 days or more (50 days was the length of the experiment) [82]. In fact, white wines packaged in clear glass bottles can be affected after just 3.3 hours exposure or, to the same degree, in green bottles after just 31 hours, when the wines were placed at a distance of 35 cm from two 40 W fluorescent lights [83]. Sparkling wines will be affected after 3.4 hours exposure in clear glass, or 18 hours in green glass. Of course, it is unlikely that bottled wine would be stored so close to fluorescent light, even in a retail or restaurant environment. As well as the duration of the exposure and the properties of the bottle, the level of damage will depend upon the spectrum of the light source, together with its intensity [84]. Wines exposed to direct sunlight would receive several thousand times the amount of UV-A radiation as that from fluorescent lights. The topic of the impact of the colour, thickness, and other aspects of the bottle is further discussed in Chapter 17.

6.11.3.2 The Role of Riboflavin, Sulfur-Containing Amino Acids, and Iron Tartrate

Certain VSCs are produced from riboflavin and amino acids by a photochemical reaction with light. Riboflavin (vitamin B_2) is a very photosensitive compound and has strong absorption of UV light. It occurs naturally in grapes, at a concentration of approximately around 50–70 µg/l, and further amounts are produced during the alcoholic fermentation by yeast and released during autolysis. Following fermentation, the level may reach 110–250 µg/l. If the wine is aged on the lees, the level will increase further. An average level in white wines is perhaps 100 µg/l. A level of riboflavin below 80–100 µg/l could limit the development of lightstrike [85]. In the process of intermolecular photoreduction, riboflavin will take up two electron equivalents from a donor, such as an amino acid, thiol, or alde-hyde. Photogeneration produces hydrogen sulfide, dimethyl disulfide, dimethyl sulfide, and methanethiol from the sulfur-containing amino acid methionine and cysteine. Wine naturally contains between 1 and 4 g/l of methionine and cysteine [85]. During the bottle ageing of traditional method and transfer method sparkling wines, prior to disgorging, sulfur-containing amino acids are freed from yeast cells as part of the process of autolysis, and if exposed to UV light, the production of DMDS in such wines is particularly likely. To make matters worse, the bubbles amply the sensory perception of the off-aromas [86]. The photodegradation of iron tartrate to glyoxylic acid and formic acid also contributes to the light-struck effect [87].

6.11.4   Prevention of Lightstrike

There are some measures that can be undertaken to help reduce the risk of lightstrike.

6.11.4.1   Vinification Measures

- Using low riboflavin-producing yeast strains during alcoholic fermentation. The study by Fracassetti et al. [85] noted that one of the strains of *S. cerevisiae* produces 170 μg/l of riboflavin – a level that could induce lightstrike.
- Fining with bentonite or charcoal. Bentonite reduces levels of riboflavin, but 1 g/l is required, and this may have a significant organoleptic impact [85].
- The addition of tannins. Some tannins, from grape skins, seed, and oak wood, have redox properties that may prevent reactions between methionine and riboflavin. Gallic tannins bind with VSCs and quinones.
- The addition of GSH. The structure of GSH has a thiol group to which methanethiol can bond.

The addition of copper is not recommended. It can form complexes with methanethiol and degrade to dimethyl disulfide [88].

6.11.4.2   Packaging Measures

Clear glass should not be used for bottling wines that are likely to be exposed to UV light for prolonged durations. This topic is discussed in Chapter 17.

6.11.4.3   Storage Measures

Wine should be stored away from light, particularly light in the range of 325–450 nm. I have found the highest incidences of light-struck wines in restaurants that have wines on display, and are lacking the rapid turnover as is often the case in supermarkets. The exposure to light also depletes sulfur dioxide levels, and increases oxygen uptake [89].

6.11.5   Treatment

There is no treatment. The fault is irreversible.

6.12 Laboratory Analysis for Sulfur Dioxide, Hydrogen Sulfide, and Volatile Sulfur Compounds

6.12.1 Sulfur Dioxide

There are four traditional laboratory methods generally used for determining SO_2:

- The aeration oxidation method, also known as aspiration/titration and the Rankine/Pocock method.
- The Ripper method
- The modified Ripper method (Ripper method with iodate)
- Automated flow method

The aeration oxidation method, also known as aspiration/titration and the Rankine/Pocock method is approved by OIV, and is still regarded as the 'gold standard' in sulfur dioxide analysis. This method, commonly used by specialists and external laboratories, is discussed in Chapter 18.

Automatic titration apparatus, using the Ripper Method, makes measurement easy and accurate. Single- or multi-parameter mini-titrators are available (e.g. Hanna Instruments) that utilise an electrochemical indicator (oxygen reduction potential [ORP] electrode), which is more accurate than a visual indicator.

A simple, if less accurate, test for free and total SO_2 can be performed using a spectrophotometer. The accuracy for sulfur dioxide is perhaps ±4 mg/l. The equipment, such as CDR WineLab®, is very user-friendly and can perform multiple analyses.

6.12.2 Hydrogen Sulfide and Volatile Sulfur Compounds

6.12.2.1 A Quick and Easy Test for H₂S Using Copper Sulfate

A quick and easy test for the presence H_2S in must or wine is to add a couple of drops of copper sulfate to a sample of must or wine, swirl or shake, and nose the wine. If the typical H_2S odour of bad eggs dissipates and the fruitiness returns, the problem is hydrogen sulfide. CAUTION: The wine must not be tasted – copper sulfate is poisonous.

6.12.2.2 The Copper/Cadmium Test for H₂S, Mercaptans (–SH Group), Disulfides, and Dimethyl Sulfide

There is a relatively easy test that can distinguish between H_2S, mercaptans (–SH group), disulfides, and dimethyl sulfide using solutions of copper sulfate and cadmium sulfate. CAUTION: These compounds are poisonous, and the wines with solutions added may be nosed but NOT tasted. The method described is adapted from Wine Analysis and Production, by Bruce Zoecklein et al. [90].

Solutions:

Reagent 1: 1% weight/volume (w/v) copper sulfate solution (1 g $CuSO_4 \cdot 5H_2O$/100 ml deionised water or 10% ethanol solution)

Table 6.1   Interpretation of results of copper/cadmium test.

Glass 1 'Control'	Glass 2 'Copper'	Glass 3 'Cadmium'	Glass 4 'Ascorbic acid + copper'	Fault indicated
Off-odour	No off-odour	No off-odour	No off-odour	Hydrogen sulfide
Off-odour	No off-odour	Off-odour	No off-odour	Mercaptans
Off-odour	No off-odour	Reduced off-odour	No off-odour	Hydrogen sulfide + mercaptans
Off-odour	Off-odour	Off-odour	No off-odour	Disulfides
Off-odour	Off-odour	Off-odour	Off-odour	Dimethyl sulfide

Reagent 2: 1% w/v cadmium sulfate solution (1 g of $CdSO_4 \cdot 8H_2O$/100 ml deionised water or 10% ethanol solution)

Reagent 3: 10% w/v ascorbic acid (10 g ascorbic acid/100 ml deionised water or 10% ethanol solution)

Procedure:

1) Put 50 ml of the wine to be analysed into each of four wine glasses.
2) Mark Glass 1 as 'Control', Glass 2 as 'Copper', Glass 3 as 'Cadmium', and Glass 4 as 'Copper + Ascorbic Acid'.
3) Add 1 ml of Reagent 1 (copper sulfate) to Glass 2 (marked Copper). Mix.
4) Add 1 ml of Reagent 2 (cadmium sulfate) to Glass 3 (marked Cadmium). Mix.
5) Add 0.5 ml of Reagent 3 (ascorbic acid) to Glass 4 ('Copper + Ascorbic Acid'). Mix and wait for three minutes. Add 1 ml of reagent 1 (copper sulfate) to the glass.
6) Nose each of the glasses in the order Glass 1, Glass 2, Glass 1, Glass 3, Glass 1, Glass 4. Compare with glass 1 each time for the presence of the original off-odour
7) Interpret the results according to Table 6.1

6.12.2.3   Gas Chromatography

VSCs can be determined by headspace analysis coupled with gas chromatography. There are several variants including the following:

- Static headspace analysis by solid-phase micro-extraction with gas chromatography for analysis (G-C SPME). This is the most widely used method, with good sensitivity.
- Headspace–Solid-Phase Micro-Extraction–Gas Chromatography–Tandem Mass Spectrometry (HS-SPME-GC-MS/MS).
- Dynamic headspace purge and trap (GC-DH-PT) involves purging the headspace with a volume of inert gas that removes most of the volatile compounds, which are then concentrated in a trap. This process has good sensitivity but can give rise to errors.
- Gas chromatography coupled with sulfur chemiluminescence detection (GC-SCD) is accurate, precise, robust, and sensitive.

6.13 Final Reflections

My introduction to the world of mercaptans in wine was as a 24 year-old wine novice when tasting wines from Sancerre that were green as grass and almost town-gas stinky. To me, they were so outlandishly bad, rather like that first cigar, that I couldn't wait to taste again and soon fell in love. I grew to love the Pouilly Fumés made by the late and great Didier Dagueneau, colloquially known as 'the wild boy of Pouilly', which were sometimes 'reduced' in youth, even though they were fermented in wood, but had a life beyond any expectation. The 1996 *Pur Sang* scored 19.5 out of 20 when tasted by Richard Hemming MW for www.jancisrobinson.com in 2017 and described as extra extraordinary [91]. Recently, I was reminded of my early experiences with Sancerre when tasting a range of wines from Weingut Tement, based in Austria's Steiermark region, in February 2020. A 2017 Sauvignon Blanc Ried Grassnitzberg 'ISTK' (sealed under a 'Vinolok' closure) exuded spectacular and delightful sulfide flint and smoke aromas. Armin Tement certainly believes that great wine is made in the vineyard, and it is not his role to intervene: 'Our work is 10 months in the vineyard. We use only natural yeast, no nutrients, no DAP and do not fine or filter. When you fine with bentonite you stop enzymes from working, and limit aromas' [personal discussion]. The 2017 Ried SULZ Morillon STK Premier Cru from Tement is deliciously mineral (Morillon is a Styrian synonym for Chardonnay). I reflect on a comment from Cliff Royle, for many years chief winemaker of Margaret River's iconic Voyager Estate and now winemaker at Flametree Wines, who espoused, 'If sulphides are now a wine fault, then the greatest wines I've ever drunk are faulty – RIP Leflaive, Coche-Dury, Roulot...' [92].

But there is a message from the dark side. It's now July 2020, and I take from my dark cellar my last bottle of 2018 Craggy Range Sauvignon Blanc, from the capitol of the variety, Marlborough, New Zealand. I've tasted 23 bottles of this wine over the last two years and enjoyed each one. I twist the saran-tin lined screw cap, pour the wine, and raise the glass – cabbage, brussel sprouts, and garlic. I had been really looking forward to a drink.

References

1 Siebert, T.E., Solomon, M.R., Pollnitz, A.P., and Jeffery, D.W. (2010). Selective determination of volatile sulfur compounds in wine by gas chromatography with sulfur chemiluminescence detection. *Journal of Agricultural and Food Chemistry* 58: 9454–62. https://doi.org/10.1021/jf102008r.

2 Fracassetti, D. and Vigentini, I. (2018). Occurrence and analysis of sulfur compounds in wine. In: *Grapes and Wines – Advances in Production, Processing, Analysis and Valorization* (ed. A.M. Jordão). Intech Open. https://www.intechopen.com/books/grapes-and-wines-advances-in-production-processing-analysis-and-valorization/occurrence-and-analysis-of-sulfur-compounds-in-wine.

3 Kotseridis, Y., Ray, J., Augier, C., and Baumes, R. (2000). Quantitative determination of sulfur containing wine odorants at sub-ppb levels. 1. Synthesis of the deuterated analogues. *Journal of Agricultural and Food Chemistry* 48: 5819–5823. https://doi.org/10.1021/jf0004715.

4 Ribéreau-Gayon, P., Dubourdieu, D., Donèche, B., and Lonvaud, A. (2006). Conditions of yeast development. In: *Handbook of Enology, The Microbiology of Wine and Vinifications*, 2e, vol. 1, 79–113. Chichester: Wiley.

5 Huang, C.-W., Walker, M.E., Fedrizzi, B. et al. (2017). Hydrogen sulfide and its roles in *Saccharomyces cerevisiae* in a winemaking context. *FEMS Yeast Research* 17 (6): fox058. https://doi.org/10.1093/femsyr/fox058.

6 Smith, M.E., Bekker, M.Z., Smith, P.A., and Wilkes, E.N. (2015). Sources of volatile sulfur compounds in wine. *Australian Journal of Grape and Wine Research* 21 (S1): 705–712. https://doi.org/10.1111/ajgw.12193.

7 Leske, P.A., Sas, A.N., Coulter, A.D. et al. (1997). The composition of Australian grape juice: chloride, sodium and sulfate ions. *Australian Journal of Grape and Wine Research* 3 (1): 26–30. https://doi.org/10.1111/j.1755-0238.1997.tb00113.x.

8 Jastrzembski, J.A., Allison, R.B., Friedberg, E., and Sacks, G.L. (2017). Role of elemental sulfur in forming latent precursors of H_2S in wine. *Journal of Agricultural and Food Chemistry* 65 (48): 10542–10549. https://doi.org/10.1021/acs.jafc.7b04015.

9 Jastrzembski, J.A. and Sacks, G.L. (2016). Sulfur Residues and Post-bottling Formation of Hydrogen Sulfide. *Research Focus Cornell Viticulture and Enology* 3a. https://grapesandwine.cals.cornell.edu/sites/grapesandwine.cals.cornell.edu/files/shared/Research%20Focus%202016-3a.pdf (accessed 16 June 2020).

10 Danilewicz, J.C. (2003). Review of reaction mechanisms of oxygen and proposed intermediate reduction products in wine: central role of iron and copper. *American Journal of Enology and Viticulture* 54: 73–85.

11 Danilewicz, J.C. (2012). Review of oxidative processes in wine and value of reduction potentials in enology. *American Journal of Enology and Viticulture* 63: 1–10. https://doi.org/10.5344/ajev.2011.11046.

12 Amerine, M.A. and Roessler, E.B. (1983). *Wines: Their Sensory Evaluation* (revised edn.). New York: W.H. Freeman & Co.

13 Coetzee, C., Buica, A., and du Toit, W.J. (2018). Research note: the use of SO_2 to bind acetaldehyde in wine: sensory implications. *South African Journal of Enology and Viticulture* 39 (2): 2018. https://doi.org/10.21548/39-2-3156.

14 du Toit, W.J., Marais, A., Pretorius, I.S., and du Toit, M. (2006). Oxygen in must and wine: a review. *South African Journal of Enology and Viticulture* 27 (1): 76–94. https://doi.org/10.21548/27-1-1610.

15 Specht, G. (2020). Yeast fermentation management for improved wine quality. In: *Managing Wine Quality*, vol. 2 (ed. A.G. Reynolds), 3–33. Cambridge: Woodhead Publishing.

16 Australian Wine Research Institute (2017). Project 3.3.2 – influencing wine style through management of oxygen during winemaking. Final Report To Australian Grape And Wine Authority. Principal Investigator: Dr Paul Smith. https://www.wineaustralia.com/research/projects/influencing-wine-style-through-managemen (accessed 28 May 2020).

17 Case, J.H. (2017). The science of wine oxidation. *SevenFiftyDaily*. 7 September. https://daily.sevenfifty.com/the-science-of-wine-oxidation (accessed 30 April 2020).

18 Jones, P.R., Kwiatkowski, M.J., Skouroumounis, G.K. et al. (2004). Exposure of red wine to oxygen post-fermentation — if you can't avoid it, why not control it? *Wine Industry Journal* 19 (3): 17–24.

19 Lesica, M. and Kosmerl, T. (2009). Microoxygenation of red wines. *Acta Agriculturae Slovenica* 93 (3): 327–336. https://www.researchgate.net/publication/265240511_Microoxygenation_of_red_wines.

20 Theron, C. (2017). Oxygen management during winemaking. *WineLand Media.* www.wineland.co.za/oxygen-management-during-winemaking (accessed 22 May 2020).

21 Smith, C. (2017). Issues and innovations in oxygen management. *Wine Business Monthly* February. https://www.winebusiness.com/wbm/index.cfm?go=getArticle&dataid=179533 (accessed 30 April 2020).

22 del Alamo-Sanza, M., Cárcel, L.M., and Nevares, I. (2017). Characterization of the oxygen transmission rate of oak wood species used in cooperage. *Journal of Agricultural and Food Chemistry* 65: 648–655. https://doi.org/10.1021/acs.jafc.6b05188.

23 Ribéreau-Gayon, P., Glories, Y., Maujean, A., and Dubourdieu, D. (2006). *Handbook of Enology: The Chemistry of Wine: Stabilization and Treatment*, 2e, vol. 2. Chichester: Wiley.

24 Zoeklein, B.W., Fugelsang, K.C., Gump, B.H., and Nury, F.S. (1990). *Production Wine Analysis*. New York: Van Nostrand Reinhold.

25 Tominaga, T., Guimbertau, G., and Dubourdieu, D. (2003). Contribution of benzenemethanethiol to smoky aroma of certain *Vitis vinifera* L. wines. *Journal of Agricultural and Food Chemistry* 51: 1373–1376. https://doi.org/10.1021/jf020756c.

26 Parr, W.V., Ballester, J., Peyron, D. et al. (2015). Perceived minerality in sauvignon wines: influence of culture and perception mode. *Food Quality and Preference* 41: 121–132. https://doi.org/10.1016/j.foodqual.2014.12.001.

27 Waterhouse, A.L., Sacks, G.L., and Jeffery, D.W. (2016). *Understanding Wine Chemistry*. Chichester: Wiley.

28 Matthews, M.A. (2015). *Terroir and Other Myths of Winegrowing*. Oakland: University of California Press.

29 Parr, W.V., Maltman, A.J., Easton, S., and Ballester, J. (2018). Minerality in wine: towards the reality behind the myths. *Beverages* 4: 77. https://doi.org/10.3390/beverages4040077.

30 Szabo, J. (2020). *Minerality: A new definition.* Canopy https://www.internationalwinechallenge.com/Canopy-Articles/minerality-a-new-definition.html (accessed 30 May 2020).

31 Rodrigues, H., Sáenz-Navajas, M.-P., and Franco-Luesma, E. (2017). Sensory and chemical drivers of wine minerality aroma: an application to Chablis wines. *Food Chemistry* 230: 553–562. https://doi.org/10.1016/j.foodchem.2017.03.036.

32 Siebert, T.E., Bramley, B., and Solomon, M.R. (2009). Hydrogen sulfide: aroma detection threshold study in red and white wine. *AWRI Technical Review* 183: 14–16.

33 Solomon, M.R., Geue, J., Osidacz, P., and Siebert, T.E. (2010). Aroma detection threshold study of methanethiol in white and red wine. *AWRI Technical Review* 186: 8–10.

34 AWRI (2020). *Low Molecular Weight Sulfur Compounds.* AWRI www.awri.com.au/commercial_services/analytical_services/analyses/lmws (accessed 12 June 2020).

35 Rauhut, D. (2017). Usage and formation of sulfur compounds. In: *Biology of Microorganisms on Grapes, in Must and in Wine* (eds. H. König, G. Unden and J. Fröhlich). Xham, Switzerland: Springer International Publishing AG.

36 Swiegers, J. and Pretorius, I. (2007). Modulation of volatile sulfur compounds by wine yeast. *Applied Microbiology and Biotechnology* 74: 954–960. https://doi.org/10.1007/s00253-006-0828-1.

37 Müller, N. and Rauhut, D. (2018). Recent developments on the origin and nature of reductive sulfurous off-odours in wine. *Fermentation* 4 (62) https://doi.org/10.3390/fermentation4030062.

38 Araujo, L.D., Vannevel, S., Buica, A. et al. (2017). Indications of the prominent role of elemental sulfur in the formation of the varietal thiol 3-mercaptohexanol in Sauvignon blanc wine. *Food Research International* 98: 79–86. https://doi.org/10.1016/j.foodres.2016.12.023.

39 Spiropoulos, A., Tanaka, J., Flerianos, I. et al. (2000). Characterization of hydrogen sulfide formation in commercial and natural wine isolates of *Saccharomyces*. *American Journal of Enology and Viticulture* 51: 233–248.

40 Kinzuric, M., Herbst-Johnstone, M., Gardner, R.C., and Fedrizzi, B. (2015). Evolution of volatile sulfur compounds during wine fermentation. *Journal of Agricultural and Food Chemistry* 63 (36): 8017–8024. https://doi.org/10.1021/acs.jafc.5b02984.

41 Goode, J. (2015). *New Yeasts for Reduction Management*, 70–76. Vineyard & Winery Management.

42 Renaissance Yeast (2020). *Low H$_2$S Yeast Technical Brief*. Renaissance Yeast https://www.renaissanceyeast.com/en/products/compare (accessed 22 June 2020).

43 Coelho, C., Perrine Julien, P., Nikolantonaki, M. et al. (2018). Molecular and macro-molecular changes in bottle-aged white wines reflect oxidative evolution – impact of must clarification and bottle closure. *Frontiers in Chemistry* 6 (95) https://doi.org/10.3389/fchem.2018.00095.

44 Considine, J.A. and Frankish, E. (2014). *Quality in Small-Scale Winemaking*. Kidlington: Academic Press.

45 Viviers, M.Z., Smith, M.E., Wilkes, E., and Smith, P. (2013). Effects of five metals on the evolution of hydrogen sulfide, methanethiol, and dimethyl sulfide during anaerobic storage of chardonnay and Shiraz wines. *Journal of Agricultural and Food Chemistry* 61 (50): 12385–12396. https://doi.org/10.1021/jf403422x.

46 Lopes, P., Silva, M.A., Pons, A. et al. (2009). Impact of oxygen dissolved at bottling and transmitted through closures on the composition and sensory properties of a Sauvignon Blanc wine during bottle storage. *Journal of Agricultural and Food Chemistry* 57 (21): 10261–10270. https://doi.org/10.1021/jf9023257.

47 Ugliano, M., Kwiatkowski, M., Vidal, S. et al. (2011). Evolution of 3-mercaptohexanol, hydrogen sulfide, and methyl mercaptan during bottle storage of Sauvignon Blanc wines. Effect of glutathione, copper, oxygen exposure, and closure-derived oxygen. *Journal of Agricultural and Food Chemistry* 59 (6): 2564–2572. https://doi.org/10.1021/jf1043585.

48 Viviers, M. (2014). Effects of metals on the evolution of volatile sulfur compounds in wine during bottle storage. *Australian and New Zealand Grapegrower & Winemaker* 600: 49–51.

49 Hudelson, J. (2011). *Wine Faults: Causes, Effects, Cures*. San Francisco, CA: The Wine Appreciation Guild.

50 Kreitman, G.Y., Elias, R.J., Jeffery, D.W., and Sacks, G.L. (2018). Loss and formation of malodorous volatile sulfhydryl compounds during wine storage. *Critical Reviews in Food Science and Nutrition* 59 (11): 1728–1752. https://doi.org/10.1080/10408398.2018 .1427043.

51 Bekker, M.Z., Day, M.P., Holt, H. et al. (2016). Effect of oxygen exposure during fermentation on volatile sulfur compounds in Shiraz wine and a comparison of strategies for remediation of reductive character. *Australian Journal Grape Wine Research* 22: 24–35.

52 Dahabieh, M., Swanson, J., Kinti, E., and Husnik, J. (2015). Hydrogen sulfide production by yeast during alcoholic fermentation: mechanisms and mitigation. *Wine and Viticulture Journal*: 23–28.

53 Cowey, G. (2008). Excessive copper fining of wines sealed under screwcaps – identifying and treating reductive winemaking characters. *The Australian & New Zealand Grape-grower & Winemaker* April, pp. 48–56.

54 Clark, A.C., Grant-Preece, P., Cleghorn, N., and Scollary, G.R. (2015). Copper(II) addition to white wines containing hydrogen sulfide: residual copper concentration and activity. *Australian Journal Grape Wine Research* 21: 30–39. https://doi.org/10.1111/ajgw .12114.

55 Vela, E., Hernández-Orte, P., Franco-Luesma, E., and Ferreira, V. (2017). The effects of copper fining on the wine content in sulfur off-odors and on their evolution during accelerated anoxic storage. *Food Chemistry* 231: 212–221. https://doi.org/10.1016/j .foodchem.2017.03.125.

56 Franco-Luesma, E. and Ferreira, V. (2014). Quantitative analysis of free and bonded forms of volatile sulfur compounds in wine. Basic methodologies and evidences showing the existence of reversible cation-complexed forms. *Journal of Chromatography A* 1359: 8–15. https://doi.org/10.1016/j.chroma.2014.07.011.

57 Franco-Luesma, E. and Ferreira, V. (2016). Reductive off-odors in wines: formation and release of H_sS and methanethiol during the accelerated anoxic storage of wines. *Food Chemistry* 199: 42–50.

58 Zhang, X., Kontoudakis, N., Blackman, J.W., and Clark, A.C. (2019). Copper(II) and sulfur dioxide in chardonnay juice and Shiraz must: impact on volatile aroma compounds and Cu forms in wine. *Beverages* 5 (70) https://doi.org/10.3390/beverages5040070.

59 Theron, C. (2010). *Vineyard and Cellar Update: The Use of Copper Citrate to Remove Sulphur Compounds*. WineLand Media www.wineland.co.za/vineyard-and-cellar-update-the-use-of-copper-citrate-to-remove-sulphur-compounds (accessed 10 June 2020).

60 International Organisation of Vine and Wine (OIV) (2020). *International Code of Oenological Practices*. Paris: OIV. http://www.oiv.int/public/medias/7213/oiv-international-code-of-oenological-practices-2020-en.pdf.

61 EUR-Lex. (2019) Commission Delegated Regulation (EU) 2019/934 of 12 March 2019 Supplementing Regulation (EU) No 1308/2013 of the European Parliament and of the Council as regards wine-growing areas where the alcoholic strength may be increased, authorised oenological practices and restrictions applicable to the production and conservation of grapevine products, the minimum percentage of alcohol for by-products and their disposal, and publication of OIV files. https://eur-lex.europa.eu/legal-content/GA/TXT/?uri=CELEX:32019R0934 (accessed 25 may 2020).

62 e-CFR. (2020) Code of Federal Regulations, Title 27, Chapter I, Subchapter A, Part 4. https://www.ecfr.gov/cgi-bin/text-idx?c=ecfr;sid=79589a2ef2d093ed0b73152fc7935f1b; rgn=div5;view=text;node=27%3A1.0.1.1.2;idno=27;cc=ecfr (accessed 30 June 2020).

63 Zoecklein, B. (2007). *Factors Impacting Sufur-like Off Odours in Wine and Winery Options*. Virginia Tech https://www.apps.fst.vt.edu/extension/enology/downloads/ SLOFactorsFinal.pdf (accessed 25 June 2020).

64 Segurel, M.A., Razungles, A.J., Riou, C. et al. (2005). Ability of possible DMS precursors to release DMS during wine aging and in the conditions of heat-alkaline treatment. *Journal of Agricultural and Food Chemistry* 53 (7): 2637–2645. https://doi.org/10.1021/ jf048273r.

65 Dagan, L. and Schneider, R. (2014). Controlling dimethyl sulphide levels in bottled wines. In: *Understanding Varietal Aromas during Alcoholic and Malolactic Fermentations. Proceedings of the XXIVes Entretiens Scientifiques Lallemand*, 39–48. Lisbon, Portugal: Lallemand. http://lallemandwine.com/wp-content/uploads/2014/07/Cahier- 2013-2014-final.pdf.

66 Lopes, P. (2012). *Closure Impact on Post-Bottling Wine Development*. VinCE Budapesten http://www.portugalglobal.pt/PT/PortugalNews/RevistaImprensaNacional/ Macroeconomia/Documents/Closure%20impact%20on%20post-bottling.pdf (accessed 25 June 2020).

67 Davis, P.M. (2012). The effect of wine matrix on the analysis of volatile sulfur compounds by solid-phase microextraction-GC-PFPD. M.Sc. Thesis. Oregon, USA: Oregon State University.

68 Nishibori, N., Kuroda, A., Yamada, O., and Goto-Yamamoto, N. (2017). Factors affecting dimethyl trisulfide formation in wine. *Food Science and Technology Research* 23 (2): 241–248. https://doi.org/10.3136/fstr.23.241.

69 Limmer, A. (2005). The chemistry and possible ways of mitigation of post-bottling sulphides. *New Zealand Wine* 1: 34–37.

70 Taber, G.M. (2007). *To Cork or Not to Cork*. New York: Scribner, a division of Simon and Schuster Inc.

71 Bekker, M.Z., Smith, M.E., Smith, P.A., and Wilkes, E.N. (2016). Formation of hydrogen sulfide in wine: interactions between copper and sulfur dioxide. *Molecules* 21 (9): 1214. https://doi.org/10.3390/molecules21091214.

72 Ferreira, V. and Franco-Luesma, E. (2016). Understanding and managing reduction problems. *Internet Journal of Viticulture and Enology* 2/1.

73 Viviers, M., Smith, M., and Wilkes, E. (2014). The role of trace metals in wine 'reduction'. *Wine & Viticulture Journal* 29 (1): 38–40. www.awri.com.au/wp-content/uploads/ 2014/04/1589-Viviers-et-al-WVJ-29-1-2014.pdf (accessed 9 July 2020).

74 Bekker, M.Z., Wilkes, E.N., and Smith, P.A. (2018). Evaluation of putative precursors of key 'reductive' compounds in wines post-bottling. *Food Chemistry* 245: 676–686. https:// doi.org/10.1016/j.foodchem.2017.10.123.

75 Ferreira, V., Franco-Luesma, E., Vela, E. et al. (2018). Elusive chemistry of hydrogen sulfide and mercaptans in wine. *Journal of Agricultural and Food Chemistry* 66 (10): 2237–2246. https://doi.org/10.1021/acs.jafc.7b02427.

76 Kontoudakis, N., Mierczynska-Vasilev, A., Guo, A. et al. (2019). Removal of sulfide-bound copper from white wine by membrane filtration. *Australian Journal of Grape and Wine Research* 25: 53–61. https://doi.org/10.1111/ajgw.12360.

77 Easton, S. (2013). Reduction - winemaker or closure? http://www.winewisdom.com/articles/techie/reduction-winemaker-or-closure (accessed 26 June 2020).

78 Ugliano, M., Dieval, J.B., Siebert, T.E. et al. (2012). Oxygen consumption and development of volatile sulfur compounds during bottle aging of two Shiraz wines. *Influence of pre- and postbottling controlled oxygen exposure. Journal of Agricultural and Food Chemistry* 60: 8561–8570. https://doi.org/10.1021/jf3014348.

79 Maujean, A., Haye, M., and Feuillat, M. (1978). Contribution à l'étude des "goûts de lumière" dans le vin de Champagne. II. Influence de la lumière sur le potentiel d'oxydoréduction. Correlation avec la teneur en thiols du vin. *Connaissance Vigne et Vin* 12 (4): 277–290. (in French). https://doi.org/10.20870/oeno-one.1978.12.4.1427.

80 Maujean, A. and Seguin, N. (1983). Contribution à l'étude des goûts de lumière dans les vins de Champagne. 3. Les réactions photochimiques responsables des goûts de lumière dans le vin de Champagne. *Sciences des Aliments* 3: 589–601.

81 Hartley, A. (2008). The effect of ultraviolet light on wine quality. www.wrap.org.uk/sites/files/wrap/UV%20&%20wine%20quality%20May%2708.pdf (accessed 1 July 2020).

82 Arapitsas, P., Dalledonne, S., Scholz, M. et al. (2020). White wine light-strike fault: a comparison between flint and green glass bottles under the typical supermarket conditions. *Food Packaging and Shelf Life* 24: 100492. https://doi.org/10.1016/j.fpsl.2020.100492.

83 Dozon, N.M. and Noble, A.C. (1989). Sensory study of the effect of fluorescent light on a sparkling wine and its base wine. *American Journal of Enology and Viticulture* 40 (4): 265–271.

84 Grant-Preece, P., Barril, C., Leigh, M. et al. (2017). Light-induced changes in bottled white wine and underlying photochemical mechanisms. *Critical Reviews in Food Science and Nutrition* 57 (4): 743–754. https://doi.org/10.1080/10408398.2014.919246.

85 Fracassetti, D., Gabrielli, M., Encinas, J. et al. (2017). Approaches to prevent the light-struck taste in white wine. *Australian Journal of Grape and Wine Research* 23 (3): 329–333. https://doi.org/10.1111/ajgw.12295.

86 Hunt, A. (2016). The (increasing) problem of lightstrike. https://www.jancisrobinson.com/articles/the-increasing-problem-of-lightstrike (accessed 9 March 2020).

87 Clark, A.C., Prenzler, P.D., and Scollary, G.R. (2007). Impact of the condition of storage of tartaric acid solution on the production and stability of glyoxylic acid. *Food Chemistry* 102 (3): 905–916. https://doi.org/10.1016/j.foodchem.2006.06.029.

88 Maujean, A. and Seguin, N. (1983). Contribution a l'étude des "goûts de lumière" dans les vins de Champagne. 4. Approches a une solution oenologique des moyens de prevention des "goûts de lumière". (Sunlight flavours in the wines of Champagne. 4. Study of an oenological solution to prevent sunlight flavour). *Science des Aliments* 3 (4): 603–613.

89 Singleton, V.L. (1987). Oxygen with phenols and related reactions in musts, wines, and model systems: observations and practical implications. *American Journal of Enology and Viticulture* 38: 69–77.

90 Zoecklein, B.W., Fugelsang, K.C., Gump, B.H., and Nury, F.S. (1995). *Wine Analysis and Production*. New York: Chapman and Hall.

91 https://www.jancisrobinson.com/tastings?search-full=%22didier%20dagueneau%22 {behind paywall} (accessed 22 June 2020).

92 Ahmed, S. (2019). *Decanter* buyer's guide: Australian chardonnay. *Decanter April*: 20–30.

7

Excessive Volatile Acidity and Ethyl Acetate

In this chapter, two of the most feared wine faults are discussed: excessive levels of volatile acidity and ethyl acetate. At high levels, volatile acidity has odours and the taste of vinegar; ethyl acetate has those of nail varnish remover and solvent. These faults may co-occur and also be present with other faults: high volatile acidity is usually also found in wines impacted by *Brettanomyces* and may be present in oxidised wines – overexposure to oxygen and inadequate sulfur dioxide management being common contributory factors.

7.1 Introduction

All wines contain volatile acids – indeed, it is not possible to produce wine without them. At low levels, these may lift some desirable aromas and flavours and increase the complexity of the wine, perhaps giving a little balsamic edge. However, at high levels, the wine will smell vinegary, exhibit a loss of fruit, and be thin and excessively sharp on the palate. The presence of high levels of volatile acidity (VA) may also be accompanied by a significant quantity of ethyl acetate (EA), which has a characteristic nose of nail varnish and nail varnish remover. Elevated levels of ethyl acetate give wine a very harsh and acidic finish and may even give a burning sensation at the back of the mouth. Significant levels of acetaldehyde may also be present in high VA wines. Elevated levels of volatile acidity are often found in oxidised wines and those impacted by the yeast *Brettanomyces*.

There are several possible sources and pathways for high levels of volatile acidity, particularly acetic acid, in wine. These include the following:

- It is a natural product of *Saccharomyces* yeast metabolism during the first half of the alcoholic fermentation (AF).
- It can be formed by acetic acid bacteria (AAB) or 'wild' yeasts on and in the cap of fermented wine in an open fermenter.
- It is produced from citric acid and any unfermented sugars by lactic acid bacteria (LAB) during and following the AF, and malolactic fermentation (MLF), if undertaken.
- It is formed by AAB as wine picks up oxygen during barrel maturation, and sometimes also during micro-oxygenation (MOX).
- It may be formed in an oxidative environment, such as on the surface of wine in ullaged tanks or barrels, by aerobic yeasts, or by AAB. Ethyl acetate may also be formed in these circumstances, which represent a major reason for excess VA and EA production.

Wine Faults and Flaws: A Practical Guide, First Edition. Keith Grainger.
© 2021 John Wiley & Sons Ltd. Published 2021 by John Wiley & Sons Ltd.

- It can be extracted into wine by a physiochemical process from oak wood during barrel maturation. The toasting process used in the production of oak barrels produces acetic acid from wood polysaccharides.

Sources and pathways for high levels of ethyl acetate include the following:

- A chemical interaction between acetic acid and ethanol.
- Microbial production by spoilage yeasts, particularly *Pichia anomala* and *Candida krusei*.
- Produced by LAB and AAB.

7.2   Volatile Acidity and Ethyl Acetate

The level of volatile acidity in a wine is often stated to be an important criterion of wine quality. However, it is perhaps more pertinent to consider it as an indicator of a wine's sanitary condition and whether it is fit for sale having regard to sensory characteristics and legal constraints. Laboratory measurements of volatile acidity are normally noted as grammes per litre (g/l) of acetic acid. From a sensory perspective, most light wine would be regarded as unmarketable at levels of volatile acidity above 1 g/l, although many fortified and/or sweet wines exceed this. The legal limits are detailed below, and perhaps, surprisingly, these are all greater than 1 g/l. It should be noted that the level of volatile acidity in a wine may well increase as it ages. We might also note here that some authors consider ethyl acetate to constitute to volatile acidity, although in fact EA is not an acid. However, increases in levels of VA are very often accompanied by the development of EA, and each of the compounds contributes to the negative sensory characteristics.

7.3   The Controversy of High Levels of Volatile Acidity

7.3.1   Are Elevated Levels of Volatile Acidity Sometimes Acceptable?

The question as to when an elevated level of VA becomes a fault or flaw is not necessarily a simple or straightforward one. There is no doubt that in the case of dry white or rosé wines, high levels of VA make the wine most unpleasant on both the nose and palate. Sweet whites that have been made from fruit infected by 'noble rot', *Botrytis cinerea* in its beneficial form, are likely to have significant VA that can add to complexity and also give the wines a little 'edge' and prevent them being cloying. However, for red wines, the style, grape variety, and, particularly, origin will impact upon the acceptability and perhaps even the desirability of relatively high levels of VA. For example, many Italian red wines have levels, which, in wines from other countries, might be considered to take the wine into the realms of flawed or even faulty, but which contribute to the very 'Italian' character of the wines. Barolo, a wine made from the indigenous Nebbiolo variety in the north-western region of Piemonte, can have a very high VA content that does not detract, partially on account of the high levels of tannin also present. Chiantis and other Tuscan wines made largely from the Sangiovese variety often exude distinctive balsamic notes. Amarone and Recioto Valpolicellas, which are appassimento wines from the north-eastern region of Veneto, often also have high

VA, which perhaps adds a little bite to balance the amounts of residual sugar these wines possess. Outside of Italy, hints of pickles may be found on some wines made from Pinot Noir, including many fine Burgundies. Sauternes including the illustrious Château d'Yquem can show heady notes of high VA. Chateau Musar, the legendary wine from Lebanon's Bekaa valley is well documented for its swingeing levels of VA (and very often *Brettanomyces*) – the individuality and style are prized by aficionados, but the wines from older vintages are often regarded as flawed or even faulty by purists. The late Serge Hochar, owner and winemaker, loved living on the edge: 'I once produced a wine that was technically perfect, but it lacked the charms of imperfection' [1]. Fortified wines, including Madeira and Oloroso Sherry of the highest quality, can exhibit spectacularly high VA levels. Over in the baking heat of Portugal's Douro region, many aged Tawny and Colheita Ports show high VA that can cut through the aromas of nuts and raisins. I recently tasted a truly sensational Oloroso V.O.R.S Sherry with an average age of 45 years from Bodegas Tradición, with a VA of 1.87 g/l. This wine was wine scored 95/100 by Robert Parker Wine Advocate [2] and achieved similar high scores from other critical sources.

The ability to limit the levels of volatile acidity in wine is relatively recent, commencing in the late 1960s and resulting from many factors, including increased understanding, the advent of temperature control for fermentations, and the widespread deployment of fermentation vats made from inert materials, particularly stainless steel. However, elevated levels of VA in wines are increasing again, partially due to higher pH levels in fruit, as producers delay picking until 'phenolic ripeness', and warmer growing seasons, a consequence of climate change.

7.3.2   1947 Château Cheval Blanc – The Greatest Faulty Wine Ever Made

Of the millions of individual wines produced in the twentieth century, there are perhaps a couple of dozen that have become legends. One of these and described by some critics as the greatest wine of all time is the 1947 vintage of Château Cheval Blanc from the Bordeaux district of Saint-Émilion. The summer that year was searingly hot; the grapes shrivelled, and of course, the wine was made in a warm cellar before the advent of temperature-controlled fermentations. Robert Parker, although now retired but still regarded by many as the most influential critic in the history of wine, notes 'the huge nose of fruitcake, chocolate, leather, coffee, and Asian spices is mind-boggling. The unctuous texture and richness of sweet fruit are amazing…after 55 years the wine is still remarkably fresh, phenomenally concentrated and profoundly complex.' He scored the wine 100/100 [3]. Jancis Robinson MW describes it as 'her last chosen wine on earth' (https://www.jancisrobinson.com/articles/1947-a-magically-faulty-vintage). However, the wine shows such a stratospheric level of volatile acidity (together with 5 g of residual sugar resulting from a stuck fermentation) that, if it were produced today, it would be designated as faulty and almost certainly could not be marketed. Mike Steinberger tasted the wine in December 2007 and proclaims 'the '47 Cheval I drank that night now ranks as the greatest wine of my life, a title I doubt it will relinquish. The moment I lifted the glass to my nose and took in that sweet, spicy, arresting perfume, my notion of excellence in wine, and my understanding of what wine was capable of, was instantly transformed…the residual sugar and volatile acidity were readily apparent, but…the flaws inexplicably became virtues' [4].

In October 2010, an Imperial (equating to eight standard bottles) was sold in Christie's in Geneva for US\$ 304 375 [5]. In October 2019, a magnum was sold by Sotheby's in London for £54 450 [6]. Specialist fine wine merchant Jan-Erik Paulson who, with others, tasted five bottles of this vintage in September 2017 described one bottle as 'fantastic' and another as 'perfection and spellbinding' (https://www.jancisrobinson.com/articles/1947-a-magically-faulty-vintage). However, he too agrees that it is 'a wine that would be classed as faulty by today's standards, with volatile acidity and residual sugar levels enough to fire any winemaker producing such a wine' (https://www.jancisrobinson.com/articles/1947-a-magically-faulty-vintage). So, the sobering, devastatingly depressing realisation is that one of the greatest wines of all time is technically faulty on account of excessive volatile acidity, and such a wine will never be produced again.

7.4 Fixed Acids and Volatile Acids

7.4.1 Total Acidity

The total acidity of a wine is the combination of non-volatile or fixed acids and volatile acids, which are those that can be separated by steam. Put simply, if a wine is 'boiled', the fixed acids stay in the liquid, but the volatile acids are released as vapour. As the name suggests, volatile acidity is the wine acid that can be detected on the nose, at 'room temperature'. All other acids are only sensed on the palate. It is perhaps pertinent at this point to examine the various fixed and volatile acids that are found in grapes and wine.

7.4.2 Fixed Acids

The important fixed acids found in wine are as follows:

7.4.2.1 Tartaric Acid
Molecular formula is $C_4H_6O_6$. This is the most important acid contained in grapes. Indeed it is not found in any other fruits of European origin, and the presence of its residue on archaeological finds is accepted as reliable evidence of viticulture, and almost certainly winemaking, in the location and period in question. However the acid is also present in tamarinds. There will usually be higher concentrations of tartaric acid in grapes from cooler climates. The concentration of tartaric acid in musts may range from 3 g/l (hot climate grapes) to 7 g/l or more (cool climate grapes). The concentration remains largely unchanged during alcoholic and MLFs. It is usually the most prevalent acid found in wine and adequate levels are required to preserve wine during the maturation and ageing processes. On very rare occasions tartaric acid may be degraded during winemaking or storage and the fault known as 'tourne' may result (see Chapter 11).

7.4.2.2 Citric Acid
Molecular formula is $C_6H_8O_7$. This acid is found in grapes, although the concentration is less than 10% of the total grape acid content. It is also produced by yeasts during the alcoholic fermentation, up to a concentration of 0.4 g/l. Elevated levels of citric acid can cause problems: if the wine undergoes MLF, citric acid may be broken down by LAB to produce diacetyl (see Chapter 11) and acetic acid.

7.4.2.3 Malic Acid

Molecular formula is $C_4H_6O_5$. This is the second most prevalent acid in grapes, its content decreasing as the grapes ripen. The concentration in musts ranges from 1–2 g/l (hot climates) up to 6.5 g/l in some musts from cool climate vineyards. Some strains of *Saccharomyces cerevisiae* can reduce its levels during alcoholic fermentation. However, the MLF, if undertaken and completed (as it almost always will be in the case of red wines), will convert all the malic acid into lactic acid. This will reduce the total titratable acid content by between 0.5 and 3 g/l. For white wines, the MLF may or may not be undertaken depending upon production region, grape variety, and of course the style of wine required.

7.4.2.4 Lactic Acid

Molecular formula is $C_3H_6O_3$. This acid is not present (other than occasionally in very small amounts) in healthy grapes, but may be present, often at high levels, in damaged fruit. During the alcoholic fermentation, a small quantity, perhaps 0.2 g/l or so, may be produced by yeast actions. During MLF, if undertaken, the actions of certain LAB convert malic acid into lactic acid. The level of lactic acid subsequent to this transformation will be approximately 50% of that of the malic acid pre-MLF. The lactic acid contents of finished wine can range from 0 up to 3 g/l.

7.4.2.5 Succinic Acid

Molecular formula is $C_4H_6O_4$. This is another acid that is not present in grapes, but up to 1 g/l is produced during the alcoholic fermentation.

7.4.3 Volatile Acids

Volatile fatty acids found in wine consist primarily of short-chain (tails of less than 6 carbons) and medium-chain fatty acids (tails with 6–12 carbons).

Generally thought of as simply as acetic acid, volatile acidity is in fact composed of acetic acid and, to a much lesser extent, butyric, isobutyric, carbonic, formic, hexanoic, isovaleric, and propionic acids. Acetic acid is the main volatile acid in wine; the other volatile acids rarely occur above their odour detection threshold.

7.4.3.1 Acetic Acid

Molecular formula is CH_3COOH, which may also be expressed as CH_3CO_2H, or condensed to $C_2H_4O_2$. Acetic acid usually comprises over 93% of a wine's VA [7]. Of course, vinegar is the best known use for acetic acid, with most vinegars containing between 4% and 18% of acetic acid by volume.

7.4.3.2 Butyric Acid

Molecular formula is $C_4H_8O_2$. This is an extremely unpleasant smelling acid, with aromas of very ripe soft cheese, or vomit. The compound has been associated with bacterial growth on grapes and also rapid MLF. The sensory detection threshold is 173 µg/l.

7.4.3.3 Isobutyric Acid

Molecular formula is $C_5H_{10}O_2$. Isobutyric acid is also known as 3-methylbutanoic acid, and β-methylbutyric acid. The odour is rancid often with notes of stale sweat. The sensory detection threshold is 33 μg/l. Isovaleric acid is discussed in greater detail in Chapter 4, as it is one of the products of the action of *Brettanomyces*.

7.4.3.4 Propionic Acid

Molecular formula is CH_3CH_2COOH. Although this volatile acid may be present in wine, it is usually at very limited levels. As with isobutyric acid, the nose is that of soft cheese and body odour. Champagne is perhaps the only wine that can contain propionic acid at levels above sensory threshold. The sensory detection threshold is in the region of 160 ng/l.

7.4.3.5 Hexanoic Acid

Molecular formula is $C_6H_{12}O_2$. This may be present in wine, but at very limited levels. The nose is that of sweat, sometimes slightly rank. The sensory threshold is 420 μg/l.

7.4.3.6 Sorbic Acid

Molecular formula is $C_6H_8O_2$. This acid is also volatile, and although not naturally present in grapes or produced in winemaking processes, it is sometimes added in the form of potassium sorbate as a fermentation inhibitor immediately before bottling to inexpensive sweet and semi-sweet wines. The compound has a faint smell of kerosene. The odour detection threshold for experienced tasters is 130 μg/l.

7.4.3.7 Sulfurous Acid

Molecular formula is H_2SO_3. Sulfurous acid is produced naturally during fermentation and also by reactions due to additions of SO_2. It is technically a volatile acid, but is not considered as a contributor to the volatile acidity of wine, and steps should be taken to remove this before laboratory analysis for volatile acidity.

7.5 Sensory Characteristics and Detection of Volatile Acidity

Excessive volatile acidity is particularly associated with a vinegar-like odour. The sensory detection threshold of total volatile acidity generally varies between 0.5 and 0.7 g/l (500–700 mg/l) depending, inter alia, upon grape variety, residual sugar, and level of alcohol. In wines from some varieties, the detection threshold may be as high as 0.9 g/l. Above threshold levels, a wine may illustrate an acrid, even bitter finish. At levels above 1.2 g/l, the nose of the wine will generally be unpleasant, and above 1.5 g/l, there will usually be a pronounced vinegar aroma and flavour. Sensory detection of acetic acid is not necessarily a straightforward matter, since invariably ethyl acetate will also be present, and the amounts and combination of both these compounds will contribute to the sensory effects. In some instances, the detected concentration of acetic acid can be under the 0.7 g/l threshold with a concentration of ethyl acetate over 100 mg/l, which contributes to the 'high VA' characteristics of the wine [8].

7.6 Legal Limits

The maximum permitted levels of volatile acidity under EU regulations are 1.08 g/l for white and rosé wines and 1.2 g/l for reds. These limits are expressed in the regulations as 18 mequiv/l for white and rosé wines and 20 mequiv/l for red wines. Commission Delegated Regulation (EU) 2019/934 [9] provides that member states may grant derogations for certain wines bearing a protected designation of origin or a protected geographical indication when they have been aged for a period of at least two years, or where they have been produced according to particular methods, and also or wines with a total alcoholic strength of at least 13 % alcohol by volume (abv). There are derogations in some EU countries for 'noble late harvest' wines, up to 1.8 g/l, although it is perhaps questionable whether any quality conscious producer would wish to market wines at such a level. Within the EU, the limits in France are lower: 0.9 mg/l for both reds and whites.

In the United States, as specified by Standards of Identity in the Code of Federal Regulations (27 CFR) [10], generally, the limits are 1.2 g/l for white and rosé wines and 1.4 g/l for red, although in California the limits are 1.1 and 1.2 g/l, respectively. In the United States, for late-harvest white wines produced from juice with a must-weight of 28 Brix or more, the permitted maximum VA is 1.5 g/l; for late-harvest red wines produced must of 28 Brix or more, the permitted maximum VA is 1.7 g/l. In Canada, the legal limits are 1.3 g/l for table wine and 2.1 g/l for ice wines. The Ontario Wine Appellation Authority respects the following limits for 'vintners quality alliance (VQA)' approval: ice wine and totally botrytis affected, 2.1 g/l; special select late harvest and botrytis affected, 1.8 g/l; late harvest and select late harvest, 1.5 g/l; and all other wines, 1.3 g/l [11]. The sensory detection threshold in ice wine is reported as being 3.185 g/l [12]. The actual VA levels in Canadian ice wines are reported as being in the range 0.49–2.29 g/l [13]. In Australia, the legal limit is 1.5 g/l for all wines (Standard 4.5.1) [14]. The legal levels in South Africa are 1.2 g/l for both red and white wines. There are no legal limits for the presence of ethyl acetate – volatile acidity at levels above legal maxima will invariably co-occur if the EA level is excessive.

7.7 Acetic Acid Bacteria

AAB, botanical family name *Acetobacteraceae,* are Gram-negative obligate aerobes. The cells are ovoid or rod shaped, and the bacterium size is usually 0.6–0.8 μm × 1.0–4.0 μm. Due to many instances of reclassification, there is some dissention as to how many genera there are in the *Acetobacteraceae* family: in 2014, Komagata et al. list 32 genera [15], but a 2018 review by Gomes et al. lists just 19 [16]. Back in 1984, just two genera and five species had been identified. AAB are ubiquitous in nature and are found in many plants, flowers, herbs, numerous fruits, and grapes. AAB also live in symbiosis with many insects, and species of the bacteria are found in the digestive tracts of these. Vinegar and fruit flies of the genus *Drosophila*, particularly *Drosophila suzukii* and *Drosophila melanogaster* host several species of AAB, especially those of the genera *Acetobacter* and *Gluconobacter*. The bacteria proliferate in the gut and are also present on the surface of the host insects [17]. These two species of fly and these two genera of bacteria are particularly implicated in acetic

Figure 7.1 *Drosophila suzukii*. Source: Courtesy Dr Gevork Arakelian.

acid production in wine, and every effort should be made to exclude them from fruit and wineries. *D. suzukii* is illustrated in Figure 7.1 and *D. melanogaster* in Figure 7.2.

AAB multiply very rapidly in conditions favourable to their development. As obligate aerobes, the bacteria cannot grow without the presence of oxygen, although they can survive in quasi-anaerobic conditions. The bacteria require an adequate carbon source to sustain growth – this is typically ethanol, which AAB convert into acetic acid [18].

The main two species implicated in the production of acetic acid in wine are *Acetobacter aceti* and *Gluconobacter oxydans*, but up to 15 other species may also be involved. AAB, mostly *Gluconobacter*, are present in grapes. On healthy grapes, the quantity is in the range of 100–10 000 CFU/ml of juice, but on damaged grapes, the quantity is perhaps 1–10 million CFU/ml [19]. On grapes infected by *B. cinerea*, the level can reach 100 million CFU/ml. They can rapidly degrade grape sugars into acetic acid. As expected, *G. oxydans* is usually the dominant species in must and the early stages of fermentation, but it is rarely found in wines as it prefers glucose (which it oxidises) for its growth [19]. *A. aceti* is the most common AAB species in the late stages of fermentation and wine [18] – *Acetobacter* prefer ethanol for growth [19], which is oxidised to acetic acid. In fact, AAB have an important ability to oxidise a wide range of substrates and to accumulate metabolic products in different media whilst avoiding toxicity to themselves [18].

The bacteria can grow in wine at any stage of winemaking, as detailed below. However, there are some particularly dangerous circumstances. If a red wine is fermented in an open top vat, the warm cap of grape skins that has been pushed to the top and exposed to the atmosphere presents an ideal environment for the growth of AAB. Growth in wine that is stored in ullaged vats or barrels is commonplace, and AAB need oxygen to convert ethanol to acetic acid. Careful use of sulfur dioxide at the appropriate stages of winemaking will inhibit or limit the growth of AAB, together with other bacteria and spoilage yeasts.

1 mm

5465223

Figure 7.2 *Drosophila melanogaster*. Source: Bugwood.orgCourtesy Ken Walker Museum of Victoria

The winery and equipment can harbour substantial colonies of AAB. They can colonise cuveries (drains can be a major source), chais, and bottling halls, on vinification equipment, and especially previously used wooden barrels. The presence in significant numbers of the bacteria on structures and equipment is consequential to poor winery hygiene.

7.8   Production of Acetic Acid in Wine

As we have seen, acetic acid is by far the most common of the volatile acids in wine. Whilst all wines contain some acetic acid, the higher the pH of the must, the greater the amount that may be produced during both the alcoholic and MLFs. The oxidation of ethanol through acetaldehyde to acetic acid is well understood and documented, wines with a pH above 3.6 having the greatest risk of high levels of production.

7.8.1   Microbiological Production

7.8.1.1   Production by Yeasts
Grapes contain many yeasts and microflora. Apiculate yeasts including *Kloeckera apiculata* and its teleomorph *Hanseniaspora uvarum* may predominate, but spherical or elongated

yeasts including *Metschnikowia pulcherrima* and its anamorph *Candida pulcherrima* are of major importance. In fact, *S. cerevisiae*, which usually predominates in alcoholic fermentations, accounts for less than 1% of the flora, whilst *H. uvarum* generally comprises over 50% of the native yeasts. Unless inhibited, *H. uvarum* is likely to have a strong presence during the early stages of the alcoholic fermentation, perhaps until a level of 4% or 5% abv has been achieved. *S. cerevisiae*, being resistant to alcohol and wine acids, will continue the fermentation from 4% or 5% abv, up until completion. The use of some non-*Saccharomyces* yeasts, particularly in association with *Saccharomyces* cultures, is gaining popularity as many winemakers seek broader and more diverse aromas and flavours than offered by many commercial *Saccharomyces* strains. By way of example, *M. pulcherrima* can produce many positive aromas, including those of citrus, and has a role in reducing the production of alcohol.

Both *Saccharomyces* and non-*Saccharomyces* yeasts produce acetic acid. *P. anomala* in particular has the ability to produce considerable quantities of acetic acid and ethyl acetate, but other 'wild' yeasts can also produce high amounts of both compounds. *H. uvarum* will deplete nitrogen, vitamins, and nutrients in the must, contributing to general yeast stress and promoting acetic acid production even when *Saccharomyces* takes over the fermentation process. During the alcoholic fermentation, especially in the most vigorous period, *S. cerevisiae* can produce at least 0.1 g/l and sometimes 0.3–0.4 g/l of acetic acid. The amount produced will depend, inter alia, on the strains of *S. cerevisiae* involved [20]. The paradox is that acetic acid is toxic to *S. cerevisiae*, and so, a significant presence (in the region of 0.7–0.9 g/l) in the fermenting juice may lead to a highly undesirable stuck fermentation. This can happen even if the fermentation temperature has not spiked at 35 °C (95 °F) or thereabouts that generally inhibits yeast action, as AAB produce yeast-inhibiting substances.

Certain speciality wine styles may pose particular problems. For example, high acetic acid production by yeasts is common during the fermentation of ice wines. The yeasts suffer high osmotic stress and adapt by exuding glycerol, which prevents the movement of water from the yeast cell into the must [21]. Glycerol is formed via an enzymatic reaction. The redox balance is changed, which is corrected for by the production of acetic acid, moving the redox potential back to equilibrium [21].

The yeast *Brettanomyces* can produce high levels of volatile acidity as well as the trademark ethyl phenols (see Chapter 4). Whilst this yeast can grow at any stage of winemaking, maturation in barrels is a particularly dangerous time, especially if there is any ullage and/or appropriate levels of molecular SO_2 are not maintained.

7.8.1.2 Production by Bacteria, Including Acetic Acid Bacteria

AAB are able to convert both glucose and ethanol into acetic acid. *Acetobacter* and *Gluconobacter* can oxidise ethanol to acetic acid enzymatically using alcohol dehydrogenase (first oxidised to acetaldehyde then to acetate with aldehyde dehydrogenase) [22]. Also, acetyl phosphate may be oxidised to produce acetic acid. AAB usually produce significant levels of acetic when there are high concentrations of oxygen and low levels of alcohol, but in wine environments with low oxygen levels and medium to high alcohol, AAB tend to produce acetaldehyde (see Chapter 5). Acetic acid is also produced by spoilage bacteria (including heterofermentative species of *Lactobacillus* and by *Oenococcus*) before and during the alcoholic fermentation, with high formation likely in stuck or sluggish fermentations. There is also a real risk of this happening during pre-fermentation cold soaking, if undertaken, and particularly during the yeast lag phase before the onset of active fermentation.

The presence of oxygen together with warm fermentation temperatures encourages the growth of AAB. Even in the range of 15–25 °C (59–77 °F), acetic acid formation may increase incrementally. LAB grow rapidly at 17–21 °C (63–70 °F), such temperatures usually used for white wine fermentations. As the fermentation rate decreases towards the end of the alcoholic fermentation, there may be a rapid growth of LAB that can use sugar, or citric acid, to produce levels of acetic acid that may cause the fermentation to stick. The degradation of sugars by LAB is via the phosphoketolase process (the way in which bacteria can break down residual sugar) [22]. Alternatively, acetic acid can simply be produced as part of the citric acid cycle. Should the temperature hit 35 °C (95 °F) or above, it is also likely that the fermentation will stick due to yeast inhibition. If this happens, there is a risk of even greater quantities of acetic acid being produced by this action of LAB on the considerable unfermented sugars. Other bacteria, including AAB, will also continue to multiply at this high temperature, and the quantity produced can quickly result in total wine spoilage. Acetic acid production may take place during and subsequent to the MLF, by heterofermentative LAB, including *Oenococcus oeni*, metabolising citric acid and any small amounts of sugar.

Even in a well-made and balanced wine, microbial action by yeasts and AAB and other bacteria may bring the total level of VA, prior to any barrel or other maturation, to 0.2–0.4 g/l. Further quantities will be produced subsequently. It should be borne in mind that even a wine fermented to 'total' dryness will contain at least 0.1 g/l and more commonly 0.3–0.5 g/l of residual sugar, some of which will be in the form of trehalose, which is manufactured solely by yeast autolysis. Of course, the modern trend, particularly in wine destined for so-called 'entry level' commercial brands, is to have residual sugar levels often as high as 5 g/l, and this can present a real danger. Some species of LAB, particularly of the genus *Lactobacillus*, degrade glycerol (present in all wines), and this results in an increase in volatile acidity, and the process can produce bitter-tasting compounds (see Chapter 11).

Production of wines in hot climates and/or hot growing seasons can present particular challenges. Super-ripe fruit resulting in high must-weights can lead to a longer lag time before the onset of fermentation, a slow yeast growth, and increased osmotic pressure. In these circumstances, the yeast is stressed, and this results in an increased growth of AAB. A must at 32 °Brix (17.7° Baumé) might result in a wine with a level of acetic acid in the region of 0.6 g/l, compared with a level of 0.4 g/l from a 22 °Brix (12.2° Baumé). High-density musts can also be low in nitrogen and other nutrients also resulting in stressed yeasts.

7.8.2 Physiochemical Production

In addition to the microbiological production of acetic acid, it may also be extracted from wood, either during the oaking of wines in barrel or by the use of chips, granules, etc. in tank. This is a physiochemical reaction. During the toasting stage of oak barrel manufacture, acetic acid is both formed and evaporated. The respected barrel manufacturer Seguin Moreau states that ageing in a new barrel would account for a maximum of 0.15 g/l increase due to extraction [23], although a figure of up to 0.2 g/l has also been quoted [24]. However, there may be further increases in acetic acid during barrel ageing as there is, of course, an oxygenation of wine taking place that may lead to dissolved oxygen (DO) levels that are favourable to microbial growth. The wine will absorb approximately 20–45 mg/l of oxygen per year, during normal barrel ageing in a 225 l barrique [25]. Racking and fining operations will result in further oxygen uptake, in the region of 0.5–3.5 mg/l. In a well-sealed barrel, a partial vacuum will develop in the barrel due to the wine shrinking as the water content of the wine evaporates through the staves and to a lesser extent the heads. There is a greater quantity of liquid evaporation in low-humidity cellars and warm cellars. This partial vacuum is lost when the bung is removed for racking and fining operations and also when samples are withdrawn for tasting or laboratory analysis. The ideal maturation cellar temperature does not exceed 16 °C (61 °F) – warmer conditions will favour the growth of acetic acid and other bacteria. The acetic acid content of a wine after 12 months in new oak may increase by 0.06–0.120 g/l, but there may be a greater increase if the barrels are second or third fill, or older. For wines that have extended barrel ageing, such as Gran Reserva Riojas, the increase may well reach 0.2 g/l. However, subsequent to barrel ageing, the VA in well-made red wines should not exceed 0.6 g/l, and in white wines that usually have shorter period of barrel ageing, a figure of 0.5 g/l may be regarded as an accepted maximum. In his seminal work '*Wine Science: Principles and Applications,* 5e', Ronald Jackson (2020) [26] states that acetic acid may be produced during barrel ageing due to hydrolysis of hemicellulose, but this is challenged by many coopers, who state that insignificant amounts of hemicellulose and/or cellobiose produced during the barrel toasting process are released.

7.9 Ethyl Acetate

Molecular formula is $CH_3COOC_2H_5$, often simplified to the condensed formula of $C_4H_8O_2$. The compound, also known as ethyl ethanoate, is the most prevalent ester in wine, accounting for over 50% of the ester content. High levels of ethyl acetate in wine are usually accompanied by high levels of acetic acid.

7.9.1 Production of Ethyl Acetate in Wine

Ethyl acetate (EA) may be produced before, during, and after the alcoholic fermentation. In white musts, typical scenarios for pre-fermentation ethyl acetate production include long settling periods and extended skin contact; in the case of red musts, a long red pre-fermentation maceration (cold soak) represents a danger of production [27]. The greatest production of EA is usually during the alcoholic fermentation. Generally speaking, the cooler the fermentation temperatures, the more esters, including ethyl acetate, will be produced and retained in the wine. Esters are of course volatile, and warmer, more vigorous fermentations are likely to drive off much of the ethyl acetate with the rapidly rising carbon dioxide. Many of the other aromatic esters are highly desirable, particularly in white wines that are brought to market early and whose secondary aromas are a major part of the taste profile. Even if not particularly cool, lengthy alcoholic fermentations are also more likely to lead to high levels of ethyl acetate.

The simplified conversion by esterification from acetic acid and alcohol may be shown as

$$CH_3CH_2OH + CH_3COOH \rightarrow CH_3COOC_2H_5 + H_2O$$

Ethanol + Acetic acid $\rightarrow$ Ethyl acetate + Water

7.9.2 Microbiological Production

7.9.2.1 Production by Yeasts

If not inhibited, the 'wild' yeast species *Candida krusei*, *Hanseniaspora uvarum*, (*Kloeckera apiculata*), *Metschnikowia pulcherrima* and *Pichia anomala* may be present and may grow in must and in the early stages of fermentation. These have the propensity to produce significant quantities of EA (and acetic acid) before and during the *Saccharomyces*-led alcoholic fermentation [28–30]. However, some strains of cultured *Saccharomyces* yeasts have the tendency to produce relatively high levels of ethyl acetate. Table 7.1 shows typical levels of ethyl acetate produced by some important yeast species.

Anecdotal evidence also suggests that 'natural yeast' fermentations often produce wines with higher-than-average levels of ethyl acetate. *P. anomala* is often present in high quantities on grapes infected by *B. cinerea* in both 'grey rot' and 'noble rot' forms. Tones of ethyl acetate on high-quality sweet wines, made from 'nobly rotten' grapes are not uncommon, as the compound may be produced early in the fermentation process before *Saccharomyces* becomes dominant. It should be noted that 'ice wines' (Canada) and eiseweine (Germany) are produced from healthy, non-*botrytised* grapes and, although levels of VA can be very high, EA levels are more modest.

Ethyl acetate can also be formed during wine storage, particularly by film-forming yeasts growing on the surface of wine in an ullaged tank or barrel. The film formed is due to yeasts budding in reproduction, but the bud or daughter fails to detach, leading to the yeast forming a network of chains that eventually cover the surface. Figure 7.3 shows the yeast *Candida vini* on the wine surface of a vat that has been left with a small ullage.

Table 7.1 Typical levels of ethyl acetate produced by various yeast species.

Yeast species	Indicative range of ethyl acetate production (mg/l)
Saccharomyces cerevisiae	10–100
Kloeckera apiculata/ Hanseniaspora uvarum	25–375
Candida krusei	220–730
Pichia anomala	137–2150

Source: Adapted from Fleet [31].

Figure 7.3 Film-forming yeast (*Candida*) showing chains.

7.9.2.2 Production by Acetic Acid Bacteria

The presence of AAB, particularly of the genus *Acetobacter* in both fermenting and finished wine, is a major cause of ethyl acetate production. Unlike VA, it is not produced by LAB. Typical ratios of acetic acid to ethyl acetate in wine with high levels of VA are between 5 : 1 and 10 : 1, although very occasionally a wine may be found with high levels of ethyl acetate but low levels of acetic acid.

7.9.3 Chemical Production

Although the majority of a wine's ethyl acetate content is likely to be as a result of the action of yeasts or AAB, it can also be formed simply by the slow esterification of ethanol and acetic acid. This process can continue during maturation and storage.

7.9.4 New Barrels Do Not Contain Ethyl Acetate

Unlike acetic acid, ethyl acetate is not present in new barrels, contrary to perceptions amongst some winemakers who have often blamed cooperage for an increase in levels during maturation. Any increase in the level during ageing in new barrels is on account of the presence of acetic bacteria in the wine, which can then manufacture the compound from acetic acid, including that extracted from the wood, and ethanol. New barrels do not harbour acetic or LAB, lacking the necessary nutrients for survival. Any surface bacteria on the barrels will be killed in the high temperatures of the toasting process, although there is the possibility of contamination due to incorrect storage. Of course, used barrels may well be heavily contaminated with AAB.

7.9.5 Sensory Characteristics and Detection of Ethyl Acetate

High levels of ethyl acetate can be recognised by an odour of pear drops, bananas, nail varnish and nail varnish remover, acetone, polish, spray paint, or glue. Indeed, ethyl acetate is a component of many commercial solvents. On the palate of wine, there is a burning sensation and/or hardness. As high levels of ethyl acetate are usually accompanied by high levels of volatile acidity (although high VA wines do not always have high levels of EA), stating a sensory detection threshold is not straightforward. The perception threshold for the characteristic 'nail varnish remover' aromas of ethyl acetate is perhaps 50–175 mg/l, although some authors quote thresholds of between 100 and 200 mg/l. The sensory detection threshold for EA as a contributor to wine aromas is much lower, perhaps as low as 12–14 mg/l [31]. At 150 mg/l, ethyl acetate may be particularly detectable on the back palate. High levels of ethyl acetate will also increase the perception of volatile acidity. As with acetic acid, detection thresholds vary from wine to wine depending, inter alia, on grape variety, style of wine, and its matrix, including the alcoholic strength.

In fact, the threshold at which ethyl acetate has an impact upon wine aromas begins as low as 10–14 mg/l, but it is often regarded as making a positive contribution when levels are below 60 mg/l. At such levels, the presence of ethyl acetate can give elements of fruitiness, a hint of cherries, richness, or even slight sweetness to a wine. When accompanied by volatile acidity at a level below 0.6 g/l, low levels of ethyl acetate can also add to wine complexity. Some commentators claim there can be a positive effect up to 80 mg/l ethyl acetate. However, at levels below sensory detection threshold but above 60 mg/l, ethyl acetate may impact negatively upon aromas, making the wine seem flat and demonstrating a loss of fruitiness – a characteristic it shares with 2,4,6-trichloroanisole (TCA).

Table wine with ethyl acetate at a level of 220 mg/l or above will certainly be unsaleable, but levels above 160 mg/l are likely to have such a negative sensory impact to make, putting the wine to market most inadvisable. Vintessential Laboratories, which have laboratories in New South Wales, Queensland, Western Australia, and Tasmania, state the average level of ethyl acetate in all wines tested in their laboratory over a period of five years was 127 mg/l, with 34% of the wines having a level higher than 150 mg/l [32]. Some sweet wines can have very high levels of EA, generally without negative sensory impact. For example, ethyl acetate concentration in Canadian ice wines ranges from 86 to 369 mg/l, with an average of 240 mg/l [13].

7.10 Prevention of Excessive Volatile Acidity and Ethyl Acetate

Levels of volatile acidity can increase throughout the wine production process, mostly as a result of microbial activity. However, the prevention of excess formation of VA and EA may be largely achieved by the implementation of basic winery hygiene and winemaking practices.

7.10.1 Fruit Selection and Processing

Only clean fruit should be used. Damaged grapes and (unless in the form of noble rot as for the production of sweet wines) grapes infected with *B. cinerea* should be excluded as these are a primary source of AAB. Clusters affected by sour rot (a complex of fungi and bacteria) must be avoided at all costs. *Zygosaccharomyces* spp. and *H. uvarum* are often prevalent in affected grapes, which can lead to high levels of acetic acid production.

Fruit should be processed quickly. Yeasts and bacteria multiply very quickly, and the warmer the temperature, the more rapid the multiplication. One single cell of a spoilage organism can become many thousands in less than 24 hours.

7.10.2 Exclusion of Vectors

Vinegar flies and common fruit flies of the genus *Drosophila*, particularly *D. suzukii* and *D. melanogaster*, are microbial vectors and should be excluded from fruit, by pruning out infected grapes or clusters and winery facilities. Alarmingly, *D. suzukii*, which is very damaging to grapes, appeared for the first time in the Bordeaux region in 2014. Fly traps and/or insect light traps should be utilised. *D. melanogaster*, particularly the females, also release foul-odoured enzymes into the grape juice. The addition of potassium metabisulfite to grapes before processing has drawbacks, but covering the bins with dry ice will help keep flies at bay and also reduce bacterial growth.

7.10.3 Cleaning

Fastidious cleaning of the winery building should be routine, including unseen areas such as drains and crevices in the building structure, which may harbour AAB. AAB can form biofilms. All equipment, including crushers, fermentation and storage vats, press, lines, hoses, and filters must be kept scrupulously clean and should be sanitised before use. All waste, particularly including stems, leaves, skins, lees, and grape pomace, should be taken at least 100 m from the winery for composting/processing or for disposal.

7.10.4 Avoiding Cross-Contamination

Rigorous steps should be taken to avoid cross-contamination. Pipettes and barrel thief should be sanitised before and after each use. All vessels, including buckets and drums, should be cleaned and sanitised before and after using for wine, additives, or processing aids.

7.10.5 Must Adjustment Including Nutrient Additions

Must should be protected with SO_2: 30–50 mg/l, depending upon pH and sanitary condition.

The pH of must for red wines should be ideally below 3.7 (but acetic spoilage can occur at pH 3.4 or above). If acidity adjustments are required, this should be undertaken at must stage, and only tartaric acid should be used. The higher the pH, the greater the production of acetic acid in both alcoholic and MLFs [33].

Fermenting wines should have sufficient nutrients to avoid yeast stress. Low nitrogen levels can lead to sluggish fermentations and increase the risk of fermentation sticking. A good provision of saturated and unsaturated fatty acids will stimulate fermentation [34], and in the case of musts rich in sugar, VA production is inverse to the amount of yeast assimilable nitrogen (YAN) (in any event, high-density musts tend to result in wines with higher levels of acetic acid). YAN comprises ammonium salts and amino acids. The best time to add YAN to minimise the production of acetic acid is at the beginning of the fermentation [35]. Diammonium phosphate (DAP) at the rate of 200 mg/l and B group vitamins, perhaps in the form of yeast hulls, may be added if necessary to nutrient-deficient musts. However, such additions may promote the growth of unwanted yeasts, and the winemaker is particularly cautioned against adding nutrients to musts containing wild yeasts, including *H. uvarum*.

7.10.6 Addition of Lysozyme

Lysozyme can be used as a preventative enzyme. Its use has pros and cons. It inhibits LAB including *Lactobacillus*, *Pediococcus*, and *Oenococcus*, by breaking cell walls. Lysozyme has now been authorised for use in must and wine by International Organisation of Vine and Wine (OIV) [36] and the European Union [9]. The enzyme may be added to the must at a rate of approximately 250 mg/l (maximum legal addition is 500 mg/l) to help prevent acetic acid production by LAB before and during the early and middle stages of the alcoholic fermentation and particularly to avoid stuck fermentations due to LAB production of acetic acid. *Lactobacillus* spp. are heterofermentative, and some species are regarded as spoilage bacteria. However, strains of *Lactobacillus plantarum* have become popular as MLF starter cultures, performing well in high pH conditions and with a high resistance to alcohol and SO_2 and possessing enzymes that produce desirable aromas [37]. *Pediococcus* spp. are homofermentative and generally regarded as spoilage bacteria – they are known producers of high levels of acetic acid (and biogenic amines) but, as in the case of Lactobacillus, possess a great number of enzymes that generate compounds responsible for desirable wine aromas.

Of course, if lysozyme is used in must, simultaneous AF and MLF are not possible, and starter cultures will subsequently be required to initiate the MLF. The product may also be added at a lower rate – 100–150 mg/l in order to kill LAB prior to barrel maturation of the wine. It should be noted that the enzyme will not inhibit very high levels of LAB. If added in high concentrations post-alcoholic fermentation, it will bind with polyphenols and thus can lighten the colour of red and rosé wines. Further, as a protein, it may increase protein instability (see Chapter 10), and it cannot be used in wines containing bentonite.

7.10.7 Cold Soaking (Red Wines)

Care should be taken if cold soaking is undertaken. The process should be used only for the most healthy grapes and soaking temperatures kept low; otherwise, the must can quickly become a bacterial melting pot. At temperatures of 4–10 °C, *K. apiculata* are far more active than *Saccharomyces* species.

7.10.8 Inoculation with Desired Strains of *Saccharomyces cerevisiae*

Unless a natural yeast fermentation, or *Saccharomyces*/non-*Saccharomyces* co-fermentation is desired, so-called 'wild' yeasts – particularly *K. apiculata* and *P. anomala* and *Pichia membranifaciens* – should be inhibited by the addition to the must of sulfur dioxide of at least 30 mg/l as detailed above. The must should then be inoculated with the desired culture of *Saccharomyces* with the aim of producing a rapid onset of alcoholic fermentation that will decrease the AAB population. *Saccharomyces* strains produce differing amounts of VA, and it is prudent to select a yeast with a low VA production potential.

7.10.9 Alcoholic Fermentation

During the yeast growth phase of the alcoholic fermentation, sufficient oxygen should be supplied to help the yeast resist the osmotic shock and limit acetic acid production. Some 6–10 mg/l of oxygen are needed at this time. High must-weights are at particular risk of acetic acid (and glycerol) production.

Fermentation temperatures should be monitored carefully – remembering that the top of vat may be considerably warmer than that at the point of the thermostat. If the temperature is too cold, it may lead to yeast stress and increased ester production including ethyl acetate; if it is too hot, there may be yeast stress and increased microbial growth. If the temperature rises to 34–35 °C, yeasts may be inhibited, and the fermentation may stick. However, acetic (and lactic acid) bacteria can tolerate such high temperatures resulting in the formation of acetic acid. The alcoholic fermentation should be completed to leave less than 1 g/l of residual sugar. A higher amount of glucose and fructose (as well as being a nutrient for possible growth of *Brettanomyces*) may result in heterofermentative LAB converting the sugars to a significant amount of acetic acid and lactic acid.

In the case of open tank fermentations, the cap should be kept submerged by punch downs, pump-overs, or other means. Grape skins in a warm, oxygen-rich atmosphere present a particular danger zone and attract vinegar *Drosophila* flies, especially when the fermentation is slowing and there is less of a protective layer of carbon dioxide. The headspace in a vat that has nearly completed or completed fermentation presents a real danger as the percentage of oxygen increases as the manufacture of CO_2 ceases, and AAB have an ample supply of carbon from wine's ethanol content to produce acetic acid. Figure 7.4 shows a dangerous raised cap in a small vat.

7.10.10 Malolactic Fermentation

For MLF, the pH should be ideally below 3.5 (but above pH 3.1, below which there is likely to be a problem with MLF taking place).

The MLF should be commenced and completed as soon as possible after the alcoholic fermentation. There is a growing trend for co-inoculation of musts with yeasts and malolactic bacteria, in order, inter alia, to reduce the lag time between the end of alcoholic fermentation and the onset of MLF. However, this does result in the presence of considerable quantities of lactic bacteria in the sugar-rich fermenting juice that, as we have seen above, poses a risk of acetic acid formation.

Figure 7.4   Dangerous raised cap.

7.10.11   Addition of Sulfur Dioxide and Maintaining Appropriate Molecular SO$_2$

Wine should be sulfured immediately following completion of the MLF. Sufficient free SO$_2$ levels, relative to the pH, should be maintained throughout ageing. The amount of molecular (active) SO$_2$ needed to inhibit bacteria is generally in the range of 0.5–0.625 mg/l. At a (low) pH of 3.3, this equates to 16–20 mg/l of free SO$_2$, but at a (high) pH of 3.8, 49–62 mg/l of free SO$_2$ is required. Bearing in mind the bound SO$_2$ component, the latter figure may take the total SO$_2$ content to levels that are not only detectable upon nosing but also approaching legal maxima. Some species and strains of AAB have a higher tolerance for SO$_2$ – for example, *Acetobacter pasteurianus* Strain A8 has been shown to survive under anaerobic conditions in wine in the presence of 0.35 and 0.8 mg/l of molecular sulfur dioxide [38].

7.10.12   Avoiding Ullage in Vats and Barrels

Vats and barrels must be topped up as soon as they become ullaged and always kept completely full. Not only AAB but also film-forming spoilage yeasts grow on the surface of wine, in the presence of oxygen. If wine is stored in a tank that is not completely full, it should be blanketed with inert gas. CO$_2$ should be used only for white wines; nitrogen or argon may be used for reds. When withdrawing necessary samples from barrels for analysis, they should be topped up immediately. If taking samples from a sealed tank, it is better not to this top up – the withdrawal will create a very slight vacuum, and the act of opening the tank to top up would result in oxygen ingress. Wine used for topping up must not kept on ullage and should be stored in small containers. Wine used for topping should be nosed and tasted before addition.

7.10.13 Micro-oxygenation

Micro-oxygenation may stimulate the growth of AAB [39, 40]. It is important to ensure that the rate of addition is less than consumption and to regularly monitor for growth and for acetic acid production.

7.10.14 Cleaning and Sanitising Barrels that Show Signs of Acetic Acid

Meticulous care should be taken with barrel cleaning and sanitation. The purchase of used barrels should be avoided if possible – unless they have been scrupulously cleaned after use, it is very likely that AAB will be present.

Any barrels that show traces of acetic acid should be saturated with an alkaline solution. A recommended procedure is to fill the barrel with cold water and leave for 24–48 hours. The barrel should then be completely drained, and refilled with 10 l of water containing 50 g of caustic soda (sodium hydroxide) or 10 g of dissolved potash. It is important that the chemicals are completely dissolved as any dry crystals will immediately scorch the wood. The barrel should be rolled and turned in order for the liquid to contact all surfaces. The solution should be drained and the barrel rinsed thoroughly with cold water. Ten litres of water containing 50 g of citric acid should then be put in the barrel (to neutralise the sodium hydroxide), which is again rolled and turned. This is drained, and the barrel is repeatedly rinsed with cold water. It is most important that all traces of citric acid are removed; otherwise, this might provide a substrate for acetic acid production. Finally, clean water should be added, the barrel rolled, and a pH test undertaken to ensure that the barrel has been neutralised. The barrel may then be drained and disinfected by steaming, followed by the burning of a sulfur candle.

The method of cleaning suggested above is different to those detailed in Section 4.8.6.3: the above method is particularly applicable to barrels showing acetic acid.

7.10.15 Cautious Use of Citric Acid

Citric acid should not be used for winemaking purposes. As we have seen, some citric acid is found in grapes, and small amounts may be produced during the alcoholic fermentation. The bacterium *O. oeni*, one of the bacteria responsible for MLFs, can metabolise citric acid to produce acetic acid. Citric acid is often used to neutralise sodium hydroxide (caustic soda), which is widely used as an effective cleaning agent. This should present no problem providing clean water is used afterwards to remove all traces of the acid. A little citric acid is sometimes added to a solution of potassium metabisulfite used for sanitising purposes, as this increases the effectiveness.

7.10.16 Avoiding the Growth of *Brettanomyces*

The 'rogue' yeast *Brettanomyces bruxellensis* is a major producer of acetic acid, particularly if oxygen is present. *Brettanomyces* can grow in an anaerobic environment, but its production of acetic acid is low in these conditions (see Chapter 4). Steps should be taken to monitor for its presence and, if found, to inhibit its growth. If *Brettanomyces* is discovered in wine before transferring to barrels, the wine should be filtered through 0.8 or 0.6 μm medium, which will capture the cells that are usually between 2 and 7 μm in size. Fine filtration will also remove other microorganisms and LAB, which, like *Brettanomyces*, can also grow in anaerobic conditions.

7.10.17 Minimising Oxygen Uptake During Cellar Operations

All handling operations, including rackings and additions of oenological agents, should be undertaken in a manner to minimise oxygen uptake. These should be limited in number, or combined, by careful planning. The transfer and racking of wine after completion of the alcoholic fermentation or during maturation introduces oxygen into the wine that can lead to a 100–1000-fold increase in AAB CFU/ml [39] – the population can reach 100 million CFU/ml.

When racking a wine that has completed its alcoholic fermentation, the receiving tank should be filled from the bottom to avoid splashing. NB: If the wine is still fermenting, tanks must NOT be filled from the bottom. Ideally, the receiving tank should be purged with inert gas.

7.10.18 Barrel Ageing

There will be an increase in volatile acidity during ageing in oak barrels. During barrel making, the toasting process produces acetic acid by hydrolysis from wood polysaccharides [41, 42]. The increase in the volatile acidity of wines aged in new barrels is usually around 0.1–0.15 g/l, which, added to other sources of production, may lead to an organoleptic impact.

7.10.19 Monitoring and Analysis

Wines should be regularly monitored and analysed for VA and for the products of *Brettanomyces* (as well as other routine analyses including pH, free SO_2, and DO). Analysis should take place:

- At must stage.
- Post-alcoholic fermentation.
- Post MLF.
- During maturation in the cellar, on a regular basis – every three of four weeks is recommended.
- Immediately pre-bottling.

If, during barrel ageing, higher-than-desired levels of VA are detected, the wine should transferred to vats, where control is easier in a non-oxidative and 'smooth walled' environment. Both *Brettanomyces* and *Acetobacter* can penetrate several millimetres into the rough inner surface of barrels – a major contributor to the challenges of inhibiting their growth.

7.10.20 Bottling

Some species and strains of AAB can produce acetic acid during storage of bottled wines. Wine spoilage in bottle by AAB has been noted, particularly due to *A. pasteurianus* [43, 44]. As sterile filtration has an organoleptic impact, it is important to note that during the bottling operation, white wines should not pick up more than 0.2 mg/l oxygen and reds not more than 0.5 mg/l. Total package oxygen (TPO levels) after bottling should be less than 1–1.25 mg/l for red wines and less than 0.5–0.6 mg/l for white wines.

7.11 Laboratory Analysis

7.11.1 Volatile Acidity and Acetic Acid Bacteria

The level of volatile acidity in wine may be measured by cash still steam distillation/titration, enzymatic assay of acetic acid, high-performance liquid chromatography (HPLC), gas chromatography, or segmented flow analysis (SFA). The cash still, as illustrated in Figure 7.5, is a relatively inexpensive and easy to operate piece of equipment for VA analysis.

AAB are very difficult to correctly identify at species levels based only on biochemical and physiological characteristics [16]. However, recently, various methods based on molecular techniques of DNA extraction and identification by polymerase chain reaction (PCR) have proved successful at identification of the genera, species, and even strains of AAB [16]. PCR with restriction fragment length polymorphism (RFLP), amplifying part of the ribosomal DNA, is accurate. Real-time PCR is a fast and reliable method to identify and quantify species without the need to culture. It has been successfully used to enumerate populations of AAB in wines [45].

7.11.2 Laboratory Analysis – Ethyl Acetate

As ethyl acetate is not an acid, it cannot be measured by steam distillation, the method often used for volatile acidity. Gas chromatography is the usual method of measuring levels of ethyl acetate.

7.12 Treatments

7.12.1 Excessive Volatile Acidity

Of course, the winemaker should take all possible steps to avoiding the production of excessive volatile acidity during winemaking and maturation. However, faced with a wine with unacceptable and possibly illegal high levels, there are several possible ways of 'rescuing'

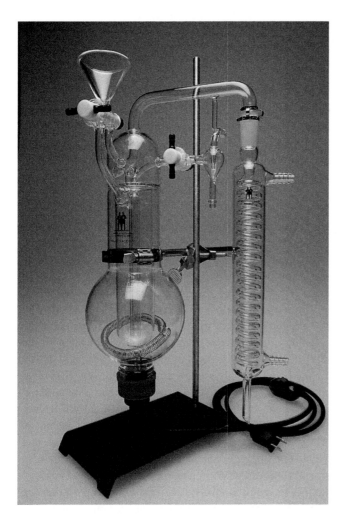

Figure 7.5   Volatile acidity (cash) still. Source: Courtesy of George Chittenden, Adams and Chittenden Scientific Glass.

the wine in question and avoiding a total loss. All of these have cost implications; these may be considerable when it is necessary to engage the services of third-party service providers.

7.12.1.1   Dealing with Rising Levels of VA

If a vat of wine is showing VA levels that are rising but still within the boundary of accept-ability, the vat should be chilled to below 7 °C and the wine sterile filtered and then sulfured to an appropriate molecular sulfur level.

7.12.1.2   Blending

Blending of the problem wine with a wine that has levels of VA well below sensory threshold is a course of action that is often undertaken. However, the winemaker should be beware of creating a greater quantity of high VA wine. It is essential that the high VA wine is sterile fil-tered before the blending; otherwise, acetic acid and other bacteria will produce further VA.

7.12.1.3 Fining

Fining with bentonite or activated charcoal may reduce the VA levels. Both have an organoleptic impact, reducing aromas and flavours and in the case of activate charcoal paling the colour too.

7.12.1.4 Nanofiltration Coupled with Ion Exchange

Nanofiltration (NF) is able to remove particles 0.002–0.005 µm in diameter. NF uses lower pressure than reverse osmosis. The process, which is permitted by USA Federal Regulations, can remove 30–40% of volatile acidity in one pass. The ion exchange column does retain some wine aromas, and operating costs are high.

7.12.1.5 Reverse Osmosis

Reverse osmosis and anion exchange has proved a successful way of VA reduction, although VA removal can never be total – the molecules of acetic acid are very small and can still pass through RO membranes. There are several patented systems in use. The first system, developed by 'post-modern winemaking guru' Clark Smith, has been effectively used since 1992. Smith's Californian Company, Vinovation Inc., has treated many millions of litres of wine from over a thousand clients from eight countries [46]. The company states that every year treated wines are included in the top 100 wines in the influential US published magazine 'Wine Spectator' [46]. Smith describes its use as 'a valuable safety net for low sulfite wine production and in support of cellar strategies pursuing microbial balance' [47]. The treatment involves pressurised wine being passed through a reverse osmosis (RO) plant that separates the permeate containing water, alcohol, and acetic acid and ethyl acetate, leaving pigmented wine concentrate as retentate. This contains all the aroma and flavour compounds. In the anion exchange column, acetic acid is removed from the permeate, adsorbing onto charged resin. The treated permeate is then mixed with the retentate and usually returned to tank, from which it may be recirculated through the system. It is claimed that the process can also remove over 30% of the VA in single pass, so several passes may be necessary to reduce very high VA to required levels. It is generally accepted that the polyamide membrane in the RO plant and epoxy resin in the anion exchange column do not taint the wine, although some aromas may be modified. There will also be a reduction in total acidity and maybe an increase in pH.

There are many possible variations on the basic process, and of particular relevance is the porosities of the membranes and operating pressure of the process. Each processing company espouses the advantages of its own particular plant and operation. For example, the Australian Company VAF Memstar state that as their patented system operates at very high pressure (up to 70 bar), the membranes operate very selectively and almost none of the wanted wine components reach the anion exchange column [48]. The company states that the VA reduction is a minimum of 30% each pass [48]. The system used by Winetech, another California-based company, is a low pressure plant, with loose thin-film membranes and pure resins. This company also claims that reduction in VA is at least 30% per pass (providing the pH is below 4). The company states that there is no significant shift in wine pH after passage through the system [49].

It is most important to note that the reverse osmosis anion exchange system does not filter the retentate (wine concentrate). Therefore, unless the wine has been previously filtered immediately prior to the process, the retentate will contain AAB and possibly other bacteria that will result in future acetic acid production. Accordingly, treated wine should be sterile filtered or, if appropriate, pasteurised to prevent recurrence of the problem. As with all RO operations, the wine, pretreatment, should be racked and free from solids including oenological products (e.g. bentonite) and also free from gas. Of course, RO treatment is expensive, but compared with the loss that a batch of unsaleable wine would entail, the price may well be worth paying.

7.12.1.6 Future Possibilities of Bio-reduction of VA Levels

A biological method of treatment has been shown to be possible, using the S26 strain of *S. cerevisiae*. A high VA wine was re-fermented with freshly crushed grapes or marc using this strain. This process has resulted in successful reduction in wines with VA of over 1.44 g/l VA, and it is claimed that there are no negative impact on aromas [50, 51], although the process is less effective in high-alcohol wines. Alice Vilela and colleagues have also shown that the use of the commercial S26 strain *S. cerevisiae* immobilised in double-layer alginate–chitosan beads could reduce the volatile acidity of a wine with 1.1 g/l acetic acid and 12.5% abv, by 28% within 72 hours and 62% within a seven-day period. There was a slight decrease in ethanol concentration (0.7%) [52].

7.12.2 Ethyl Acetate

High levels of ethyl acetate represent a fault that is perhaps more difficult to treat, and there is some dissent as to the effectiveness of reverse osmosis and ion exchange in removing the compound. California-based Winesecrets and Winetech claim that the resins used in the ion exchange column are very effective.

7.13 Final Reflections

The opening sentence of this book reads: 'Wines are produced today in over 65 countries, and it is often stated that production standards are higher than at any time in the 8000 or more years of vinous history.' However, although standards are generally high, wines produced in different continents, countries, regions, and districts have in many ways become more the same, more globalised and standardised. To put it bluntly as if (with apologies to Evelyn Waugh) I ever put it any other way, wine has become more boring. Gone are the days when the great red wines would develop over 20 or more years and were structured to last half a century. I well recall drinking, just before the turn of the century a bottle of 1928 Château Yon-Figeac from Saint-Émilion. This was a vintage that exuded hard tannins that took decades to soften and for the wines to become harmonious. The wine I tasted was from a bottle that was ullaged to mid-shoulder, and my expectations were low. It was superb. The VA was high but the complexity indescribable. Like the 1947 Château Cheval Blanc, such a wine will not be made again, especially by market-led wine producers who put approachability first.

References

1 Howard, P. (2017). Chateau Musar, Legend of Lebanon. www.winealchemy.co.uk/chateau-musar-legend-lebanon (accessed 29 June 2020).

2 Parker, R. (2013). Bodegas Tradición Oloroso VORS. Robert Parker's Wine Advocate, August 2013. Issue 208. www.robertparker.com.

3 Parker, R.M. Jr., (2003). *Bordeaux: A Consumer's Guide to the World's Finest Wines.* New York: Simon & Schuster.

4 Steinberger, M. (2008). The greatest wine on the planet. How the '47 Cheval Blanc, a defective wine from an aberrant year, got so good. Slate. Webpage. https://slate.com/human-interest/2008/02/how-the-47-cheval-blanc-got-to-be-the-greatest-wine-ever-made.html (accessed 28 June 2020).

5 Reuters Life. (2010). Cheval Blanc sets world record at Christie's wine sale. Reuters. Webpage. https://www.reuters.com/article/us-auction-wine/cheval-blanc-sets-world-record-at-christies-wine-sale-idUKTRE6AF4UA20101116 (accessed 28 June 2020).

6 Millar, R. (2019). Cheval Blanc 1947 tops London sale. *The Drink's Business* 18 October. https://www.thedrinksbusiness.com/2019/10/cheval-blanc-1947-tops-london-sale (accessed 19 June 2020).

7 Buick, D. and Holdstock, M. (2003). The relationship between acetic acid and volatile acidity. *AWRI Technical Review* 143: 39–43.

8 Gardner, D.M. (2015). Is Your Wine Slowly Turning Into Vinegar? Basic Information About Volatile Acidity. *Penn State Extension Wine & Grapes U*. https://psuwineandgrapes.wordpress.com/2015/02/20/is-your-wine-slowly-turning-into-vinegar-basic-information-about-volatile-acidity (accessed 30 June 2020).

9 EUR-Lex. (2019) Commission Delegated Regulation (EU) 2019/934 of 12 March 2019 supplementing Regulation (EU) No. 1308/2013 of the European Parliament and of the Council as regards wine-growing areas where the alcoholic strength may be increased, authorised oenological practices and restrictions applicable to the production and conservation of grapevine products, the minimum percentage of alcohol for by-products and their disposal, and publication of OIV files. https://eur-lex.europa.eu/legal-content/GA/TXT/?uri=CELEX:32019R0934 (accessed 25 may 2020).

10 Code of Federal Regulations. (2020) Title 27, Chapter I, Subchapter A, Part 4. https://www.ecfr.gov/cgi-bin/text-idx?c=ecfr;sid=79589a2ef2d093ed0b73152fc7935f1b;rgn=div5;view=text;node=27%3A1.0.1.1.2;idno=27;cc=ecfr (accessed 30 June 2020).

11 Ontario Wine Appellation Authority. (2019). Wine Standards –Specifications for grapes and wine making. Ontario Wine Appellation Authority. https://www.vqaontario.ca/Regulations/Standards (accessed 30 June 2020).

12 Cliff, M.A. and Pickering, G.J. (2006). Determination of odour detection thresholds for acetic acid and ethyl acetate in ice wine. *Journal of Wine Research* 17 (1): 45–52.

13 Nurgel, C., Pickering, G.J., and Inglis, D. (2004). Sensory and chemical characteristics of Canadian ice wines. *Journal of the Science of Food and Agriculture* 84: 1675–1684. https://doi.org/10.1002/jsfa.1860.

14 Australia New Zealand Food Standards Code – Standard 4.5.1 – Wine production Requirements (Australia Only). www.legislation.gov.au/Details/F2017C01001 (accessed 30 June 2020).

15 Komagata, K., Lino, T., and Yamada, Y. (2014). The family *Acetobacteraceae*. In: *The Prokaryotes* (eds. E. Rosenberg, E.F. DeLong, S. Lory, et al.), 3–78. Berlin Heidelberg: Springer.

16 Gomes, R.J., de Fatima Borges, M., de Freitas Rosa, M. et al. (2018). Acetic acid bacteria in the food industry: systematics, characteristics and applications. *Food Technology and Biotechnology* 56 (2): 139–151. https://doi.org/10.17113/ftb.56.02.18.5593.

17 Crotti, E., Rizzi, A., Chouaia, B. et al. (2010). Acetic acid bacteria, newly emerging symbionts of insects. *Applied and Environmental Microbiology* 76 (21): 6963–6970. https://doi.org/10.1128/aem.01336-10.

18 Mas, A., Torija, M.J., González, A. et al. (2007). Acetic acid bacteria in oenology. *Contributions to Science* 3 (4): 511–521. https://doi.org/10.2436/20.7010.01.27.

19 Ribéreau-Gayon, P., Dubourdieu, D., Donèche, B., and Lonvaud, A. (2006). *Handbook of Enology – The Microbiology of Wine and Vinifications*, vol. 1. Chichester: Wiley.

20 Millan, C. and Ortega, J.M. (1988). Production of ethanol, acetaldehyde, and acetic acid in wine by various yeast races: role of alcohol and aldehyde dehydrogenase. *American Journal of Enology and Viticulture* 39: 107–112.

21 Moss, R. (2015). How volatile fatty acids and sulphurous compounds impact on key aromas. *Wineland Media*. www.wineland.co.za/how-volatile-fatty-acids-and-sulphurous-compounds-impact-on-key-aromas (accessed 23 June 2020).

22 Swiegers, J.H., Bartowsky, E.J., Henschke, P.A., and Pretorius, I.S. (2005). Yeast and bacterial modulation of wine aroma and flavour. *Australian Journal of Grape and Wine Research* 11: 139–173. https://doi.org/10.1111/j.1755-0238.2005.tb00285.x.

23 Seguin Moreau (2011). Panorama, Volatile Acidity. http://en.seguin-moreau.fr/Advice/PANORAMA (accessed 13 August 2020).

24 Jacon, V. and Chatonnet, P. (2004). *The Barrel – Selection, Utilization, Maintenance*. Bordeaux: Vigne & Vin Publications Internationales.

25 Ribéreau-Gayon, P., Glories, Y., Maujean, A., and Dubourdieu, D. (2006). *Handbook of Enology – The Chemistry of Wine: Stabilization and Treatment*, 2e, vol. 2. Chichester: Wiley.

26 Jackson, R.S. (2020). *Wine Science: Principles and Applications*, 5e. San Diego, CA: Academic Press.

27 Romano, P. (2005). Proprietà tecnologiche e di qualità delle specie di lieviti vinari. In: *Microbiologia del vino* (eds. M. Vicenzini, P. Romano and G. Farris), 101–131. Milano: Ambrosiana.

28 Romano, P., Suzzi, G., Comi, G., and Zironi, R. (1992). Higher alcohol and acetic acid production by apiculate wine yeasts. *Journal of Applied Bacteriology* 73: 126–130. https://doi.org/10.1111/j.1365-2672.1992.tb01698.x.

29 Plata, C., Millán, C., Mauricio, J.C., and Ortega, J.M. (2003). Formation of ethyl acetate and isoamyl acetate by various species of wine yeasts. *Food Microbiology* 20 (2): 217–224. https://doi.org/10.1016/S0740-0020(02)00101-6.

30 Fleet, G. (1993). *Wine Microbiology and Biotechnology*. London: Taylor & Francis.

31 Margalit, Y. (2012). *Concepts in Wine Chemistry*, 3e. San Francisco, CA: The Wine Appreciation Guild.

32 Byrne, S. and Howell, G. (2016). *Ethyl Acetate – A Misunderstood Fermentation Problem*. Vintessential Laboratories. Webpage. www.vintessential.com.au/ethyl-acetate-a-misunderstood-fermentation-problem (accessed 22 July 2019).

33 Hudelson, J. (2011). *Wine Faults: Causes, Effects, Cures*. San Francisco, CA: The Wine Appreciation Guild.

34 Delfini, C. and Costa, A. (1993). Effects of grape must lees and insoluble materials on the alcoholic fermentation rate and the production of acetic acid, pyruvic acid, and acetaldehyde. *American Journal of Enology and Viticulture* 44: 86–92.

35 Bely, M., Rinaldi, A., and Dubourdieu, D. (2003). Influence of assimilable nitrogen on volatile acidity production by *Saccharomyces cerevisiae* during high sugar fermentation. *Journal of Bioscience and Bioengineering* 96 (6): 507–512. https://doi.org/10.1016/S1389-1723(04)70141-3.

36 International Organisation of Vine and Wine (OIV) (2021). *International Code of Oenological Practices*. Paris: OIV. https://www.oiv.int/public/medias/7713/en-oiv-code-2021.pdf.

37 Gil-Sánchez, I., Suáldea, B.B., and Moreno-Arribas, M.V. (2019). Malolactic fermentation. In: *Red Wine Technology* (ed. A. Morata), 85–98. London: Academic Press.

38 du Toit, W.J., Pretorius, I.S., and Lonvaud-Funel, A. (2005). The effect of sulphur dioxide and oxygen on the viability and culturability of a strain of *Acetobacter pasteurianus* and a strain of *Brettanomyces bruxellensis* isolated from wine. *Journal of Applied Microbiology* 98: 862–871. https://doi.org/10.1111/j.1365-2672.2004.02549.x.

39 du Toit, W.J., Marais, J., Pretorius, I.S., and du Toit, M. (2006). Oxygen in must and wine: a review. *South African Journal of Enology and Viticulture* 27 (1): 76–94. https://doi.org/10.21548/27-1-1610.

40 du Toit, W.J., Lisjak, K., Marais, J., and du Toit, M. (2006). The effect of micro-oxygenation on the phenolic composition, quality and aerobic wine-spoilage microorganisms of different South African red wines. *South African Journal of Enology and Viticulture* 27 (1): 57–67. https://doi.org/10.21548/27-1-1601.

41 Vivas, N. (2000). *Manuel de tonnellerie à l'usage des utilisateurs de futaille*. Bordeaux: Féret Éditions.

42 Zamora, F. (2003). *Elaboración y crianza del vino tinto: aspectos científicos y prácticos*. Editorial Mundi-Prensa. Madrid: AMV Ediciones.

43 Bartowsky, E.J., Xia, D., Gibson, R.L. et al. (2003). Spoilage of bottled red wine by acetic acid bacteria. *Letters in Applied Microbiology* 36: 307–314. https://doi.org/10.1046/j.1472-765X.2003.01314.x.

44 Bartowsky, E.J. and Henschke, P.A. (2008). Acetic acid bacteria spoilage of bottled red wine – a review. *International Journal of Food Microbiology* 125 (1): 60–70. https://doi.org/10.1046/j.1472-765X.2003.01314.x.

45 González, A., Hierro, N., Poblet, M. et al. (2006). Enumeration and detection of acetic acid bacteria by real-time and nested polymerase chain reactions. *FEMS Microbiology Letters* 254: 123–128. https://doi.org/10.1111/j.1574-6968.2005.000011.x.

46 Vinovation. Removal of volatile acidity from wine. http://www.vinovation.com/acidreduction.htm (accessed 30 June 2020).

47 Smith, C. (2014). *Postmodern Winemaking*. Berkeley: University of California Press.

48 https://www.vafmemstar.com.au/service/volatile-acidity-reduction/ (accessed 31 December 2020).

49 https://12f17fe5-0048-f74d-c863-1f30ab4c8c8f.filesusr.com/ugd/e4ddad_7e1741cc6cf64ac4b01568964cdd3e57.pdf (accessed 1 July 2020).

50 Vilela-Moura, A., Schuller, D., Mendes-Faia, A., and Côrte-Real, M. (2008). Reduction of volatile acidity of wines by selected yeast strains. *Applied Microbiology and Biotechnology* 80: 881–890. https://doi.org/10.1007/s00253-008-1616-x.

51 Vilela-Moura, A., Schuller, D., Falco, V. et al. (2010). Effect of refermentation conditions and micro-oxygenation on the reduction of volatile acidity by commercial *S. cerevisiae* strains and their impact on the aromatic profile of wines. *International Journal of Food Microbiology* 141 (3): 165–172. https://doi.org/10.1016/j.ijfoodmicro.2010.05.006.

52 Vilela, A., Schuller, D., Mendes-Faia, A., and Côrte-Real, M. (2013). Reduction of volatile acidity of acidic wines by immobilized *Saccharomyces cerevisiae* cells. *Applied Microbiology and Biotechnology* 97 (11): 4991–5000. https://doi.org/10.1007/s00253-013-4719-y.

8

Atypical Ageing (ATA) – Sometimes Called Untypical Ageing (UTA)

In this chapter, I cover one of the less discussed wine faults. This fault only impacts upon white and rosé wines. Atypical ageing (ATA) has been reported to affect up to 20% of white wines in most countries in certain years. The fault manifests with a significant loss of varietal character, a substantial decrease in fruity and floral aromas, and the possible presence of odours of wet dishcloth, wax, and wet wool.

8.1   Introduction

ATA is perhaps something of a broad descriptor relating to white and rosé wines that age rapidly in an unwanted and unexpected manner, losing varietal characteristics and illustrating undesirable aromas and flavours. Red wines are not affected by ATA, for reasons discussed below. The defect can develop from as early as a few weeks and up to 12 months after the first post-fermentation addition of sulfur dioxide (SO_2). The fault is sometimes referred to as 'untypical ageing' or, in Germany where it has been extensively studied, as 'Untypische Alterungsnote' or UTA (pronounced oo-tar). ATA was first noticed in some German wines in 1988, In some years it has resulted in 20% of wines submitted for certification for 'quality wine status' being rejected [1]. Since the 1990s, it has been identified in wines from other European and several New World countries, although there is some dispute as to its possible presence in Australian and New Zealand wines. It has been estimated that up to 20% of USA white wines might be affected by the problem [2], which has been described as 'one of the most serious quality problems in white winemaking in nearly all wine producing countries' [3]. The fault remains controversial and is little discussed outside of scientific circles, although a substantial body of research has been conducted. However, there can be little doubt that ATA is widespread and has a serious sensory impact upon affected wines. Wines affected by ATA suffer a major loss of varietal character together with a substantial decrease in fruity aromas. They may exhibit odours of wet dishcloth, wax, furniture varnish, and washing detergent. On the palate, the wine will be thin-bodied and bitter. The odours and other negative effects on the palate will increase as the wine ages.

8.2 The Atypical Ageing Controversy

There is no doubt that some faults discussed in this book must be regarded as serious defects if they are present in finished wine at concentrations above sensory detection thresholds. These include the haloanisole taints, discussed in Chapter 3. Compounds associated with *Brettanomyces* (discussed in Chapter 4) and Volatile Sulfur Compounds (discussed in Chapter 6) might only be regarded as faults once the concentration of the relevant compounds reaches a certain level, which varies according to grape variety and other aspects of a wine's matrix. Even then, the level may be the subject of dispute. However, in the case of ATA, there is some dissention as to whether it is a fault. In some countries, the sensory characteristics exhibited by affected wines are considered as nothing other than an 'extension of terroir' [4]. There can also be little doubt that many experienced tasters, writers, and critics confuse the fault with 'premox'. This topic was discussed in Chapter 5. To further complicate matters, the results of the research have sometimes been inconclusive or conflicting. Some researchers regard the issue as one of premox. However, premox was not 'discovered' until 2002, some 14 years after ATA was first noticed. The fault is barely written about in popular wine literature, and my attempts at discussing the topic with numerous winemakers from Australia, New Zealand, and USA have often been met with blank expressions or a rapid change of the line of conversation. Winemakers from Austria, Germany, and Italy have been more forthcoming, but many are keen to pronounce that ATA is a problem that only affects other wineries, regions, or even countries. Some, such as Morten Hallgren (winemaker at Ravines Wine Cellars on New York's Keuka Lake) are more open. Speaking of his geographical area, Morten says, 'Some people still deny it's a problem, some people accept it, but we definitely have it here' [5].

ATA has been described as the 'most intractable problem in white winemaking in modern times' [Volker Schneider, personal conversation]. Whilst researching for this book I have encountered many white wines exhibiting very pronounced characteristics of ATA, including several examples from the Graves district of Bordeaux (particularly 2016 vintage), the southern Italian region of Campania (again 2016 vintage), the German regions of Mosel and Rheinhessen, and Spain's La Mancha. Having referred to tasting notes accumulated over many years, there is now little doubt in my mind that many poorly performing white wines were victims of the fault, often exhibiting 'waxy' and 'mothball' aromas. I have noted a high incidence of occurrence with wines made with certain grape varieties, including Müller–Thurgau, Kerner, Colombard, and Ugni Blanc. However, this was likely the result of over-cropping. This is often the case with these varieties farmed with little thought to quality, rather than any genetic susceptibility to the fault.

8.3 The Causes of Atypical Ageing and Formation Pathways

8.3.1 The Role of 2-Aminoacetophenone (2-AAP)

The fault of ATA occurs predominantly in wines produced in hot and dry years, from vines that are suffering hormonal stress. Hydric stress is the major causal factor, particularly when vines endure this during the period from 10 days before and up until 10 days after veraison, together with general physiological and hormonal stress. Over-cropping is also an important factor. Harvested grapes contain tryptophan and its metabolites, particularly the phytohormone indole-3-acetic acid (IAA), an odourless acid (molecular formula $C_{10}H_9NO_2$). IAA degrades in the winemaking process to produce N-formyl-2-aminoacetophenone which decomposes to 2-aminoacetophenone (2-AAP) [6], which has odours of mothballs, soap, and the other aromas listed in Section 8.4. The presence of 2-AAP is the major reason for the ATA defect [7]. 2-AAP (molecular formula C_8H_9NO) is sometimes referred to as o-aminoacetophenone or *ortho*-aminoacetophenone.

It should be noted that the amount of 2-AAP produced is not related to the amount of IAA [7]. In other words, it is the mere presence of IAA can lead to 'high' levels of 2-AAP production. A concentration of more than 1 µg/l of 2-AAP is the chemical marker that a wine is a victim of ATA. Its presence at or above this level is also responsible for adverse sensory characteristics [8]. Incidentally, 2-AAP is one of the main compounds responsible for the so-called 'foxy' nose and taste of wines made from some varieties of the American *Vitis labrusca* species, including Niagara, Isabella, and Concord.

8.3.2 The Formation of 2-AAP

The degradation of IAA to 2-AAP has two formation pathways: biological and chemical.

- *Biological formation*: Wild (non-*Saccharomyces*) yeasts, if stressed by nutrient deficiency in the early stages of the alcoholic fermentation, can produce AAP. It is generally accepted that *Saccharomyces* yeasts do not degrade IAA [9], but the compound may be produced by a chemical pathway in wines where *Saccharomyces* species have been the only yeasts involved. Grape must may begin a spontaneous fermentation due to the activity of non-*Saccharomyces* yeasts.
- *Chemical formation*: IAA may be present in must, but in its bound form, i.e. in this state it cannot react. Reactions during the fermentation process may convert IAA into the free form. When sulfites oxidise, oxygen-free radicals are formed, which are highly chemically reactive with certain other elements and compounds. In the case of red wines, these radicals are scavenged by tannins and other phenols. However, in white and rosé wines, in the absence of sufficient phenols, the radicals break down free IAA and produces N-formyl-2-aminoacetophenone which further degrades to 2-aminoacetophenone. Sulfite oxidation may occur immediately after the first addition of SO_2 following the completion of the alcoholic fermentation, in wines where the malolactic fermentation (MLF) is not desired, or after the addition of SO_2 following the MLF. The presence of both free IAA and free SO_2 is necessary for the production of 2-AAP. Thus ATA can appear from the time of the first sulfurisation. However, it may not be noticed upon tasting at this early stage, as esters and other aromatics produced by the fermentation may predominate and mask the ATA characteristics. As the aromas and flavours of the esters wane, the unwanted characteristics of ATA may come to the fore. Subsequent careless or unconsidered cellar operations can dissipate fermentation aromas that, particularly in some wine styles, the producer would wish to retain, thus exposing the presence of 2-AAP and the onset of ATA.

There needs to be some, albeit a very low amount of dissolved oxygen (DO) in the wine for the free radicals noted above to be formed. However, at the first addition of SO_2, the wine usually contains suspended yeasts or is still on yeast lees, both of which are acknowledged scavengers of DO. The amount of DO required to form the free radicals leading to ATA is tiny. It is stated that 0.15–0.50 mg/l of DO are sufficient to produce characteristics of ATA [4]. This amount of DO is much less than is required to give a wine any notes of oxidation.

8.3.3 The Role of Other Chemical Compounds

In addition to 2-AAP, other chemical compounds have been suggested as contributing to ATA. These include skatole (3-methyl-indole) [10]. Skatole in low concentrations has a floral aroma, often redolent of jasmine, but in high concentrations, it is very unpleasant with strong odours of faeces.

The Involvement of Sotolon, Phenylacetaldehyde The involvement of sotolon, which is a volatile compound, in the ATA of dry white wine is a topic which is also subject to much dissent. Sotolon is the common name for 4,5-dimethyl-3-hydroxy-2,5-dihydrofuran-2-one (3-hydroxy-4,5-dimethylfuran-2(5H)-one). It may be formed when 2-ketobutyric acid produced during the alcoholic fermentation, reacts with acetaldehyde. It is a marker of oxidation. It may also be produced by oxidative degeneration of ascorbic acid (which is often added to wines as an antioxidant and, paradoxically, to delay the onset of ATA) [11]. Sotolon is often present at high levels in some fortified wines such as Madeira but is said to cause an irreversible (ATA) defect when it exceeds 7–8 µg l in dry white wines [12]. Sotolon can give odours of unrefined sugar.

8.3.4 Distinguishing ATA from Premature Oxidation (Premox), and Reduction

Wines suffering from premature oxidation (premox) contain high levels of DO. However, ATA can occur in wines with very low levels of DO, and the fault should be distinguished from oxidation, including premox. I have discussed ATA with some producers whose wines have exhibited the fault. Their immediate response was that the wines were not kept on ullage and that every step was taken to avoid exposure to oxygen. Appropriate additions of SO_2 post alcoholic fermentation or MLF, and the maintenance of adequate free SO_2 levels during maturation, are cellar operations undertaken to help protect wine from oxidation (and microbial growth). But, as we have noted, it is the addition of SO_2 that can trigger the onset of ATA.

ATA aromas and flavours have also often been confused with those of reduction. Wines can suffer from both ATA and reduction [13] as vine hydric stress and yeast stress due to a nitrogen deficiency can lead to the production of volatile sulfur compounds that are present in reduced wine (see Chapter 6). Dominant volatile sulfur compounds can mask the aromas of ATA which will become apparent should the former be removed from the wine by chemical (copper sulfate) treatment or ageing with controlled oxygenation.

8.4 Sensory Detection

8.4.1 Sensory Characteristics

Wines suffering from ATA can be unexpectedly light in colour. On the nose, there is a major loss of varietal characteristics, a substantial decrease in general fruity and floral aromas, and a loss of so-called 'minerality'. Some of the more pronounced aromas present maybe those of wet dishcloth, mothballs, camphor, wax, wet wool, lime blossom, putty, furniture varnish, and floor polish. Other aromas may include those of washing detergent, soap, acacia, jasmine, lemon blossom, curry, sulfur, and urine. On the palate, the wine will be thin-bodied, and the finish may exude a bitterness, which is not tannin related, together with metallic notes. After some time in tank or bottle, these aromas and tastes increase. The fault can be distinguished from premature oxidation (premox) when nosed and tasted as wines suffering from premox maintain their varietal character to a large degree. However, it will be overlaid by oxidative notes.

8.4.2 Sensory Perception Threshold of 2-AAP

ATA may be detected on wines containing 0.5–1.5 µg/l of 2-AAP. The actual threshold within this range would appear to depend upon the aroma intensity of the wine, with 0.7–1 µg/l being figures accepted by many researchers. However, in sensory examinations of wines with ATA, there is a low correlation between the level AAP and the level of ATA characteristics [4]. Further, when sound wines have had AAP added in trials, they fail to illustrate all the sensory characteristics of ATA [14]. The perception thresholds are highly matrix dependent [15].

8.5 Laboratory Detection

There are several laboratory options for 2AA detection:

- *Gas chromatography and mass selective (GC/MS) detection*: 2-AAP is extracted from wines by liquid–liquid, solid–liquid, or solid phase microextraction (SPME);
- *High-performance thin-layer chromatography with fluorescence detection (HPTLC-FLD)*: This method can detect 2-AAP down to 0.2–0.3 µg/l, well below the minimum sensory threshold of at least 0.5 µg/l [16];
- *Heart-cut multidimensional gas chromatography coupled with selective mass spectrometric detection (H/C MDGC-MS-MS)*: This is an effective method to detect 2-AAP at a level considerably below the sensory threshold [17];
- Ultrasound-assisted SPME coupled with GC-MS [18].

8.6   The Main Viticultural Causes of ATA

ATA is a problem that begins in the vineyard. It is exclusively attributed to vineyard management [3]. Viticultural factors that can result in ATA affected wines include:

- *Vine stress*: White grapes harvested from physiologically stressed vines, may be deficient in desirable aroma compounds, lacking in oxygen radical scavengers, and deficient in yeast assimilable nitrogen (YAN). The grapes may contain a high IAA concentration together with a low level of phenolic oxygen radical scavengers;
- *Inadequate grape ripeness*: ATA may be a consequence of poor physiological ripeness, even in cases where the sugar content of grapes is satisfactory. Wines made from late-harvested grapes are less affected than those from the same vineyard blocks but harvested early [7, 19]. Quality-conscious growers generally avoid early harvesting for many reasons, including the threat of ATA. As Markus Molitor, of the eponymous Mosel wine estate, says, 'a too early harvest might cause an atypical early ageing of the wines, and they often get some bitter flavours, too' [personal discussion];
- *High cropping levels*: High yields, particularly if coupled with, premature harvest, give a propensity for ATA;
- *Low nutrient levels*: This is an area where research has given inconsistent results, but it is generally accepted that low nitrogen levels and deficiencies in nutrients in the vineyard give a propensity for ATA in resultant wines.

The producers I have spoken to who were prepared to discuss ATA all feel that the problem results entirely from a lack of knowledge, care, and attention in the vineyard. Götz Blessing of Weingut Göttelmann in the Nahe region of Germany states, 'it's not something that will be found in the wine from high quality producers other than in a hot, drought year such as 2003. With good vineyard nutrition and sensible restriction on the amount of crop, it should never be a problem' [personal discussion]. However, although there has been much research, it would seem that further explanations are needed for the viticultural causes of ATA formation [20].

8.7   Prevention

8.7.1   In the Vineyard

As we have seen, ATA is born in the vineyard, and prevention is largely down to appropriate vineyard management. The following measures might be appropriate:

- If legal (which is not the case without derogations in many regions of the European Union), irrigation pre-, during and post-veraison should be undertaken to avoid severe hydric stress of vines;
- Adequate soil nitrogen and nutrient levels should be maintained [2]. Both fertigation and foliar nitrogen additions are usually beneficial. Foliar nitrogen sprays have been shown to increase YAN, the value of which is discussed below [21]. Soil nitrogen additions of 60 kg/ha have been recommended – higher amounts leading to a predilection to ultimate ATA production [22]. The addition of farmyard manure has also been shown to be

effective in the effects of ATA. A humus concentration of at least 2.5% should be maintained;

- The level of green cover crops, which will compete for water and nitrogen, should be reduced, particularly early in the season, perhaps by ploughing alternate rows [23]. It is interesting to note that the widespread implementation of green cover crops in some countries immediately preceded the discovery of ATA affected wines;
- Yield restriction by summer pruning and green harvesting is desirable as high crop levels can lead to delayed ripening and a propensity for 2-AAP production;
- Under-ripe fruit must be avoided – the harvest should be delayed until the grapes have achieved aromatic ripeness. A simple way of evaluating the fruit for physiological ripeness is the skins should be translucent, and they should be easily separated (peeled) from the berry pulp. The seeds should also be brown, not green, in colour.

8.7.2 In the Winery

The following should be undertaken, particularly if the yield is high and/or the crop has suffered from hydric stress:

- Fruit should be processed quickly;
- Whole cluster pressing should be considered, although there is much dissent as to how beneficial this is. When complete grapes are pressed, there is a limited phenolic uptake, and less IAA extracted. Many advocates suggest gentle pressing techniques should be used, and the fractions from each pressing vinified separately. However, other researchers have found that skin contact can reduce the concentration of AAP by up to 35% [24]. The theory is that the increased amounts of phenols extracted from the grapes scavenge the oxygen radicals responsible for the chemical conversion of IAA into AAP. A skin contact period of 6–12 hours may also be beneficial in extracting a greater quantity of primary fruit aromas, that may help mask the aromas of 2-AAP. To further complicate, skin contact and/or the addition of tannins to add sufficient phenols to the wine with the intention of avoiding the formation of 2-AAP are likely to increase bitterness and astringency of the wine, and promote deterioration due to oxidation;
- Must should be clarified, by settling (débourbage), gas-flushed centrifuge, or other means. Post clarification the turbidity should be no greater than 100 nephelometric turbidity units (NTU) as the solids may contain bound IAA;
- Maintaining sufficient yeast nutrients is most important. Yeast nutrients contain radical scavengers. A level of at least 250 mg/l of YAN is advisable (a minimum of 150 mg/l is usually sufficient to ensure trouble-free fermentation). Both diammonium phosphate (DAP) and a yeast nutrient containing free amino acids, sterols, vitamins B_1 (thiamine), B_3 (niacin), and B_9 (folate), and magnesium sulfate might be added following the yeast lag phase, but care should be taken to restrict additions to the minimum necessary;
- There is a wide difference in the ability of various commercial yeast strains to produce IAA, and a strain or strains with low production potential should be utilised [25];
- MLF may be beneficial in delaying the formation 2-AAP. Of course, if MLF is undertaken, the first addition of SO_2 is not usually made until this has been completed;
- Storage on lees in barrel has been shown to reduce the concentration of AAP [26].

8.7.3 Ascorbic Acid Addition

The addition of ascorbic acid (vitamin C) at a rate of 100–150 mg/l is the only largely effective means of addressing the future formation of ATA. Such addition may delay the onset of ATA by two years [27]. The addition is ideally made simultaneously with, or within a few days of, the first sulfurisation of the fermented wine. It is the addition of SO_2 at this time that triggers the 2-AAP formation, resulting in ATA. The level free SO_2 should be checked regularly, and further additions made as necessary. There is a view that the ascorbic acid should be added when the level of free SO_2 in the wine ceases to fall, i.e. remains constant, but the formation of 2-AAP will have already commenced. However, this later addition will inhibit further formation and thus may successfully limit the ATA components to below the sensory threshold. To summarise, the addition of ascorbic acid may delay the onset, or reduce the sensory effects of ATA, but will not prevent its eventual appearance.

It should be noted that the addition of ascorbic acid is a preventative, not a curative. Further, the addition of ascorbic acid may present several issues. If the wine contains other than a tiny amount of DO, the ascorbic acid may be spontaneously oxidised. Appropriate measures should be taken to avoid both controlled oxygenation and rapid oxygen pick-up. Wines containing ascorbic acid should not be stored in barrels, tanks should be kept full, pumping operations should be minimised, and pumps and hoses should be gas flushed. The feeder tank for bottling operations should also be gas blanketed. In wines that are prone to reduction, ascorbic acid may increase the reduction characteristics. Wines should be adjusted as necessary, with particular attention paid to wines that are to be bottled with a screw cap closure and which may become victim to so-called post-bottling reduction. Of course, storage in cool conditions slows down all chemical reactions, including the formation of 2-AAP.

8.8 Treatments

There are no effective treatments. The chemical reactions leading to the formation of ATA related compounds are irreversible. The addition of ascorbic acid is ineffective at treating ATA compounds already formed, and these cannot be removed by fining, filtration, reverse osmosis or any other means. Wines suffering from the fault should be consumed as soon as possible (if at all) as with the waning of primary fruit, the unwanted characteristics of ATA will become more apparent.

8.9 Final Reflections

Wet dishcloth, wet wool, wax, putty, floor polish, and zero in the way of fruit, I have all the evidence of ATA in my glass. It's April 2020, and COVID-19 lockdown has brought its challenges, but this wine makes me smile. For once I'm thankful that I can't invite friends to the house. The wine is a 2019, just seven months old. And it's my second ATA impacted wine of the week.

References

1 Christoph, N., Bauer-Christoph, C., Geßner, M., and Köhler, H.J. (1995). Die "Untypische Alterungsnote im Wein", Teil I: Untersuchungen zum Auftreten und zur sensorischen Charakterisierung der "Untypischen Alterungsnote". *Rebe und Wein* 48: 350–356.

2 Henick-Kling, T., Gerling, C., Martinson, T. et al. (2009). Studies on the origin and sensory aspects of atypical aging in white wines. 14–16 April 2008, Trier, Germany. *Proceedings of the 15th International Enology Symposium, International Association of Enology, Management and Wine Marketing*, Zum Kaiserstuhl 16, 79206 Breisach, Germany.

3 Schneider, V. (2014). Primer on atypical aging. *Wines and Vines* 4: 45–51.

4 Schneider, V. (2014). Atypical aging defect: sensory discrimination, viticultural causes, and enological consequences. A review. *American Journal of Enology and Viticulture* 65 (3): 277–284.

5 Patterson, T. (2009). Godawful wines made in the vineyard. *Wines & Vines* July 2009. https://www.winesandvines.com/columns/section/24/article/65511/Godawful-Wines-Made-in-the-Vineyard (behind paywall) (accessed 26 February 2018).

6 Christoph, N., Gessner, M., Simat, T.J., and Hoenicke, K. (1999). Off-flavor compounds in wine and other food products formed by enzymatical, physical, and chemical degradation of tryptophan and its metabolites. *Advances in Experimental Medicine and Biology* 467: 659–669.

7 Hoenicke, K., Simat, T.J., Steinhart, S. et al. (2002). 'Untypical aging off-flavor' in wine: formation of 2-aminoacetophenone and evaluation of its influencing factors. *Analytica Chimica Acta* 458 (1): 29–37. https://doi.org/10.1016/S0003-2670(01)01523-9.

8 Rapp, A., Versini, G., and Ullemeyer, H. (1993). 2-Aminoacetophenone: causal component of 'untypical aging flavour,' 'naphthalene note,' 'hybrid note' of wine. *Vitis – Journal of Grapevine Research* 32: 61–62.

9 Dollmann, B., Schmitt, A., Köhler, H. et al. (1996). Formation of the 'untypical aging off-flavour' in wine: generation of 2-aminoacetophenone in model studies with *Saccharomyces cerevisiae*. *Wein-Wissenschaft* 51 (2): 122–125.

10 Hühn, T., Sponholz, W.R., and Grossmann, M. (1999). Release of undesired aroma compounds from plant hormones during alcoholic fermentation. *Vitic. Enol. Sci.* 54 (4): 105–113.

11 Robinson, A.L., Boss, P.K., Solomon, P.S. et al. (2014). Origins of grape and wine aroma. Part 1. Chemical components and viticultural impacts. *American Journal of Enology and Viticulture* 65 (1) https://doi.org/10.5344/ajev.2013.12070.

12 Gabrielli, M., Fracassetti, D., and Tirelli, A. (2014). UHPLC quantification of sotolon in white wine. *Journal of Agricultural and Food Chemistry* 62 (21): 4878–4883. https://doi.org/10.1021/jf500508m.

13 Rauhut, D., Shefford, P.G., Roll, C. et al. (2003). Effect on diverse oenological methods to avoid occurrence of atypical aging and related off-flavours in wine. In: *7th International Symposium of Oenology* (eds. A. Lonvaud-Funel, G. de Revel and P. Darriet), 376–379. Londres, Paris: Tec &. Doc, Lavoisier.

14 Cheng, L., Lakso, A., Henick-Kling, T. et al. (2004). Conclusions for three years of study on the effect of drought stress and available nitrogen on the formation of atypical aging flavor defect in wine. In: *Proceedings of the 33rd Annual New York Wine Industry Workshop* (ed. T. Henick-Kling), 68–71. Geneva/NY: Cornell University.

15 Perry, D.M. and Hayes, J.E. (2016). Effects of matrix composition on detection threshold estimates for methylanthranilate and 2-aminoacetophenone. *Food* 5 (2): 35. https://doi .org/10.3390/foods5020035.

16 Schmarr, H.G., Keiser, J., and Krautwald, S. (2016). An improved method for the analysis of 2-aminoacetophenone in wine based on headspace solid-phase microextraction and heart-cut multidimensional gas chromatography with selective detection by tandem mass spectrometry. *Journal of Chromatography A* 1477: 64–69. https://doi.org/10.1016/j .chroma.2016.11.029.

17 Horlacher, N. and Schwack, W. (2016). Determination of 2-Aminoacetophenone in wine by high-performance thin-layer chromatography-fluorescence detection. *Journal of Chromatography A* 1432: 140–144. https://doi.org/10.1016/j.chroma.2015.12.081.

18 Mihaljević Žulj, M., Maslov, L., Tomaz, I., and Jeromel, A. (2015). Determination of 2-aminoacetophenone in white wines using ultrasound assisted SPME coupled with GC-MS. *Journal of Analytical Chemistry* 70 (7): 814–818. https://doi.org/10.1134/ S1061934815070102.

19 Simat, T.J., Hoenicke, K., Gessner, M., and Christoph, N. (2004). Metabolism of tryptophan and indole-3-acetic acid formation during vinification and its influence on the formation of 2-aminoacetophenone. *Mitteilungen Klosterneuburg* 54: 43–55.

20 Linsenmeier, A., Rauhut, D., Kürbel, H. et al. (2007). Ambivalence of the influence of nitrogen supply on o-aminoacetophenone in 'Riesling' wine. *Vitis* 46 (2): 91–97.

21 Hühn, T., Cuperus, S., Pfliehinger, M. et al. (2003). Influence of environment on the composition of substrates and related atypical ageing in white wines. In: *Oenologie. 7e Symposium International d'OEnology* (eds. A. Lonvaud-Funel et al.), 139–143. Paris: Lavoisier Tec & Doc.

22 Linsenmeier, A., Rauhut, D., Kürbel, H. et al. (2007). Untypical ageing off-flavour and masking effects due to long-term nitrogen fertilization. *Vitis* 46 (1): 33–38.

23 Schwab, A.L. and Peternel, M. (1997). Investigation about the influence of a long term green cover on must and wine quality with special consideration of Franconian pedological and climatic conditions. *Vitic. Enol. Sci.* 52 (1): 20–26.

24 Bach, H.P. (2005). Untypischen Alterungston vermeiden. *Winzer-Zeitschrift* 9: 32–34.

25 Mihaljević, Ž.M., Korenika, A.M.J., Puhelek, I. et al. (2012). Influence of different yeast strains on metabolism of tryptophan and indole-3-acetic acid as the precursors of "Untypical Aging off-flavor" (UTA) in wine. *Book of Abstracts of the International Symposium for Agriculture and Food*, Skopje, pp. 55–56.

26 Lavigne-Cruege, V., Cutzach, I., and Dubourdieu, D. (2000). Interprétation chimique du vieillissement aromatique défectueux des vins blancs. Incidence des modalités d'élevage. In: *Oenologie 99. 6e Symposium International d'Oenology* (eds. A. Lonvaud-Funel et al.), 433–438. Paris: Lavoisier Tec & Doc.

27 Henick-Kling, T. (2012). A Note on Atypical Aging in White Wine. *Viticulture and Enology Extension News – Washington State University* Fall, 9.

9

Fermentation in Bottle

In this chapter, the serious fault of an unwanted fermentation of a wine after it has been bottled is briefly discussed. This will result in, at very least, aroma, flavour, and texture modifications or more seriously cloudy appearance, fizziness, sediments, and raised corks. In a worst scenario, bottles can burst due to the pressure of the carbon dioxide gas formed, causing injury. Bottle fermentation is consequential to the presence in the bottled wine of viable yeast (which may result in an alcoholic fermentation) or lactic acid bacteria (LAB) (which may result in a malolactic fermentation [MLF]), together with a growth substrate. The presence of LAB may also result in other faults that are discussed in Chapters 11 and 14.

9.1   Introduction

9.1.1   Bottle-Fermented Sparkling Wines

Before discussing unwanted fermentation in bottle, it is pertinent to remind ourselves that most high-quality sparkling wines undergo a controlled fermentation in the bottle in which the wine is sold. This is usually a second fermentation, the prime purpose of which is to make the wine sparkling but also to impart the aromas and flavours of autolysis, from the decomposition of the yeast lees that settle on the side of the bottle during initial cellar maturation. These yeast lees will be removed by a process of riddling (*remuage*) and disgorging (*dégorgement*). However, there are also high-quality sparkling wines that are made by a second or continuous fermentation in a pressure tank (autoclave) followed by a filtration under pressure before being bottled. These include the finest Proseccos, e.g. Conegliano Valdobbiadene Superiore di Cartizze DOCG.

9.1.2   Pétillant-Naturels

One of the newest wine styles to make waves in the last few years is, in reality, one of the oldest styles. Pétillant-naturels, commonly abbreviated to pét-nats, are wines that are bottled before they have completed their primary fermentation. For many years, there have been a few sparkling wines available made by the *Méthode Ancestrale*, and pét-nats are taking this 'back' to an even more natural foundation. The term 'pét-nat' was first used back in the 1990s by growers in the Loire and Loir valleys, France, but now the wines are made in

Wine Faults and Flaws: A Practical Guide, First Edition. Keith Grainger.
© 2021 John Wiley & Sons Ltd. Published 2021 by John Wiley & Sons Ltd.

several countries, including (northern) Italy and the United States. Many producers use no added yeast, and riddling and disgorging may, or may not, take place. Undisgorged wines are served cloudy, for even if yeast has settled, it will be roused as the bottle is opened. James Christopher Tracy, winemaker at Channing Daughters in Long Island, New York, explained the wines as 'exactitude within a non-exact style' [1]. Artisan beer makers too are happy for the yeasts in their 'bottle-conditioned' beers to work their wonders after leaving the brewery, although the yeasts for conditioning are usually added at bottling, and are chosen for their flocculation characteristics, and remain in the bottles if the beer is carefully poured.

Historically it was not uncommon for 'still' wines from some, usually cooler, regions to undergo a slight bottle fermentation if the alcoholic or MLFs had not been completed by bottling time. Indeed, a few wines were expected to have a slight pétillance as a consequence of the MLF taking place in bottle. These included Vinho Verde, from the northwest of Portugal. Writing in 1969, British author Raymond Postgate described the wine as naturally semi-sparkling with a slight prickle derived from the MLF finishing in bottle [2]. Of course, such bottle activity was, at best, a pretty unreliable activity. 'It's a great wine locally but really doesn't travel' was the historical oft-heard expression from travellers who, having encountered a regional delight, found that once back home it suffered from bottle fermentation or bacterial spoilage.

9.1.3 Unwanted Bottle Fermentation

With the advances in winemaking knowledge and technology in recent decades, the possibility of an alcoholic fermentation or MLF or refermentation taking place in bottle or other packaging seems, prima facie, an event that would only occur in a winemaker's worst nightmare.

In reality, although thankfully a rare event, unwanted fermentation of wine in bottle is again on the increase. Manuel Malfeito-Ferreira notes that according to his experience, 'refermentation problems have increased in the last years in red wines because of the addition of concentrated grape juices to make softer wines in accordance to modern market demands' [3]. As noted in Chapter 4, Australian red wines saw an average residual sugar increase from 0.5 g/l in 1998 to 2.1 g/l in 2008 [4], and today a concentration of 5 g/l is far from uncommon, particularly in so-called entry-level wines. Of course, in such cases, if any viable yeasts find their way into the bottle, refermentation is likely to occur. A pre-bottling sterile filtration will trap any yeasts and bacteria from the wine, although a great many producers of high-quality wines do not undertake this, believing it negatively modifies aromas and flavours and compromises quality. In any case, there are several areas of possible exposure post filtration. Wine bottles are delivered to wineries on shrink-wrapped palates, and such wrapping is easily and frequently damaged. Whilst ideally winery bottling facilities are isolated from production areas, yeast or bacteria contamination of empty bottles and corks or other closures frequently happens. With an inadequate pre-bottling rinsing, the seeds for disaster are sown. Bottling equipment too is easily contaminated with yeast or bacteria, and careful maintenance and thorough cleaning are essential.

9.2 Sensory Detection

There are many indicators of a fermentation taking place or having taken place in bottle. These include:

- *Raised corks, due to the production of carbon dioxide (CO$_2$)*: it should be noted that raised corks can also be a consequence of heat damage – see Chapter 14.
- *Bubbles or pétillance*: this is often the main sign of a bottle fermentation. However, some wines may have a slight natural pétillance, such as wines from the Mosel region, Germany, where a little CO$_2$ from the primary fermentation remains in the wine (the MLF does not take place for these wines). Also some wines are bottled under a blanket of CO$_2$ or the bottles flushed with gas, and a tiny amount will dissolve in the wine. This can give added freshness to a crisp white or rosé wine.
- *Cloudiness*: a milkiness or turbidity, due to the suspension of working, live cells or the agitation of dead cells: it should not be confused with hazes (see Chapter 10), which are lacking in sold matter.
- *Sludgy sediments*: these may be pale grey, or mushroom in colour, and should not be confused with the tannin sediments that naturally occur in mature red wines or tartrate crystals that may appear in certain white wines (see Chapter 15).

9.3 Alcoholic Fermentation in Bottle

An alcoholic fermentation may take place in bottle if yeasts are present together with a fermentable sugar. This may be due to their presence in the wine or contamination of bottles or equipment before or during the bottling process. The yeasts most likely to cause problems are *Saccharomyces* spp., *Schizosaccharomyces pombe*, *Zygosaccharomyces bailii*, *Saccharomycodes ludwigii*, and *Brettanomyces* spp. These yeasts have a high resistance to sulfur dioxide (SO$_2$) – *Zygosaccharomyces* in particular is incredibly resistant to alcohol (up to 15% abv) and alarmingly resistant to SO$_2$, tolerating a staggering 3 mg/l of molecular SO$_2$ [5]. At a pH of 3.6, to achieve this molecular SO$_2$ level, the free SO$_2$ would need to be 188 mg/l, which by itself would take a dry red wine over the legal limit for SO$_2$ in the European Union! Film-forming yeasts may also cause issues, although the alcohol tolerance for these is usually low. *Pichia* spp. and *Candida* spp. may form a film or a ring of cells that, in the presence of oxygen, sticks to the glass in the neck of the bottle if the level of free SO$_2$ is too low [3]. This is more likely if the closure does not prevent the diffusion of oxygen, as might be the case with saran–tin-lined screw cap.

If, at bottling, a wine contains 1.6 g/l of glucose or fructose, this could produce 800 mg/l of CO$_2$ that is sufficient to give a noticeable spritz on the palate [6]. The presence of CO$_2$ higher than the saturation concentration of about 1400 mg/l (at 20 °C) could result from the fermentation of 2.8 g/l of sugars – this would be sufficient to raise the bottle cork.

9.4 Malolactic Fermentation (MLF) in Bottle

When making white, rosé, or sparkling wines, depending upon the style required, the wine may or may not undergo complete or partial MLF. In the case of red wines, almost without exception, a complete MLF is required. The MLF, in fact, is a transformation caused by the action of some strains of LAB – *Oenococcus oeni* is the species usually preferred by wine-makers. MLF can be induced by warming the vats or inoculating with strains of LAB and nutrients. The ideal conditions for the fermentation to take place include a temperature of approximately 20 °C and SO_2 levels close to zero. There is a movement amongst winemakers to undertake the alcoholic fermentation and MLF simultaneously. However, there are instances when the MLF stops short of completion, leaving some malic acid remaining in the wine. A wine that has completed the MLF will contain less than 300 mg/l of malic acid. One gramme of malic acid produces 0.67 g of lactic acid and 0.33 g of CO_2. Accordingly 300 mg/l of malic acid could produce 100 mg/l of CO_2. Whilst an MLF of this amount could produce cloudiness, it would take some 2.4 g/l of malic acid to ferment before there would be a noticeable spritz on the palate [6]. However, it is vital that if malic acid is present in the wine at bottling, LAB are not. In the case of red wines, even if the malic acid does not ferment, there will be a negative impact upon the palate. Of course, the presence of malic acid together with viable LAB exposes the wine to the risk of the MLF resuming after packaging.

There are several possible causes for an incomplete MLF including:

- pH too low;
- SO_2 too high;
- Temperature too low;
- Lack of nutrients for the malolactic bacteria.

Further discussion of faults caused by lactic bacteria, including the species that do not conduct MLF, will be discussed in Chapters 11 and 14.

9.5 Prevention: Preparing Wine for Bottling and the Bottling Process

Fermentation in bottle may be avoided by effective winery and equipment cleaning and sanitation, pre-bottling and bottling adjustments, and cellar operations and procedures.

9.5.1 Preparing Wine for Bottling

9.5.1.1 Fining and Other Clarification

The clarification of wine post MLF, or post alcoholic fermentation if MLF is not undertaken, will decrease the microbiological load at each operation. Simple settling will reduce the microbial content, and high-speed centrifugation can achieve practically sterile wines shortly after fermentation [7]. Wines should be fined using appropriate agents before final preparation for bottling. Bentonite may be used to remove proteins in white wine and help with 'heat' stability, but there is a loss of desirable aromas and flavours.

9.5.1.2 Adjustment of Free and Molecular SO$_2$

The level of free and molecular SO$_2$ should always be adjusted in wines immediately before packaging. The amount of free SO$_2$ required to prevent any MLF taking place in container depends, inter alia, upon the pH of the wine. A molecular SO$_2$ of 0.8 mg/l may generally speaking be regarded as effective at inhibiting LAB. This equates to a level of free sulfur dioxide between 25 mg/l for a wine with a pH of 3.3 and 40 mg/l and for a wine with a pH of 3.5. Many winemakers mistakenly believe the growth of LAB can be prevented by adjusting free sulfur dioxide to 30 or 35 mg/l at bottling time, but in wines with a pH above 3.4, this may be insufficient. Furthermore, the level of free SO$_2$ will fall immediately post bottling, as some of the compound becomes 'bound' in the wine.

9.5.1.3 Addition of a Fermentation Inhibitor

The addition of a fermentation inhibitor for wines containing any residual sugar or malic acid will effectively prevent bottle fermentation. There are three inhibitors in general use:

- Lysozyme;
- Dimethyl dicarbonate (DMDC) marketed as Velcorin®;
- Potassium sorbate.

Sodium benzoate is also a fermentation inhibitor, but it is not permitted for use by the International Organisation of Vine and Wine (OIV) or for wines produced or marketed in the EU.

Lysozyme The growth of LAB bacteria may be effectively inhibited by the addition of lysozyme. This is a low-molecular-weight protein derived from egg whites that brings about lysis of the cell wall of Gram-positive bacteria including LAB species. Its use is permitted by the OIV International Code of Oenological Practices up to a maximum dose of 500 mg/l [8]. As the dosing equipment is expensive, it is generally out of reach of small producers, unless mobile facilities are available. Lysozyme should always be used along with sulfur dioxide.

Dimethyl Dicarbonate (DMDC) (Velcorin) DMDC with a molecular formula of $C_4H_6O_5$ is a dimethyl ester of dicarbonic acid used as 'cold sterilant' in the food industry. Other than in wines, it is mainly used in carbonated or non-carbonated juice beverages, iced teas, and flavoured waters. DMDC, which is marketed under the trade name Velcorin, may be added shortly before or during the bottling process to obtain microbiological stability of bottled wine containing fermentable sugars and to prevent the development of unwanted yeast and LAB. It is most effective at low pH, high alcohol content, and low microbial population [9]. The maximum addition according to OIV International Code of Oenological Practices is 200 mg/l, and the wine must not be marketed whilst its presence is detectable [8]. It is generally accepted that the use of DMDC does not result in any organoleptic impact.

DMDC works by penetrating into yeast cells and binds with and deactivates certain amino acid residues that are located at the catalytic centre of key cellular enzymes [10], particularly alcohol dehydrogenases and glyceraldehyde-3-phosphate dehydrogenase. The amount of DMDC required to deactivate yeast varies according to the species, the strain, and the number of colony forming units (cfus) present and is also influenced by the pH of wine.

Bacteria, including LAB, are more resistant than yeasts, and so this preservative should not be regarded as a 'catch-all' when used alone [11]. The Australian Wine Research Institute (AWRI) noted that an addition of 100 mg/l of DMDC appears to be sufficient to kill yeast in wine if there are less than 500 CFU/ml [12]. The manufacturer of Velcorin (LANXESS) stated that the usual dosage is between 125 and 200 mg/l [13]. However, some strains of *Brettanomyces bruxellensis* (which can be responsible for fermentation in bottle) may only be inhibited at DMDC levels above 150 mg/l [14]. Accordingly, to be sure of Brett inhibition, the recommended dosage of DMDC for this purpose is 200 mg/l, which is the OIV limit for wines [8].

The effectiveness of DMDC is dependent upon several factors that need to be taken into account when determining the dosage of DMDC [10]. Some other non-*Saccharomyces* yeasts have a very high tolerance – by way of example in one study, *Z. bailii* (at a concentration of 1000 CFU/ml) required a minimum of 250 mg/l to be inhibited, and two strains of *Pichia anomala* (at a similar concentration) required a minimum of 400 mg/l of DMDC for inhibition [14]. These concentrations are above those legally permitted in the European Union (EU) and by OIV [8]. However, the concentration of viable yeast or bacteria should always be reduced to below 500 cfu/ml before the addition of the compound. The use of DMDC as a preservative (E242) is permitted for wine in the EU under Regulation 2019/934 [15] (which amends previous regulations). It is approved as an additive for wine in the United States by FDA 21 Code of Federal Regulations, providing 'the viable microbial load has been reduced to 500 microorganisms per millilitre (500 CFU/ml) or less by current good manufacturing practices such as heat treatment, filtration, or other technologies prior to the use of dimethyl dicarbonate' [16]. It is also approved as a *processing aid* for use in wine in Australia and New Zealand under Food Standard Code 1.3.3. [17]. DMDC must not be present in the food when sold. It fully decomposes into carbon dioxide and methanol, after which no antimicrobial effect remains [18]. The level of methanol must not exceed 400 mg/l for red wines and 250 mg/for white and rosé wines when these are marketed [8]. If wine is treated with the maximum permitted level of 200 mg/l, this 'would result in an additional 98 mg/l of methanol to wine, which already contains an average of about 140 mg/l from natural source' [18]. It will be seen that in the case of white wines, the total will be close to the OIV limit.

DMDC must always be used in synergy with sulfur dioxide. It should be noted that DMDC is quickly hydrolysed into methanol and carbon dioxide (CO_2) [19], so it is important to maintain appropriate SO_2 levels. DMDC must be completely homogenised into wine and so requires special dosing apparatus, as illustrated in Figure 9.1.

The dosing units may be purchased, rented, or already incorporated into mobile bottling facilities. It is recommended that Velcorin is added post-final filtration before bottling.

Potassium Sorbate Sorbic acid, in the form of potassium sorbate, may be added to wine containing residual sugar prior to bottling as a fermentation inhibitor. The rate of addition is 100–200 mg/l. It should be noted that sorbic acid does not actually kill yeast but merely inhibits the activity of *Saccharomyces cerevisiae*, other species of *Saccharomyces*, and some other yeast genera. It is not effective against film-forming yeasts, including *Candida* spp., nor against *Brettanomyces* spp. or *Zygosaccharomyces*. It is should not be considered to be a replacement for sulfur dioxide, as it does not have antibacterial properties. Sorbic acid only

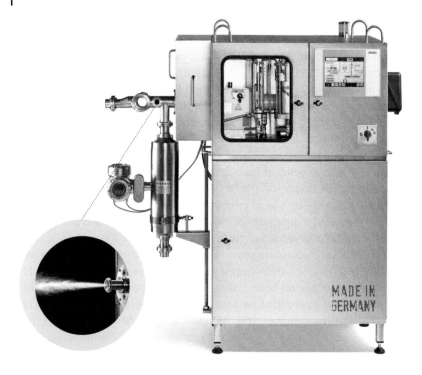

Figure 9.1 Velcorin dosing unit. Source: Courtesy of LANXESS.

remains effective for a relatively short period of time. It degrades to ethyl sorbate that is present in all wines that have had sorbic acid additions. This can have negative organoleptic implications as discussed in Chapter 14.

9.5.1.4 Heat Treatments

Flash pasteurisation by which the wine is heated and cooled in plate heat exchangers will kill yeasts and LAB – the wine may be sterile-filtered before bottling to avoid recontamination. An alternative, which is less popular than in the past, is in hot bottling, or thermolisation, by which wine is heated and bottled at the desired temperature and then cooled after bottling. This does give a problem with fill heights as the wine shrinks post bottling.

9.5.1.5 Membrane Filtration

Filtration by diatomaceous earth is often carried out during wine ageing, and the tightest earths can reduce microbial numbers drastically [7]. Other forms of nominal or depth filtration will remove 99.9% of bacteria and yeast cells [20] and may be undertaken pre- and post-maturation. Providing wine has been adequately pre-filtered to avoid membrane fouling, membrane filtration immediately prior to bottling using a sterile (0.45 m or less) medium will capture any remaining yeasts and bacteria (but will not prevent contamination from the bottling line, bottles, or corks). Figure 9.2 shows a bank of membrane filters.

Figure 9.2 Bank of membrane filters.

9.5.2 The Bottling Process

9.5.2.1 The Final Production Source of Contamination

The bottling and closing process is the last production source of contamination of wine, and although wines can become contaminated in bottle, e.g. with TCA (see Chapter 3), it is generally accepted that, other than in the event of contaminated, damaged, or compromised packaging, yeast or bacteria cannot enter a sealed container. Usually, the contamination of wine at this stage results simply from lack of hygiene and, particularly, inadequate disinfection of all the bottling facility, including surfaces in contact with the cork or other closure [3].

9.5.2.2 The Bottling Line and Operation

Whilst ideally winery bottling facilities are isolated from production areas, yeast or bacteria contamination of empty bottles and corks or other closures can happen. With an inadequate pre-bottling rinsing, the seeds for disaster are sown. In fact, the bottle rinser and rinsing water are frequent sources of contamination [3, 21, 22].

Bottling equipment is easily contaminated with yeast or bacteria. The bottling line should be sanitised prior to the bottling run by flushing with water at over 80 °C (276°) for at least 20 minutes [23] or by the use of steam with caustic cleaning agents. Care should be taken that there are no 'cold spots' in the bottling unit. Steam disinfection as an alternative to the use of chemicals has the advantage of possibly killing yeasts and bacteria without directly contacting the cells. However, steam cleaning can result in a negative pressure within the filler when cooling takes place, which leads to sucking in air that may contain spoilage yeasts in suspension and result in immediate contamination of the filler. This issue can be resolved by introducing sterile compressed air during cooling [3]. It must be noted that biofilms can be resistant to hot water sterilisation and *Z. bailii*, often implicated in refermentation in bottle, easily forms biofilms. However, inadequate sanitation can result from not being able to correctly apply the cleaning and disinfection programs, for example, the design of the filler and corker, poor maintenance, and the microbiological quality of the ambient atmosphere [3]. In fact usually the main sources of contamination are the filler and corker, and the danger increases as the equipment ages or in the case of inadequate maintenance [24]. Many parts of the system, including surfaces concealed by o-rings and the valves of isobarometric fillers, are very difficult to reach with sanitising agents.

There is a modern trend, particularly in so-called 'entry-level' wines, for them to be produced or finished with a few grammes of residual sugar, even though the wines are usually labelled as 'dry'. This poses a real danger of fermentation in bottle. Manuel Malfeito-Ferreira noted that when grape juice concentrate is used in production, yeasts of the genus *Zygosaccharomyces*, particularly *Z. bailii*, are frequently detected in the fillers [3].

The corking unit is also a frequent source of contamination as the surfaces in contact with the sides of the corks can suffer heavy contamination [3]. These areas include the feeding hopper, the cork transport tube, and the corker jaws. Colonisation of yeasts and bacteria can result from the formation of water droplets and condensation with the consequence of contamination and colonisation by yeasts suspended in the air. This can be prevented by the heating the jaws up to lethal temperatures, which is possible with modern corkers. Even the situation of enclosing the bottling line in a room isolated from the winery to reduce contamination from the winery ambient air can be dangerous. With good design, which includes overpressure, wet air evacuation and an enclosed bottling room can assist in proper bottling. However, as noted by Donnelly, when poorly designed, such bottling rooms can be microbial incubators, and they are dangerous sources of airborne and general contamination [25].

Minimising oxygen uptake at bottling will assist in maintaining the level of free SO_2 in the wine post-bottling. Regular measurements of dissolved oxygen (DO) in the wine and headspace oxygen (HSO) should be made during the bottling process [26]. An empty bottle prior to preparation for filling contains 750 ml of air, which includes an oxygen content of over 200 mg. If this is left unpurged, this could increase the DO in the wine by 0.3–0.7 mg/l, depending on the filling technology adopted [26]. Pre-filling purging and headspace oxygen evacuation will help minimise oxygen additions during bottling, reducing the potential for microbial spoilage.

9.6 Treatment

Faced with a wine that is undergoing or has undergone an alcoholic fermentation or MLF in bottle, the producer has little choice other than to de-bottle the wine and put in tank to complete the fermentation, followed by sulfurisation, fining, filtering, and, if the wine is sound, re-bottling. However, there is likely a considerable aroma and flavour modification.

9.7   Final Reflections

The wine world has changed almost out of recognition since my first tentative sips or, if I recall correctly, gulps as a young student. I soon learned to look out for wines that were fermenting in bottle and after a sip would relegate these to the cooking pot. As I finish writing this chapter with a glass of unpretentious pét-nat in hand, a wine style that is pure fun and extremely enjoyable, I begin to wonder if I have discarded some magic over the years.

References

1 Postgate, R. (1969). *Portuguese Wine*. London: J. M. Dent & Sons Ltd.
2 https://www.winemag.com/2018/08/07/pet-nat-wine-guide (accessed 4 July 2020).
3 Malfeito-Ferreira, M. (2019). Spoilage yeasts in red wines. In: *Red Wine Technology* (ed. A. Morata), 219–235. London: Academic Press.
4 Godden, P. and Muhlack, R. (2010). Trends in the composition of Australian wine, 1984–2008. *The Australian and New Zealand Grapegrower & Winemaker* 558: 47–61.
5 Fugelsang, K.C. and Edwards, C.G. (2007). *Wine Microbiology – Practical Applications and Procedures*, 2e. New York: Springer.
6 Butzke, C. (2010). *Preventing Refermentation*. Purdue Extension https://www.extension.purdue.edu/extmedia/FS/FS-56-W.pdf (accessed 16 July 2020).
7 Malfeito-Ferreira, M. (2011). Yeasts and wine off-flavours: a technological perspective. *Annals of Microbiology* https://doi.org/10.1007/s13213-010-0098-0.
8 International Organisation of Vine and Wine {OIV} (2021). *International Code of Oenological Practices*. Paris: OIV. https://www.oiv.int/public/medias/7713/en-oiv-code-2021.pdf.
9 Escott, C., Loira, I., Morata, A. et al. (2017). Wine spoilage yeasts: control strategy. In: *Yeasts Industrial Applications* (ed. A. Morata). Intech Open. https://doi.org/10.5772/intechopen.69942. https://www.intechopen.com/books/yeast-industrial-applications/wine-spoilage-yeasts-control-strategy.
10 Portugal, C., Sáenz, Y., Rojo-Bezares, B. et al. (2014). Brettanomyces susceptibility to antimicrobial agents used in winemaking: in vitro and practical approaches. *European Food Research and Technology* 238 (4): 641–652. https://doi.org/10.1007/s00217-013-2143-2.
11 Costa, A., Barata, A., Malfeito-Ferreira, M., and Loureiro, V. (2008). Evaluation of the inhibitory effect of dimethyl dicarbonate (DMDC) against microorganisms associated with wine. *Food Microbiology* 25: 422–427.
12 AWRI (2020). *Microbiological stabilisation*. Fact Sheet. AWRI www.awri.com.au/industry_support/winemaking_resources/storage-and-packaging/pre-packaging-preparation/microbiological-stabilisation (accessed 23 July 2020).
13 LANXESS (2020). *Velcorin – Spectrum of Use*. LANXESS https://velcorin.com/velcorin/spectrum-of-use/velcorin-wine (accessed 20 July 2020).

14 Renouf, V., Strehaiano, P., and Lonvaud-Funel, A. (2008). Effectiveness of dimethlydi-carbonate to prevent *Brettanomyces bruxellensis* growth in wine. *Food Control* 19 (2): 208–216.

15 EUR-Lex. (2019) Commission Delegated Regulation (EU) 2019/934 of 12 March 2019 supplementing Regulation (EU) No 1308/2013 of the European Parliament and of the Council as regards wine-growing areas where the alcoholic strength may be increased, authorised oenological practices and restrictions applicable to the production and con-servation of grapevine products, the minimum percentage of alcohol for by-products and their disposal, and publication of OIV files. https://eur-lex.europa.eu/legal-content/GA/TXT/?uri=CELEX:32019R0934 (accessed 25 May 2020).

16 Code of Federal Regulations. (2020) Title 27, Chapter I, Subchapter A, Part 24. https://www.ecfr.gov/cgi-bin/text-idx?c=ecfr&sid=506cf0c03546efff958847134c5527d3&rgn=div5&view=text&node=27:1.0.1.1.19&idno=27 (accessed 14 June 2020).

17 Australia and New Zealand Food Standard Code 1.3.3. (2016) www.foodstandards.gov.au/code/Documents/1.3.3%20Processing%20aids%20v157.pdf (accessed 8 April 2020).

18 European Food Safety Authority {EFSA} (2015). Scientific opinion on the re-evaluation of dimethyl dicarbonate (DMDC, E 242) as a food additive. *EFSA Journal* 13 (12): 4319. https://efsa.onlinelibrary.wiley.com/doi/pdf/10.2903/j.efsa.2015.4319.

19 Milheiro, J., Filipe-Ribeiro, L., Vilela, A. et al. (2017). 4-Ethylphenol, 4-ethylguaiacol and 4-ethylcatechol in red wines: microbial formation, prevention, remediation and overview of analytical approaches. *Critical Reviews in Food Science and Nutrition* 59 (9): 1367–1391.

20 Horton, D. and Clark, M. (2016). Wine maker's nightmare: Refermentation and unstable sweet wines, and how to avoid. *University of Minnesota Grape Breeding and Enology Monograph.* https://enology.dl.umn.edu/news/wine-makers-nightmare%C2%A0-re-fermentation-and-unstable-sweet-wines-and-how-avoid (accessed 16 July 2020).

21 Donnelly, D. (1977). Elimination from table wines of yeast contamination by filling machines. *American Journal of Enology and Viticulture* 28: 182–184.

22 Neradt, F. (1982). Sources of reinfections during cold-sterile bottling of wine. *American Journal of Enology and Viticulture* 33: 140–144.

23 Tran, T., Wilkes, E., and Johnson, D. (2015). Microbiological stability of wine packaging in Australia and New Zealand. *Wine and Viticulture Journal*, March/April: 46–49.

24 Loureiro, V. and Malfeito-Ferreira, M. (2003). Yeasts in spoilage. In: *Encyclopedia of Food Sciences and Nutrition*, 2e (eds. B. Caballero, L. Trugo and P. Finglas), 5530–5536. London: Academic Press.

25 Donnelly, D. (1977). Airborne microbial contamination in a winery bottling room. *American Journal of Enology and Viticulture* 28: 176–181.

26 Letaief, H. (2016). Key points of the bottling process. *Wines & Vines* May. https://winesvinesanalytics.com/features/article/168227/Key-Points-of-the-Bottling-Process (accessed 30 April 2020).

10

Hazes

In this chapter, the causes of formation of hazes in bottled wine and how these may be prevented are discussed. Hazes may occur on account of unstable proteins, microorganisms, or the presence of metals, particularly copper or iron. They are most noticeable in white or rosé wines. After time in bottle, hazy wines may precipitate a sediment. Hazes are usually just a fault of appearance and have no other sensory impact. The steps to be taken during vinification to avoid haze formation are discussed.

10.1 Introduction

There can be little doubt that clarity in wine, particularly in white and rosé wine, is important to the consumer, and the presence of any haze usually leads to the conclusion that the wine is faulty. Interestingly, consumers are prepared to accept or even expect hazes or cloudiness in many other drinks and in many cases regard this as a sign of quality or purity. Such drinks include 'white' or wheat beers, some fruit juices, cloudy lemonade and alcoholic lemonade, limoncello, and lemon vodka, together with ouzo, pastis, absinthe, and raki (which take on a cloudiness when water is added). Even with wine, the new (old) style of fizzy pét-nats that are bottled before they have completed their primary fermentation are growing in popularity.

For the producer, the onset of a haze in white or rosé wine post-bottling can prove very expensive. If the wines have not left the warehouse and are deemed to be correctable, the costs will include uncorking, decanting to vat, haze treatment, racking, stabilising, filtering, washing and sterilising bottles, rebottling, labelling, and packing. If the wines have been despatched, then the economic cost in product recall can be massive, especially if the wines have been exported, to which costs must be added the detriment to the producer's reputation. It is fortunate that incidences of hazes are less common than was once the case, but to counter this, consumers are now undoubtedly less tolerant of visual defects. Hazes and deposits account for 31% of the investigations conducted by the Australian Wine Research Institute (AWRI) [1].

There are several sources of hazes in wine:

- Protein
- Microbes: yeast or bacteria
- Metals

Wine Faults and Flaws: A Practical Guide, First Edition. Keith Grainger.
© 2021 John Wiley & Sons Ltd. Published 2021 by John Wiley & Sons Ltd.

Of these, protein haze is the most prevalent and therefore constitutes the majority of the discussion in this chapter. The hazes can also form deposits or sediments after some time in bottle. Other sources of deposits are tannic matters that precipitate as a natural consequence of wine ageing, and tartrates (see Chapter 15).

10.2 Protein Haze

10.2.1 The Formation and Impact of Protein Hazes

Protein haze is now a fairly rare fault that may affect white or rosé wines. The fault is particularly likely to occur if the wine is subjected to elevated temperatures during transport or storage. The wine will appear dull and oily, as a result of positively charged dissolved grape proteins denaturing and being massed together in insoluble light-dispersing aggregates. The wine will also appear hazy, and the particles may eventually form unattractive sediments [2]. There is no modification or deterioration in aromas or flavours (Figure 10.1).

Proteins are present in grapes, with some varieties, e.g. Grüner Veltliner, having particularly high concentrations. Generally speaking, the riper the grapes, the higher the protein content. Proteins are also contained in yeast and may be imparted into wine by autolysis, but these proteins may have a negative effect regarding the formation of hazes. Protein haze formation is predominantly due to an excess of two groups of pathogenesis-related

Figure 10.1   Protein haze. Source: Courtesy of ETS Laboratories.

proteins: thaumatin-like proteins and chitinases [3]. These proteins are small and compact with globular structures [4]. There are other types of protein present in must and wine that are not haze forming. The total protein content of wine is usually between 15 and 300 mg/l. The mechanisms of protein haze formation are still not fully understood, but it is known that the presence of procyanidins is needed for haze formation [5]. Phenolic compounds increase haze development by crosslinking denatured proteins to stimulate aggregate development [6]. It is also certain that the sulfate anion is necessary for haze formation [7]. The concentration of sulfate concentration and the overall ionic strength of wines play a part, as the presence of sulfates and other ions in sufficient quantities favours protein aggregation [8]. Proteins may aggregate post-bottling, particularly during storage or transport at elevated temperatures. When the aggregates crosslink, a haze is formed. Sedimentation may, or may not, occur.

Proteins may be either positively (cation) or negatively (anion) charged depending upon the wine pH. Thaumatin-like proteins and chitinases are nearly always positively charged, at the range of pH found in wine. The isoelectric point is the pH at which a molecule carries no positive or negative electrical charge, i.e. the net charge is zero. The isoelectric point of a protein is the point that the protein is least soluble and therefore unstable – above and below the isoelectric point, the protein will be soluble. It is only unstable proteins that need removing from wine. The pH of wine may change at times during the winemaking process, and accordingly, the isoelectric point is also changed. For this reason, protein (heat) stabilisation should be undertaken after the completion of all wine operations that may have an effect upon pH.

10.2.2 Prevention of Protein Haze Formation

10.2.2.1 Pre-fermentation Reduction in Haze-Forming Proteins

The total protein content, including haze-forming proteins, may be reduced by the taking of appropriate actions pre-fermentation:

- Grapes should be harvested before becoming overripe.
- Must may be treated with pectolytic enzymes with proteolytic activity [5].
- Pre-fermentation maceration with extraction enzymes should be avoided if the grapes are protein rich. In the case of some varieties, skin contact can increase unstable protein by up to 50% [5].
- Dry ice may be used to protect harvested grapes instead of sulfur dioxide reducing sulfate concentration.
- The tannin from stems can combine with unstable proteins in crushing and pressing; hand-picking is perhaps preferable to mechanical harvesting (at least from a haze formation perspective) [9].
- Settling and racking the must can decrease protein content by 10–15%.
- Fining musts with bentonite (see Section 10.2.2.4) can reduce amino acid content by up to 30% – care must be taken that sufficient nutrients are retained for yeast development and health.
- Pre-fermentation addition of tannin can precipitate proteins but may have a negative effect on mouthfeel.

10.2.2.2   Post-fermentation Reduction in Haze-Forming Proteins

- Following the alcoholic fermentation and malolactic fermentation (MLF) (if undertaken), the wine should be racked to separate it from remaining phenolic compounds along with gross lees.
- The mannoproteins contained in fine lees (or added to wine) can act as protective colloids and help prevent unstable grape proteins from precipitating.
- The addition of lysozyme to inhibit the growth of lactic acid bacteria (LAB) adds substantial amounts of non-grape protein to wine. The addition necessary to avoid MLF is 250–500 mg/l, which is 2–50 times the amounts naturally present in grapes, and may result in protein instability [10]. To complicate matters, lysozyme, although potentially unstable, does not respond well to the heat tests described below [10].

10.2.2.3   Testing for Protein Stability

As we have seen, the stability of proteins in wine depends, inter alia, upon the pH. Accordingly, stability tests should be undertaken after all treatments have been completed. Cold stabilisation, as commonly used to precipitate tartrates, will affect the pH (and total acidity). There are two types of test that may be used to indicate protein stability:

- Heat tests
- Reagent tests

Heat (Stability) Tests   Heat/protein stability tests are the traditional and still most used way to determine if a wine is at risk of forming protein haze. Although they are probably the most reliable method to indicate haze/sediment formation in the bottle during storage [5], nevertheless, these do not always accurately predict haze formation, particularly during long-term storage [7]. One of the drawbacks of heat stability tests is that all proteins tend to be precipitated and the amount of fining agent required can be therefore overestimated. Another disadvantage is the time required for completing the test [5], which ranges from 5 to 22 hours, according to the heating and cooling temperatures and regime utilised, and the particular process undertaken.

A heat test extensively used involves the wine being heated to 80 °C (176 °F) for six hours and then cooled for three hours. The turbidity of the wine is measured, either by sight or ideally with a turbidimeter (also known as nephelometer) – the measurements should be made both before and after the procedure and compared with a control wine. If measuring by sight, a focussed light should be passed through the samples and the beam of light observed at right angles. Of course, a turbidimeter is more accurate for this purpose. To 'pass' the heat test the increase in turbidity for a white wine must be less than two Nephelometric Turbidity Units (2 NTUs), although some winemakers prefer to rely on a lower level such as 1 NTU. The AWRI notes that recent research shows that heating for two hours to this temperature is sufficient [11] – a fact sheet '*Measurement of heat stability of wine*' is published by AWRI and gives a detailed procedure for undertaking a heat test [11]. In a 2009 study by Mireia Esteruelas et al., a 'fast' heat test involving heating the wine to 90 °C (194 °F) for one hour in a thermostatic water bath followed by chilling at a temperature of 4 °C (39 °F) for six hours in a refrigerator was found to be the most similar to the natural precipitate and perhaps be the most appropriate stability test [12].

Reagent-Based Tests There are also reagent-based test kits available such as Bentotest®, Proteotest®, and Prostab®, which are relatively simple to use. The Prostab test requires the availability of a spectrophotometer.

Other possible protein stability tests include the *tannin* test and *ethanol* test. In reality none of these are good predictors of the amount of fining agent required to stabilise the wine, and accordingly are not discussed here.

10.2.2.4 Fining with Bentonite

Bentonite is a fine clay, formed by the decomposition of volcanic ash that was deposited millions of years ago. It is mined from the ground and has many uses. These include a lubricant for oil drilling, cat litter, a thickener in paints, a component in sunscreen, and a binder for animal feeds – it has an ability to adsorb mycotoxins [13]. The US state of Wyoming has 70% of the known deposits in the world [14]. Bentonite is hydrated aluminium silicate that comprises mostly of montmorillonite.

Bentonite is the standard fining material for removing haze-forming proteins in wine. It is negatively charged and thus neutralises positively charged proteins in wine, acting as a cation exchanger. Although bentonite fining will not stabilise negatively charged proteins, i.e. if the isoelectric point of a protein is lower than the wine's pH, the proteins present in wine are invariably positively charged. Fining with bentonite binds unstable proteins, which are adsorbed onto the material. However, particularly if the wine is already protein stable, bentonite can bind to aroma and flavour compounds. The addition of bentonite has been known to strip colour, particularly from red and rosé wines [10]. Winemakers should always check the wine for protein stability prior to treating with bentonite and are reminded that heat stability tests should only take place after any cold stabilisation or other treatments, such as ion exchange or electrodialysis, used to precipitate tartrates (see Chapter 15).

There are several different types of bentonite. In the wine industry, sodium bentonite (Na bentonite) and calcium bentonite (Ca bentonite) are the two types generally used. Sodium bentonite swells more than calcium bentonite, creating a larger surface area for adsorption. Sodium bentonites create a larger sediment (and potential wine loss). Calcium bentonites can be activated with sodium carbonate at 80°C (176°F) resulting in an exchange of sodium for calcium, producing sodium-activated bentonite (Ca–Na or Na–Ca bentonite). This increases adsorption properties. Bentonites must be hydrated with water and/or wine, forming a slurry, before use.

It is generally accepted amongst winemakers and well documented by researchers that bentonite not only has a negative impact on colour but also diminishes desirable aromas and flavours. It is probable that most aroma compounds are removed as an indirect effect of protein removal; some hydrophilic odour-active compounds are weakly hydrogen bound with protein surfaces, and more hydrophobic aromatic molecules can bind to interior protein sites that have a stronger affinity for hydrophobic substances [13]. Bentonite can also remove phenolic compounds [15], and whilst white wines usually contain less phenolics than reds, such loss can have an impact upon mouthfeel, as well as flavours. The study by He et al. showed that the total phenolics of 107.8 mg/l in a Chardonnay before bentonite fining were decreased to 94.8, 98.9, 95.3, and 95.3 mg/l, depending upon the type of bentonite used [3]. In the same study, the concentration of total phenolics in a skin-fermented Sauvignon Blanc was 414.8 mg/l before treatment, but after fining with bentonite, the concentration of total phenolics decreased to 409.4, 340.4, 398.9, and 394.3 mg/l, depending upon the type of bentonite. Whilst such reductions may have the positive effect of reducing astringency, alteration of other aspects of mouthfeel can be negative.

Apart from the loss of aromas and flavours, there are further downsides to the use of bentonite. There is a reduction in the quantity of wine, due to lees formation. This can amount to between 3% and 10% (or even more) of the wine volume. Some wine can be recovered from bentonite sedimentation, and this may be achieved by utilising a lees filter or rotary vacuum filter, but such processing has a negative impact upon quality [6, 16], including damaging oxygen uptake. It has been estimated that the annual global value of wine loss due to bentonite additions was about 1 billion Australian dollars [17]. Any remaining traces of bentonite in wine can rapidly foul membrane filters. Further, the used bentonite cannot be recycled and requires disposal, which may represent an additional cost and environmental impact. Accordingly, much research is taking place on bentonite alternatives [6]. Alternatives investigated include ultrafiltration (UF), addition of proteolytic enzymes, flash pasteurisation, other adsorbents (silica gel, hydroxyapatite, and alumina), zirconium oxide, natural zeolites, chitin and chitosan, carrageenans, and the use of mannoproteins [5]. Carrageenans, naturally occurring polysaccharides extracted from *Rhodophyta* red seaweeds, have been shown to be effective at protein stabilisation with more intense flavour and aroma profiles than those stabilised with bentonite [18].

Fining Trials Before bentonite is used, fining trials should be undertaken to determine the amount required, which usually ranges from 200 mg/l to 1 g/l. A series of six measuring jars each containing 200 ml of wine should be set up. The first jar should have no bentonite added, the second 0.5 g/l, and the third 1 g/l, and the additions to the others continue in similar increments. As the bentonite needs to be added in a solution (a 5% solution is appropriate), the additions required are 0, 2, 4, 6, 8, and 10 ml of (5%) suspension in order to give 0, 0.5, 1, 1.5, 2, and 2.5 g/l bentonite to the series of 200 ml samples [19]. The samples should be allowed to rest overnight, the top layer of each removed and filtered through a 0.45 µm membrane, and then subject to heat or reagent stability tests.

Bentonite Additions Whilst bentonite may be added at any stage of wine production, must, fermenting wine, or wine, it would appear that adding during the early stages of fermentation requires the least additions [4]. However, determining the amount of addition at this stage can be imprecise, and subsequent additions may well be required. Undertaking fermentation with bentonite present also increases the rate of fermentation, which may be advantageous [2]. The wine can be racked off the bentonite solids post fermentation, and the detrimental impact upon aroma and flavour may be less than if the earth is added to wine.

The different types of bentonite have similar efficiencies in removing haze-related proteins, and all result in significant decrease in total phenolic concentration [3]. However, there is a variance in phenolic composition depending on the type of bentonite. In the study by He et al. [3], fining with Ca–Na bentonite resulted in the lowest concentrations of caftaric acid and flavanols, particularly epicatechin gallate, gallocatechin, catechin, and epicatechin. This could lead to alteration in the mouthfeel of the fined wine.

After fining with bentonite, the wine should undergo a further protein stability test to see if an additional treatment is required.

10.2.2.5 Alternatives to Bentonite

Ultrafiltration   UF can be used to remove certain proteins by separating molecules according to their molecular weight. However, colour and desirable aroma and flavour compounds are removed in the process. A 2018 study by Mariana Rodrigues de Sousa et al. showed that the UF operation lead to the clarification of white wines, with a decrease of turbidity of over 80% in the treated wine compared with the base wine [20].

The Use of Chitosan and Chitin–Glucan   Both chitosan and chitin–glucan of fungal origin can remove certain proteins from wine and be used to carry out a treatment to prevent protein haze. The amount to be used should be determined by preliminary testing. However, the recommended dose used should be less than or equal to 100 g/hl. In a study by Simone Vincenzi et al., the addition of 1 g/l of chitin to wine decreased the haze induced by the heat test by 50%, and the addition of 20 g/l of chitin decreased the wine haze by almost 80%. This haze reduction was directly related to the removal of certain grape chitinases [21]. The compounds are known to reduce the levels of tartaric acid.

10.2.2.6 Possible Future Alternatives to Bentonite

Laboratory-, pilot-, and industry-scale trials of the use of proctase have been successfully carried out in Australia [8] and would seem to indicate that proctase (enzymes) could be used as an alternative to bentonite. The product is currently not permitted for use in the European Union (EU), or for wines sold in the EU.

10.3 Microbial Hazes

The presence of yeasts or bacteria in bottle can lead to the formation of hazes and sediments, as discussed in Chapter 9. *Saccharomyces cerevisiae*, other *Saccharomyces* spp., *Zygosaccharomyces* spp., and *Brettanomyces bruxellensis* can ferment any sugars in the bottled wine leading, inter alia, to cloudiness/haze and sediments. *Zygosaccharomyces* spp. is often found in wine where grape juice concentrate has been blended to wine to give a little sweetness. It can tolerate high alcohol levels and is very likely to cause refermentation in bottle. The presence of LAB can lead to MLF, haze formation, and sedimentation in bottle. Finally, although they are aerobic, endospore-producing *Bacillus* spp. have been found in bottled wine occasionally, also resulting in haze and sedimentation.

10.4 Metal Hazes

The two most common metal hazes in wine are copper and iron.

10.4.1 Copper Haze

10.4.1.1 Formation of Copper Haze

The presence of copper in wine can stem from several sources including soil, fertilisers, fungicides, pesticides, fining agents, and filtration materials. The use of copper as a vineyard fungicide, usually in the form of 'Bordeaux mixture' comprising copper sulfate and lime, remains controversial. Copper is permitted in certified organic vineyards, and the growth in organic production is perhaps one reason why the amount of copper in wine is increasing. As well as being undesirable in wine due to being toxic, copper can cause the fault of copper haze, also known as copper casse.

Copper haze can occur when copper levels in wine are above 0.5 mg/l. It is only really noticeable in white and rosé wines, imparting a red haze and often throwing deposits [22]. During the last 15 years, there has been an increased usage of copper sulfate pentahydrate ($CuSO_4 \cdot 5H_2O$ or $CuH_{10}O_9S$), as a fining treatment for wines containing volatile sulfur compounds such as H_2S, or exhibiting other reductive characteristics. The maximum addition according to the International Organisation of Vine and Wine (OIV) International Code of Oenological Practices is 1 g/h/l [23]. In the years following the introduction of screw-cap closures, the number of wines receiving copper treatments increased substantially. Very occasionally, wine may come in contact with copper during vinification processes. Happily, the use of valves made of brass (an alloy of copper and zinc) is a largely a thing of the past, but brass ferrules and fittings are still sometimes encountered; I have noted these on several occasions when inspecting wineries, mostly on thermostats and sight tubes. It is also not unknown for winemakers to deliberately allow wine to contact copper or brass, in order to counter reductive tendencies.

Wines should be screened for copper prior to bottling. The maximum concentration recommended to avoid the risk of copper haze is 0.5 mg/l, and the maximum copper residue in wine (other than certain fortified wines) according to the OIV is 1 mg/l [23].

10.4.1.2 Prevention/Removal of Copper Haze

There are several methods available to prevent the formation of and to remove copper haze. The processes summarised here are permitted at the time of writing by the OIV [23], but may not be legal in all countries. 'Blue fining' using cyanide compounds, usually potassium ferrocyanide, can be added to the wine, which then forms an insoluble complex with copper ions that precipitate the copper [24]. Gum arabic (*Acacia verek*) is commonly used on fined and filtered wines prior to bottling to prevent problems with hazes due to copper. It is approved by OIV and EU (E414). The dose is 10–30 g/hl, to be determined by trial. Although gum arabic prevents haze formation, it does not eliminate the copper [24]. The maximum permitted concentration is 0.3 g/l [23]. Polyvinylimidazole–polyvinylpyrrolidone (PVI/PVP) copolymers may be used to reduce copper at a maximum amount of 500 mg/l. They must be removed by filtration no later than two days after the addition [23]. Chitosan or chitin–glucan of fungal origin may also be used to remove copper. The maximum dose for this purpose is 100 mg/l [23].

10.4.2 Iron Haze

10.4.2.1 Formation of Iron Haze

The presence of iron in wine may stem from several sources including soil, yeast supplements, bentonite, fining agents, filtration materials, and metal materials [25]. Iron haze (casse) is now a very rare fault. It is usually caused when wine comes in physical contact with iron. Historically, fermentation and storage tanks were often constructed of iron or mild steel, and it was essential that these were coated or lined. Enamel was a popular and effective coating but was prone to damage, exposing wine to the metal. Epoxy resin is a less easily damaged coating and was extensively used. Such lined tanks are still occasionally encountered, particularly in small wineries. At high concentrations, above 6 mg/l, the presence of iron can increase the risk of forming an iron casse. This is most likely to be visible in white wines in the form of a white or grey haze or sediment.

10.4.2.2 Prevention/Removal of Iron Haze

A screening should take place pre-bottling. There are several methods to correct excessive concentrations of iron. These include the addition of gum arabic, citric acid, tannin, or calcium phytate; fining with PVI/PVP; or the use of potassium ferrocyanide [23].

10.5 Final Reflections

Misty Alfonsi of Fenn Valley Vineyards, Lake Michigan, is unashamed. She noted that in a Sauvignon Blanc, a protein haze had appeared as a very light fog, usually in bottles that had been exposed to heat for a period of time. It had been decided not to fine the wine with bentonite, which would not only have removed the protein but also have stripped some of the aroma from the wine [25]. 'We opted not to inflict this damage on the wine', she says, 'we were more concerned about retaining the character of the wine than preventing the formation of a slight haze …' [25]. I am reminded that perceptions of quality, which in a white wine include clarity and brightness, are often built upon illusions.

References

1 Th. (2017). Winemaking Education, Common Wine Faults and Their Impacts. 7th Wine Regulatory Forum Ha Noi, Viet Nam 11–12 May 2017. http://mddb.apec.org/Documents/2017/SCSC/WRF/17_scsc_wrf_013.pdf (accessed 10 May 2020).

2 Marangon, M., Pocock, K.F., and Waters, E.J. (2012). The addition of bentonite at different stages of white winemaking: effect on protein stability. *Grapegrower & Winemaker* 580: 71–73.

3 He, S., Hider, R., Zhao, J., and Tian, B. (2020). Effect of bentonite fining on proteins and phenolic composition of chardonnay and Sauvignon Blanc wines. *South African Journal of Enology and Viticulture* 41 (1): 113–120. https://doi.org/10.21548/41-1-3814.

4 Marangon, M., Van Sluyter, S.C., Waters, E.J., and Menz, R.I. (2014). Structure of haze forming proteins in white wines: vitis vinifera thaumatin-like proteins. *PLoS One* 9 (12): e113757. https://doi.org/10.1371/journal.pone.0113757.

5 Cosme, F., Fernandes, C., Ribeiro, T. et al. (2020). White wine protein instability: mechanism, quality control and technological alternatives for wine stabilisation – an overview. *Beverages* 6: 19. https://www.mdpi.com/2306-5710/6/1/19/htm.

6 Van Sluyter, S.C., McRae, J.M., Falconer, R.J. et al. (2015). Wine protein haze: mechanisms of formation and advances in prevention. *Journal of Agricultural and Food Chemistry* 63 (16): 4020–4030. https://doi.org/10.1021/acs.jafc.5b00047.

7 Pocock, K.F., Salazar, F.N., and Waters, E.J. (2012). The effect of bentonite fining at different stages of white winemaking on protein stability. *Australian Journal of Grape and Wine Research* 17 (2): 280–284.

8 Robinson, E., Scrimgeour, N., and Marangon, M. (November/December2012). Beyond bentonite. *Wine & Viticulture Journal November/December*: 24–30.

9 Zoecklein, B. (2020). Wine proteins and protein stability. Monograph *Virginia Tech*. https://www.apps.fst.vt.edu/extension/enology/downloads/wm_issues/Wine%20Proteins%20and%20Protein%20Stability.pdf (accessed 30 June 2020).

10 Butzke, C. (2010). Fining with Bentonite. Monograph, *Purdue Extension*. https://www.extension.purdue.edu/extmedia/FS/FS-53-W.pdf (accessed 30 June 2020).

11 The Australian Wine Research Institute (2020). Measurement of heat stability of wine. Monograph. www.awri.com.au/industry_support/winemaking_resources/laboratory_methods/chemical/heat_stab (accessed 25 July 2020).

12 Esteruelas, M., Poinsaut, P., Sieczkowski, N. et al. (2009). Comparison of methods for estimating protein stability in white wines. *American Journal of Enology and Viticulture* 60 (3): 302–311.

13 Lambri, M., Colangelo, D., Dordoni, R. et al. (2016). Innovations in the use of bentonite in oenology: interactions with grape and wine proteins, colloids, polyphenols. In: *Grape and Wine Biotechnology* (ed. A. Morata). IntechOpen. https://doi.org/10.5772/64753. https://www.intechopen.com/books/grape-and-wine-biotechnology/innovations-in-the-use-of-bentonite-in-oenology-interactions-with-grape-and-wine-proteins-colloids-p.

14 Sutherland, W.M. (2014). Wyoming Bentonite Summary Report September 2014. *State of Wyoming*. Monograph. https://www.wsgs.wyo.gov/products/wsgs-2014-bentonite-summary.pdf (accessed 27 July 2020).

15 Ghanem, C., Taillandier, P., Rizk, M. et al. (2017). Analysis of the impact of fining agents types, oenological tannins and mannoproteins and their concentrations on the phenolic composition of red wine. *LWT – Food Science and Technology* 83: 101–109. https://doi.org/10.1016/j.lwt.2017.05.009.

16 Waters, E.J., Alexander, G., Muhlack, R. et al. (2005). Preventing protein haze in bottled white wine. *Australian Journal of Grape and Wine Research* 11: 215–225. https://doi.org/10.1111/j.1755-0238.2005.tb00289.x.

17 Majewski, P., Barbalet, A., and Waters, E. (2011). $1 Billion hidden cost of bentonite fining. *Australian & New Zealand Grapegrower & Winemaker* (569): 58–62.

18 Ratnayake, S., Stockdake, V., Grafton, S. et al. (2019). Carrageenans as heat stabilisers of white wine. *Australian Journal of Grape and Wine Research* 25: 439–450. https://doi.org/10.1111/ajgw.12411.

19 Pocock, K.F., Waters, E.J., Herderich, M.J., and Pretorius, I.S. (2008). Protein stability tests and their effectiveness in predicting protein stability during storage and transport. *Wine Industry Journal* 23 (2): 40–44. https://www.researchgate.net/publication/323108602_Protein_stability_tests_and_their_effectiveness_in_predicting_protein_stability_during_storage_and_transport_A_W_R_I_R_E_P_O_R_T.

20 de Sousa, M.R., de Pinho, M.N., and Catarino, S. (2018). Clarification and protein stabilization of white wines by ultrafiltration. Book of Abstracts: Euromembrane 2018, July 9, Valencia, Spain.

21 Vincenzi, S., Polesani, M., and Curioni, A. (2005). Removal of specific protein components by chitin enhances protein stability in a white wine. *American Journal of Enology and Viticulture* 56 (3): 246–254.

22 Clark, A.C., Wilkes, E.N., and Scollary, G.R. (2015). Chemistry of copper in white wine: a review. *Australian Journal of Grape and Wine Research* 21: 339–350. https://doi.org/10.1111/ajgw.12159.

23 International Organisation of Vine and Wine {OIV} (2019). *International Code of Oenological Practices*. Paris: OIV. http://www.oiv.int/public/medias/7213/oiv-international-code-of-oenological-practices-2020-en.pdf.

24 Claus, H. (2020). How to deal with uninvited guests in wine: copper and copper-containing oxidases. *Fermentation* 6 (1): 38. https://doi.org/10.3390/fermentation6010038.

25 https://www.fennvalley.com/2014/09/24/haze-in-sauvignon-blanc-2013 (accessed 10 July 2020).

11

Lactic Acid Bacteria-Related Faults

In this chapter, lactic acid bacteria (LAB) and the positive and negative roles they may play in wine production and quality will be discussed. Their role in the malolactic fermentation (MLF), which is considered essential in nearly all red wines and desirable in many white wines, is explained. On the negative side, LAB can cause several faults of appearance (including ropiness), off-aromas, and off-flavours (including acrolein, excessive diacetyl, and mousiness). Some of these constitute serious faults. It can also be involved in the formation of unwanted volatile sulfur compounds (see Chapter 6) and the production of high levels of volatile acidity (see Chapter 7) and can be responsible for an unwanted fermentation in packaged wine (see Chapter 9). Some LAB are also involved in the production of biogenic amines and ethyl carbamate, as discussed in Chapter 14.

11.1 Introduction to Lactic Acid Bacteria

LAB are present on grapes and leaves, in must, and in wine in varying amounts at the different stages of vinification. LAB are generally either cocci (round or oval shape) or bacilli (rod shape).

The four most important genera of LAB found in grape must and wine are the following:

- *Lactobacillus*: homofermentative, facultative heterofermentative, and heterofermentative bacilli – usually found singly or in chains;
- *Oenococcus*: heterofermentative cocci – usually found as long chains;
- *Pediococcus*: homofermentative cocci – often grouped in tetrads (clusters of four);
- *Leuconostoc*: heterofermentative cocci – occurring in pairs or chains [1].

Only some of the species of LAB are responsible for MLF, but all species can bring about most unwelcome alteration in aroma or flavour profile or impact upon texture. Even *Oenococcus oeni*, the species most commonly used for MLF, can in certain circumstances be responsible for faults. LAB-related changes can result in a serious loss in quality, and some LAB-associated faults render the wine totally undrinkable and spoiled. An image of *Oenococcus oeni*, magnified 5 000 times, is shown as Figure 11.1.

All species of LAB are chemotrophs, that is, they oxidise chemical compounds in order to obtain energy for their metabolism. Whilst LAB may oxidise many compounds, the oxidation of sugars is the principal pathway to produce their energy [2]. Most species

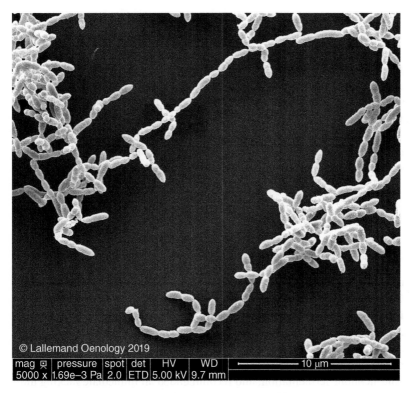

mag 囚 | pressure | spot | det | HV | WD | ——————— 10 μm ———————
5000 x | 1.69e–3 Pa | 2.0 | ETD | 5.00 kV | 9.7 mm

© Lallemand Oenology 2019

Figure 11.1 *Oenococcus oeni* 5000×. Source: Courtesy of Lallemand.

are heterofermentative or facultative heterofermentative, that is, they produce differing end products depending upon the growth substrate. There are some species that are homofermentative, and when a sugar of the hexose group, usually glucose, is the substrate (the source of carbon), this will be converted into (mostly) lactic acid. Pentoses are another group of sugars found in very small quantities in must and wine that can also be degraded by LAB. If malic acid is the substrate, some species of LAB will convert this into (mostly) lactic acid in the MLF process. However, certain species of LAB will, if the carbon substrate is a six-carbon sugar (glucose or fructose) or citric acid, produce acetic acid and adenosine triphosphate (ATP). This process may very well result in unacceptably high levels of volatile acidity, as discussed in Chapter 7.

11.2 Lactic Acid Bacteria and Their Natural Sources

LAB are usually present in relatively low quantities on sound grapes and also on leaves. On dehydrated or damaged fruit, the quantity may be considerably higher, especially if the grape skins are split and the bacteria can obtain nutrients from the grape pulp [3]. Fruit that has been mechanically harvested may also expose the bacteria to nutrients within the berries as the machines rip the grapes off the pedicels, so any delay in transport or processing may lead to significant increase in the quantities present. Colonies of LAB may also be present in the winery and on inadequately sanitised equipment. The population of LAB is usually low in must obtained from healthy grapes, and this will decline further during the alcoholic fermentation (AF) as the bacteria compete with *Saccharomyces cerevisiae*. Following the completion of the AF, and if not inhibited, populations grow rapidly, and some of the species then conduct the MLF, provided the wine conditions are suitable as detailed below. Although *O. oeni* is the predominant species for the MLF, it has rarely been isolated from grapes in the vineyard.

LAB require less oxygen to grow than is contained in the atmosphere, i.e. they are microaerophilic. This means that they can grow in wine even when care is taken to exclude extraneous sources of oxygen. The bacteria are able to grow in full, sealed tanks.

11.3 Malolactic Fermentation (MLF)

11.3.1 History of the Understanding MLF

The seeds of knowledge of what is now termed the MLF were sown in the early nineteenth century. In 1837, Freiherr von Babo described the phenomenon of a 'second fermentation' in wines, coinciding with the rise in temperatures during late spring [4, 5]. This second fermentation released carbon dioxide (CO_2) and was responsible for renewed cloudiness in the new wines. Although Antonie Van Leeuwenhoek had first observed bacteria in 1676, von Babo did not consider bacteria to be responsible, but related the activity to 'the melting of the grease of the alcoholic fermentation, and recommended an immediate racking into a new barrel with the addition of sulfur dioxide (SO_2), fining, temperature reduction, all followed by a second racking and stabilisation with another addition of SO_2' [4, 5]. Louis Pasteur, during his studies on wine spoilage, isolated the first bacteria from wine [6]. By 1891, Hermann Müller-Thurgau perhaps best known for his work in breeding the grape variety that bears his name, had theorised that the acid reduction that takes place could be due to bacterial activity [5]. In 1913, Müller-Thurgau and Osterwalder published their seminal investigation into LAB in wine and the bacterial degradation of malic acid to lactic acid and CO_2 [7]. However, in the early to mid-twentieth century, most winemakers still regarded the process as a mystery. Particularly in cooler regions, a natural MLF would usually not take place until the rise of temperatures in spring. The alcoholic fermentations would often not have been completed until late autumn or early winter by which time the ambient temperature of the chai or cellar would be too cold for the MLF to proceed. When the wine began to bubble again in spring with the warmer weather, winemakers drew parallels with the sap rising in the vines, and the wine coming to life again. It is only since the early 1960s that the bacteria involved have been understood in any detail as expounded by Émile Peynaud and Domercq [8] and since the late twentieth century that strains of MLF cultures by laboratories have enabled a large degree of control by winemakers.

11.3.2 The MLF Process

MLF is almost always desirable in the production of red wines and may, or may not, be undertaken for white, rosé, and sparkling wines. LAB of the species *O. oeni* are mainly responsible for this process, during which malic acid (and some sugars, particularly fructose) is converted into lactic acid. Malic acid has apple-like flavours (it is named after the genus *Malus*, most species of which are eating or cooking apples).

The simple equation using condensed formulae is

Lactic acid bacterla

$$C_4H_6O_5 \quad \rightarrow \quad C_3H_6O_3 + CO_2$$

Malic acid $\quad \rightarrow \quad$ Lactic acid + Carbon dioxide

The process is in fact a transformation of the dicarboxylic malic acid into the monocarboxylic acid lactic acid. This can be shown as

Lactic acid bacterla

$$HOOC.CH_2.CHOH.COOH \quad \rightarrow \quad CH_3.CHOH.COOH + CO_2$$

Malic acid $\quad \rightarrow \quad$ Lactic acid + Carbon dioxide

Malic acid contains two carboxylic acid (COOH) groups whilst lactic acid contains just one. Accordingly, a major effect of the MLF is reducing the acidity of the wine that undergoes this process, with a corresponding increase in pH. Wines that have not undergone an MLF usually have a fresh, crisp texture, sometimes even with a slight prickle on the tongue, characteristics that generally are undesirable in red wines. Conversely, wines that have had a complete MLF have a lower total acidity, softer texture, and roundness or smoothness, often with creamy notes, although this latter characteristic can also due to the undertaking of the practice of bâtonnage. However, the MLF process is more complex than a simple decarboxylation of malic acid to lactic acid. Numerous grape and alcoholic fermentation-derived wine components are consumed by LAB, metabolising these into new compounds. Distinctive desirable aromas and flavours may be metabolised by *O. oeni*, enhancing the nose and palate profile and elevating the complexity. The process of MLF increases the presence of many ethyl esters and reduces acetate esters. For wines that have barrel or other oak ageing, MLF helps lift the oak characteristics, due to *O. oeni* releasing oak lactones from the wood [9]. The process of MLF without doubt improves quality in white wines from certain grape varieties and regions – with white Burgundies and other Chardonnays produced almost the world over being prime examples. The completion of MLF also improves stability and reduces the risk of subsequent microbial growth, as a source of carbon, in the form of malic acid, is removed.

Nowadays, it is common to inoculate wines with the desired strains of bacteria, but even this is no guarantee of a successful initiation or completion of the MLF. There is also a growing trend amongst winemakers to co-inoculate must with homofermentative bacteria together with a desired strain of *S. cerevisiae* to initiate co-fermentation, i.e. the alcoholic fermentation and MLF taking place simultaneously. This may result in the presence of considerable quantities of lactic bacteria in the sugar-rich fermenting juice that can result in unwanted flavour modifications and increased production of acetic acid. However, it does permit the addition of SO_2 immediately after completion of the co-fermentation and may minimise the risk of problems with *Brettanomyces* – the period between the conclusions of the alcoholic fermentation on onset of the MLF is a particularly dangerous time for the growth of Brett (see Chapter 4).

A successful MLF is subject to the wine being within parameters with regard to five key aspects:

- Temperature;
- SO_2;
- pH;
- Alcohol;
- Nutrients.

However, it is not possible to state precise parameters for any of these, as a particularly favourable value for one may compensate of a somewhat unfavourable value of another.

11.3.2.1 Temperature

The temperature range for MLF lies between 18 and 25 °C (64–77 °F), with 20 °C (68 °F) perhaps being the optimum.

11.3.2.2 Sulfur Dioxide (SO₂)

LAB generally will not grow when the level of free SO_2 is above ± 10 mg/l. Again this figure is far from precise and is, inter alia, very much pH dependent: the level of molecular SO_2 is key. Differing species and strains have varying molecular SO_2 tolerances. Whilst the death of LAB often occurs between 25 and 45 mg/l of free SO_2, alarmingly, some strains of *Pediococcus damnosus* implicated in 'ropy' wine are not inhibited at a free SO_2 concentration as high as 50 mg/l.

11.3.2.3 pH

Although the growth of LAB is possible at a pH as low as 3, such growth will be very slow, and the onset of the MLF may not happen at all below a pH of 3.2. Growth becomes faster as the pH goes higher and may be particularly rapid when the pH is 3.8–3.9, in which figures represent the absolute upper limit for any winemaker would wish to have in their wine, even when the vineyards are in a hot climate. The MLF may take just 14 days in a wine with a pH of 3.83 but over five months when the pH is 3.15 [10]. Differing species and strains of LAB have individual pH levels below which they will not be active, which vary depending on the substrate converted. Strains of *Oenococcus* may transform malic acid if the pH is as low as 3.23 but only convert sugars at a pH above 3.51. Of course, sugars may also be converted into unwelcome acetic acid (see Chapter 7). However, lactobacilli may convert sugars if the pH is above 3.32 and malic acid at a pH above 3.38. Thus, the presence of lactobacilli may present a danger of increased volatile acidity and off-flavours, as discussed below.

11.3.2.4 Alcohol

Different genera and species of LAB have varying tolerances to high alcohol levels. *O. oeni* will cease to grow at levels above 13–14% abv, and conversely *Lactobacillus fructivorans*, *Lactobacillus brevis,* and *Lactobacillus hilgardii* may survive at levels up to 20% abv.

11.3.2.5 Nutrients

As with other bacteria, LAB need a source of carbon, nitrogen, and nutrients in order to grow. All of these are usually present in sufficient quantities in wine. As we have seen, the carbon may come from sugars or malic, citric, or other organic acids. Nutrients may have been depleted following the alcoholic fermentation. Assimilable nitrogen is supplied mostly by amino acids, and a deficiency in these may, on occasions, lead to problems in the initiation or completion of the MLF. The bacteria also need a supply of minerals, particularly magnesium (Mg), manganese (Mn), potassium (K), sodium (Na), and phosphorus (P). B group vitamins are generally not synthesised by LAB; of course, these are often added to musts to provide nutrients for *Saccharomyces.*

11.4   Undesirable Aromas, Off-Flavours, and Wine Spoilage Caused by Lactic Acid Bacteria

11.4.1   Off-Odours Associated with Lactic Acid Bacteria

Specific types of wine off-odours, fault, and spoilage are associated with certain species of LAB. The bacteria can partially or completely metabolise aroma precursors, aromatic compounds, and phenolic compounds, and the metabolised products alter the aroma and taste profiles. As most species of LAB will not really grow in a wine with a pH of 3.2 or less, the incidence of lactic acid-related faults is rare in cool climate white wines. Particular off-odours may result from the growth of bacteria such as *L. brevis*, *L. hilgardii*, *Lactobacillus plantarum*, *Leuconostoc mesenteroides*, *Pediococcus* spp., and *Streptococcus mucilaginosus*. These taints are summarised in Table 11.1 and detailed below.

It should be noted that some species of Lactic bacteria, including *L. hilgardii*, *L. fructivorans*, *L. plantarum,* and *O. oeni* can remain active at alcohol levels of up to 22% abv, and accordingly the related aroma and flavour modifications can affect fortified wines.

11.4.2   Acrolein

Acrolein has a chemical formula of CH_2CHCHO or C_3H_4O. The compound is also known as propenal. The fault is often referred to as 'amertume', which is a French word for bitter.

All wines contain glycerol (glycerine), usually in a concentration of 5–9 g/l, which has a positive impact upon taste. It is one of the many products formed by yeasts during the early stages of alcoholic fermentation. Glycerol can be degraded by bacteria, and the resultant reduction in concentration has a negative impact upon quality and produces off-aromas and off-flavours. Some species of LAB, particularly *Lactobacillus casei*, *L. fructivorans*, and *L. hilgardii* (and possibly *Pediococcus* spp.), convert glycerol into 3-hydroxypropionaldehyde (3-HPA) [11], also known as 3-hydroxypropanal or reuterin. This is the precursor to acrolein that, when combined with polyphenols, imparts a bitter taste. Acrolein bitterness, loss of glycerol, and causal rod-shaped bacteria were described by Pasteur in 1873 [6]. Due to the

Table 11.1   LAB-related off-odour and wine spoilage compounds.

Name	Compound	Sensory characteristics	LAB responsible
Amertume	Acrolein	Bitter	*Lactobacillus cellobiosus* *L. hilgardii* *Leuconostoc mesenteroides* *Pediococcus parvulus*
Diacetyl	Diacetyl	Butter Rancid butter Butterscotch	*Lactobacillus plantarum* *Oenococcus oeni* *Pediococcus* spp.
Mannitol	Mannitol	Bittersweet taste	*Lactobacillus brevis*
Ropiness graisse	β-D-Glucan (exopolysaccharide)	Oily, slimy viscous appearance	*Leuconostoc mesenteroides* *Pediococcus damnosus* *Pediococcus pentosaceus*
Mousiness	2-Acetyltetrahydropyridine (ATHP), 2-ethyltetrahydropyridine (ETHP)	Odours of mice on retro-olfaction	*Lactobacillus brevis* *Lactobacillus cellobiosus* *Lactobacillus hilgardii*
Tourne	Acetic and succinic acids	Dull, brown, turbid appearance; lactic smell	*Lactobacillus brevis* *L. plantarum*
Indole	Indole	Plastic smell	*Lactobacillus linderii* Various yeasts – see Chapter 14

role of polyphenols, it is mainly red wines, particularly those with high tannin levels that are affected, although the problem can be found in some whites, notably if they have been aged on lees. High pH wines are most likely to be affected. The degradation of glycerol by LAB in the presence of glucose can also result in an increase in volatile acidity.

Sensory characteristics: Acrolein is mostly sensed on the palate, especially the back palate, as a bitter taste. Although acrolein can have a piercing and acrid smell reminiscent of burnt cooking fat, this is not generally found in wine, due to the compound combining with polyphenols.

Sensory perception threshold: 10 mg/l of acrolein.

11.4.3   Excess Diacetyl

Diacetyl is also known as 2,3-butanedion, with a chemical formula of $C_4H_6O_2$ or $CH_3COCOCH_3$. It is a diketone whose presence at low levels (up to 4 mg/l) can add pleasant aromas and complexity to wine [12]. At higher levels, it gives notes of butter, nuts, and/or butterscotch, but at excess concentrations, it can dominate with overt and unwelcome rancid butter aromas and flavours. Diacetyl is metabolised from citric acid and some sugars. Although a small amount, perhaps 0.2–0.3 mg/l, can be produced by *S. cerevisiae* during the alcoholic fermentation, the vast majority will be metabolised by various species of LAB, mainly of the genera *Lactobacillus* and *Pediococcus*. Diacetyl can also be produced by *O. oeni*, the bacterium usually favoured by winemakers for the MLF. To minimise diacetyl production, it is recommended to use a neutral bacterial strain for the MLF, with a high rate of inoculation, and to let the wine have extended contact with yeast and bacteria lees [13].

Cysteine and methionine are sulfur containing amino acids, and diacetyl can react with these to produce volatile sulfur compounds (see Chapter 6).

Sensory characteristics: Aromas of butter, nuts and/or butterscotch. At high concentrations there are rancid butter aromas and flavours.

Sensory perception threshold: The threshold for diacetyl flavours depends on the wine type and matrix. Martineau et al. found that the flavour detection threshold was 0.2 mg/l for a lightly aromatic Chardonnay wine, 0.9 mg/l for a low tannic aromatic Pinot Noir wine, and 2.8 mg/l in a full-flavoured, full-bodied Cabernet Sauvignon wine [13]. Concentrations of diacetyl in excess of 5–7 mg/l may be regarded detrimental to wine quality as a result of off-flavours, and high levels may cause spoilage.

11.4.4 Mannitol

Mannitol is a polyol (sugar alcohol) carbohydrate that is found in most fruits and vegetables, with a chemical formula of $C_6H_{14}O_6$. It is formed by enzymatic reduction of fructose. *L. brevis* is the main genus of bacteria implicated in its production, but it can also be produced by *Lactobacillus kunkeei*, which is a strong inhibitor of yeasts [14], and *O. oeni*. As it is mostly metabolised from fructose (catalysed by mannitol dehydrogenase), it is sometimes found in wines with residual sugar, particularly those with low pH and which have undergone MLF. Mannitol production from fructose is accompanied by propanol, butanol, lactic acid, and high levels of acetic acid – wines with excessive mannitol usually exude excessively high levels of volatile acidity.

Sensory characteristics: A bittersweet taste (particularly when accompanied by high levels of volatile acidity) and can give an aggressive, caustic, corrosive, and viscous texture. On the nose, there are often vinegary and ester aromas, largely on account of the co-occurrence of high volatile acidity.

Sensory perception threshold: This has not been accurately determined in wine, varying considerably according to the matrix. However, in water, it is 7.3 g/l [15].

11.4.5 Ropiness

Known in France as 'graisse' (fat), ropiness gives wine a slimy, oily appearance. The name ropiness is given on account of the 'chains' of bacteria often present. Ropiness mainly occurs in white wines and is rare in red wines because the bacteria responsible do not grow so well in the presence of tannins. The fault is caused by strains of the bacteria

P. damnosus (usually the main species involved), *Pediococcus pentosaceus, O. oeni*, and *Le. mesenteroides* metabolising glucose and possibly dextrose (if it has been used for must enrichment) to produce the polysaccharides glucan and dextran [16]. It has been shown that so-called 'ropy' strains of *P. damnosus* can have a population of between 10^4 and 10^6 CFU/ml in bottled wines, even with a 'high' free SO_2 level of 50 mg/l [17]. The species has a high alcohol tolerance and is largely resistant to SO_2 [2]. *L. plantarum, Lactobacillus sanfrancisco*, and even the yeast *S. cerevisiae* are also able to metabolise glucose and dextrose into polysaccharides. Ropiness can be controlled by lowering the pH to 3.5 or less [18]. Problems with ropiness may occur in tank following fermentation but unfortunately are most likely to appear in bottle [2].

Sensory characteristics: A murky, slimy, oily appearance with high viscosity. On the palate, there is a most unpleasant oily or slimy texture.

11.4.6 Mousiness

Mousiness has already been discussed in Chapter 4, as it is often metabolised by *Brettanomyces*. However, the fault can also be caused by LAB, particularly *L. brevis, L. fermentum, L. hilgardii*, and *L. plantarum* and *Pediococcus* spp. Even *O. oeni*, usually favoured by winemakers for the MLF, can produce the fault that, as with other types of bacterial spoilage, is most likely in high pH wines. There are several nitrogen-heterocyclic imines in which its presence can result in the off-flavour. These include 2-acetyl-3,4,5,6-tetrahydropyridine (ATHP) (this is the most likely compound), 2-acetyl-1-pyrroline (APY), and 2-ethyl-3,4,5,6-tetrahydropyridine (ETHP). The simultaneous presence of at least two of these compounds is necessary for the off-flavour to be perceptible [19]. 2-Acetyl-1,4,5,6-tetrahydropyridine may also be involved. These compounds are derived from the α-amino acids lysine and ornithine. A high pH (above 3.6) and low SO_2 levels facilitate the growth of the bacteria involved in the production of these compounds. However, very few studies have analysed the origin of this fault, and little is known about its repercussion on wine quality due to the complexity of the process but also because it co-occurs with other faults [20].

The fault of mousiness can also be of chemical origin, when hydrogen peroxide has been used to remove SO_2 from must or wine [21]. It is also possible that 2-acetyl-3,4,5,6-tetrahydropyridine (ATHP) can be formed from methylglyoxal and proline via a Maillard reaction [22]. In practical terms, inadequate levels of SO_2, a high pH, and exposure to oxygen put a wine at risk of developing mousiness. There is no treatment.

Sensory characteristics: Mousiness does not really manifest itself on the nose of a wine (the compounds responsible are not volatile) but is apparent after the wine has been swallowed, and the saliva in the mouth diluted the wine acids and raised the pH. The flavours may be described as mouse urine, dirty mouse cage, popcorn, and crackers. These are particularly noticeable on retro-olfaction and on the finish of the wine [23]. Some tasters also mention characters of dried sausage skins, vomit, or dirty mops [24].

Sensory perception threshold: According to Snowden et al. [25], the sensory perception threshold for ATHP is 1.6 µg/l, 2-acetylpyrroline is 100 ng/l, and 2-ethyl-3,4,5,6-tetrahydropyridine is 150 µg/l. However, the Australian Wine Research Institute (AWRI) states that there is no defined sensory threshold [3] for mousy taint. About 30% of people are anosmic to mousiness [21].

11.4.7 Tourne

Tourne is a fault that was recognised and named by Pasteur [6]. It is caused by the degrada-tion of tartaric acid to mostly acetic and succinic acids by LAB. The main species responsible are *L. plantarum* and *L. brevis*. When tartaric acid degrades, there is a decrease in total acidity, as well as an increase in volatile acidity, and an increase in microbial growth that may cause cloudiness in the wine. The wine becomes insipid and weak, and red pigments become less intense.

Sensory characteristics: A dull, brown, turbid appearance and a 'lactic' smell. It is often flat and flabby on the palate. Succinic acid exhibits a strong sour and salty character, lingering on the palate [23].

Sensory perception threshold: There are no reliable figures for sensory perception thresh-old of tourne, but acetic acid will be unpleasant in wines at levels above 1.2 g/l. When succinic acid exceeds 0.5 g/l, it tastes sour and salty, rendering the wine unbalanced [23].

11.4.8 Indole

The chemical formula of indole is C_8H_7N. *Lactobacillus linderii* is the main bacterium associated with the production of indole, although it can be produced by many bacteria. Non-*Saccharomyces* yeasts can also produce high levels of indole during a sluggish fermentation. The formation is by the degradation of the amino acid tryptophan.

Sensory Characteristics: At very low concentrations, indole has a flowery aroma – it is a con-stituent of orange blossom. However, it occurs naturally in human faeces, and at medium concentrations, it has an intense faecal smell. It also gives a plastic-like aroma to wine.

Sensory perception threshold: The sensory perception threshold of indole ranges from 15 to 330 µg/l, depending upon the wine matrix.

This compound is discussed further in Chapter 14.

11.4.9 Geranium Taste

Sorbic acid is often used as a yeast inhibitor and to ensure stability of wines containing residual sugar. It can be reduced by LAB, particularly *O. oeni* into 2,4-hexadien-1-ol (sorby-lalcohol), which can spontaneously form 2-ethoxyhexa-3,5-diene.

Sensory characteristics: A pronounced nose of crushed geranium leaves.
Sensory perception threshold: 2-ethoxyhexa-3,5-diene has an olfactory threshold of 100 ng/l.

This topic, together with other off-odours related to the use of sorbic acid, is discussed in more detail in Chapter 14.

11.5 Prevention of Lactic Acid Bacteria-Related Faults

The following measures will prevent the occurrence of LAB-related faults:

- Maintaining good hygiene in the winery and maturation cellar, with all equipment thor-oughly cleaned and sanitised or sterilised. During the harvest period, picking bins should be thoroughly cleaned before each *trie* to limit the increase in populations of LAB (and other bacteria).

- The addition of lysozyme and/or sterile filtration will inhibit LAB. Lysozyme is an enzyme (protein) obtained from egg whites that destroys the membranes and thus kills Gram-positive bacteria, including those of the genera *Oenococci, Lactobacilli,* and *Pediococci.* Yeasts are unaffected by lysozyme, as are moulds and Gram-negative bacteria, including acetic acid bacteria. As lysozyme is a protein, phenolics decrease its activity. Accordingly, lysozyme is less active in red wines than white wines. The product is approved for use in the United States at a dose up to 500 mg/l and by the International Organisation of Vine and Wine (OIV) and European Union at the same dosage.
- If MLF is desired, fresh bacteria cultures from dependable sources should be added.
- A low pH, below 3.5, restricts the growth of *Lactobacillus* spp. and *Pediococcus* spp. Where legal, pH may be adjusted by the addition of tartaric acid.
- Vats and barrels should be completely full, or the wine blanketed with inert gas.
- Careful SO_2 management, both in must and post-MLF, will inhibit growth of LAB. The AWRI suggests that immediately post MLF, an addition of 40–50 mg/l of SO_2 is made [3]. However, the amount of the addition required will depend upon the pH of the wine and of course the level of any SO_2 already present. Care must be taken with regard to any sensory impact, and to ensure that, allowing for such additions as may be necessary before bottling, the wine remains within legal maxima. The AWRI noted that a level of 0.8 mg/l of molecular SO_2 will inhibit growth. This is also the level that will generally inhibit growth of *Brettanomyces*. Pre-bottling SO_2 should be adjusted as required: a post-bottling level of 0.6 mg/l is suggested, and accordingly the pre-bottling level depends upon bottling procedures.
- Wine should be stored below 18 °C (65.5 °F).
- Sterile filtration (0.45 μm) will capture LAB and may be considered for 'at-risk' wines. These include any wine to which sorbic acid is added.

11.6 Analysis

The detection of LAB may be made by the following:

- Light (optical) microscopy – this is somewhat unreliable. The morphology of most LAB is from 0.5 to10 μm.
- Plating – using selective or differential media.
- Fluorescent microscopy – this enables reliable immediate microscopic evaluation of samples. This technique uses fluorescent dyes to mark cells.
- Quantitative polymerase chain reaction (qPCR) also known as real-time PCR. This process uses molecular biology techniques that amplify and monitor the amplification of targeted ribosomal RNA and DNA.

The analysis of LAB-metabolised compounds can be undertaken by high-performance liquid chromatography (HPLC) or by gas chromatography–mass spectroscopy (GC-MS).

11.7 Final Reflections

During the last 150 years, since the groundbreaking work of Pasteur, the incidence of individual wine faults has been something of a roller coaster ride: increases have been followed by falls, only to be often followed by another increase. In 1945, William Vere Cruess, faculty member and prolific researcher at University of California, Berkeley, wrote with regard to LAB spoiled wines, 'At present, 1945, it is very difficult to find such wines, as the very general use of SO_2 and cooling have practically eliminated such spoilage' [26] – thanks to Lucy Joseph for pointing to this. In 2015, the AWRI noted that the incidence of mousy (and Brett) characters in white wines submitted to their helpdesk was on the rise [21]. At recent wine tastings, I have found several wines with excess diacetyl and just a few with the bittersweet taste of mannitol. However, as the rise in popularity of no-added sulfur 'natural' wine continues, with constant pressures on winemakers generally to reduce levels of sulfites, some bacterial faults that seemed to have been confined to history are perhaps making a most unwelcome reappearance.

References

1 Inês, A. and Falco, V. (2018). Lactic acid bacteria contribution to wine quality and safety. In: *Generation of Aromas and Flavours* (ed. A. Vilela). IntechOpen. https://doi.org/10.5772/intechopen.81168. https://www.intechopen.com/books/generation-of-aromas-and-flavours/lactic-acid-bacteria-contribution-to-wine-quality-and-safety.

2 Ribéreau-Gayon, P., Dubourdieu, D., Donèche, B., and Lonvaud, A. (2006). *Handbook of Enology: The Microbiology of Wine and Vinifications*, vol. 1, 140. Chichester: Wiley.

3 AWRI (2020). Avoiding spoilage caused by lactic acid bacteria. www.awri.com.au/wp-content/uploads/2011/06/Avoiding-spoilage-from-LAB.pdf (accessed 12 July 2020).

4 von Babo, F. (1837). Die Mängel und Krankheiten des Weines und deren verbesserung. In: *Kurye Belehrung über die zweckmäßige Behandlungsart der eingekellerten Weine*. Heidelberg: Heidelberg.

5 Krieger, S. (2005). The history of malolactic bacteria in wine. In: *Malolactic Fermentation in Wine* (eds. R. Morenzoni and K.S. Specht). Montréal: Lallemand.

6 Pasteur, L. (1873). *Etudes sur le vin: Ses maladies, causes qui les provoquent, procédés nouveaux pour les conserver et pour les vieillir*. Paris: Imprimerie Royale.

7 Müller-Thurgau, H. and Osterwalder, A. (1913). Die Bakterien im Wein und Obstwein und die dadurch verursachten Veränderungen. *Zentralblatt für Bakteriologie II* 36: 129–338.

8 Peynaud, E. and Domercq, S. (1961). Étude sur les bacteriés lactiques du vin. *Annales de Technologie Agricole* 10: 43–60.

9 Bartowsky, E.J., Costello, P.J., Abrahamse, C.E. et al. (2009). Wine bacteria – friends and foes. *AWRI Wine Industry Journal* 24: 2. www.winebiz.com.au.

10 Bousbouras, G.E. and Kunkee, R.E. (1971). Effect of pH on malolactic fermentation in wine. *American Journal of Enology and Viticulture* 22: 121–126.

11 Bauer, R., Cowan, D.A., and Crouch, A. (2010). Acrolein in wine: importance of 3-hydroxypropionaldehyde and derivatives in production and detection. *Journal of Agricultural and Food Chemistry* 58 (6): 3243–3250. https://doi.org/10.1021/jf9041112.

12 Bartowsky, E.J. (2009). Bacterial spoilage of wine and approaches to minimize it. *Letters in Applied Microbiology* 48 (2): 149–156. https://doi.org/10.1111/j.1472-765X.2008.02505.x.

13 Martineau, B., Acree, T.E., and Henick-Kling, T. (1995). Effect of wine type on the detection threshold for diacetyl. *Food Research International* 28 (2): 139–143. https://doi .org/10.1016/0963-9969(95)90797-E.

14 *Lactobacillus kunkeei*. UC Davis Viticulture and Enology https://wineserver.ucdavis.edu/ industry-info/enology/wine-microbiology/bacteria/lactobacillus-kunkeei (accessed 12 July 2020).

15 Waterhouse, A.L., Sacks, G.L., and Jeffery, D.W. (2016). *Understanding Wine Chemistry*. Chichester: Wiley.

16 Lonvaud-Funel, A. (1999). Lactic acid bacteria in the quality improvement and deprecia-tion of wine. *Antonie Van Leeuwenhoek* 76: 317–331.

17 Lonvaud-Funel, A. and Joyeux, A. (1988). *Une altération bactérienne des vins: la maladie des vins filants. Science des Aliments* 8: 33–34.

18 du Toit, M. and Pretorius, I.S. (2000). Microbial spoilage and preservation of wine: using weapons from nature's own arsenal – a review. *South African Journal of Enology and Viticulture* 21 (1): 74–97.

19 Costello, P., Lee, T.H., and Henschke, P.A. (2001). Ability of lactic acid bacteria to pro-duce N-heterocycles causing mousy off-flavour in win. *Australian Journal of Grape and Wine Research* 7: 160–167. https://doi.org/10.1111/j.1755-0238.2001.tb00205.x.

20 Costello, P.J. and Henschke, P.A. (2002). Mousy off-flavor of wine: precursors and biosynthesis of the causative N-heterocycles 2-ethyltetrahydropyridine, 2-acetyltetrahydropyridine, and 2-acetyl-1-pyrroline by *Lactobacillus hilgardii* DSM 20176. *Journal of Agricultural and Food Chemistry* 50: 7079–7087. https://doi.org/10 .1021/jf020341r.

21 AWRI (2015). Avoid mousy, off-flavours. *Grape Grower and Winemaker* 613: 50. www .awri.com.au/wp-content/uploads/2018/04/s1694.pdf.

22 Künzler, L. and Nikfardjam, M.P. (2013). Investigations into the formation of 2-acetylpyridine and the mousy off-flavor in wine. *Mitteilungen Klosterneuburg* 63 (4): 187–198. https://www.researchgate.net/publication/296742145_Investigations_ into_the_formation_of_2-acetylpyridine_and_the_mousy_off-flavor_in_wine/link/ 5bf2780ba6fdcc3a8de0f27b/download {in German}.

23 Palacios, A. (2005). Organoleptic defects caused by uncontrolled malolactic fermen-tation. In: *Malolactic Fermentation in Wine* (eds. R. Morenzoni and K.S. Specht). Lallemand: Montréal.

24 Tempère, S., Chatelet, B., de Revel, G. et al. (2019). Comparison between standardized sensory methods used to evaluate the mousy off-flavor in red wine. *OENO One* 53 (2) https://doi.org/10.20870/oeno-one.2019.53.2.2350.

25 Snowden, E.M., Bowyer, M.C., Grbin, P.R., and Bowyer, P.K. (2006). Mousy off-flavor: a review. *Journal of Agricultural and Food Chemistry* 54 (18): 6465–6474. https://doi.org/ 10.1021/jf0528613.

26 Cruess, W.V. (1946). Spoilage and other microorganisms of wine. *The Fruit Products Journal and American Food Manufacturer*: 229–236. April.

12

Smoke Taint and Other Airborne Contaminations

Making wines that are free from fault and flaw is usually within the control of the producer. The process begins with ripe grapes that are free from pest or disease and continues with care and diligence through all the stages of vinification and packaging. However, the issue of smoke taint in wine, which is increasing in some New World countries, stems from circumstances outside of the producer's control. In this chapter, I consider the causes that give-rise to smoke taint, the limited steps that may be taken to prevent the formation of off-odours and off-flavours, and how they may be treated. I also briefly consider other airborne contaminations, including eucalyptol taint.

12.1   Introduction

Smoke tainted grapes and the wines made from them is usually a consequence of wild bush-fires or controlled scrub-burning. It may also be a result of other aerial pollution. Whilst old world vineyards are not immune; it is in regions and countries of the New World including, but not limited to Australia, South Africa, Chile, and California that the problem is most likely. Impacted grapes will have modified composition and adverse sensory characteristics. On the nose, affected wines can have aromas of smoke, ash, smoked cheese, smoked fish, and disinfectant. On the palate, the characteristics include ash, burnt flavours, tastes of chorizo, and an 'ashtray' finish. The unpleasant aspects are particularly noticeable retro-nasally.

The issue of smoke taint in wines is relatively new. It has been an increasing problem in Australia in the last 20 years or so. According to the Australian Wine Research Institute (AWRI), smoke taint first became a big problem in March 2003 (over 1.3 million hectares of land burnt), and since then has been a regular occurrence [1]. In the Australian state of Victoria, where controlled burns are mandatory, some 40% of the wine grape crop was affected in the 2009 harvest. It is estimated that in that year, the financial impact amounted to some $368 million (ASD). Australian harvests have, to an extent, been impacted every year from 2013 to 2020, a particularly disastrous year. In 2020, bush fires destroyed some vineyards, and smoke taint affected several regions, particularly the Hunter Valley, the Mudgee in New South Wales, and the King Valley and Gippsland in Victoria. Bruce Tyrrell of the Hunter Valley producer Tyrrell's Wines, estimated that the smoke impacted about 75% of Hunter Valley's 2020 vintage premium wines, and that Sémillon, for which the

Wine Faults and Flaws: A Practical Guide, First Edition. Keith Grainger.
© 2021 John Wiley & Sons Ltd. Published 2021 by John Wiley & Sons Ltd.

region is particularly famous, Chardonnay and Shiraz, would not be bottled on account of smoke taint [2]. The department of Primary Industries in Victoria state that the problem is increasing, and by the year 2050 the risk of bushfires and consequent exposure of vines will be in the range of 15–70% annually [3]. In recent years, South Africa, particularly the Stellenbosch and Constantia regions, have been affected. In California, the fires of 2017 and 2018 had a severe impact in several areas. Fortunately, the disastrous fires of 2019 occurred after the harvest was in the wineries. In August 2018 in California's Lake County, the growers declared that that Constellation Brands, the second-largest wine-producing company in the world, had rejected all of the Sauvignon Blanc it had contracted to buy in the county, estimated to be the equivalent of approximately 100 000 cases of wine [4], on account of perceived smoke taint. In September, growers in Oregon's Rogue valley had grapes with a contracted value of about US$ 4 million rejected by another buyer [4].

12.2 Smoke Taint Compounds in the Atmosphere

Smoke taint compounds are relatively unstable in the atmosphere, varying in composition and quantity according to atmospheric conditions. They are affected by moisture, temperature, photooxidation, and wind direction [5]. Smoke taint compounds are usually worse in fresh smoke that is directly downwind of a fire. Smoke contains volatile organic compounds (VOCs) which are formed from lignin sources. Chemical transformation of organic compounds also generates additional particulate matter, secondary organic aerosol (SOA), which accounts for a substantial fraction of aerosol in the troposphere. However, a quantitative and predictive understanding of SOA formation does not exist [6].

There are two main ways the taint compounds get onto vines and into grapes:

- They are absorbed into grapes directly via the waxy cuticle. It is generally accepted that this is the major pathway to modify grapes;
- They are absorbed by the stomata and waxy cuticles of leaves and may translocate to the grapes [7] or remain in the leaves [5]. There is some dissent as to whether smoke taint compounds can be translocated to grapes via leaves.

As might be expected, the concentration of smoke-taint-related compounds will be highest on the grape skins.

The compounds that produce smoke taint can be reliably measured by using Tenax sorbent tube air sampling, followed by gas chromatographic analysis. This will determine the individual compounds present and their concentrations. The duration of exposure, and intensity of the smoke, impact upon the uptake of volatile phenols in grapes. Smoke-like aromas and flavours are pronounced in wines produced from grapes exposed to a high smoke density (20% obscuration) for a short duration, and a low smoke density (2.5% obscuration) for a long duration [7]. Although the following year's grapes will not be modified, the smoke exposure may reduce the yield in that year [8].

Grapes may also become tainted by pollution other than bushfires, i.e. general aerial pollution, which is the source of 'smoke taint' characteristics in some European wines. *m*-Cresol and *p*-cresol taints have been found in Swiss wines. The source has been attributed to pollution from factories close to the vineyards.

12.3 Critical Times in the Growing Season and Duration of Exposure for Smoke Taint to Impact

Up until flowering time, there is a very low risk of there being smoke uptake into the yet-to-be-formed grapes, although the total risk period commences from when the vine's shoots are 10 cm in length. From the time the berries begin to develop up until veraison, the risk is low to medium. The period of high risk commences around 7–10 days, following veraison and right up to the grapes being harvested [9, 10]. There is a risk of grapes becoming tainted following even a short exposure to smoke, but the risk becomes high if the period lasts for 30 minutes or more. If the smoke is low-density, several days may be required for absorption into grapes [5]. There is evidence that the taint will build up following several, individual periods of exposure.

12.4 The Volatile Phenols Responsible for Smoke Taint; Their Odours and Flavours and Sensory Detection Thresholds

The volatile phenols responsible for smoke taint characteristics in wines are:

- Guaiacol (2-methoxyphenol) – organoleptic characteristics: smoke, smoked cheese, smoked fish, smoky bacon crisps (chips), and ash. Guaiacol, as with most of the smoke taint compounds, is also a product of pyrolysis of oak wood consequential to the toasting of barrels. Therefore it can be present in oak-aged wines that have not been subjected to vineyard smoke. It is a metabolite of *Brettanomyces* (see Chapter 4) and can be found in so-called 'cork tainted' wines (see Chapter 3). Guaiacol is an important smoke taint marker. It was found above the odour detection threshold (ODT) in 8 out of 12 South African red wines that had received vineyard smoke exposure in a study by McKay et al. [11]. The aroma sensory threshold of guaiacol is 23 μg/l [12], and it is 27 μg/l on the palate. In their 2018 study, McKay et al. confirmed the orthonasal threshold to be 23 μg/l [13];
- Syringol (2,6-dimethoxyphenol) – organoleptic characteristics: charred, charcoal. Guaiacol and syringol are often the two most abundant volatile phenols in smoke [5]. However, syringol has a high sensory threshold and has limited impact in odours and palate, and there is some dispute as to its importance as a smoke taint marker;
- 4-Methyl-guaiacol (2-methoxy-4-methylphenol), – organoleptic characteristics: sweet, spicy, clove, leather-like, smoky. Sensory detection threshold: 65 μg/l. 4-Methyl-guaiacol is also a main smoke taint marker and is often present in the highest concentrations;
- 4-Ethylguaiacol – see Chapter 4;
- 4-Ethylphenol – see Chapter 4;
- 4-Methyl-syringol (2,6-dimethoxy-4-methylphenol) – organoleptic characteristics: charred, charcoal. As with syringol, the sensory threshold is high;
- Eugenol – organoleptic characteristics: spice, cloves. Sensory threshold: 130 μg/l;
- *m*-Cresol (*meta*-cresol also known as 3-methylphenol) – organoleptic characteristics: plastic, poo (faecal), Elastoplast, and ash. Sensory threshold: 20 μg/l [12];
- *o*-Cresol (*ortho*-cresol, also known as 2-methylphenol) – organoleptic characteristics: plastic, burnt, and medicinal. Sensory threshold: 62 μg/l [12, 14];

- *p*-Cresol (*para*-cresol, also known as 4-methylphenol) – organoleptic characteristics: smoky, faecal, stables, and tar. Sensory threshold: 64 µg/l [12]. Cresols are derived from lignin and may be present in toasted oak barrels and oak alternatives.
- Furfural – organoleptic characteristics: caramel toffee. Sensory threshold: 60 µg/l.

12.5 Smoke Taint in Wines

The presence of volatile phenols in wine as listed above may result from direct pathways from the must produced from affected grapes. When smoke taint compounds are absorbed in the grapes, a small proportion of the volatile phenols remain in their free (odorous) form. However, a large proportion becomes immediately bound to sugars to form glycosides, i.e. the vine will bind or glycosylate each of these volatile phenols to form phenolic glycosides, which are non-volatile [15]. During the winemaking process, these bound forms are released from the sugar molecule [15]. However, phenolic glycosides act as precursors for volatile phenols in wines, which can be released as a result of hydrolysis catalysed by enzymes or acids. Such release can take place before and during fermentation, also during maturation and/or storage.

12.6 Other Sources of Guaiacol and 4-Methyl-guaiacol in Wines

Guaiacol and methylguaiacol may also be present in wines as a result of oak ageing. The consequential concentration of guaiacol and methylguaiacol is in the region 10–100 µg/l and 1–20 µg/l, respectively. Guaiacol concentrations of approximately 20 µg/l may be regarded as giving positive sensory attributes. Guaiacol can also be extracted into wine from cork closures [16]. *o*-Cresol and *m*-cresol may also be found in oak-aged wines. However, phenolic glycosides do not generally appear in wine as a result of oak ageing, and their presence is regarded as a reliable indicator of grapes impacted by smoke.

12.7 Laboratory Testing

Analysis for the appropriate range of phenolic glycosides may be undertaken by high performance liquid chromatography-mass spectrometry (HPLC-MS) methods, which give precise quantification of the compounds present. As alternatives, gas chromatography-mass spectrometry (GC-MS) and MS/MS (triple quadrupole) methods accurately determine free and glycoside-bound guaiacol and 4-methyl guaiacol, the two primary markers of smoke taint in wines.

12.8 Prevention of Development of Smoke-Related Volatile Phenols from Affected Grapes

Preventing grapes from being tainted in the vineyard is usually beyond the control of the grower. If the crop has been heavily impacted, serious consideration is given, whether to harvest at all. As we have seen, the vine transforms some of the smoke taint compounds into bound non-volatile phenols, i.e. in glycoside forms. In bound form, the compounds may not be apparent upon sensory analysis of grapes or must. If laboratory grape analysis shows high levels of glycoconjugated phenols, techniques to remove these will not be effective in the long term [17].

However, some several operations and techniques may limit the consequential impact – many of these are cumulative.

- If grapes have been exposed to smoke, they should be harvested by hand. The integrity of fruit should be maintained by picking into small 12–16 kg lug boxes;
- All materials other than grapes (MOGS), particularly leaves, should be excluded from the crusher. This is relatively easy to achieve by manual bunch sorting on selection tables in the vineyard and/or at the winery. Berries may be manually selected on a belt or vibrating table, or by optical or mechanised selection process such as Tribaie;
- Grapes should be crushed as cool as possible and the must kept at a temperature below 10 °C (50 °F) before adjusting and raising the temperature for the onset of fermentation;
- When making white wines, whole-cluster pressing is preferable, and skin contact pre-fermentation should be avoided. A study by Kelly et al. found that (Sauvignon Blanc) wines, made with fruit crushed before pressing extracted 39% of total glycoconjugated phenols, which was approximately double the extraction (±18% of total glycoconjugated phenol) of wines made from whole-cluster pressed (Chardonnay) [18];
- Cold soaking should be avoided;
- Purified enzymes may be used for settling white must, but no enzyme should be used for red must as they will extract the volatile phenols;
- It is often recommended that for red wines, the total maceration period should be reduced and post-fermentation maceration avoided. This advice is based upon the fact that guaiacol and guaiacol glycoconjugates accumulate largely in grape skins. However, the full extraction of smoke derived compounds is usually completed within two to three days, usually before full extraction of anthocyanins, and certainly before the full extraction of tannins [19]. In 2015 Renata Ristic et al. attributed more apparent smoke-related aromas and flavours in red wines, due to the increased duration of skin contact undertaken with red winemaking, which increased extraction of smoke-derived volatile phenols and their glycoconjugates [20];
- The use of high vitality and high aroma yeast is recommended;
- Fermentation temperatures should be kept at the lower end of the required scale;
- Pressing of the pomace should be at low pressure. The pressing of each of the fractions should be kept separate. and these should be analysed for volatile phenols. If high levels are detected in any fraction, it should be discarded. There is less extraction of phenolic contaminants in the first 400 l/tonne [21].
- If malolactic fermentation (MLF) is required, this should be initiated by inoculation;
- Post-alcoholic fermentation, if there are signs of taint, an application of a specific inactivated yeast could help to reduce some of the smoky characters.

Pre-processing washing of grapes would appear to be largely ineffective at removing the taint compounds [1]. In any case, water retained on grape-skins will dilute juice concentration and reduce the potential alcohol, and acidity.

12.9 Treatments

There are treatments available for smoke tainted wines, but none of these provides a complete solution. Non-volatile glycosides may remain in the wine, and after some time, the smoke taint may return as a result of acid or enzyme catalysed hydrolysis.

12.9.1 Activated Carbon

Activated carbon (aka activated charcoal) will remove much of the smoke taint, but it is a nonselective treatment. It will simultaneously remove many desirable aromas. Synthetic mineral fining may be used as an alternative to activated carbon.

12.9.2 Other Fining Treatments

The use of other fining agents can lead to small improvements by reducing the ash-like aromas of smoke taint, but they have considerable drawbacks by reducing desirable aromas. A study by Fudge et al. found that the majority of the commercial fining agents evaluated as potential remedial treatments for smoke-affected wine showed poor selectivity towards the volatile phenols associated with smoke taint [22].

12.9.3 Reverse Osmosis (RO) and Solid Phase Adsorption

Reverse osmosis (RO) and solid phase adsorption techniques can be used separately or combined. They can initially reduce associated volatile phenols by up to 33%, perhaps taking these below the sensory thresholds. Using RO and a polystyrene-based adsorbent resin, guaiacol, 4-methylguaiacol, 4-ethylguaiacol, and 4-ethylphenol may be successfully removed [23]. However, desirable aromas may also be removed, and alarmingly, the taint may eventually return to the wine as the remaining phenols continue to develop. The challenge hindering permanent removal of the compounds is that bound glycosylated phenols may be hydrolysed into guaiacol and 4-methylguaiacol during the winemaking process. RO does not remove bound phenols. With the ageing of the wine, hydrolysis will convert these glycoconjugates back to their state of odorous compounds [24, 25].

12.9.4 The Addition of Oak Products to Affected Wines

The perceived complexity of an affected wine may be enhanced by the additions of tannins and oak chips [25]. These may reduce the perception of smoke-related characteristics. However, as oak products contain guaiacol, trials are done to ensure that such additions do not make matters worse.

12.9.5 Blending

Blending smoke-tainted wine with the un-impacted product can give a wine with odours below the sensory detection thresholds, at least in the short-term.

12.9.6   Limiting Time in Bottle Pre-sale

Whilst not a treatment, if tainted wine is bottled, it should be sold quickly into a market where it will be consumed early. Although much of the taint was removed by previously discussed methods, it may well reappear after some months in bottle. The producer may be advised to use an alternative or 'cleanskin' label.

12.10   Other Airborne Contaminations

12.10.1   Spray Drift

Spray drift of pesticides and herbicides that are not appropriate for viticulture can be a consequence of careless (and particularly aerial) spraying of neighbouring crops. In Australia, in 2005, the wine was found to be contaminated as a result of spray drift. The herbicide used contained 2,4-dichlorophenoxyacetic acid.

12.10.2   Eucalyptol – 1,8-Cineole

Eucalypt aroma and taste characteristics may be found in wines from some districts in Australia, Chile, and California. The compound responsible is 1,8-cineole, commonly known as eucalyptol. Affected wines will illustrate aromas of eucalyptus or mint, together with tones of camphor. It is red wines, particularly those produced from Cabernet Sauvignon, that are most likely to exhibit the characteristics. The sensory threshold is from 2.2 to 3.3 µg/l depending on the grape variety. However, a lower threshold of 1.1 µg/l has been reported, studying a Merlot wine from California and 1.3 µg/l in a Tannat wine [26]. Eucalypt character, like *Brettanomyces*, remains a controversial topic, with many lauding the extra dimension in the aroma and palate matrix of affected wines. But, as with the products of *Brettanomyces*, individual consumers will have differing levels of taint that they regard as unacceptable. Wines with strong eucalyptus odours may contain more than 20 µg/l of eucalyptol. A level of 27.5 µg/l is perhaps a consumer rejection threshold [27]. Generally, it is only red wines that may contain significant amounts of 1,8-cineole [28].

Laboratory experiments conducted by California's ETS Laboratories first demonstrated that eucalyptol could move from eucalyptus leaves to grape berries, without any direct contact [29]. Eucalyptol may also be transferred through vine leaves and petioles (MOGS), and from eucalyptus leaves and nuts from trees that may end up in harvested grapes. The taint may be found in wines from vines grown near eucalypt trees [30]. There will be airborne transmission of 1,8-cineole, which is a major component of the oils secreted by eucalyptus trees, and this adsorbs onto the grape skin, leaves, and stalks. The highest level is usually found in the vine leaves and the second highest in the stems. Eucalyptus trees shed bark regularly, and this, as well a whole and part leaves, can become lodged in the vine canopy, making a major contribution to the taint.

There are reports that 1,8-cineole can arise solely from terpene precursors within some grapes, and wines with a eucalyptus character can be made from grapes grown in vineyards far away from eucalyptus tree cultivations [26].

12.10.2.1 Prevention

There are a few preventative steps that can be taken in winemaking to help reduce the impact of 1,8-cineole:

- Vines situated very close to blue gum trees are particularly susceptible. Depending on the level of contamination, fruit from a number of the rows nearest to the boundary of the trees should be harvested and processed separately, and possibly just used in low quantities in blends;
- MOGS should be excluded at crushing and pressing time. This is easier if the fruit is hand-harvested, but sorting using belt, vibrating and optical sorting tables can all be effective. It is most important to exclude impacted vine leaves from cold soaks and fermentations;
- 1,8-Cineole is extracted from skins during maceration and fermentation. Accordingly, skin contact should be reduced as far as practicable.

12.11 Final Reflections

Smoke taint can be devastating for grape growers and wine producers. However, the impact of bushfires upon residents, communities, properties, land, and the environment can be catastrophic. In a typical 'fire season' in Australia's New South Wales, some 300 000 ha will be burned. In the summer of 2019/2020, some 5.4 million hectares burned in the State and a total of 12.6 million hectares in Australia [31]. The fires were responsible for the release of 434 million tonnes of carbon dioxide, and an estimated 1 billion animals were killed [31].

References

1 Høj, P., Pretorius, I., and Blair, R.J. (eds.) (2003). The Australian Wine Research Institute Annual Report. Urrbrae, South Australia: The Australian Wine Research Institute. www.awri.com.au/wp-content/uploads/2003_AWRI_Annual_Report.pdf (accessed 27 July 2020).

2 Gough Henly, S. (2020). After the fires, Australian wineries assess the damage. *Wine Spectator*, 3 March. https://www.winespectator.com/articles/after-the-fires-australian-wineries-assess-the-damage? (accessed 12 March 2020).

3 Whiting, J. and Krstic, M. (2007). *Understanding the Sensitivity to Timing and Management Options to Mitigate the Negative Impacts of Bush Fire Smoke on Grape and Wine Quality – Scoping Study*. Knoxfield, Victoria: The Department of Primary Industries.

4 Mobley, E. (2018). Wineries, vineyards clash over how to handle grapes affected by wildfire smoke. *San Francisco Chronicle* 5 October. https://www.sfchronicle.com/wine/article/Wineries-vineyards-clash-over-how-to-handle-13285829.php (accessed 14 March 2020).

5 Agriculture Victoria (2016). Smoke taint in wine Information sheet 2 of a series of 5. www.hin.com.au/__data/assets/pdf_file/0009/164628/Smoke-taint-risk-and-managment-in-vineyards.pdf (accessed 20 June 2020).

6 Krstic, M.P., Johnson, D.L., and Herderich, M.J. (2015). Review of smoke taint in wine: smoke-derived volatile phenols and their glycosidic metabolites in grapes and vines as biomarkers for smoke exposure and their role in the sensory perception of smoke taint. *Australian Journal of Grape and Wine Research* 21: 537–553. https://doi.org/10.1111/ajgw.12183.

7 Brodison, K. (2013). Effect of smoke in grape and wine production. Bulletin 4847. Perth, Western Australia. Department of Agriculture and Food Western Australia. http://citeseerx.ist.psu.edu/viewdoc/download; jsessionid=7DCE37912829907FD8A627BFB05F4822?doi=10.1.1.392.6887&rep=rep1& type=pdf (accessed 10 July 2020).

8 O'Kennedy, K. (undated). Managing smoke taint in wines. Monograph. Institute of Grape and Wine Sciences, Stellenbosch University. https://www.sasev.org/wp-content/uploads/2018/08/85_managing-smoke-taint-in-wine.pdf (accessed 29 June 2020).

9 Kennison, K.R., Wilkinson, K.L., Pollnitz, A.P. et al. (2009). Effect of timing and duration of grapevine exposure to smoke on the composition and sensory properties of wine. *Australian Journal of Grape and Wine Research* 15: 228–237. https://doi.org/10.1111/j .1755-0238.2009.00056.x.

10 Kennison, K.R., Wilkinson, K.L., Pollnitz, A.P. et al. (2011). Effect of smoke application to field-grown Merlot grapevines at key phenological growth stages on wine sensory and chemical properties. *Australian Journal of Grape and Wine Research* 17: S5–S12. https://doi.org/10.1111/j.1755-0238.2011.00137.x.

11 McKay, M., Bauer, F.F., Panzeri, V.V. et al. (2019). Profiling potentially smoke tainted red wines: volatile phenols and aroma attributes. *South African Journal of Enology and Viticulture* 40 (2): 2221–2236. https://doi.org/10.21548/40-2-3270.

12 Parker, M., Osidacz, P., Baldock, G.A. et al. (2012). Contribution of several volatile phenols and their glycoconjugates to smoke-related sensory properties of red wine. *Journal of Agricultural and Food Chemistry* 60 (10): 26229–22637. https://doi.org/10.1021/jf2040548.

13 McKay, M., Bauer, F.F., Panzeri, V., and Buica, A. (2018). Testing the sensitivity of potential panelists for wine taint compounds using a simplified sensory strategy. *Food* 7 (11) https://doi.org/10.3390/foods7110176.

14 Parker, B.M., Baldock, G., Hayaska, Y. et al. (2013). Seeing through smoke. *Wine and Viticulture Journal* 28: 42–46.

15 Agriculture Victoria (2016). Smoke taint in wine Information sheet 1 of a series of 5. www.hin.com.au/__data/assets/pdf_file/0008/164627/Smoke-taint-in-wine.pdf (accessed 20 June 2020).

16 Simpson, R.F., Amon, J.M., and Daw, A.J. (1986). Off-flavour in wine caused by guaiacol. *Food Technology in Australia* 38: 31–33.

17 Wilkinson, K.L., Ristic, R., Pinchbeck, K.A. et al. (2011). Comparison of methods for the analysis of smoke related phenols and their conjugates in grapes and wine. *Australian Journal of Grape and Wine Research* 17: S22–S28. https://doi.org/10.1111/j.1755-0238 .2011.00147.x.

18 Kelly, D., Zerihun, A., Hayasaka, Y., and Gibberd, M. (2014). Winemaking practice affects the extraction of smoke-borne phenols from grapes into wines. *Australian Journal of Grape and Wine Research* 20 (3): 386–393. https://doi.org/10.1111/ajgw.12089.

19 Hervé, E., Price, S., and Burns, G. (2019). Relationship between smoke markers (free guaiacols) in grapes and red wines: observations from recent vintages in California. *Œno2019 11th International Symposium of Enology/IVAS 2019 11th edition of In Vino Analytica Scientia*. Monograph available at: http://ets-public-assets.s3-us-west-1.amazonaws.com/website-article-pdfs/Smoke+Red+Ferments+Poster+IVAS+2019+EH052919.pdf (accessed 31 July 2020).

20 Ristic, R., Boss, P.K., and Wilkinson, K.L. (2015). Influence of fruit maturity at harvest on the intensity of smoke taint in wine. *Molecules* 20 (5): 8913–8927. https://doi.org/10.3390/molecules20058913.

21 Simos, C. and Krstic, M. (2020). I can smell smoke — now what? *Grapegrower & Winemaker* 672: 28–31. www.awri.com.au/wp-content/uploads/2020/01/s2131.pdf.

22 Fudge, A.L., Schiettecatte, M., Ristic, R. et al. (2012). Amelioration of smoke taint in wine by treatment with commercial fining agents. *Australian Journal of Grape and Wine Research* 18: 302–307. https://doi.org/10.1111/j.1755-0238.2012.00200.x.

23 Fudge, A.L., Ristic, R., Wollan, D., and Wilkinson, K.L. (2011). Amelioration of smoke taint in wine by reverse osmosis and solid phase adsorption. *Australian Journal of Grape and Wine Research* 17 (2): S41–S48. https://doi.org/10.1111/j.1755-0238.2011.00148.x.

24 Singh, D.P., Chong, H.H., Pitt, K.M. et al. (2011). Guaiacol and 4-methylguaiacol accumulate in wines made from smoke-affected fruit because of hydrolysis of their conjugates. *Australian Journal of Grape and Wine Research* 17: S13–S21. https://doi.org/10.1111/j.1755-0238.2011.00128.x.

25 Ristic, R., Osidacz, P., Pinchbeck, K. et al. (2011). The effect of winemaking techniques on the intensity of smoke taint in wine. *Australian Journal of Grape and Wine Research* 17: S29–S40. https://doi.org/10.1111/j.1755-0238.2011.00146.x.

26 Hervé, E., Price, S., and Burns, G. (2003). Eucalyptol in wines showing a Eucalyptus" aroma. Poster Paper. *Bordeaux, Oenologiques 2003" VIIme Symposium International d'Oenologie* (19–21 June 2003).

27 Saliba, A.J., Bullock, J., and Hardie, W.J. (2009). Consumer rejection threshold for 1,8-cineole (eucalyptol) in Australian red wine. *Food Quality and Preference* 20: 500–504. https://doi.org/10.1016/j.foodqual.2009.04.009.

28 Capone, D.L., Van Leeuwen, K., Taylor, D.K. et al. (2011). Evolution and occurrence of 1,8-cineole in Australian wine. *Journal of Agricultural and Food Chemistry* 59 (3): 953–959. https://doi.org/10.1021/jf1038212.

29 ETS Laboratories (2015). Controlling eucalyptus flavor in wine. https://www.etslabs.com/library/14 (accessed 20 June 2020).

30 Farina, L., Boido, E., Carrau, F., and Dellacassa, E. (2005). Terpene compounds as possible precursors of 1,8-cineole in red grapes and wines. *Journal of Agricultural and Food Chemistry* 53: 1633–1636. https://doi.org/10.1021/jf040332d.

31 Werner, J. and Lyons, S. (2020). The size of Australia's bushfire crisis captured in five big numbers. ABC Science, 4 March 2020. www.abc.net.au/news/science/2020-03-05/bushfire-crisis-five-big-numbers/12007716 (accessed 2 July 2020).

13

Ladybeetle and Brown Marmorated Stink Bug Taints

In this chapter, I briefly examine the problems that certain species of ladybeetle and the brown marmorated stink bug (BMSB) may cause in wine. If harvested with grape clusters and present in the grape crush, even in small numbers, each of these may impart off-odours and off-flavours to wine. Although thankfully rare at present, it would appear that incidences of the problems are increasing.

13.1   Introduction

There are, of course, many pests that may affect the grapevine and its fruit. Ground pests such as nematodes may have a serious impact on the health of the vine, and some root-feeding pests including *Phylloxera vastatrix* (*Daktulosphaira vitifoliae*) and *Margarodes vitis* can kill *Vitis vinifera* vines. Pests are also the vectors of some vine diseases. Of particular concern to vine growers in the United States and South America is Pierce's disease, caused by the bacterium *Xylella fastidiosa*, which is spread by xylem-feeding leafhoppers known as sharpshooters, particularly the glassy-winged sharpshooter (*Homalodisca vitripennis*). Various pests may impact upon the quantity and quality of fruit: as well as being disease vectors, they cause damage and can bring unwanted bacteria and the precursors of compounds that contribute to wine faults. Flies, particularly the spotted winged drosophila (*Drosophila suzukii*) and common fruit fly (*Drosophila melanogaster*), can be responsible for the onset of sour rot in grapes and elevated levels of volatile acidity in wine.

However, there are three vineyard pests of which the mere presence in grape clusters, irrespective of damaged fruit, can be responsible for off-odours and off-flavours in wines. These are the seven spot ladybeetle (*Coccinella septempunctata*), the multicoloured Asian ladybeetle (MALB) (*Harmonia axyridis*), and the BMSB (*Halyomorpha halys*). The taint imparted to wine by ladybeetles is caused by certain methoxypyrazines and that of the BMSB certain aldehydes, particularly *trans*-2-decenal (T2D), also known as (*E*)-2-decenal.

13.2   Methoxypyrazines

Methoxypyrazines are compounds that are key to the aroma and taste characteristics of green peppers, asparagus, and peas having herbaceous and grassy organoleptic profiles.

Wine Faults and Flaws: A Practical Guide, First Edition. Keith Grainger.

They are found in some grape varieties and impart green, grassy, and catty aromas and flavours, which may also include overt bell pepper, and gooseberry notes. If present at high levels, they can give such unripe tones that wines can be aggressive and distinctly unpleasant. Of red varieties, Cabernet Sauvignon, Cabernet Franc, and Carmenère can often show some of the associated grassiness. These notes may add a pleasant foil to the predominant cassis aromas and give a little edge and lift to a wine. Wines from cooler climates are most likely to show these, slightly green, characteristics. Of course, if the fruit is barely ripe, perhaps on account of early picking, the grassy character can predominate, but this issue has waned in the last couple of decades. Of course, the variety that really exudes overt methoxypyrazine characteristics is Sauvignon Blanc, and some wine regions, particularly Marlborough in the north of New Zealand's South Island, have built a reputation upon honing associated aromas and flavours to a fresh and exciting perfection.

3-Isobutyl-2-methoxypyrazine (IBMP) is usually the most dominant of the pyrazine compounds in Sauvignon Blanc and other pyrazine containing grapes and, as such, is assertive on both the nose and palate. The aroma threshold is 1–2 ng/l. There are other methoxypyrazines that may also be present in wine. These include the following:

- *2-Isopropyl-3-methoxypyrazine (IPMP)*: This is also found in some grape varieties and can have an aroma of peas.
- *2-Methoxy-3,5-dimethylpyrazine (MDMP)*: This gives rise to the so-called fungal must taint. The sensory characteristics include odours of mushroom, mould, earth, potato, green hazelnut, and dust and the odour of an engineering workshop. MDMP is an extremely potent compound. The perception threshold in white wine is noted as 2 ng/l.
- *2-sec-Butyl-3-methoxypyrazine*: This may be present in tiny amounts. The aroma is of peas and ivy-leaves.

IPMP has been identified as the primary cause of ladybeetle taint [1] although MDMP is also implicated [2].

13.3 Ladybeetles (Also Known as Ladybirds and Ladybugs)

13.3.1 Background

The seven spot ladybird (*C. septempunctata*) is the most common ladybeetle in Europe and may infest grape clusters on the vine, particularly within seven days of the harvest. The insects may be present on harvested bunches, and unless removed during grape sorting, pre-crushing can be responsible for imparting IPMP and other methoxypyrazines to must and wine. The methoxypyrazines are present in the haemolymph (the blood equivalent) of the creature. There is anecdotal evidence of wines from some regions in some vintages (e.g. Burgundy 2004 and 2011) being badly affected by related taints.

The MALB (*H. axyridis*) is now common in the United States, Canada, France, the United Kingdom, and South America having been introduced as a means of biocontrol tools for aphids. The first widely reported linked ladybeetle taint was in North America in 2001, when some winemakers reported 'burnt peanut butter' odours similar to 'crushed ladybugs' [3]. Large numbers of MALB had been noted in the vineyards that year [3]. In eastern

Figure 13.1 *Harmonia axyridis.* Source: Courtesy of Gyorgy Csoka, Hungary Forest Research Institute, Bugwood.org

Canada and northern United States, ladybeetle taint has been responsible for the spoilage of significant quantities of wine [4].

H. axyridis is shown in Figure 13.1, and *C. septempunctata* is shown as Figure 13.2.

In a study by Susanne Kögel et al., IPMP was detected in the haemolymph of all specimens of both *C. septempunctata* and *H. axyridis*. The frequency of IBMP in *C. septempunctata* was greater (60%) than in *H. axyridis* [5].

13.3.2 Infested Grapes

To increase their chances of survival during the winter, some ladybirds, especially *H. axyridis*, feed on grapes in late autumn [5]. They often infest the bunches close to harvest time, when the grapes are sugar rich. The number of bugs feeding on a cluster varies year by year, but it is not unknown for 20 or more to be present. Of course, ladybugs are voracious consumers of aphids, and heavy infestations would seem to coincide with a large presence of aphids in vegetation.

13.3.3 Sensory Characteristics, Detection, and Consumer Rejection Thresholds

13.3.3.1 Sensory Characteristics

The sensory characteristics of ladybug taint may include one or more of the following:

- Peanut.
- Asparagus (in varieties where this is not a desired aroma or flavour).

Figure 13.2 *Coccinella septempunctata*. Source: Courtesy of Susan Ellis, Bugwood.org

- Bell pepper (in varieties where this is not a desired aroma or flavour).
- Herbaceous odours (in varieties where this is not a desired aroma or flavour).
- Earthiness.

Tainted wines will also have a reduction in desired varietal characteristics.

13.3.3.2 Sensory Detection Thresholds and Consumer Rejection Thresholds

Detailing accurate odour detection thresholds (ODTs) and taste detection thresholds (TDTs) of wines affected by related taints has proved challenging, with considerable variation in stated figures. A study by Pickering et al. [6], using ascending forced choice method of limits, concluded 'best estimation thresholds' for IPMP range from 0.32 to 2.29 ng/l, depending upon the wine style and evaluation mode. Although other methoxypyrazines play supporting roles, there is no doubt that IPMP is the most important contributor to ladybug taint [7, 8].

13.3.4 Prevention

The spraying of insecticide upon the grapes on the vines so close to harvest should not be undertaken. Pickers may be instructed to eliminate as many ladybugs as possible, but there is no doubt that rigorous sorting by hand or with an optical sorter prior to crushing/pressing is the most effective method to exclude ladybeetles from the must.

13.3.5 Treatments

The green aromas and flavours in both red and white wines may be masked by oak ageing. The 2006 study by Pickering et al. [4] found that the use of oak chips reduced the levels of IPMP taint in both red and white wines. Of course, such treatment is not appropriate for many wine styles and qualities. IPMP remains relatively stable during bottle ageing. The use of activated charcoal can be effective in reducing methoxypyrazine concentrations in white wines [4], but colour, desirable aromas, and flavours are diminished. Research has been taking place as to the possible use of food-grade polyethylene-, polypropylene-, and silicon-based polymers, which has shown it may be possible to remove up to 40% of the taint compounds, with minimal sensory changes [9]. Any commercial use of this technology perhaps lies in the future.

13.4 Brown Marmorated Stink Bug (*Halyomorpha halys*) Taint

13.4.1 Background

The BMSB *Ha. halys* is a mottled brown, shield-shaped bug, between 12 and 17 mm long. The adjective 'marmorated' stems from the Latin for marbled, which is the tell-tale appearance of the creature's back. It is a pest to many species of plant, including corn, soya beans, hazel, and pecan nuts. Crop losses can be up to 90%.

A native of eastern Asia, particularly China, Korea, and Japan, the BMSB first appeared in the United States in Pennsylvania, where it was identified in 1998. Since then, it has spread rapidly, initially through the north-east, and by 2010, it was being described as a large-scale problem in Mid-Atlantic vineyards [10]. By September 2017, it been identified in some 44 states, including (and perhaps most alarmingly) in California, Washington State, and the Willamette Valley in Oregon. It is also present in four Canadian provinces. It was first reported in Europe in 2007 and is now widespread in Italy and Hungary.

Its spread over distance is usually by 'hitchhiking' on commodities or plant material. Shipments of electrical components and bricks to Australia from Italy have been found to contain the bugs. Strident efforts have been made to keep the pest out of New Zealand [11]. Shipments of cars from Japan have been found to contain the bug, and in early 2018, three ships holding 10 000–12 000 cars were sent back to sea by New Zealand port authorities. New Zealand Biosecurity Minister Damien O'Connor noted in October 2019, 'In the past

5549916

Figure 13.3 *Halyomorpha halys*. Source: Courtesy of Kristie Graham, USDA ARS, www.Bugwood.org.

few months we've increased the number of additional BMSB risk countries from 16 to 33 and we've upped our mandatory pre-arrival treatment requirements for targeted vehicles, machinery and parts' [12].

An image of *Ha. halys* is shown as Figure 13.3.

In countries outside of Asia, the bug has no natural predators. The level of crop damage already being incurred and potential future damage as the bug hitch-hikes to new territories has led to considerable scientific research on BMSB, with an increasing number of papers published each year since 2012, and over 40 in 2017, 26 in 2018, and another 26 in 2019 [13].

All of the five stages of the life cycle of BMSB have been observed in vineyards, and the bug is able to live on a diet comprising only grapes. BMSB may cause physical damage to grapes and vines, including necrosis and berry shrivelling, lowering both yield and quality. It is estimated that the loss in yield can be as high as 37% [14]. The loss to other crops can be as high as 90%. The New Zealand Institute of Economic Research has estimated that if the pest becomes established there, the economic loss in horticultural exports could amount to $4.2 billion a year [15].

The BMSB taint imparted into wine can have serious organoleptic impact. It is red wines that are particularly affected. There can be no doubt that incidences of BMSB taint will increase during the next decade. Together with atypical ageing, which may affect white wines (see Chapter 8) and ladybeetle taint, discussed above, BMSB taint is one of the most important 'new' faults to affect wine.

There are several other species of stink bug including the consperse stink bug, red shouldered stink bug, and rough stink bug. The BMSB can be distinguished from these by the presence of two white bands on the antennae [16].

13.4.2 Infested Grapes

Wines made from grapes infested by BMSB contain the following aldehydes:

- (E)-2-Decenal, also known as *trans*-2-decenal (formula $C_{10}H_{18}O$). This compounds gives the characteristic BMSB taint.
- Tridecane (formula $C_{13}H_{28}$).
- *trans*-2-Undecen-1-ol (formula $C_{11}H_{22}O$).
- Dodecane (formula $C_{12}H_{26}$).

These compounds are particularly secreted when bugs, present in grape clusters, become stressed during winemaking processes. The bugs often migrate to the centre of the cluster close to harvest [10] and may not be apparent, even in the sorting of whole bunches. The inclusion of three live bugs per cluster is sufficient to raise the trans-2-decenal level in the resultant wine to above consumer rejection threshold (CRT) [17]. However, back in 2011, Dr. Joseph A. Fiola, specialist in viticulture at Western Maryland Research and Education Center in Keedysville, Maryland, added controlled numbers of MBSBs to 25 lb lug boxes of grapes. He noted that he could detect a crushed cilantro (coriander) off-aroma in the juice when as few as 5–10 bugs were added per lug box, although in this case the resultant wines did not show the off-aromas [18]. Pallavi Mohekar et al. found that adding three BMSB per grape cluster gave levels of 614 µg/l of tridecane and 2 µg/l of (E)-2-decenal [19].

13.4.3 Sensory Characteristics, Detection, and Consumer Rejection Thresholds

13.4.3.1 Sensory Characteristics
(E)-2-Decenal is the most odorous of the BMSB-related compounds found in must and wine, although tridecane may be the most prevalent. These two compounds account for 70% of the BMSB stress-related compounds in red wine. (E)-2-Decenal has very strong 'green' aromas, tainting red wines and reducing quality. Impacted wines have coriander (cilantro), herb-like, green, and musty odours and flavours. There is generally a loss of black fruit, red fruit, and floral characteristics in the wine. Tridecane has a faint fuel-oil odour, *trans*-2-undecen-1-ol is floral or rose-like, and dodecane has a faint but objectionable paraffin odour.

13.4.3.2 Sensory Detection Thresholds and Consumer Rejection Thresholds (CRT)
The sensory detection threshold for (E)-2-decenal is 0.5 (µg/l) [19]. In red wine, the CRT is 4.8 µg/l [19, 20], and CRT is 13.0 (µg/l). Another study by Pallavi Mohekar, examining the detection and rejection thresholds of (E)-2-decanal in a Pinot Noir wine, found that 28% of all participants had CRTs below 0.05 µg/l, 65% of participants did not reject (E)-2-decanal-containing wine until the concentration exceeded 12 µg/l, and 7% of the participants had CRT above 30 µg/l T2D [21].

13.4.4 Prevention

The pressing stage in particular is a critical point for the release of the compounds. Tridecane has a faint fuel-oil odour but is not implicated in off-aromas and flavours, although it may have a (presently unknown) effect on wine quality. As the pressing stage for white wines takes place prior to fermentation, the stress compounds released are blown off by carbon dioxide, so it would seem that there is a low risk of BMSB taint affecting white wines.

13.4.5 Treatments

Removal to any great extent of BMSB taint from affected red wines is pretty much impossible.

- All methods of fining, including bentonite, gelatine, casein, and activated charcoal, are ineffective at removing BMSB taint. Such fining agents bind with non-volatile compounds and proteins, and it is suggested that their ineffectiveness in removing (*E*)-2-decenal is due to their having weak binding ability to the compound [17].
- Reverse osmosis has limited effect, removing perhaps 10% of the taint.
- Barrel ageing and other oak treatments, particularly the addition of oak chips, help mask the taint.
- Taints will be reduced after a period of bottle ageing, and extended bottle ageing may be a natural and practical solution to remove BMSB taint from wines. The ageing process involves many compound modifying reactions, including hydrolysis, degradation, condensation, and reduction. In reductive conditions, aldehydes, including tridecane and (*E*)-2-decenal, are converted into alcohols.

13.5 Final Reflections

The history of invasive species causing damage to crops and flora is full of good intentions, carelessness, and outright recklessness. The MALB was imported to the United States from Russia, Japan, and Korea as part of a United Stated Department of Agriculture (USDA) biocontrol strategy [22]. The insects eat tree aphids, aphids that cause damage to soy crops, and coccids. The species became established in the 1990s and was certainly successful for its primary purpose. The insects do feed on some fruits, but with the exception of raspberries, it is mostly fruit that has already been subjected to damage. From 1995, the insect was also introduced as biocontrol in Europe and is now established in 13 European countries. It is very possible that its role in wine taint is, of now, embryonic.

The BMSB hitch-hiked its way into the Western countries and is now a major pest to many annual and perennial crops in the territories it has invaded [23]. The physical damage the bug inflicts on fruits can be enormous. It is a pest of peaches, pears, apricots, plums, apples, tomatoes, etc. The estimated economic damage to US crops alone exceeds US$ 21 billion [24]. Again, it is very possible that its role in wine taint is, of now, embryonic.

Whilst sipping a glass of Mad Dog Shiraz from Australia's Barossa Valley, I am reminded of the textbook example of the damage caused by the well-intentioned introduction of an alien species. In 1935, 100 cane toads were introduced in Queensland, Australia, with the intention of eradicating the cane beetles that were destroying crops of sugarcane. The toads did eat large numbers of cane beetles. However, disastrously cane toads have glands on the back of their neck that emit bufotoxin. Their tadpoles are also highly toxic. Bufotoxin is poisonous to both dogs and cats and many native animal species. The initial cane toad population of 100 has grown to an estimated 2 billion [25].

References

1 Kőgel, S., Botezatu, A.I., Hoffmann, C., and Pickering, G.J. (2015). Methoxypyrazine composition of Coccinellidae-tainted Riesling and Pinot noir wine from Germany. *Journal of the Science of Food and Agriculture* 95 (3): 509–514. https://doi.org/10.1002/jsfa .6760.

2 Botezatu, A.I. (2013). Methoxypyrazines and ladybug taint in wines. PhD thesis. Brock University.

3 Pickering, G.J., Lin, J.Y., Riesen, R. et al. (2004). Influence of *Harmonia axyridis* on the sensory properties of white and red wine. *American Journal of Enology and Viticulture* 55 (2): 153–159. https://www.researchgate.net/publication/275963541_Influence_of_ Harmonia_axyridis_on_the_Sensory_Properties_of_White_and_Red_Wine.

4 Pickering, G.J., Lin, J.Y., Reynolds, A. et al. (2006). The evaluation of remedial treatments for wine affected by *Harmonia axyridis*. *International Journal of Food Science and Technology* 41 (1): 77–86. https://doi.org/10.1111/j.1365-2621.2005.01039.x.

5 Kögel, S., Gross, J., Hoffmann, C., and Ulrich, D. (2012). Diversities and frequencies of methoxypyrazines in hemolymph of *Harmonia axyridis* and *Coccinella septempunctata* and their influence on the taste of wine. *European Food Research and Technology* 234: 399–404. https://doi.org/10.1007/s00217-011-1646-y. https://www.researchgate .net/publication/230881734_Diversity_and_frequencies_of_methoxypyrazines_in_ hemolymph_of_Harmonia_axyridis_and_Coccinella_septempunctata_and_their_ influence_on_the_taste_of_wine.

6 Pickering, G.J., Karthik, A., Inglis, D. et al. (2007). Determination of ortho- and retronasal detection thresholds for 2-isopropyl-3-methoxypyrazine in wine. *Journal of Food Science* 72 (7): 468–472. https://doi.org/10.1111/j.1750-3841.2007.00439.x.

7 Botezatu, A.I., Kotseridis, Y., Inglisa, D., and Pickering, G.J. (2013). Occurrence and contribution of alkyl methoxypyrazines in wine tainted by *Harmonia axyridis* and *Coccinella septempunctata*. *Journal of the Science of Food and Agriculture* 93: 803–810. https://doi.org/10.1002/jsfa.5800.

8 Pickering, G.J., Lin, Y., Reynolds, A. et al. (2005). The influence of *Harmonia axyridis* on wine composition and aging. *Journal of Food Science* 70 (2): S128–S135. https://doi .org/10.1111/j.1365-2621.2005.tb07117.x.

9 Pickering, G.J., Inglis, D., Botezatu, A. et al. (2014). New approaches to removing alkyl-methoxypyrazines from grape juice and wine. *Scientific Bulletin, Series F: Biotechnologies* XVIII: 130–134. http://biotechnologyjournal.usamv.ro/pdf/2014/Art22.pdf.

10 Timmer, J. (2017). Will the Brown Marmorated Stink Bug be a Problem in Wine and Juice? *Wines & Grapes U*. March 2017. Penn State Extension. https://psuwineandgrapes

.wordpress.com/2017/03/10/will-the-brown-marmorated-stink-bug-be-a-problem-in-wine-and-juice (accessed 10 June 2020).

11 Haye, T., Gariepy, T., Hoelmer, K. et al. (2015). Range expansion of the invasive brown marmorated stinkbug, *Halyomorpha halys*: an increasing threat to field, fruit and vegetable crops worldwide. *Journal of Pest Science* 88 (4): 665–673. https://doi.org/10.1007/s10340-015-0670-2.

12 O'Connor, D. (2019). Government on high alert for stink bugs. http://Beehive.govt.nz. The official website of the New Zealand Government. https://www.beehive.govt.nz/release/government-high-alert-stink-bugs (accessed 28 June 2020).

13 StopBMSB.org (2020). *Management of Brown Marmorated Stink Bug in US Specialty Crops*. Scientific publications http://www.stopbmsb.org/about-us/scientific-publications (accessed 3 March 2020).

14 Smith, J.R., Hesler, S.P., and Loeb, G.M. (2014). Potential impact of *Halyomorpha halys* (*Hemiptera*: *Pentatomidae*) on grape production in the Finger Lakes Region of New York. *Journal of Entomological Science* 49 (3): 290–303. https://doi.org/10.18474/0749-8004-49.3.290.

15 New Zealand Institute for Economic Research (2017). Quantifying the economic impacts of a Brown Marmorated Stink Bug incursion in New Zealand. www.hortnz.co.nz/assets/UploadsNew/Quantifying-the-economic-impacts-of-a-Brown-Marmorated-Stink-Bug-Incursion.pdf (accessed 5th February 2020).

16 Agriculture and Natural Resources, University of California (2015). Pest Alert! Brown Marmorated Stink Bug *UC IPM* May. http://ipm.ucanr.edu/pestalert/pabrownmarmorated.html (accessed 24 July 2020).

17 Mohekar, P., Osborne, J., and Tomasino, E. (2018). Effects of fining agents, reverse osmosis and wine age on Brown Marmorated Stink Bug (*Halyomorpha halys*) taint in wine. *Beverages* 4: 17. http://www.mdpi.com/2306-5710/4/1/17.

18 Fiola, J.A. (2011). Brown Marmorated Stink Bug (BMSB) Part 3-Fruit Damage and Juice/Wine Taint. *Timely Viticulture*. University of Maryland extension. https://extension.umd.edu/sites/extension.umd.edu/files/_docs/programs/viticulture/TimelyVitBMSB3R.pdf (accessed 15 August 2020).

19 Mohekar, P., Osborne, J., Wiman, N.G. et al. (2017). Influence of winemaking processing steps on the amounts of (*E*)-2-decenal and tridecane as off-odorants caused by brown marmorated stink bug (*Halyomorpha halys*). *Journal of Agricultural and Food Chemistry* 65 (4): 872–878. https://doi.org/10.1021/acs.jafc.6b04268.

20 Mohekar, P., Lim, J., Lapis, T., and Tomasino, E. (2014). Consumer rejection thresholds of *trans*-2-decenal in Pinot noir: linking wine quality to threshold segmentation. *Proceedings of the Institute of Food Technology Annual Meeting*, New Orleans, LA, USA (22–24 June 2014).

21 Mohekar, P., Lapis, T.J., and Wiman, N.G. (2017). Brown marmorated stink bug taint in Pinot noir: detection and consumer rejection thresholds of *trans*-2-decenal. *American Journal of Enology and Viticulture* 68 (1): 120–126. https://doi.org/10.5344/ajev.2016.15096.

22 Boggs, J. and Jones, S.C. (2014). Multicoloured Asian lady beetle – agriculture and natural resources fact sheet. Monograph. Ohio State University Extension. https://

ohiograpeweb.cfaes.ohio-state.edu/sites/grapeweb/files/imce/pdf_factsheets/Lady%20Beetle.pdf (accessed 30 June 2020).

23 Lee, D.H. (2015). Current status of research progress on the biology and management of *Halyomorpha halys.* (*Hemiptera: Pentatomidae*) as an invasive species. *Applied Entomology and Zoology* 50: 277–290. https://doi.org/10.1007/s13355-015-0350-y.

24 Lupi, D., Dioli, P., and Limonta, L. (2017). First evidence of *Halyomorpha halys* (Stål) (*Hemiptera Heteroptera, Pentatomidae*) feeding on rice (*Oryza sativa* L.). *Journal of Entomological Research* 49 (1) https://doi.org/10.4081/jear.2017.6679.

25 RSPCA Queensland (2020). Cane toads: an introduced menace. www.rspcaqld.org.au/blog/wildlife-conservation/cane-toads-an-introduced-menace (accessed 6 July 2020).

14

Sundry Faults, Contaminants, Including Undesirable Compounds from a Health Perspective and Flaws Due to Poor Balance

In this chapter, I cover in brief some sundry, flaws, taints, and contaminants that may affect wine. The presence of undesirable compounds that have a possible impact on human health is discussed. These include biogenic amines, ethyl carbamate, ochratoxin A (OTA), dibutyl phthalate, and other phthalates. With the possible exception of biogenic amines, these compounds do not generally have an organoleptic impact. However, other faults discussed can have serious sensory implications, and in the worst cases, render affected wines undrinkable. These include styrene, indole, geosmin, and heat damage. Finally, I briefly consider the topic of balance in wines and how unbalanced wines, although not faulty, may be seriously flawed.

14.1 Mycotoxins, Particularly Ochratoxin A

Mycotoxins may be present on harvested grapes, being present in the vineyard on vines and grapes suffering from fungal diseases. Whilst they have not been identified as producing off-odours and flavours, they pose risks to human health.

14.1.1 The Presence of Ochratoxin A

OTA is the main mycotoxin found in wine [1]. OTA can cause irreversible damage to the kidneys and liver and is a known carcinogen in some animals and a possible human carcinogen by the International Agency for Research on Cancer (IARC) (Group 2B) [2]. In the European Union (EU), wine is estimated to be the second greatest dietary source of OTA (cereals being the first). Other foods and drinks that may include the toxin include beer, coffee, cocoa, nuts, and spices. The maximum limit of the OTA allowed according to EU regulations for wine (including sparkling wine), excluding liqueur wine and wine with an alcoholic strength of not less than 15% alcohol by volume (abv) is 2 μg/l [3]. China has recently introduced a similar maximum (National Health and Family Planning Commission GB 2761-2017) [4]. At the time of writing, no other country has a regulatory limit. Some 1.2% of wine samples analysed on a global basis have been shown to exceed this limit [1], but a high percentage of wines show detectable levels (over 50% in some countries) [5].

Wine Faults and Flaws: A Practical Guide, First Edition. Keith Grainger.
© 2021 John Wiley & Sons Ltd. Published 2021 by John Wiley & Sons Ltd.

A recent study found 89% of the USA wines examined had detectable levels of OTA, and the average level (of all wines examined) was 1.2 μg/l, and two wines were over the EU legal limit [6]. One historic study found detectable levels in 100% of wines in Turkey [7]. The presence in wine of OTA is more likely in reds, and particularly sweet whites, than in dry white or rosé wines.

14.1.2 The Origins of Ochratoxin A in Wines

OTA originates in the vineyard. It may be present in the skin and flesh of grapes that have been affected by species of the fungi *Aspergillus* and/or *Penicillium* [8]. *Aspergillus carbonarius* has been identified as the main species responsible for OTA contamination in grapes [9], but *Aspergillus niger* and *Penicillium verrucosum* are also implicated. *Aspergillus* is frequently found in warmer and wetter regions, and *Penicillium* in colder and drier regions. The fungi are only able to penetrate grapes due to skin damage, which may be caused by insects, including larvae of the ubiquitous European Grapevine Moth (*Lobesia botrana*), and birds. Pathogenic fungi including powdery mildew (*Uncinula necator*) and *Botrytis cinerea* also damage grapes and allow penetration by *Aspergillus* and/or *Penicillium* species. Some products commonly used as fungicides in vineyards have been found to promote fungal growth and/or OTA production. These include copper [10] and carbendazim [11].

OTA is minimised in wine by eliminating mouldy and diseased grapes during the picking and sorting processes. The reduction in OTA levels by the rejection of mouldy grapes can be as much as 98% [12]. The risk of potential contamination of wine may be reduced by limiting skin contact of affected grapes and by must clarification. Red wines often have extended periods of pre-fermentation skin maceration, sometimes at elevated temperatures and in aerobic conditions. During this period, grape must may pick up OTA from the grapes [13].

The concentration of OTA changes during winemaking processes [1, 14]. During fermentation, the levels will decrease considerably in the generally anaerobic conditions that inhibit fungal growth [15]. OTA is reduced during alcoholic fermentation by adsorption to solids, such as grape skins and yeast cell walls (yeast hulls) [16]. Red wine stored in tanks followed by draining show a significant decrease of OTA of about 55%. The malolactic fermentation (MLF) will also reduce OTA levels [14].

14.1.3 Prevention of, or Minimising, Contamination with OTA

The following actions will help minimise OTA production:

- establish and implement a hazard analysis critical control point (HACCP) system in vineyard and winery with OTA designated as a hazard;
- steps should be taken to avoid grape contamination by toxigenic fungal species;
- fruit should be carefully harvested and transported to minimise damage;
- cool storage of grapes before processing;
- harvested fruit should be subjected to rigorous sorting, including rejection of mouldy and diseased grapes. Optical sorters such as Tribaie, as illustrated in Figure 14.1 are an efficient means of excluding mouldy, rotten, and damaged fruit.

Figure 14.1 Tribaie sorting grapes. Source: Courtesy of Amos Industrie.

If the harvest has a high risk of contamination:

- extended macerations should be avoided;
- low volume, low pressure, and fast pressings should be used;
- must filtration should be undertaken;
- yeast hulls should be added to must at a recommended dose of 20–40 g/hl;
- wines should be stored in an inert tank and carefully racked off sediments;
- treatment with chitosan or chitin–glucan should be undertaken as discussed below.

The International Organization of Vine and Wine (OIV) has published a Code Of Sound Vitivinicultural Practices In Order To Minimise Levels Of Ochratoxin A In Vine-Based Products [17].

14.1.4 Analysis

OTA may be effectively determined in wines by immunoaffinity columns (IAC) clean-up, followed by high-performance liquid chromatography coupled with fluorescent detection (HPLC–FD). Unfortunately, the method is somewhat expensive.

14.1.5 Treatment of Affected Wines

There are several methods available for reducing the levels of OTA in affected wines. These include fining with egg albumen, gelatine, bentonite, activated charcoal, or chitosan, or membrane filtration.

Ertan Anli found that OTA levels in red wine are most effectively reduced by fining with isinglass or egg albumen, which removed 21.3% and 19.7% of OTA, respectively [14]. In the case of white wines, bentonite was the most effective fining agent removing 20.3% of OTA when utilised at a rate of 50 g/hl [14]. Fining of white wine with casein did not play a significant role in the removal of OTA, removing only 2% of OTA [14]. Fining with gelatine may contribute to the removal of up to 58% of OTA from red wine [13]. The International Code of Oenological Practices suggests the treatment using chitosan and chitin–glucan to reduce OTA in wine, the maximum doses being 500 g/hl [18]. Fining with charcoal can also reduce level, but this will strip both colour and desirable aromas. Filtration, particularly using a small pore or sterile membrane, is effective. Gambutti et al. reported an 80% decrease in OTA in white wine after filtration through 0.45 µm membranes [19].

14.2 Dibutyl Phthalate and Other Phthalates

Phthalates are believed to be hormone disruptors in humans. They are widely used as plasticisers but are not chemically bound to the materials they are used in. Wines may become contaminated with dibutyl phthalate (DBP) or other phthalates during production. Research conducted by Pascal Chatonnet et al. revealed that 59% of the (French) wines analysed contained significant quantities of DBP and only 17% did not contain any detectable quantity of at least one of the reprotoxic phthalates [20]. Alarmingly, the research showed that over 11% of the wines analysed did not comply with European Union (EU)-specific migration limits (SML) for materials in contact with food. The EU SML for DBP is 0.3 mg/g [20].

Polymer materials often contain high quantities of phthalates which may be released over time into wine by contact. Chatonnet et al.'s study analysed a variety of materials frequently present in wineries and found that vats coated with epoxy resin were the major source of contamination and the authors recommended that the use of these should be terminated [20].

14.3 Ethyl Carbamate

Ethyl carbamate (EC) is a known carcinogen in animals. It is classified as 'a probable human carcinogen' (Group 2A) by the IARC [2], and several countries have set limits for its concentration in wine. It is present in all alcoholic beverages, usually in concentrations of 12–15 ppm [21], with the highest concentration usually found in distilled spirits.

14.3.1 Production of Ethyl Carbamate in Wines

The two major precursors for EC production are urea and citrulline [22]. Urea is released by yeasts during alcoholic fermentation and reacts with alcohol to produce ethyl carbamate. There are cultured yeast strains available that have been modified to limit the production of the compound. The amino acid arginine is quantitatively one of the major amino acids in wine. Certain bacteria can degrade arginine into citrulline, which is also a source of EC production. These include commercially available cultured strains of lactic acid bacteria (LAB). In a study published in 2000, all the 27 commercial LAB strains tested excreted citrulline from the degradation of arginine. Some of them were able to reutilise previously excreted citrulline [23].

14.3.2 Prevention of Ethyl Carbamate in Wines

The most effective means of reducing EC concentration in wine is to reduce the level of precursors. In the vineyard, excessive fertilisation with urea, ammonia, and other nitrogen-based fertilisers should be avoided. Excessive levels of nitrogen, in particular arginine, in musts can lead to EC production, and this is one of the many reasons why additions of diammonium phosphate (DAP) and other yeast nutrients need to be undertaken with caution. The levels in must of arginine, which is metabolised by yeast to produce urea, should be kept below 1 g/l. The use of a yeast strain that limits urea extraction (and EC production) is advised. After the completion of MLF, the possible growth of *Oenococcus oeni* and hetero-fermentative LAB should be inhibited.

The addition of an active acid urease enzyme produced from *Lactobacillus fermentum* will reduce the level of urea (a major precursor of EC), and help avoid the formation of ethyl carbamate during ageing. Urease transforms urea into ammonia and carbon dioxide. The process is permitted by the OIV [18]. Following treatment, urease should be removed by filtration.

High alcohol levels, long storage times, particularly at elevated temperatures and light exposure favour the production of ethyl carbamate – levels may be very high in heat-damaged wines. Accordingly, cool, dark storage conditions should be maintained. High temperatures over an extended period are often reached if wine is shipped without temperature control. A study by Roberto Larcher et al. showed that less than five days at 40 °C could be considered sufficient to produce from 15 to 30 µg/l EC, in wines with more than 20 mg/l urea when bottled [24].

14.3.3 Analysis

The most reliable and fastest method of extraction is solid-phase microextraction (SPME). Somewhat less reliable due to background interference is solid-phase extraction (SPE) [25]. Identification may be made by flame ionisation detection (FID), alkali FID (AFID), or selective ion monitoring (SIM). GC/MS may effectively carry out quantification.

14.4 Biogenic Amines

Biogenic amines are toxic at high concentrations, and some people have an intolerance at the levels very often found in wine. They can cause anaphylactic responses. Other possible effects include headaches, facial flushing, nausea, and respiratory distress. The content of amines in wines varies from trace levels up to 130 mg/l. Red wines usually contain up to 50 mg/l, with a lesser concentration in whites [26]. Biogenic amines can have a 'blood-like' odour and taste, particularly in the case of cadaverine which at very high concentrations can have a smell of rotting flesh. However, it is the impact upon human health that should be the greatest concern to winemakers.

Over 20 biogenic amines that may be found in wines have been identified. The most important are:

- Histamine – the most toxic of the biogenic amines found in wines;
- Tyramine;
- Putrescine;
- Cadaverine.

These amines are mainly produced from microbial decarboxylation of the amino acids histidine, tyrosine, ornithine, and lysine, respectively. They are produced mainly by LAB, but also by some yeasts, including *Brettanomyces bruxellensis*. Production of biogenic amines is generally higher in wines with a high pH, and those kept on yeast lees [27]. The addition of DAP as a yeast nutrient late in the fermentation process facilitates production. The presence of biogenic amines may be an indication of poor winemaking practices and a 'natural' MLF. Using a commercial strain of LAB can help minimise production.

14.5 Ethyl Sorbate and Sorbyl Alcohol (Geraniol) Off-Odours

14.5.1 The Use of Potassium Sorbate and Conversion to Compounds Giving Off-Odours

Sorbic acid, in the form of potassium sorbate, may be added to wine containing residual sugar before bottling as a fermentation inhibitor. It should be noted that sorbic acid does not actually kill the yeast, but merely inhibits the activity of *Saccharomyces cerevisiae*, other species of *Saccharomyces*, and certain other yeast genera. It is generally not effective against film-forming yeasts nor *Brettanomyces* spp. or *Zygosaccharomyces*. It should not be considered as a replacement for sulfur dioxide, as it does not have antibacterial properties.

Sorbic acid only remains effective for a relatively short time. It degrades to ethyl sorbate which is, in fact, present in all wines that have had sorbic acid additions. Ethyl sorbate can give a candy or pineapple aroma following a period of storage. However, sorbic acid can be reduced by LAB, particularly *O. oeni* into 2,4-hexadien-1-ol (sorbyl alcohol), which can spontaneously form the ethyl ether 2-ethoxyhexa-3,5-diene. LAB use sorbic acid as a source of carbon.

Yeasts are not involved in this process, but other bacteria may also form sorbyl alcohol. Affected wines have a nose of crushed geranium leaves.

Sensory characteristics: Ethyl sorbate gives a wine a nose of celery and/or pineapple. 2-Ethoxyhexa-3,5-diene gives pronounced nose of crushed geranium leaves.

Sensory perception threshold: The threshold for ethyl sorbate is 135 mg/l. 2-Ethoxyhexa-3,5-diene has an olfactory threshold of 100 ng/l.

14.5.2 Prevention

If possible, the use of potassium sorbate should be avoided. It should only be used for wines with residual sugar, for which it is not possible to use other means to avoid the risk of fermentation in bottle. If potassium sorbate is used, care should be taken to maintain adequate levels of SO_2, which will inhibit the action of lactic acid and other bacteria. A single addition of the necessary amount of potassium sorbate should be undertaken no more than a day or two, prior to bottling. The compound releases 74% sorbic acid. Ethyl sorbate should not be added if the wine is to be held in storage for extended periods, due to the degradation described above.

The use of sterile filtration followed by cold, sterile bottling, or pasteurisation are possible nonchemical alternative methods of removing/inhibiting LAB and negating the risk of fermentation in a bottle of residual sugars.

14.6 Paper-Taste

A wet 'paper-taste' or odour and taste of wet cardboard is a rare fault in recent times. It is most likely caused by passing the wine through a sheet filter or other type of filtration without cleansing the pads/media, including diatomaceous earth. Improper storage of filter materials can lead to them absorbing odorous substances from the environment. This is also the case with fining materials such as bentonite and activated carbon. The problem may be prevented by cleansing the filter system before use with a mild citric acid solution, followed by flushing with clean water to remove acid residues.

14.7 Plastic Taints – Styrene

'Plastic' taints in wine are usually due to styrene (also known as phenylethylene or vinyl-benzene) contamination and may stem from many sources including plastic tanks, vessels, and tank linings. Styrene is used in the production of plastic, and if vessels are damaged, poorly maintained or aggressively cleaned, the compound can be released into the wine. The most likely time of pick-up is during storage. Styrene monomers can also migrate into wine from plastic-lined transport containers. Non-glass packaging materials have also been implicated in plastic odours in wines, including, 'bag-in-box', certain linings of cans, and bottles. The sensory perception threshold of styrene in wine is 50–210 µg/l.

14.8 Indole

14.8.1 The Presence of Indole in Wine

The chemical formula of indole is C_8H_7N. Indole is a normal wine component, formed as a by-product of tryptophan metabolism – tryptophan is an amino acid. The concentration in sound wines ranges from 1 to 10 µg/l [28]. At very low concentrations, indole can add complexity to wine aromas and flavours, but at higher levels, it can give off-aromas and flavours. High levels are usually produced during sluggish fermentation when yeast assimilable nitrogen (YAN) levels are low. Although it can be produced early in the fermentation process by *S. cerevisiae*, non-*Saccharomyces* yeasts are likely to be the major producers, with the 'wild' yeast *Candida stellata* able to produce particularly high amounts (1033 μg/l) [29]. *Lactobacillus*, particularly *L. linderii,* is another known vehicle of production. The formation is by the degradation product of the amino acid tryptophan. It is generally the case that in a normal, efficient, alcoholic fermentation that any indole produced is removed by the process of catabolic metabolism. However, if the fermentation is sluggish, excess indole, produced by non-*Saccharomyces* species may still present by the end of fermentation.

14.8.2 Sensory Perception

At high concentrations, Indole gives a plastic-like off-flavour in wine. At a medium concentration, it can give a faecal aroma. The sensory threshold is 15–330 μg/l, depending upon the wine matrix. Coulter et al. determined a level of 23 μg/l in white wine, but a third of the panellists used for this threshold determination had best estimate thresholds of 9 μg/l or lower [30]. However, it is estimated that 50% of the population are anosmic to indole in wine (https://www.aroxa.com/wine/wine-flavour-standard/indole).

14.9 Geosmin

14.9.1 Sources of Geosmin in Wine

Geosmin is also known as *trans*-1,10-dimethyl-*trans*-9-decalol (and *trans*-1,10-dimethyl-*trans*-9-decanol). The name geosmin means 'odour (osmo) of earth (geo)' and is derived from the Greek. The chemical formula is $C_{12}H_{22}O$. Geosmin has been identified as a compound responsible for earthy off-odours in red and white wines [31]. The fault has been historically found in wines from many regions, particularly in wet vintages. However, in recent years, it has been particularly noted in some wines from the Loire Valley, Alsace, Bordeaux, and northern and southern Burgundy, including Beaujolais. Geosmin is very stable in the fermentation and ageing of wine.

The compound is produced by some algae and soil bacteria, including *Streptomyces*. It is a metabolite of *B. cinerea* and *Penicillium expansum* [32]. It is introduced into wine from rotten grapes which contain the fungi. Other *Penicillium* species, including *P. brevicompactum, P. canescens,* and *P. thomii,* can also produce geosmin [33]. It is worth noting that 59 different species of *Penicillium* have been isolated from grapes in vineyards around the world [9]. Geosmin is often found in tap water and water that has been stored in tank. There is a danger of wine contamination when such water is used to mix oenological additives, e.g. DAP and bentonite. In some countries outside of the EU, water may be legally added to wine, inter alia to reduce the alcohol level. This poses a risk of geosmin taint (along with the possibility of several other microbial and chemical contaminants).

14.9.2 Organoleptic Detection and Sensory Perception Thresholds

The geosmin fault is identified on both the nose and palate. There is a pronounced aroma of beetroot, often accompanied by earthy, damp soil odours [34]. Some tasters liken the nose to the smell of rain hitting soft earth. The nose can occasionally have some hints of mushroom, but care must be taken to distinguish this from the characteristics exhibited by wines contaminated with haloanisoles (see Chapter 3). The nose of Geosmin infected wines has also been likened to a dirty fish tank.

The perception threshold is often quoted as 25 ng/l or even higher. A recent review by Maria Carla Cravero notes a range of 30–45 ng/l [35]. However, using a panel of six expert tasters, Georg Weingart et al. determined a group olfactory (odour) threshold level of just 1.1 ng/l with individual tasters varying from 0.1 ng/l up to 22.6 ng/l [36]. Gustatory (taste) detection threshold was pretty much identical. This research which considered the determination of geosmin and TCA in white and red Austrian wines by headspace SPME-GC/MS and comparison with sensory analysis yielded some fascinating results. Geosmin was above the SPME-GC/MS limit of detection (0.5 ng/l) in 93% of the wines analysed, and over 2 ng/in 55% of the wines. 2,4,6-Trichloroanisole was detected only in 20% of all investigated wines and above 2 ng/l in 4% of all wines. The researchers state that their results show that geosmin is a much more prevalent cause of 'off-flavours' than TCA [36].

14.9.3   Treatment

Numerous treatments have been studied for their effect in reducing geosmin concentrations in wine. In 2014, Maria Lisanti et al. tested activated charcoal, bentonite, polyvinylpolypyrrolidone (PVPP), yeast cell walls, potassium caseinate, zeolite and grapeseed oil for efficacy and effect on wine aromas [37]. Only the grapeseed oil was able to substantially decrease the concentration of geosmin in both red and in white wine (by 83% and 81%, respectively). Potassium caseinate had limited effect in red wine and activated charcoal in white wine, reducing the geosmin concentration by 14% and 23%, respectively. However, all of the effective treatments resulted in a decrease in several aroma compounds [37].

14.10   2-Bromo-4-methylphenol – Iodine, Oyster Taste

Iodine or oyster taste in wine may be attributed to 2-bromo-4-methylphenol. Hydrochloric acid is used for the regeneration of ion exchange resins used for the acidification of wines. This acid contains 2-bromo-4-methylphenol, which once captured by the resin can migrate into a wine subjected to ion exchange [38].

14.11   Heat Damage

14.11.1   Transport and Storage

Wine may suffer heat damage post-packaging during storage or transport. The ideal temperature for storage of wine lies in the range of 12–16 °C (53–61 °F), although a range of 9 –19 °C (48–66 °F) is acceptable. Within this range, there is no doubt that temperature consistency is more important than the precise temperature. However, in practical circumstance represented by both large warehouses and small commercial premises such as many restaurants, this can be hard to achieve. It is during the shipping and transporting of wine that heat damage is most likely to occur. Temperature control during this time might be regarded as essential, but in practice, it is far from commonplace. Shipping containers may be stacked on deck or in a hot hold, and are often held on the dockside for extended periods before and after actual shipping. The incidence of heat-damaged wines is perhaps increasing. Ships are using low-grade fuel resulting in longer journeys. Damage can also occur to individual bottles stored on retailers' shelves under heat emitting lights, and hot shelves in bars.

14.11.2 Sensory Detection

The first sign of heat damage may be before the bottle is opened, particularly if sealed with a natural cork. The cork may be significantly raised or pushed out just a little, or wine will have soaked the sides of the cork. This is easy to see in the case of red wines. Wines may be subject to browning, with red wines paler and white wines deeper than normal.

Heat damaged wines will usually smell jammy, confected, sweet (although sweetness is an illusion for we cannot physically smell sweetness), sometimes tones of caramel, spun sugar and often with tones of raisins, nuts and fruitcake. In severe cases, affected wines will exhibit signs of oxidation, including bitter, dried out, burnt, flat, nutty, Sherry-like flavours, curry notes, all with increased and aggressive acidity.

Wine quality may be considerable reduced as a consequence of storage or transport at high temperatures, even if the usual signs of damage discussed previously are not apparent. A loss of aromatics is very likely in these circumstances. In considering the thiols present in Marlborough (New Zealand) Sauvignon Blanc, Jamie Goode notes that wine stored/shipped at 20 °C will lose half the content of 3-mercaptohexyl acetate (which gives mango, guava, and passion fruit aromas) every two months [39]. At very least, heat-damaged wines have a reduced shelf life.

14.11.3 Pathways to Heat Damage

Many chemical changes take place when wines are exposed to heat, and the reader is referred to the 2015 study by Neil Scrimgeour et al. [40]. There will be a rapid fall in free SO_2, and oxidation will begin. The relative rate of oxygen uptake is 17 times more at 30 °C than at 15 °C [40]. Acetate esters (isoamyl acetate, hexyl acetate) are more readily destroyed at increased storage temperature than are the ethyl esters (ethyl hexanoate, ethyl decanoate). However, the levels of ethyl esters will decrease. There will be a likely formation of furfural and formation of ethyl carbamate. There is evidence that high metal levels, especially copper, increase heat-related deterioration.

14.12 Matters of Balance

14.12.1 Unbalanced Wines May Be Regarded as Flawed

In some 'fault' compounds, there can be much dispute as to the concentration at which they have a negative impact upon aromas and taste. There may be dissent as to the concentration at which they comprise a fault. Much may depend upon the matrix of a wine and leaving fault compounds aside, many aspects of this make it enjoyable or not. One of the key considerations is that of balance – the relationship between the various basic individual structural components and the interplay between the compounds that contribute aromas and flavours. If a wine is very unbalanced, and particularly if one component dominates or is lacking on the palate, the wine may be considered to be flawed. A wine is unbalanced if the concentration (or lack of concentration) of key components results in disharmony. The key components of balance include:

- alcohol;
- acidity;
- body and structure;
- tannins (in the case of red wines);
- fruit and flavour concentration;
- residual sugar (if any).

I will briefly consider the role of each of these.

14.12.2   Alcohol

The alcohol content of so-called 'table' wines, i.e. those that have not been fortified, ranges from 7.5% to over 15.5% abv. Excessive alcohol in any wine numbs the fruit, inhibits freshness and can even give a burning sensation on the palate. A wine with excessive alcohol is unbalanced, less pleasant and less drinkable in other than relatively small quantities. Of course, excessive alcohol results from high must weights which are the consequence of super-ripe or over-ripe grapes. There can be little doubt that climate change has had an impact on high sugar levels in fruit, but many producers have been held hostage to the concept of phenolic ripeness, fearing that the critics and market will run shy from any hint of greenness or unripe tannins in a wine. Although many believe that high alcohol levels give stability (and long life), in truth, the opposite is the case as high levels of ethanol provide nutrients for microbes to degrade wine [41]. Thankfully, many producers are now picking earlier, and the use of yeasts non-*Saccharomyces* yeasts that produce less alcohol is gaining interest. These include *Metschnikowia pulcherrima* and *Pichia kudriavzevii,* which may be used as stand-alones or in co-fermentations with *S. cerevisiae* [42, 43].

Winemakers sometimes reduce alcohol levels in wine by adding water to the must. This process is illegal in many countries, including member states of the EU. Euphemisms are often used for the process, including 'the white horse', 'the black snake', and 'flushing out the bins very well'. The filtration of must may also reduce sugar content. There are methods of removing alcohol from over-alcoholic wines, including the use of reverse osmosis (RO) machines or spinning cone columns, but these remain controversial.

14.12.3   Acidity

A wine will have poor balance if it has excessive or insufficient acidity. Wines with excessive acidity will feel sharp and aggressive on the palate, and those with insufficient acidity will be dull and flabby. Excessive levels of acidity may be a consequence of growing season following veraison being cooler than normal, or the harvest has been undertaken prematurely. Must analysis should be conducted to indicate levels of tartaric and malic acid, and the total acidity. A low pH may inhibit the onset of the MLF if this is required. Excessive acidity levels may be remedied, where legal, preferably at the stage of must adjustment. Calcium carbonate may be used to deacidify must, the only real challenge being to calculate the amount correctly. Potassium bicarbonate may also be used to remove excess acidity from wine. The neutralised acids are precipitated as solid salts, which settle on the bottom of the vat. The wine should then be racked, and if necessary, filtered. Electrodialysis, or anion exchange treatment, may also be used to deacidify musts. Insufficient acidity may be a consequence of very ripe grapes, as sugars increase, the acidity level falls. Where legally permissible, must acidification may take place by adding tartaric, lactic or malic acid, by cation exchange or by the use of elctrodialysis.

14.12.4 Body and Structure

Body, sometimes referred to as weight or mouthfeel, is more of a tactile than a taste sensation. It is a loose term to describe the lightness or fullness of the wine in the mouth. The body of a wine is supported by its *structure,* which is a combination of acidity, alcohol, tannin (red wines), and any sweetness. The structure may be thought of as the architecture of a wine. If any of the components dominate, the wine may be considered to be out-of-balance.

14.12.5 Tannins

Wines with high levels of tannins, which are coarsely textured or very unripe, may be considered as flawed. Full-bodied red wines that are low in tannins will lack 'grip' and may be regarded as flawed. Excessive tannins are one of the main causes of unbalanced wines. Prevention may be achieved by:

- harvesting at appropriate phenolic ripeness;
- destemming, before crushing – stalks contain bitter tannins;
- undertaking pre-fermentation cold soaks for colour, aroma, and flavour extraction;
- extraction during fermentation by punch down or 'gentle' pump-overs;
- avoiding post-fermentation maceration – hard tannins are extracted readily into high alcohol juice;
- not overextending the period of barrel ageing;
- fining, particularly near the end of barrel maturation – see treatments below.

Excessive tannins may be treated by:

- *Fining*: Fining with egg whites (albumin), casein, gelatin, isinglass, bentonite, and PVPP will reduce the tannin levels. Egg whites are the preferred method for fining many quality red wines, such as Bordeaux which will be treated in barrel. Four to seven whites per barrique are generally appropriate. All fining operations will remove some desirable (and undesirable) aroma and flavour compounds, and tannins of all molecule sizes.
- *Gum arabic*: The addition of gum arabic considerably improves the balance of wines and especially those with accented acidic or tannic notes. The complex molecular structure

of gum arabic allows it to place itself among the taste buds receptors, by temporarily isolating them to delay or reduce the astringent and bitter perceptions [43]. In red wines, gum arabic may be added to improve balance, soften astringency, mask bitterness, and increase 'sweet' sensations [44]. Gum arabic is also used to improve protein and tartrate stability, as discussed in Chapters 10 and 15. The dose may vary from 10 to 100 g/hl. The compound is permitted for use in wine by OIV and in the EU.

14.12.6 Fruit and Flavour Concentration

The concentration of fruit and flavour concentration lies at the heart of the balance of a wine. Whilst wines lacking in fruit will be lean and very angular, excessive concentrations can result in so-called 'fruit bombs' which can mask complexity and leave the palate overwhelmed.

14.12.7 Residual Sugar (If Any)

As far as balance is concerned, any residual sugar must be countered by other components, particularly acidity. A sweet wine with low acidity will be flabby, cloying, and distinctly lacking in freshness.

14.13 Final Reflections

How an individual wine is perceived depends upon history, education, training, culture, and a hundred and one other factors inbuilt in the taster. To Chinese people, health aspects are a priority, and the presence of biogenic amines, OTA, or ethyl carbamate are a cause of concern to those who are aware of these. To British consumers, a lack of detectable sensory faults and a wine perceived to be balanced are the important criteria. But a wine that a British drinker considers to be excessive in acidity may be perfect for an Italian, and a very high alcohol, fruit bomb wine be unctuous to a recently retired USA wine critic.

References

1 Gil-Serna, J., Vázquez, C., González-Jaén, M.T., and Patiño, B. (2018). Wine contamination with ochratoxins: a review. *Beverages* 4 (1): 6, 1–21. https://doi.org/10.3390/beverages4010006.

2 International Agency for Research on Cancer [IARC] (2019). World Health Organisation (2019). List of classification. https://monographs.iarc.fr/list-of-classifications (accessed 1 June 2020).

3 European Commission (2006). Commission Regulation EC No 1881/2006 setting maximum levels for certain contaminants in foodstuffs. *Official Journal of the European Union* 2006,OJ L 364, 5–24. Consolidated and amended version. https://eur-lex.europa.eu/legal-content/EN/TXT/PDF/?uri=CELEX:02006R1881-20200701&from=EN (accessed 20 July 2020).

4 National Health and Family Planning Commission (NHFPC) (2017). Maximum levels of mycotoxins in food (GB 2761-2017). www.nhfpc.gov.cn/sps/s7891/201704/b83ad058ff544ee39dea811264878981.shtml summary: http://www.cnwinenews.com/html/2017/putaojiu_0420/108373.html (in Chinese) (accessed 6 March 2019).

5 Mateo, R., Medina, A., Mateo, E.M. et al. (2007). An overview of ochratoxin A in beer and wine. *International Journal of Food Microbiology* 119: 79–83. https://doi.org/10.1016/j.ijfoodmicro.2007.07.029.

6 de Jesus, C.L., Bartley, A., Welch, A.Z., and Berry, J.P. (2017). High incidence and levels of ochratoxin A in wines sourced from the United States. *Toxins* 10 (1): 1–12. https://doi.org/10.3390/toxins10010001.

7 Anli, E., Çabuk, B., Vural, N., and Başpinar, E. (2005). Ochratoxin A in Turkish wines. *Journal of Food Biochemistry* 29: 611–623. https://doi.org/10.1111/j.1745-4514.2005.00043.x.

8 Kassemeyer, H.-H. and Berkelmann-Löhnertz, B. (2009). Fungi of grapes. In: *Biology of Microorganisms on Grapes, in Must and in Wine* (eds. H. König, G. Unden and J. Fröhlich), 61–87. Berlin: Springer-Verlag.

9 Rousseaux, S., Filofteia Diguta, C., Radoï-Matei, F. et al. (2014). Non-botrytis grape-rotting fungi responsible for earthy and moldy off-flavors and mycotoxins. *Food Microbiology* 38: 104–121. https://doi.org/10.1016/j.fm.2013.08.013.

10 Bellí, N., Marín, S., Sanchis, V. et al. (2006). Impact of fungicides on *Aspergillus carbonarius* growth and ochratoxin A production on synthetic grape-like medium and on grapes. *Food Additives and Contaminants* 23 (10): 1021–1029. https://doi.org/10.1080/02652030600778702.

11 Medina, A., Mateo, R., Valle-Algarra, F. et al. (2007). Effect of carbendazim and physicochemical factors on the growth and ochratoxin A production of *Aspergillus carbonarius* isolated from grapes. *International Journal of Food Microbiology* 119 (3): 230–235. https://doi.org/10.1016/j.ijfoodmicro.2007.07.053.

12 Quintela, S., Villarán, M.C., López de Armentia, I., and Elejalde, E. (2013). Ochratoxin A removal in wine: a review. *Food Control* 30 (2): 439–445. https://doi.org/10.1016/j.foodcont.2012.08.014.

13 Lasram, S., Mani, A., Zaied, C. et al. (2008). Evolution of ochratoxin A content during red and rose vinification. *Journal of the Science Food Agriculture* 88: 1696–1703. https://doi.org/10.1002/jsfa.3266.

14 Anli, R.E., Vural, N., and Bayram, M. (2011). Removal of ochratoxin A (OTA) from naturally contaminated wines during the vinification process. *Journal of the Institute of Brewing* 117 (3): 456–461. https://doi.org/10.1002/j.2050-0416.2011.tb00493.x.

15 Blesa, J., Soriano, J.M., Moltó, J.C., and Mañes, J. (2007). Factors affecting the presence of ochratoxin A in wines. *Critical Reviews in Food Science and Nutrition* 46 (6): 473–478. https://doi.org/10.1080/10408390500215803.

16 Fernandes, A., Ratola, N., Cerdeira, A. et al. (2007). Changes in ochratoxin A concentration during winemaking. *American Journal of Enology and Viticulture* 58 (1): 92–96. https://www.researchgate.net/publication/279688781_Changes_in_Ochratoxin_A_Concentration_during_Winemaking.

17 International Organization of Vine and Wine (OIV) (2005). Code of Sound Vitivinicultural Practices in order to Minimise Levels of Ochratoxin A In Vine-Based Products. http://www.oiv.int/en/technical-standards-and-documents/good-practices-guidelines/code-of-sound-vitivinicultural-practices-in-order-to-minimise-levels-of-ochratoxin-a-in-vine-based-products-2005 (accessed 10 July 2020).

18 International Organisation of Vine and Wine {OIV} (2020). *International Code of Oenological Practices*. Paris: OIV. http://www.oiv.int/public/medias/7213/oiv-international-code-of-oenological-practices-2020-en.pdf.

19 Gambuti, A., Strollo, D., Genovese, A. et al. (2005). Influence of enological practices on ochratoxin A concentration in wine. *American Journal of Enology and Viticulture* 56 (2): 155–162.

20 Chatonnet, P., Boutou, S., and Plana, A. (2014). Contamination of wines and spirits by phthalates: types of contaminants present, contamination sources and means of prevention. *Food Additives & Contaminants A*: 1. https://doi.org/10.1080/19440049.2014.941947.

21 Butzke, C.E. and Bisson, L.F. (1997). Ethyl Carbamate Preventative Action Manual, Department of Viticulture & Enology. University of California, Davis, California for US FDA. https://www.fda.gov/downloads/Food/FoodborneIllnessContaminants/UCM119802.pdf (accessed 10 February 2020).

22 Mira de Orduña, R., Patchett, M.L.R., Liu, S.-Q., and Pilone, G.J. (2001). Growth and arginine metabolism of the wine lactic acid bacteria *Lactobacillus buchneri* and *Oenococcus oeni* at different pH values and arginine concentrations. *Applied and Environmental Microbiology* 67 (4): 1657–1662. https://doi.org/10.1128/AEM.67.4.1657-1662.2001.

23 Mira de Orduña, R., Liu, S.-Q., Patchett, M.L., and Pilone, G.J. (2000). Ethyl carbamate precursor citrulline formation from arginine degradation by malolactic wine lactic acid bacteria. *FEMS Microbiology Letters* 183 (1): 31–35. https://doi.org/10.1111/j.1574-6968.2000.tb08929.x.

24 Larcher, R., Moser, S., Menolli, A.U. et al. (2013). Ethyl carbamate formation in sub-optimal wine storage conditions and influence of the yeast starter. *Journal International des Sciences de la Vigne et du Vin* 47 (1): 65–68. https://doi.org/10.20870/oeno-one.2013.47.1.1535.

25 Jiao, Z., Dong, Y., and Chen, Q. (2014). Ethyl carbamate in fermented beverages: presence, analytical chemistry, formation mechanism, and mitigation proposals. *Comprehensive Reviews in Food Science and Food Safety* 13: 611–626. https://doi.org/10.1111/1541-4337.12084.

26 Waterhouse, A.L., Sacks, G.L., and Jeffery, D.W. (2016). *Understanding Wine Chemistry*. Chichester: Wiley.

27 Lonvaud-Funel, A. and Joyeux, A. (1994). Histamine production by wine lactic acid bacteria: isolation of a histamine-producing strain of *Leuconostoc oenos*. *Journal of Applied Bacteriology* 77: 401–407. https://doi.org/10.1111/j.1365-2672.1994.tb03441.x.

28 Capone, D.L., van Leeuwen, K.A., Pardon, K.H. et al. (2010). Identification and analysis of 2-chloro-6-methylphenol, 2,6-dichlorophenol and indole: causes of taints and off-flavours in wines. *Australian Journal of Grape and Wine Research* 16: 210–217. https://doi.org/10.1111/j.1755-0238.2009.00065.x.

29 Arevalo-Villena, M., Bartowsky, E.J., Capone, D., and Sefton, M.A. (2010). Production of indole by wine-associated microorganisms under oenological conditions. *Food Microbiology* 27 (5): 685–690. https://doi.org/10.1016/j.fm.2010.03.011.

30 Coulter, A.D., Capone, D.L., Baldock, G.A. et al. (2008). Taints and off-flavours in wine—case studies of recent industry problems. In: *Proceedings of the 13th Australian Wine Industry Technical Conference: 28 July–2 August 2007; Adelaide* (eds. R.J. Blair, P.J. Williams and I.S. Pretorius), 73–80. Wine Industry Technical Conference Inc.

31 Darriet, P., Pons, M., Lammy, S., and Dubourdieu, D. (2000). Identification and quantification of geosmin, an earthy odorant contaminating wines. *Journal of Agricultural and Food Chemistry* 48 (10): 4835–4838. https://doi.org/10.1021/jf0007683.

32 la Guerche, S., Chamont, S., Blancard, D. et al. (2005). Origin of (−)-geosmin on grapes: on the complementary action of two fungi, *Botrytis cinerea* and *Penicillium expansum*. *Antonie Van Leeuwenhoek* 88: 131–139. https://doi.org/10.1007/s10482-005-3872-4.

33 la Guerche, S., de Senneville, L., Blancard, D., and Darriet, P. (2007). Impact of the *Botrytis cinerea* strain and metabolism on (−)-geosmin production by *Penicillium expansum* in grape juice. *Antonie Van Leeuwenhoek* 92: 331–341. https://doi.org/10.1007/s10482-007-9161-7.

34 la Guerche, S., Dauphin, B., Pons, M. et al. (2006). Characterization of some mushroom and earthy off-odours microbially induced by the development of rot on grapes. *Journal of Agricultural and Food Chemistry* 54 (24): 9193–9200. https://doi.org/10.1021/jf0615294.

35 Cravero, M.C. (2020). Musty and moldy taint in wines: a review. *Beverages* 6: 41. https://doi.org/10.3390/beverages6020041.

36 Weingart, G., Schwartz, H., Eder, R., and Sontag, G. (2010). Determination of geosmin and 2,4,6-trichloroanisole in white and red Austrian wines by headspace SPME-GC/MS and comparison with sensory analysis. *European Food Research and Technology* 231: 771–779. https://doi.org/10.1007/s00217-010-1321-8.

37 Lisanti, M., Gambuti, A., Genovese, A. et al. (2014). Earthy off-flavour in wine: evaluation of remedial treatments for geosmin contamination. *Food Chemistry* 154: 171–178. https://doi.org/10.1016/j.foodchem.2013.12.100.

38 Barbe, J.C., Tempere, S., Riquier, L. et al. (2014). 2-Bromo-4-methylphenol, a compound responsible for iodine off-flavor in wines. *Journal of Agricultural and Food Chemistry* 62 (48): 11620–11627. https://doi.org/10.1021/jf504222c.

39 Goode, J. (2016). Thiols and beyond. https://wineanorak.com/thiolsandbeyond.htm (accessed 11 March 2020).

40 Scrimgeour, N., Nordestgaard, S., Lloyd, N.D.R., and Wilkes, E.N. (2015). Exploring the effect of elevated storage temperature on wine composition. *Australian Journal of Grape and Wine Research* 21: 713–722. https://doi.org/10.1111/ajgw.12196.

41 Hudelson, J. (2011). *Wine Faults: Causes, Effects, Cures*. San Francisco, CA: The Wine Appreciation Guild.

42 Rőcker, J., Strub, S., Ebert, K., and Grossmann, M. (2016). Usage of different aerobic non-*Saccharomyces* yeasts and experimental conditions as a tool for reducing the potential ethanol content in wines. *European Food Research and Technology* 242: 2051–2070. https://doi.org/10.1007/s00217-016-2703-3.

43 Ciani, M. and Comitini, F. (2019). Use of non-saccharomyces yeasts in red winemaking. In: *Red Wine Technology* (ed. A. Morata), 51–68. London: Academic Press.

44 Bichescu, C. and Stanciu, S. (2018). The sensory properties and chromatic characteristics of Feteascà Neagrà red wine after the treatment with gum arabic and alternative oak products. *Romanian Biotechnological Letters* 23 (4): 13793–13803. https://www.e-repository.org/rbl/vol.23/iss.4/9.pdf (accessed 28 April 2020).

15

TDN and Tartrate Crystals: Faults or Not?

In this chapter, two issues that are particularly divisive will be discussed. First, the odours of kerosene, or paraffin, imparted to some white wines by the presence of 1,1,6-trimethyl-1,2-dihydronaphthalene (TDN) are hated by many tasters, but lauded by others. Second, the physical presence in wine of crystals of potassium bitartrate or calcium tartrate that do not impact upon wine aroma or taste, but is often a cause of concern to consumers, and consequentially wine producers spend time, money, and particularly energy in attempting to prevent their formation. The various methods of preventing crystal precipitation are briefly discussed here.

15.1 TDN

15.1.1 What Is TDN?

Kerosene, paraffin, diesel, or petrol odours have long been associated with mature Riesling wines. The aromas, which develop in bottle, are due to the formation of the compound generally known as TDN (1,1,6-trimethyl-1,2-dihydronaphthalene), one of the group of C_{13}-norisoprenoids. These are aroma and flavour compounds that are formed by the enzymatic breakdown of carotenoid pigments [1]. TDN may be present in aged Riesling wine at a concentration of up to 250 mg/l, especially in wines produced in warm climates or vintages. The discussion as to whether TDN is a fault, a flaw, or an attribute perhaps almost warrants a book of its own. Generally speaking, winemakers and consumers in Europe are more positive about the compound than their New World counterparts, particularly those from Australasia and New Zealand. C_{13}-norisoprenoids are important aroma compounds in both red and white wines – Peter Winterhalter and Russell Rouseff noted that about 70% of the total concentration of volatile secondary metabolites in a Chardonnay juice comprised C_{13}-norisoprenoids [2].

15.1.2 The Formation of TDN

TDN precursors are formed in grapes. Carotenoids mostly begin to be synthesised at the first stage of fruit formation; the synthesis continues until veraison [3], and then the compounds break down during ripening. Carotenoids absorb light, including UV light, which would be

Wine Faults and Flaws: A Practical Guide, First Edition. Keith Grainger.
© 2021 John Wiley & Sons Ltd. Published 2021 by John Wiley & Sons Ltd.

harmful to grape cells and DNA. The production of TDN precursors is higher in warm to hot climates and and where there is a high level of sunlight exposure, due to enzymatic or acid-catalysed reactions [4]. Climate change has perhaps resulted in elevated TDN levels in recent years. With increased temperatures and more intense sun exposure, the appearance of TDN is now favoured in wine at the earlier stages of ageing and at higher concentrations [5] than was historically the case. Appropriate canopy management can perhaps mitigate the effects of such elevated temperature and sun exposure [6]. TDN can be associated with atypical ageing (ATA), which may be present when the vines have suffered from hydric or heat stress, as discussed in Chapter 8.

Carotenoid precursors are also slowly converted to TDN in the acidic wine environment. It would appear that lower pH provides a favourable environment for acid hydrolysis of the glycosylated precursors of TDN. However, wines from some grape varieties have high levels of carotenoids but low levels of TDN, even after extensive ageing.

15.1.3 Sensory Characteristics and Sensory Detection Thresholds

15.1.3.1 Sensory Characteristics
TDN is particularly detected on the nose of a wine and retro-nasally. At low levels, the compound has aromas of toast and marmalade, while at high levels, the penetrating odours of aviation fuel, kerosene, diesel, or petrol are to the fore.

15.1.3.2 Sensory Detection Thresholds
The sensory detection threshold in wine is perhaps 2–4 µg/l, although it was historically believed to be 20 µg/l [7, 8]. In a 2012 study by Gavin Sacks et al., the threshold was determined to be 2 µg/l [7]. In 2020, Andii Tarasov et al. stated a detection threshold of about 4 µg/l, a recognition threshold of 10–12 µg/l, and a rejection threshold of 71–82 µg/l [9]. This comparatively high rejection threshold perhaps illustrates the challenge facing those who perceive even moderate levels of TDN to be a fault. A 2014 study by Carolyn Ross et al. using untrained consumers found a higher detection level of 18–20 µg/l and a consumer rejection threshold of 80–160 µg/l, depending on the composition of the wine [10].

15.1.3.3 Prevalence of TDN in Riesling Wines
The 2012 Sacks et al. study [7] found that 27 of 28 Riesling wines (1–3 years of age) had suprathreshold TDN concentrations, whereas only 7 of 69 non-Riesling wines of a similar age showed levels above threshold. TDN levels have long been known to increase in wine up to a concentration of 200 µg/l over a period of 10–12 years [11], although higher levels have now been noted. A survey of 116 Riesling wines was undertaken by the Australian Wine Research Institute (AWRI) [8]. The wines were tasted by a panel and analysed by gas chromatography–mass spectrometry (GC-MS). This found TDN levels of up to 254 µg/l (the highest mean and total levels were in wines from Eden Valley, South Australia), and all the wines examined had suprathreshold levels [8].

15.1.4 Minimising TDN in Wine Production

As stated, the precursors of TDN are present in harvested grapes and are greater in hot climates and seasons. Minimising production revolves around appropriate viticultural practices. Leaf removal on both the eastern and western sides of the vine has been shown to increase the production [6]. Carotenoids can be reduced if the grapes are shaded [12], with the mid-season perhaps being the most important period [13]. Higher levels of vineyard nitrogen fertilisation perhaps lead to the presence of fewer TDN precursors in grapes [14]. This might be due to the fact that higher nitrogen levels are likely to produce greater vigour and a denser leaf canopy. Sufficient soil water is also a factor as vine stress will compact the canopy – partial drying of the rootzone will increase sunlight penetration [8]. Silke M. Gerdes et al. [15] found that TDN levels increased at exposure above 20% of full sunlight. Elevated temperature and intense sun exposure lead to both higher concentrations and earlier appearance of TDN during the ageing process [16]. In hot years, consideration might be given to earlier picking, but of course this can have generally detrimental organoleptic consequences. Certain strains of *Saccharomyces cerevisiae* produce greater quantities of TDN.

15.1.5 TDN in Bottled Wine

The level of TDN in bottled wine generally increases as the wine ages, particularly if this is at elevated temperatures. However, TDN is scalped by cork [17], so following an initial increase, levels can actually reduce after the wine has been in bottle for a few years (if the bottle is sealed with a cork closure). Almost all Australian and New Zealand Rieslings are sealed with screw cap, which is perhaps one of the reasons why high levels of TDN are often found in mature Rieslings from these countries.

15.1.6 The TDN Controversy

Whether overt TDN is perceived as a fault, flaw, or something that adds an exciting extra dimension to a mature Riesling depends as much on the culture of the taster as on the wine matrix. Winemakers can mostly agree on the source, but there any semblance of consensus ends. Steve Smith MW, acclaimed New Zealand winemaker, now of Pyramid Valley Vineyards stated, 'There is no doubt TDN starts in the vineyard with stressed vines. I don't mind a little when it occurs in aged Rieslings. However, I would try to eradicate it in winemaking, if I knew it was there' [personal discussion]. Jamie Marfell of Marlborough's Brancott Estate said, 'I quite like a little (TDN) in mature Rieslings, I see it in the fruit in hot years and think it's part of the characteristics the variety can show' [personal discussion]. Matt Thomson, another New Zealand legend who is a consultant and winemaker at Blank Canvas, was even more emphatic, 'TDN is a characteristic of the (Riesling) variety, and it stems from the vineyard in warm years. It gives an extra dimension, an extra excitement to the aged wines' [personal discussion]. Joachim Heger of Weingut Dr Heger in the Baden region of Germany was quite philosophical, 'I think it's part of the character of Riesling, something that develops with age and adds complexity. I don't want it to dominate but it's a point of difference' [personal discussion]. Not all would agree. Armin Schüttler and colleagues believe that the petrol-like aroma is not appreciated in modern Riesling wines [6]. Interestingly, in the AWRI survey of 116 Riesling wines [8], a number of wines containing high TDN levels (including the second-highest TDN concentration of 246 μg/l found in a 1998 Eden Valley Riesling) were given high-quality scores by the tasting panel [8].

15.2 Tartrate Deposits

The presence in bottled wine of tartrate crystals has long caused concern amongst many consumers. At first glance, the crystals may resemble tint pieces of broken glass or crystal sugar, although they can be amorphous or powder-like.

15.2.1 The Formation of Tartrate Deposits

The precipitated deposits comprise either of the following:

- Potassium bitartrate, also known as potassium hydrogen tartrate (KHT), with a chemical formula of $C_4H_5KO_6$;
- Calcium tartrate (CaT), with a chemical formula of $C_4H_4CaO_6$ (usually as the calcium L-tartrate isomer but very occasionally as calcium DL-tartrate).

Of these, potassium bitartrate is by far the most common. The crystals may attach themselves to the cork or fall as sediment in the bottle. Figure 15.1 shows tartrate crystals in the bottom of a bottle of Premier Grand Cru Sauternes.

In white wine, tartrates are usually colourless. In red wine, they are purple or red and may co-precipitate with tannins to form a heavy, amorphous sediment. The deposits may occur in any wine type but are particularly common in white wines (which generally have higher levels of tartaric acid than reds), which are often stored at low temperatures for an extended duration in restaurants and bars. Potassium bitartrate is much less soluble at low temperatures, favouring precipitation in these circumstances. Tartrates are tasteless and do not detract from wine quality in any way, other than their physical presence. They pose no risk to health.

Figure 15.1 Tartrates in a bottle of Sauternes.

Tartrate crystals can be found in wines of the finest provenance: for example, their presence in mature bottles of Château d'Yquem, Premier Cru Supérieur Sauternes, is common. Yet the global wine industry spends a vast sum of money every year on cold stabilisation treatments of wines to avoid consumer dissatisfaction. Cold stabilisation of wine reduces total acidity by between 0.5 and 1.5 g/l and also affects pH. Accordingly the organoleptic characteristics of a wine are modified – of course this also happens if tartrates naturally precipitate. Whilst it is true that the presence of tartrates sometimes cause unwarranted concern to consumers, it is generally the less knowledgeable buyers, including those in less mature markets, who are likely to reject wines showing deposits. Buyers of cheese accept rinds and holes in Emmental and many other types. It is my belief that producers should re-invest some of the money currently spent on tartrate stabilisation into consumer education. Put simply, if the makers of some of the world's finest wines are content for tartrates to be present in their product and deliberately avoid the interventionist steps that could be used to minimise appearance in bottle, then why not for all wines? However, I feel for the producer faced with the dilemma as to whether or not to use the interventionist techniques necessary to avoid tartrates. John Kinney of the flagship Occasio Winery in Livermore Valley, California, noted,

> In 2015, we chose to ignore them. Though the resulting wines were beautiful, among the highest rated wines in Livermore Valley that year, we had to hand-sell them from our tasting room. We couldn't wholesale the wines because, in a grocery store, no one is around to explain the reasons for the little crystals in the wine. At least for us, ignoring tartrates is no longer an option. That is why we have returned to removing them … For a procedure that is only of cosmetic benefit, I can't help but think that cold stabilization is an unnecessary step in winemaking practice [18].

Tartar and tartrates can form at any stage of winemaking from crushing through to the finished wine in bottle. Deposits may be found on crushers, presses, inside tanks, and several other items of winemaking equipment. The core of the crystals will be the deposit and growth of potassium or calcium ionised to tartaric acid [19]. As alcohol levels increase during fermentation, the solubility of potassium bitartrate decreases. Some of the tartrates will be removed during the racking operations post fermentation. Figure 15.2 shows tartrates left in a tank after the first racking, following the alcoholic fermentation of a red wine.

15.2.2 Sources of Calcium in Wines

The level of calcium in wines can range from 40 to 150 mg/l. At levels above 60 mg/l (red wines) or 80 mg/l (white wines), there is a risk of calcium tartrate precipitation. There are several possible sources for calcium in must and wines:

- Vineyard soils;
- Calcium carbonate used to de-acidify musts and wine;
- Milk or casein used for fining;
- The use of calcium bentonite for fining;
- Unlined concrete fermentation and storage vats [20].

Figure 15.2 Potassium bitartrate precipitated in vat.

The use of concrete vats, which gained in popularity from the beginning of the twentieth century until stainless steel became commonplace from the late 1970s, has again been on the increase during the last 15 years or so. At the beginning of the 1970s, only three Bordeaux properties were using stainless steel; all others were equipped with wood or concrete. Today, concrete is desired by some producers on account of its thermal inertia and, if unlined (although sometimes a light spray of cream of tartar is used), the ability to allow minute oxygen permeation. It is perhaps partially on account of the increased use of concrete vats that the precipitation of calcium tartrate crystals is becoming more common [20].

15.2.3 Calcium Tartrate Instability

The precipitation of CaT is very unpredictable and can often occur many years after bottling. CaT is 10 times less soluble in wine than KHT. Further, unlike for potassium bitartrate, chilling has little or no effect on calcium tartrate precipitation. Accordingly, two of the tests commonly used to predict crystallisation (the test of chilling the wine and the mini-contact test) are largely ineffective as a predictive tool for CaT precipitation. A more reliable test consists of seeding a sample of wine with micronised calcium tartrate and measuring the quantity of calcium precipitated after two days' storage at a cold temperature. Calcium precipitation above 25 mg/l indicates a high risk of CaT instability.

Levels of calcium above 60 mg/l (red wines) and 80 mg/l (white wines) may be regarded as constituting a danger of instability. Crystals of calcium-L tartrate are more likely to form in wines with a high pH [20], including wines that have had a complete malolactic fermentation (MLF). In any case, malic acid is an efficient calcium L-tartrate crystallisation inhibitor. Polyuronic acids of grape pectins, which are naturally found in wine, are also efficient inhibitors of calcium L-tartrate crystal formation [20].

15.2.4 Sources of Potassium in Wines

The level of potassium in wines can range from 400 to 2000 mg/l.

The sources of potassium in wines are as follows:

- *Grape skins and pulp*: Potassium is the most prevalent cation in grapes. The skins usually contain the highest concentration, perhaps (by weight) four times as much as the pulp. Grapes are by far the highest source of potassium in wine.
- The use of potassium metabisulfite as a source of sulfur dioxide.
- Oenological treatments, such as potassium hydrogen carbonate, potassium caseinate, and potassium sorbate.

15.2.5 Potassium Bitartrate Instability

Precipitation of potassium bitartrates is by far the most common source of tartrate deposits in wine. Polyphenols and polysaccharides and proteins can interact with bitartrates and prevent the nucleation necessary for crystals to grow. Red wine, in particular, contains 'protective colloids' that help inhibit precipitation, although these can be removed during filtration. Lees ageing, as often undertaken for certain white wine styles, imparts mannoproteins to wines, and these can help promote tartrate stability. Potassium bitartrate has a limited solubility in wine that will depend, inter alia, upon the alcohol content, the pH, the synergy between anions and cations, and the temperature [19]. The solubility progressively decreases as the alcohol concentration increases [21].

15.2.6 Prevention of Tartrate Formation

There are various methods that can be used during the winemaking process to reduce the risk of tartrate precipitation in bottled wines. Some methods are effective against the precipitation of all tartrates, some only for KHT, and others only for CaT. The techniques may be physical (including physiochemical) or chemical methods. Alternatively, these may be grouped into 'subtractive' and 'non-subtractive or additive' processes. Subtractive methods involve reducing the concentration of tartaric acid and/or potassium or calcium in wine. Additive methods use protective colloids or crystallisation inhibitors. Additive techniques are generally more respectful to sensory qualities and considerably more environmentally friendly, especially with regard to energy consumption. Another major advantage in the use of additive methods, particularly for the small winery, is that no investment in major plant is required, as is the case for the subtractive methods of cold stabilisation, ion exchange, or electrodialysis.

15.2.7 Prevention of Precipitation of Potassium Bitartrate Crystals in Packaged Wine

15.2.7.1 Methods Used for KHT Stabilisation

The subtractive methods that can be used to prevent potassium tartrate precipitation are as follows:

- *Cold stabilisation*: Traditional, contact process;
- Ion exchange;
- Electrodialysis.

The additive processes for the prevention of potassium tartrate precipitation are the following:

- Metatartaric acid;
- Carboxymethylcellulose (CMC);
- Mannoproteins;
- Gum arabic;
- Potassium polyaspartate.

15.2.7.2 Cold Stabilisation

The conventional methods of inhibiting the precipitation of tartrates in the finished wine are by utilising traditional cold stabilisation or the contact process shortly before bottling. In fact these processes only really work for potassium bitartrate and even in this case are not always totally effective. They must take place after all other stabilisation procedures, e.g. protein stabilisation, as any change in pH will seriously impact upon the effectiveness of the operation. Using cold stabilisation, the wine is first fined to negate the protective effect of micro-crystals. The tank of wine is then chilled to just above its freezing point: e.g. chilled to $-4.5\,°C$ ($24\,°F$) for a wine containing 12% abv or chilled to $-8.5\,°C$ ($16.7\,°F$) for fortified wines containing 20% abv. It is vital that the temperature is not lowered to below the wine's freezing point or irreparable damage to the wine will ensue.

The freezing point can be calculated as follows:

$$\frac{\% \text{ alcohol} - 1}{2} = \text{Wine freezing point}\,(\text{-}°C)$$

The wine is held, preferably in insulated tanks, at the temperature just above freezing point for 10–21 days, depending upon the temperature (colder temperatures require shorter holding time). After this operation, the wine can be racked off the crystals that will have formed and be bottled.

If the wine has a pH below 3.65 pre-treatment, during the cold stabilisation process, there will be a reduction in pH by up to 0.2 pH units. For wines at higher pH than 3.7 pre-treatment, the pH will be increased. There are other major drawbacks with traditional cold stabilisation, including the amount of time taken and the copious energy usage. It is also expensive in terms of equipment. Further, the level of natural crystallisation inhibitors will change as wines mature, leading to possible precipitation of tartrates. Put simply, cold stabilisation is not always totally effective in the long term.

A more effective, quicker, and cheaper system is that of the contact process. This involves chilling wine to −3 °C (27 °F) to +4 °C (39 °F) and adding 1–2 g/l of purified tetrahydrated crystals of potassium bitartrate, followed by a vigorous stirring for at least four hours to keep the crystals in suspension and maximise the contact area. The wine should be seeded with the compound in a very fine (less than 100 µm) grain form. Alternatively, the wine is pumped through a crystal bed. The added crystals act as nuclei and attract and accelerate further crystallisation. Crystal precipitation takes approximately 6–8 days.

Whichever method is used, it is important that the wine is not allowed to warm up prior to racking and filtration; otherwise the potassium tartrate may become re-dissolved in the wine. Oxygen uptake is a major risk at such low temperatures, so it is important that the wine in tank is blanketed with inert gas.

15.2.7.3 Ion Exchange
Ion exchange makes use of cation resins that are selective towards potassium and calcium, and accordingly the treatment lowers the concentration of potassium and calcium ions in the wine, replacing the potassium and calcium ions with hydrogen ions. It is perhaps the most cost effective method of tartrate stabilisation [22] but does affect pH and certain organoleptic characteristics. In 2009, the process was first permitted in the European Union (EU) under Regulation (EC) No. 606/2009. Under current EU Regulation 2019/934 [23], the International Code of Oenological Practices (OIV) [24] and the OIV International Oenological Codex [25], an acid regenerated cation exchange resin must be used in the process. Sodium regenerated resin is not permitted. The ion exchange process may be viewed as a rapid and effective means of tartrate prevention but is an expensive option.

15.2.7.4 Electrodialysis
A far more effective method of tartrate stabilisation than cold stabilisation processes is the use of electrodialysis, but the capital cost of the equipment puts this out of reach of many wineries, although, in some regions, mobile machines and operators can be hired. Electrodialysis uses selective membranes allowing the passage of potassium, calcium, and tartrate ions under the influence of an electrical charge. The removal of tartrates by this method is fast (a single cross-flow pass), efficient, and reliable and is adjusted to the characteristics of the wine being processed. The speed of the process makes the system particularly attractive to producers who wish to get a wine, such as a new season's Sauvignon Blanc, to market early. There is also less wine loss, perhaps 1%, compared with up to 3.5% with cold stabilisation. As the process takes place without refrigeration, there is a considerable reduction in energy costs, up to 95% compared with the 12+ days for cold stabilisation.

15.2.7.5 Metatartaric Acid Addition
Metatartaric acid is a mixture of polymers with different molecular weights comprising a polymerised compound formed by the esterification between the carboxylic group of one tartaric acid unit and an alcohol group of another molecule of tartaric acid. The addition of metatartaric acid, at a maximum rate of 10 g/hl (100 mg/l), is permitted by the EU (Regulation (EU) 2019/934) [23] and OIV [24] and is commonly used to prevent tartrate crystal

formation. However, its addition is only effective at protecting from tartrate precipitation for a limited amount of time, perhaps nine months or so, due a rapid hydrolysis of the ester group. After this period, the wine is likely to form even more crystals than would be the case if no additions were made.

15.2.7.6 Carboxymethylcellulose (CMC)

CMC may be used as an efficient and effective preventive against KHT deposits [26]. However, it is not effective against calcium L- and DL-tartrates. CMC works by coating with a protective film the crystal nuclei, which are the precursors for precipitation. This film deforms the morphology of the crystal, physically limiting nuclei growth [19]. It is possible that some young white wines may be too unstable for CMC use [27]. Being derived from cellulose, CMC is not a 'natural' fining agent. It is available in liquid or dry form – the latter must be thoroughly dissolved in wine by a rigorous stirring. Before addition of CMC, all cellar operations (other than final filtration) including the addition of any necessary oenological products or processing aids must have been completed. Prior to the addition, the wine must be protein stable (the protein lysozyme must not be present), or there is a considerable risk of the formation of protein haze. As CMC may be removed from solution and block filters (some CMCs have a dramatic initial effect on wine filterability [28]), a minimum of 48 and up to 120 hours should be allowed following addition prior to final filtration and subsequent bottling [29].

The use of CMC is permitted in the EU up to 200 mg/l [23], but *not* in still red and rosé wines as it can cause colour precipitation in previously colour-stable wines. For this reason, any use of CMC in red wines (where legal) must be always used in combination with gum arabic for colour stability.

15.2.7.7 Mannoproteins

A further alternative for the prevention of KHT crystal formation is the addition of mannoproteins, which are extracted from the cell walls of yeasts. Mannoproteins react with cellulose, and therefore any addition must be at least 48 hours prior to sheet filtration. In any event, the addition can decrease a wine's filterability. Of course, the addition of mannoproteins affects a wine's protein stability. Mannoproteins are marketed by Laffort as MANNOSTAB®. The dose is 10–30 g/hl depending upon the degree of instability. Their use is permitted in the EU and by OIV.

15.2.7.8 Potassium Polyaspartate (KPA)

Potassium polyaspartate (KPA), with a chemical formula of $C_4H_5NO_3K$, is an odourless, light brown powder with usually 90% dry matter [30]. It is completely soluble in water and wine and may be used to stabilise wines against tartrate precipitation. KPA works by enveloping tartaric molecules as a 'colloid protector' so that the tartaric acid bases never get close enough to initiate crystallisation. The compound is manufacture by Enartis and marketed as Zenith Uno, and the rates of addition are 100 ml/l. KPA was first authorised for use in wine by the EU under Regulation (EU) 2017/1961; this regulation has since been repealed and incorporated into Regulation 2019/934 [23]. Its maximum permitted dose is 10 g/hl (100 mg/l). On 10 February 2020, KPA was listed in the United States by the alcohol and tobacco tax and trade bureau (TTB) as an approved additive for tartrate stabilisation under

24.250. Wines must be protein stable before addition and not contain lysozyme. As this is a relatively new product, there are as yet no longitudinal studies upon its effectiveness, but anecdotal evidence would indicate that its performance is better than CMC, mannoproteins, and metatartaric acid.

15.2.7.9   Considerations as to the Choice of Methods of Prevention of KHT Crystallisation

The additive methods of metatartaric acid, CMC, and KPA are reactive with wine proteins due to their high negative charge. For this reason, it is imperative to check wine protein stability and be sure wine is well below the maximum stability limit – wines close to the stability limit can form haziness or sediment following the addition of metatartaric acid, CMC, and/or KPA. Metatartaric acid can be used only for short-term conservation. CMC has a low cost but can only be used in white wine. Ion exchange is effective in all wines, has a relatively low cost, but has an impact upon pH. Furthermore, it is not easy to automate. A comparison of the various methods of KHT stabilisation is shown in Table 15.1.

15.2.7.10   Gum Arabic and Co-blends of Additives

Gum arabic, also known as *Acacia verek*, is commonly used on fined and filtered wines prior to bottling to prevent problems with hazes including copper and iron hazes and pigment precipitation. It also has limited effect as a tartrate inhibitor. Gum arabic is mostly harvested from *Senegalia senegal* (also known as *Acacia senegal*) and other African *Acacia* species. It is tapped from the bark during the dry season and has a large number of uses including as a fixative for food flavours, an emulsifier, a stabiliser of dairy products and in pharmaceuticals. The compound has been approved for use by the OIV and in the EU. The maximum dose according to OIV and EU regulations is 300 mg/l.

Gum arabic is normally used in combination with other tartrate stabilisation additives, such as metatartaric acid and/or CMC. In recent years several co-blends of two or more additives that have inhibitory properties in their own right towards tartrate instability have been created by oenological product manufacturers. These are particularly valuable on

Table 15.1   Stabilisation for potassium bitartrate.

	Physical treatments				Chemical treatments			
	Cold stabilisation	Contact process	Ion exchange	Electro- dialysis	Metatartaric acid	KPA	CMC[a]	Mannoproteins
Effectiveness	Good	Good	Good	Very good	Very good	Very good	Very good	Fair
Lasting effect	Good	Good	Good	Very good	Poor	Good	Very good	Good
Reacts with proteins	N/A	N/A	N/A	N/A	Yes	Yes	Yes	No
Affects colour	No	No	No	No	No	No	Yes	No
Energy use	High	Medium	Low	Low	N/A	N/A	N/A	N/A

a)   Not permitted in EU for red and rosé still wines.

overcoming deficiencies that may be present in a single additive. For example, Stab MegaR, manufactured by Enartis, is a co-blend of CMC, gum arabic, and mannoproteins. Zenith Color is a co-blend of KPA and gum arabic, which also helps prevent the precipitation of unstable colour.

15.2.8 Prevention of Precipitation of Calcium Tartrate Crystals in Packaged Wine

15.2.8.1 Methods Permitted by the OIV

Traditional cold stabilisation is not a particularly effective technique for preventing calcium tartrates. Cold temperatures do not significantly increase or accelerate CaT precipitation, and in any event the usual 12–14 days' cooling treatment would be too short. Nevertheless, cold stabilisation remains a permitted method by the OIV for the prevention of CaT, as well as KHT [24]. Ion exchange and electrodialysis, as discussed Section 15.2.7, are effective. The chemicals permitted by the OIV and in the EU for CaT stabilisation are the additions of metatartaric acid (as discussed in Section 15.2.7.5), calcium tartrate, and DL-tartaric acid.

15.2.8.2 The Addition of Calcium Tartrate

Calcium tartrate, specifically micronised crystalline calcium tartrate, can be used for seeding wine to promote the rapid formation and precipitation of calcium tartrate crystals. The treatment can be carried out at 'cellar temperature' of 10 to 15 °C (15–59 °F), avoiding the costs of cooling the wine. Precipitation of crystals takes approximately 7–10 days, with minimal impact upon sensory characteristics. As CaT precipitation induces KHT precipitation (not the other way round), it is appropriate to stabilise for calcium CaT before KHT. The maximum dose is 200 g/hl (2 g/l).

15.2.8.3 The Addition of Racemic (DL) Tartaric Acid

The addition of high racemic tartaric acid (DL)-tartaric acid can be effective at preventing CaT precipitation, if used early in vinification, e.g. at must stage or during the alcoholic fermentation. The compound, which is marketed as Ca^{2+} Stab*, works by reacting with soluble calcium and precipitating D-calcium tartrate. The dosage, in mg/l, can be calculated by subtracting the desired calcium content in the wine from the original calcium content and multiplying by 3. Treatment must be undertaken by a qualified oenologist.

A comparison of the various methods of CaT stabilisation is shown in Table 15.2.

15.2.9 Predictive Tests

15.2.9.1 Potassium Bitartrate

A traditional predictive test for potassium bitartrate stability may be undertaken by chilling wine to −4 °C (28 °F) holding for six days and checking for crystallisation. Alternatively, a mini-contact test may be undertaken. This measures the wine conductivity variation before and after refrigeration for 30 minutes with potassium bitartrate seeding. A white wine is chilled to 0 °C (32 °F) and seeded with 10 g/l KHT crystals and held for four minutes. A red wine is seeded with 20 g/l of KHT and held for 12 minutes. Conductivity readings are then made.

Table 15.2 Stabilisation for calcium tartrate.

	Physical treatments				Chemical treatments		
	Cold stabilisation	Contact process	Ion exchange	Electro-dialysis	Metatartaric acid	Crystalline calcium tartrate	Racemic tartaric acid
Effectiveness	Poor	Fairly Poor	Fair	Very good	Good	Very good	Good
Lasting effect	Poor	Poor	Good	Very good	Poor	Very good	Good
Reacts with proteins	N/A	N/A	N/A	N/A	Yes	Yes	No
Affects colour	No	No	No	No	No	No	No
Energy use	High	Medium	Low	Low	N/A	N/A	N/A

A convenient means of stability testing is with a portable microprocessor ion selective photometer, such as Hannah Instruments HI 83748 TARTARIC ACID ISM, or with Check-Stab equipment.

15.2.9.2 Calcium Tartrate

The tests used to predict KHT instability (the cold test and mini-contact test) are not effective tools for CaT instability. Cold temperatures have little effect on the rate of CaT precipitation, and in any case the tests are not long enough [31]. A simple but effective test is to seed a wine sample of wine with ground calcium tartrate crystals. After two days in cool conditions, the amount of calcium precipitated is measured: a precipitation of more than 25 mg/l indicates CaT instability.

15.3 Final Reflections

Its April 2019, and I'm tutoring a tasting for New Zealand wines for the Lincoln (UK) Wine Society. The diverse reactions of the 40 members present to a wine showing marked TDN characteristics, a 2015 Pegasus Bay Riesling from Waipara (sealed with a saran–tin screw cap), gives an illustration of the differing perceptions of the impact the compound has upon quality. Nearly a quarter of the participants (9 out of 40) gave the wine a maximum score, with comments including 'delicious and penetrating petrol aromas', 'heady and sensual', and 'aromas that are almost addictive'. Most other participants gave the wine a high score, but about 12.5% of the group (5 out of 40) awarded the wine a low score, with observations such as 'unpleasant diesel tones' and 'odours of solvent'. In the same month as part of a Rheingau (Germany) tasting, to a group of trade students, I showed a very mature 2002 Riesling Auslese: Hattenheimer Wisselbrunnen (VDP Grosse Lage) from Weingut Hans Lang. This semi-sweet, delicate wine was loved by all, with comments such as 'the kerosene notes really add an extra dimension to a nose that exudes both green and red apples, toffee-apple, barley sugar with a hint of citrus'. This wine was sealed with a natural cork that may well have scalped some of the TDN content, thus preventing it from overwhelming the fruit and making the complexity.

References

1 Waterhouse, A.L., Sacks, G.L., and Jeffery, D.W. (2016). *Understanding Wine Chemistry*. Chichester: Wiley.

2 Winterhalter, P. and Russell L., R. (2001). Carotenoid-derived aroma compounds: an introduction. In: *Carotenoid-Derived Aroma Compounds*, ACS Symposium Series, vol. 802, 1–17. Washington, DC: American Chemical Society. https://doi.org/10.1021/bk-2002-0802.ch001.

3 Baumes, R., Wirth, J., Bureau, S. et al. (2002). Biogeneration of C_{13}-norisoprenoid compounds: experiments supportive for an apo-carotenoid pathway in grapevines. *Analytica Chimica Acta* 428 (1): 3–14. https://doi.org/10.1016/S0003-2670(01)01589-6.

4 Bakker, J. and Clarke, R.J. (2012). *Wine Flavour Chemistry*, 2e. Chichester: Wiley-Blackwell.

5 Tarasov, A., Giuliani, N., Dobrydneva, A. et al. (2019). Absorption of 1,1,6-trimethyl-1,2-dihydronaphthalene (TDN) from wine by bottle closures. *European Food Research and Technology* 245 (11): 2343–2351. https://doi.org/10.1007/s00217-019-03351-8.

6 Schüttler, A., Guthier, C., Stoll, M. et al. (2015). Impact of grape cluster defoliation on TDN potential in cool climate Riesling wines. *BIO Web of Conferences* 5: 1006. https://doi.org/10.1051/bioconf/20150501006.

7 Sacks, G.L., Gates, M.J., Ferry, F.X. et al. (2012). Sensory threshold of 1,1,6-trimethyl-1,2-dihydronaphthalene (TDN) and concentrations in young Riesling and non-Riesling wines. *Journal of Agricultural and Food Chemistry* 60: 2998–3004. https://doi.org/10.1021/jf205203b.

8 Black, C., Francis, L., Henschke, P. et al. (2012). Aged Riesling and the development of TDN. *Wine & Viticulture Journal* September/October. www.awri.com.au/wp-content/uploads/Sept-Oct-2012-AWRI-Report.pdf.

9 Tarasov, A., Giuliani, N., and Dobrydnev, A. (2020). 1,1,6-Trimethyl-1,2-dihydronaphthalene (TDN) sensory thresholds in Riesling wine. *Food* 9: 606. https://doi.org/10.3390/foods9050606.

10 Ross, C.F., Zwink, A.C., Castro, L., and Harrison, R. (2014). 1,1,6-Trimethyl-1,2-dihydronaphthalene in 1-year-old Riesling wines. *Australian Journal of Grape and Wine Research* 20: 335–339. https://doi.org/10.1111/ajgw.12085.

11 Winterhalter, P. (1991). 1,1,6-Trimethyl-1,2-dihydronaphthalene (TDN) formation in wine. 1. Studies on the hydrolysis of 2,6,10,10-tetramethyl-1-oxaspirol[4.5]dec-6-ene-2,8-diol rationalizing the origin of TDN and related C_{13} norisoprenoids in Riesling wine. *Journal of Agricultural and Food Chemistry* 39 (10): 1825–1829.

12 Steel, C.C. and Keller, M. (2000). Influence of UV-B irradiation on the carotenoid content of *Vitis vinifera* tissues. *Biochemical Society Transaction* 28: 883–885. https://doi.org/10.1042/bst0280883.

13 Kwasniewski, M.T., Vanden Heuvel, J.E., Pan, B.S., and Sacks, G.L. (2010). Timing of cluster light environment manipulation during grape development affects C_{13} norisoprenoid and carotenoid concentrations in Riesling. *Journal of Agricultural and Food Chemistry* 58 (11): 6841–6849. https://doi.org/10.1021/jf904555p.

14 Linsenmeier, A.W. and Löhnertz, O. (2007). Changes in norisoprenoid levels with long-term nitrogen fertilisation in different vintages of *Vitis vinifera* var. Riesling wines. *South African Journal of Enology and Viticulture* 28: 17–24. https://doi.org/10.21548/28-1-1455.

15 Gerdes, S.M., Winterhalter, P., and Elbeler, S.E. (2002). Effect of sunlight exposure on norisoprenoid formation in white Riesling grapes. In: *Carotenoid Derived-Aroma Compounds*, Chapter 19 (eds. P. Winterhalter and R.L. Rouseff), 262–272. Washington, DC: American Chemical Society. https://doi.org/10.1021/bk-2002-0802.ch019.

16 Dobrydnev, A., Tarasov, A., Müller, N. et al. (2020). An optimized method for synthesis and purification of 1,1,6-trimethyl-1,2-dihydronaphthalene (TDN). *MethodsX* 7: 56–61. https://www.sciencedirect.com/science/article/pii/S2215016119303449

17 Capone, D., Sefton, M.A., Pretorius, I., and Hoj, P. (2003). Flavour 'scalping' by wine bottle closures. *Wine Industry Journal* September/October 18 (5): 16–20.

18 Occasio Winery (2020). *Tartrate Stability – A New Approach on the Horizon*. Occasio Winery https://occasiowinery.com/tartrate-stability-new-approach-horizon (accessed 10 August 2020).

19 Swarts, A.C. (2017). A look of tartrate stabilisation of wine in the South African Wine Industry. Dissertation. The Cape Wine Academy.

20 AWRI (2020). *Calcium Instability*. AWRI www.awri.com.au/industry_support/winemaking_resources/fining-stabilities/hazes_and_deposits/calcium_instability (accessed 14 May 2020).

21 AWRI (2020). *Potassium instability*. AWRI www.awri.com.au/industry_support/winemaking_resources/fining-stabilities/hazes_and_deposits/potassium_instability (accessed 10 August 2020).

22 Lasanta, C. and Gomez, J. (2012). Tartrate stabilization of wines. *Trends in Food Science & Technology* 28 (1): 52–59. https://doi.org/10.1016/j.tifs.2012.06.005.

23 EUR-Lex (2019). Commission Delegated Regulation (EU) 2019/934 of 12 March 2019 supplementing Regulation (EU) No 1308/2013 of the European Parliament and of the Council as regards wine-growing areas where the alcoholic strength may be increased, authorised oenological practices and restrictions applicable to the production and conservation of grapevine products, the minimum percentage of alcohol for by-products and their disposal, and publication of OIV files. https://eur-lex.europa.eu/legal-content/en/ALL/?uri=CELEX:02019R0934-20190607 (accessed 14 May 2020).

24 International Organisation of Vine and Wine (OIV) (2019). *International Code of Oenological Practices*. Paris: OIV. http://www.oiv.int/public/medias/7213/oiv-international-code-of-oenological-practices-2020-en.pdf.

25 International Organisation of Vine and Wine (OIV) (2019). *International Oenological Codex*. Paris: OIV. http://www.oiv.int/en/technical-standards-and-documents/oenological-products/international-oenological-codex.

26 Greeff, A.E., Robillard, B., and du Toit, W.J. (2012). Short- and long-term efficiency of carboxymethylcellulose (CMC) to prevent crystal formation in South African wine. *Food*

Additives and Contaminants – Part A, Chemistry, Analysis, Control, Exposure and Risk Assessment 29 (9): 1374–1385. https://doi.org/10.1080/19440049.2012.694122.

27 Marsh, R. and Mills, S. (2013). Assessment of CMC-induced tartrate stability over 12 months. *Wine & Viticulture Journal* 28 (4): 36–37.

28 Bowyer, P., Moine, V., Gouty, C. et al. (2010). CMC: a new potassium bitartrate stabilisation tool. *Australian & New Zealand Grapegrower & Winemaker* 558: 65–68.

29 Wilkes, E., Tran, T., and Scrimgeour, N. (2013). CMCs, busting the myths! (and adding some new ones). *Proceedings of a seminar organised by the Australian Society of Viticulture and Oenology, Adelaide 2012*. Australian Society of Viticulture and Oenology.

30 Eder, R., Willach, M., Strauss, M., and Philipp, C. (2019). Efficient tartaric stabilisation of white wine with potassium polyaspartate. *BIO Web of Conferences* 15: 02036. 42nd World Congress of Vine and Wine. https://doi.org/10.1051/bioconf/20191502036. https://www.bio-conferences.org/articles/bioconf/pdf/2019/04/bioconf-oiv2019_02036.pdf (accessed 28 June 2020).

31 Enartis (2019). *Wine tartrate stability*. Enartis News https://www.enartis.com/wp-content/uploads/2019/10/Enartis-Newsletter-StabTartarica_Pacific_2.pdf (accessed 12 May 2020).

16

Must Correction, Wine Correction, and Alcohol Reduction Using Membrane Technologies

In this chapter, I discuss the use of membrane technologies in wine production. Technologies covered include reverse osmosis (RO), membrane microfiltration (MF), ultrafiltration (UF), and electrodialysis (ED). There are many occasions during vinification when the technologies may be utilised, including must correction, pH adjustment, wine filtration, wine correction – particularly reducing alcohol content, wine stabilisation, and tartrate stabilisation. Membrane equipment may also be utilised to remove, or at least reduce the levels of, fault compounds including high levels of acetic acid, and also the removal of some taints. I also include an overview of traditional methods of clarification and filtration in common use in wineries.

16.1 Introduction

16.1.1 The Use of Membranes in Stabilising Wines for Export Markets

The changes in the pattern of wine consumption of wine during the last 40 years have been dramatic. There has been spectacular growth in sales in many countries that have limited domestic production and thus import most of the wines offered for sale. Before the 1970s, only 10% of wines were exported from producing countries, including intra-Europe 'exports'. By 2016 over one third of wines globally were consumed in a different country to the one where they were made [1]. This huge rise in exports has been made possible by considerable improvements in the stability of the products sold. The journey undertaken to market by wine in both bottle and bulk may be many thousands of miles from the New World to the Old or vice versa. During such journeys, containers of wine, whether on the decks of ships or in the holds, are often subject to elevated temperatures that encourage microbial growth. The delays at the dockside and long road journeys, often in curtained trailers, also provide conditions that can lead to product deterioration. Membrane microfiltration (MF) and ultrafiltration (UF) have become important tools in ensuring clarity and stability of the product, and their use has become commonplace, at least by large producers. Other membrane processes, including RO, have enabled producers to achieve a product of reasonably consistent quality and style. Global wine sales now amount to the equivalent of 30 billion 75 cl bottles, the majority of which is low-cost wine designed for 'everyday' drinking.

Wine Faults and Flaws: A Practical Guide, First Edition. Keith Grainger.
© 2021 John Wiley & Sons Ltd. Published 2021 by John Wiley & Sons Ltd.

16.1.2 Wine as a 'Natural' Product

It is generally accepted that the quality and style of wines from any region, or any producer small or large will not be consistent from vintage to vintage. It is these variations that make individual wines so exciting to wine lovers. In other words, the quality properties of the finished product are not precisely predictable, unlike the product of other drink production technologies [2]. There is constant debate as to the respective roles of nature and nurture, and as to how much wine production is, or should be, art or science. However, there is no doubt that the use of modern winemaking techniques has led to wines moving closer in style [3], and this is decried by passionate small producers and serious and educated wine lovers. There is a desire amongst consumers for wines that have the minimum of chemical additives, including sulfur dioxide (SO_2), and the minimum of intervention. Membrane technologies sit well with the former, but not so well with the latter, at least in the eyes of the consumer.

16.2 Membrane Processes Used in the Wine Industry

Membrane processing applications relevant to the wine industry include:

- Front-end microfiltration (MF);
- Cross-flow (tangential) microfiltration (CFMF);
- Ultrafiltration (UF);
- Nanofiltration (NF);
- Reverse osmosis (RO);
- Electrodialysis (ED);
- Pervaporation.

In the last few decades, the amount of research into and development of the oenological use of membranes has been immense. In addition to the institutes undertaking broad-based research into food and drink science and technology, research has also been undertaken by institutes specialising in wine research. There is a perception that these are mostly based in so-called 'New World' wine-producing countries. The Australian Wine Research Institute (AWRI), established in 1955, and the Department of Enology of the University of California at Davis (and several other New World institutions) have produced a considerable amount of groundbreaking studies. However, 'Old World' institutions have been equally active. The work of Émile Peynaud, regarded as the grandfather of modern oenology and the first to experiment with RO as a means of must concentration [4], Jean and Pascal Ribéreau–Gayon, Denis Dubourdieu and more recently Pascal Chatonnet's work on taints and Martine Mietton–Peuchot's work on membrane technology at Université Bordeaux (2) Segalen have helped to maintain that institution's position as a world leader. However, the widespread use of membrane technologies in winemaking, particularly in small and medium-sized wineries, is not as universal and has not developed as rapidly as many authors suggest. Even at the start of the current century, the use of membrane filtration in the wine industry was mainly as a safeguard after pad filtration to ensure the complete removal of microorganisms before bottling [5].

Consumers require and expect wines that are clear and stable and, with the exception of bottle matured fine wines that comprise a tiny percentage of the total wine market, free from sediments. However, it is the so-called gatekeepers, not the consumer or producer who have the power in the industry and who prescribe product specifications. In the United Kingdom, supermarkets account for over 75% of all 'off sales' of wine. Many of these specify a complete absence of all viable microorganisms, which means that so-called sterile filtration regimes have to be used [3]. Conversely, Robert Parker recently retired, but undisputedly the most important wine critic of all time is well documented as having an aversion to all filtration. Parker's influence was so great that even Cru Classé Chateaux in Bordeaux adjusted their winemaking to suit his perceived palate. Some other influential critics decry the use of interventionist techniques and what they perceive as excessive manipulation. The websites of many producers of super premium wines boast that the wines are bottled without filtration or only after a light (pad) filtration. 'Good wine is made in the vineyard' is an oft heard expression, meaning that if top quality, healthy grapes are produced from vines that are in balance, the winemaker does not need to carry out other than basic adjustments and manipulations. Careful fruit selection on a sorting table to exclude rotten, damaged, or less than perfectly ripe fruit is now commonplace at quality-conscious producers – 30 years ago all would go into the fermentation vats. Rackings of classic red wines during their time in barrel maturation have been reduced – 30 years ago four rackings in the first year in the chai was commonplace in the Bordeaux region. Today, many producers undertake a single racking.

Although membrane processing should have a good press as methods of improving and stabilising wines that do not involve chemical additions, in practice perceptions are often negative and the processes regarded as highly manipulative, interfering with nature and stripping the heart and soul from wines. However, producers of inexpensive and 'branded' wines need to bring their products to market quickly, and the historic slow passage through the winery has been significantly reduced by using intervention, such as a single pass through a cross-flow membrane filter shortly after fermentation.

16.3 Clarification

16.3.1 'Traditional' Clarification and Filtration Methods in Common Use

There are several methods of clarification and filtration in common use in wineries around the world for the clarification of must, wine, and lees. Centrifugation, generally using a plate separator centrifuge, is effective for basic clarification, and is often used for must. Before fermentation of white wines, must needs to be largely free of solid matter to avoid the development of off-flavours. Whilst natural setting (débourbage) may take place over 24 or more hours, centrifugation eliminates this waiting period. After fermentation, centrifugation may be used at various stages, including post-fining. It is often used pre-filtration. However, centrifugation is a violent method and many wineries avoid the process. Models of the centrifuge are now available that can be flushed with inert gas to prevent the problem of oxygen uptake.

Immediately after the conclusion of the alcoholic fermentation wines will contain considerable solid matter, including grape solids and yeast hulls; with time, most of these solids will settle, and the wine may be racked off. Alternatively, the wine may undergo a depth filtration using Kieselguhr (diatomaceous earth) powder. Diatomaceous earth, which comprises crystalline silica, has been used as a filtration adjuvant since the late nineteenth century [6], but due to health concerns, its use is diminishing. It is a known carcinogen material, the wearing by operatives of personal protection equipment (PPE) is essential. If crystalline silica is inhaled, the sharp particles may become stuck in the lungs which may result in silicosis or cancers. Diatomaceous earth also presents issues regarding deposits of residue in landfill sites. The use of cellulose material, which is an organic and biodegradable alternative, has gained ground in recent years. The filtration may be undertaken using a mobile earth filter or rotary vacuum filter (which will cope with the filtration of lees and reduce wine loss). However, there is a generally unwelcome oxygen uptake during this process. Perlite, made from processed volcanic rock, may also be used for filtration of grape must and wine containing a high percentage of solids. However, it is very abrasive and can cause wear to injection pumps [6].

Although they will not cope with wines containing gross solids, sheet filters, also known as pad filters, and plate and frame filters are very versatile and extensively used in wineries of all sizes, and in small wineries may be the only type of filter available. The sheets are constructed of cellulose fibres, sometimes with the addition of granular components such as Kieselguhr, perlite, or polyethylene fibres, and sometimes also cation resins, with an electrostatic charge, to attract particles with the opposite electrostatic charge [3]. Filter sheets are manufactured in several grades with uses ranging from a basic polishing to complete sterilisation. The large internal surfaces of filter pads enable the retention of considerable volumes of turbid liquid, up to $3 l/m^2$. It is argued that such performance levels are unequalled by membranes [6]. However, the filters gradually block, necessitating a labour-intensive and taking apart and reassembly, which also results in wine loss. The filters can also be messy. It is a very rare exception to see a drip-free example! The enclosed lenticular depth filter is a more hygienic and less labour-intensive piece of equipment.

All the above filtration equipment requires design and manufacture to high standards and careful operation by trained personnel to avoid negative consequences to quality. In particular, ingress of oxygen is particularly damaging. It is to be avoided – ideally, filter equipment should be purged to evacuate air, but in practice, this only happens in the most quality-conscious wineries. Both earth and pad filters present an environmental challenge regarding the disposal of depleted filter material. However, capital costs relatively low compared with most membrane filters.

16.3.2 Membrane Filtration of Must and Wine

16.3.2.1 Clarification of Must
Cross-flow microfiltration (CFMF) is a fast and efficient method of must clarification that can be used instead of traditional methods, including diatomaceous earth filtration and centrifugation. The main issue is that of membrane fouling due to trapping of feed particles. CFMF is further discussed below.

16.3.2.2 Clarification of Wine

Membrane filtration of wine can be carried out as front end microfiltration (MF), which requires a previous clarification and filtration using one of the methods detailed in Section 16.3.1, or CFMF, which will achieve the desired clarity in a single pass.

If a wine contains any residual sugar, and has an alcoholic degree of less than approximately 15.5% abv, there is a considerable risk of a most undesirable re-fermentation taking place in the bottle (or other packaging) if any yeast colony-forming units are present. This topic was discussed in Chapter 9. Thermotic bottling, heating the pre-bottled wine to 54 °C (129 °F) is one method commonly used to guard against this risk, but it does present issues with fill heights as the wine cools. Other methods are flash pasteurisation, immediately before bottling, at a temperature of approximately 75 °C (167 °F) for 30 seconds and post-bottling tunnel pasteurisation at 82 °C (180 °F) for approximately 15 minutes. However, membrane microfiltration immediately before bottling is the method preferred by many quality-conscious producers. It avoids the flavour modification occasionally associated with pasteurisation [7]. Membrane filtration is also a sensible precaution before bottling of wines that have not undergone malolactic fermentation, as the consequences of this taking place in bottle would also result in wine spoilage. *Oenococcus oeni*, *Leuconostoc oenos*, *Pediococcus damnosus*, and *Lactobacillus brevis* are the main bacteria involved and are all removed by 0.45 μm membrane [8]. *L. brevis* (lactic bacteria) can break down compounds in wine and result in a threefold increase in volatile acidity [9]. Yeasts of the species a *Saccharomyces cerevisiae* (the most dominant yeast in most fermentations) are removed using a 0.8 μm membrane pore size. As discussed in Chapter 4, *Brettanomyces bruxellensis* may require 0.8 μm, 0.6 μm, or even 0.45 μm membrane, depending upon the strain. Membranes with pore diameters of 0.2–0.45 μm are typically used for the final filtering of white wine and 0.45–0.65 μm for reds [10]. The same author claims that microfiltration eliminates the use of fining substances. However, many winemakers still undertake traditional fining to remove colloids, even if the wine is destined for microfiltration.

16.3.3 Cross-Flow Micro Filtration

Cross-flow (tangential) filtration was first used for the clarification of wine in the late 1980s. The system will normally comprise a monoblock containing a pre-filter unit (to prevent rapid fouling) and a membrane filtration unit arranged in series. The main advantage over other filtration methods is that it is capable of clarifying wine, achieving microbiological stabilisation, and filtration in a single operation [11].

Even if the juice has an initial turbidity of 400 nephelometric turbidity units (NTUs) the filtered wine can be of brilliant clarity, with a turbidity of just 1 NTU. Cross-flow filters work best when the wine has a wide particle size distribution, which means that they are ideal for filtering wine shortly after fermentation when there is a high concentration of suspended solids. Other accepted advantages with CFMF include a reduction in labour costs compared with multistage filtration, less oxygen pick-up with a single operation and minimal wine waste – 2% or less. The economics of CFMF for lees recovery compared with

Kieselguhr can be particularly attractive to the larger producer. The colour, aroma, and flavour of wine recovered from lees are generally regarded as little affected by the membrane filtration process. If cross-flow filtration is undertaken shortly after the completion of the alcoholic fermentation, the filtrate may be used as inoculums for other alcoholic or malolactic fermentations [7].

There are valuable reasons for performing microfiltration prior to wine going into vat or barrel for maturation, and cross-flow filtration is the only cost and time-efficient procedure at this stage. Spoilage by *B. bruxellensis* was discussed in Chapter 4, but *Saccharomyces ludwigii* and other rogue yeasts are known to spoil wine in tanks. Several film-forming yeasts including *Candida* spp., *Issatchenkia orientalis*, *Pichia membranaefaciens* can also spoil wine in barrels. With good cellar hygiene and meticulous topping up of barrels, such problems are rare.

The automation aspect of cross-flow filtration is attractive to many winemakers – less chance of human error, but this is not accepted by all. Stefano Migotto, the founder of wine specialist filtration company Winetech, states, '… we removed all the automation from our systems. I couldn't trust the accuracy of the computer. Now, if I make a mistake, I make the mistake. I don't want some computer to open or close a valve because some high frequency device is passing by. Now, all of our cross-flow filtration machines are operated manually. That gives me better control of the process' [12].

When cross-flow filters were first introduced, there was a prevalent view amongst wine lovers, critics, and many oenologues that cross-flow filtration was detrimental to the quality of wine [5]. The technologies and membranes used had migrated from other industries, in particular the biopharmaceutical industry, and stripped flavours and tannins from wine. Taints sometimes occurred because of the plastic nature of the membranes, with a correspondent recommendation that membranes were conditioned with acidified hot water before use [5]. There have been huge improvements in membrane technologies in the last three decades, including the introduction of ceramic (inorganic) and wine specific hollow-fibre (organic) membranes. Spiral-wound membranes are now generally falling out of use. Nevertheless, the negative perceptions, although not as universal, linger today. On several occasions, recently when I have visited wineries with professional colleagues, upon seeing the cross-flow filter, the colleagues made remarks such as 'clearly the focus here is not on quality'.

The simple question as to whether modern cross-flow filtration negatively impacts upon aromas, flavour, and quality is perhaps still unanswered. However, Youssef El Rayess and Martine Mietton–Peuchot note that fouling may lead to 'excessive retention of some wine components, which may lead to a loss of some organoleptic characters' [11]. There is also a risk that high molecular weight carbohydrate and protein colloids may be removed during tangential filtration. These are an integral part of the wines composition, aroma, and taste [6]. Attila Rektor et al. [13] conducted sensory analyses on musts of Hungarian varieties (Kekfrankos, Furmint, and a blend) that had undergone MF treatment using hollow-fibre and multi-tube membranes and control samples. Trained assessors considered that the appearance of the micro-filtered musts was more attractive (there was a colour intensity decrease). Although to a certain extent of loss in flavour and odour intensity was observed in the treated samples, the tasters preferred the treated samples as they found the 'off-flavour' less intense. A decrease was also noted in the colour density of a Cabernet Sauvignon wine studied by Arriagada–Carrazana et al. [14] who examined the effects of membrane filtration (0.65 μm final filter) on aromatic and phenolic quality.

The researchers noted that polyphenolics were also decreased and there was a difference in some aroma compounds which they attributed to membrane adsorption. De Pinho reported that the application of MF (and UF) to wine clarification has been restricted by the lack of knowledge on the possible removal of polysaccharides and other macromolecules that may be an important contribution to the quality of wines [15].

16.4 Membrane Fouling

In the last 20 years, there has been considerable advancement in the understanding of fouling of membranes during wine filtration. In 1999, Roger Boulton et al. attributed poor membrane life with certain wines to be mainly caused by colloidal complexes of proteins, phenols, other entities, and polysaccharides [16]. By 1999, it had already been established that polysaccharides with polyphenols are the main compounds responsible for the fouling of membranes in wine clarification [17]. Jeffrey Koehler et al. [18] noted that it is well established that membrane fouling by proteins leads to large flux decline. The strongest fouling occurs on account of hydrophobic interactions between protein and membrane material in combination with conditions where protein–protein interactions are also facilitated. Urkiage et al. [2] found the best membrane material for filtering red wine to be cellulose acetate with pore sizes of 0.2 and 0.45 μm being suitable for the physio-chemical and flux reduction point of view. However, today this would usually be regarded as a too-fine porosity for a quality red wine. In the early 2000s, Aude Vernhet and Michel Moutounet made significant contributions to the understanding of membrane fouling build-up [19], and Vernhet et al. made huge advances in the understanding of the impact on fouling of wine polysaccharides and polyphenols [20]. In a study by Ulbricht et al., it was again noted that aggregates of polyphenols and polysaccharides present in red wine have a major contribution to adsorptive fouling [21]. This study indicated that fouling is strong for polyethersulfone (PES) but very weak for polypropylene membranes [21]. The low adsorption tendency of wine ingredients to polypropylene membranes results in higher fluxes and longer service life of the respective filtration modules in wine clarification. However, membrane fouling remains a major limiting factor to the wider use of this technology in the wine industry.

Wines made from grapes affected by *Botrytis cinerea*, which produces the polysaccharide β-glucan that has a molecular weight of 1 000 000 kDa, can cause severe membrane fouling.

Filtration inhibition starts from 0.6 mg/l of β-glucan with a 0.45 μm membrane filter (a figure much less than previously believed), and concentrations of 1 mg/l and above lead to complete blocking of the membrane after a filtrate quantity of only 1.5 l [22].

16.5 Must Correction, Wine Correction, and Alcohol Reduction

16.5.1 Reverse Osmosis

RO may be used for:

- must correction (pre-fermentation);
- wine correction (post-fermentation) including the removal of excessive acetic acid;
- reducing alcohol content.

The technique is highly suitable for must and wine treatments, taking place at low temperature, in the region of 15 °C (59 °F), with a maximum turbidity of 400 NTU.

16.5.1.1 Must Correction by RO – Increasing Sugar Concentration

If a must has a deficiency of sugar, which will result in an unbalanced wine having an inadequate degree of alcohol, several methods may be available to adjust this. These are chaptalisation (must enrichment), the addition of concentrated grape must or rectified concentrated must, the bleeding (saignée) of juice from vats containing crushed red grapes, cryo-concentration, vacuum evaporation, and RO. Their usage will depend on the legal restrictions within the region of production and the territory where it is sold. Traditionally, in the more northerly producing regions in Europe, the potential alcoholic degree of wines would be increased by the process of chaptalisation (the addition of sucrose). This method has lost favour, partially on account of its impact on wine balance. There has been discussion in the European Union (EU) on banning the practice, but this has not been implemented at the time of writing. The addition of concentrated grape must is occasionally practised in some countries, e.g. Italy. Cryo-concentration involves the cooling of must until it begins to freeze. The unwanted water element, in the form of ice, is removed by filtration. It is claimed that by this method, there is virtually no loss of flavour [3]. In the case of a white must of the Macabeo variety, this method comes closest to removing the water without affecting the other components [23]. Vacuum evaporators have improved somewhat in the last two decades. They now operate at acceptable temperatures, 25–30 °C (77°–86 °F), and aromas are retained in an aroma trap, condensed, and added back to the wine.

The first experiments in using RO for must concentration were made back in 1970 [4]. However, it was to be nearly 20 years (following the advent of composite membranes) before further detailed experiments took place [24]. Using the technology, the water content of the must may be reduced by between 5% and 20%. The method is particularly valuable in reducing water content when there has been rain at harvest time. However, although Ronald Jackson [7] states that RO may be used for immature berries, if the ripeness of the fruit has been restricted due to cold weather, its utilisation for must concentration is less successful. Although sugar will be concentrated, so will acidity and green tannins, resulting in an aggressive and unbalanced wine. Martine Mietton–Peuchot et al. noted that the technique should not be used in palliation of lack of fruit maturity, but that good results are obtained with mature water inflated grapes [25]. Fermentations can slow down when high sugar concentrations have been achieved by various methods, including RO [26]. The juice should be pre-filtered using a membrane of 20 μm or finer before the operation. RO for the purpose of concentration of musts to be fermented into wine was permitted in the European Union from 1999 (Council Regulation (EC) No. 1493/1999, since superseded by subsequent regulations). The maxima set by current EU Regulations and the International Organisation of Vine and Wine (OIV) in the *International Code of Oenological Practices* for must concentration by RO are a maximum 20% decrease in must volume and a maximum 2% increase in potential alcohol [27]. Interestingly, the EU authorities do not consider that this negates the concept of *terroir*. However, when two Bordeaux properties covered their vineyards with plastic sheeting to deflect rainwater during the 2000 growing season, the resulting wines were not accepted as Appellation Controlée, resulting in the properties marketing the wines as the world's most expensive Vins de Table [28].

RO has also been used to produce concentrated juice destined for making icewine-like wines [7]. Experienced tasters will immediately recognise that the wine lacks the distinctive flavours that result from vineyard over-ripening and the freezing of the grapes on the vine, or even by cryoextraction.

16.5.1.2 Must Correction by RO, Ultrafiltration/Nanofiltration – Reducing Sugar Concentration

The impact of climate change, and the harvesting of wine grapes later to ensure phenolic ripeness, has resulted in an often unwelcome increase in must density and consequent high alcohol content after fermentation. Ultrafiltration (UF) plus RO or UF coupled with NF, is one effective method for sugar reduction, which is now permitted by the OIV [27] and EU [29]. The desired reduction in must sugars does involve a slight concentration of other components [30].

16.5.2 Ultrafiltration

Ultrafiltration operates at lower pressures and larger porosities than RO. Ultrafiltration plus nanofiltration can also be used for sugar removal from must, NF giving greater flow rates than RO [31]. Ultrafiltration can be a useful tool to adjust the style of red wine after fermentation. Tannins can, if required, be selectively removed based on molecular weight. The tannins and colour are held in the retentate, i.e. producing a red wine with deeper colour and firmer tannic structure than the base wine. There is the possibility of adding these tannins and colour from the retentate to other red wines should they have a deficiency of these. Alternatively, UF can produce two separate wines that can stand alone: (i) the permeate a lighter, smoother red, or even a rosé, blush or blanc de noirs style, and (ii) the retentate a red with full body and firm tannic structure. Bitterness resulting from the inclusions of hard press fractions can also be reduced. However, it should be noted that careful handling of sound fruit should have avoided these problems arising. UF can be used to remove proteins, and therefore can offer security against protein haze formation. The use of ultrafiltration to help correct wine flaws can be particularly valuable. In the case of white wines, the removal of tint and pinking, and in the case of reds, removing oxidised phenolics that would otherwise cause premature browning. These can be selectively removed based upon molecular weight. The precursors of browning can also be removed. Also, UF may have beneficial effects in rejuvenating old wines [7].

16.5.3 Wine Correction – Reducing Alcohol Content

The mean alcohol content of wines has increased considerably in the last 30 years. As well as the impact of climate change, the industry has been held hostage to the concept of phenolic ripeness, believing that the market and some critics demand softer styles of reds than in the past [28]. In 1989, in the first edition of *Making Good Wine*, referring to Australian Cabernet Sauvignon Bryce Rankine wrote: 'A ripeness of 10–12 °C Baumé (18–21.6° Brix) is usual, which results in wine containing between about 10 and 12% alcohol by volume' [32]. Today, alcohol levels over 14.5% are commonplace, and in some regions, producers struggle to restrain alcoholic content and keep the wine balanced. A report from the AWRI details that between 1984 and 2004, average alcohol levels increased in wines made from all grape varieties and in all Australian regions, with red wines rising from an average of 12.3% to 13.9% abv [33]. An increase from 12.3% to 13.9% abv is not an increase of 1.6% alcohol, but an increase of 13% in the alcohol content. In recent years, wines from many countries, including California (USA) and Chile, have faced adverse criticism from wine experts and others on account of their perceived excessive alcohol. Apart from the obvious health and social issues raised by an increase in average alcohol levels, excessive alcohol in any individual wine numbs the fruit, inhibits freshness and can even give a burning sensation on the palate. The wine is unbalanced, less drinkable in other than relatively small quantities and may be considered to be flawed.

There are several techniques available for alcohol reduction. All of these have organoleptic impacts and other positive and negative consequences. The addition of unfermented grape juice (süssreserve) is a process once commonly used in Germany during the production of inexpensive wines, with an unfermented grape juice addition representing 10–15% of the total volume, and resulting in a finished wine that is medium-sweet with a high level of primary fruit aromas, which is the main reason for the operation. As consumer preference for dry wines has increased in recent years, the use of süssreserve has considerably declined. Processes based on high temperature (evaporation) result in strong alteration or loss of the wine aroma [34] and are not generally used in the wine industry. One technique that is illegal in most production countries (including member states of the EU) is simply adding water to the must, fermenting juice or wine. Technical processes include spinning cones, RO, RO plus perstractive membrane, and RO plus distillation. However, the use of RO for this purpose has remained controversial [28]. A study by Catarino and Mendes [35] indicates that NF and pervaporation are effective for dealcoholising wine and retaining original characteristics. Another study by Agence Nationale de La Recherche [36] looked at various methods of dealcoholising wines, including treatment of must by the combination of UF–NF already noted, and dealcoholisation of finished wine by blending after partial or total distillation, stripping, RO combined with distillation or membrane contactors, NF combined with membrane contactors and direct treatment by membrane contactors. The study found that in most cases, a treatment that eliminated 2° of alcohol, obtained by blending of the original wine with the wine dealcoholised by distillation or the combination of RO combined with membrane contactors results in products that cannot be distinguished from the control wine. This study claims that in the case of the samples of Merlot, Syrah, Chardonnay, and Sauvignon Blanc and with original pre-treatment alcohol degrees of 13–14% abv, reductions in alcohol of up to 3% were not perceptible to the average consumer (even with sensory analysis training). However, wine professionals rated reduced alcohol wines lower, citing a loss of complexity [36].

Interesting tests in dealcoholisation of red wine by direct RO with a pilot plant equipped with a polypropylene hollow-fibre membrane contractor were conducted in the study by Liguori et al. [34]. After five cycles alcohol was reduced from 12.8% to less than 0.5%, most of the reduction taking place in the first two cycles. The resultant dealcoholised 'wine' was lacking in aroma, but the authors state that the process is suitable for use in the wine industry to reduce the total alcohol content.

16.6 Fault Correction

16.6.1 Removing Acetic Acid

An excess of volatile acidity is generally regarded as a fault in wines, as discussed in Chapter 7. If the level is very high, the wine may smell of vinegar. Although other acids, including carbonic, sulfurous, butyric, and formic acid are volatile, acetic acid is the main culprit. As with ethanol, the molecules of acetic acid are very small and can pass through RO membranes. However, effective removal of volatile acidity can be achieved by RO plus ion exchange or by RO plus adsorption. However, the level of volatile acidity that is acceptable in any wine is subjective. For example, high volatile acidity is regarded by connoisseurs as part of the intrinsic character of many reds from Italy.

16.6.2 Removal of *Brettanomyces* Yeasts, and the Treatment of Volatile Phenols

With the increase in gatekeepers demanding red wines with a few grammes of residual sugar, the increase in alcohol content in most wines during recent years, and the drive by all involved in the industry including winemakers, regulatory authorities, the press and consumers to reduce the levels of sulfur dioxide in wine, it is not surprising that the instances of red wines tainted with compounds metabolised by *Brettanomyces* spp. (commonly known as 'Brett') have increased during the last decades. *Brettanomyces* is discussed in detail in Chapter 4.

A complete treatment for *Brettanomyces* and related volatile phenols is a two-stage operation: removal of the viable *Brettanomyces* cells, and removal of the odorous synthesised compounds, including 4-ethylphenol (4-EP) and 4 -ethylguaiacol (4-EG).

Brettanomyces cells are usually between 2 and 7 μm in size, but they can be smaller, depending upon the strain, and the cell size is likely to be reduced when the yeast is stressed. Accordingly, many strains of the yeast will be removed by filtration using 1.2 or 1 μm material, but others are only removed with 0.8 [37], 0.6 μm or even 0.45 μm (sterile) material. Sterile membrane filtration, using a medium with a pore size no greater than 0.45 μm, is almost always effective in removing *Brettanomyces* [38, 39]. However, it is possible that cells in a viable, but non-culturable (VBNC) state may still pass through this size. This may happen when the cells are stressed, as is the case following sulfur additions. The yeasts may become smaller or elongated. Complete assurance would be obtained with a 0.22 μm membrane, but there is little doubt that this can result in severe organoleptic modification.

The composition of the filter material is pertinent, as even with larger pore sizes some materials can be effective, and result in less impact upon the colloidal structure of the wine. In a study by Duarte et al. [40], polypropylene filters showed poor efficacy – *B. bruxellensis* colonies were too numerous to count following filtration. Using PES filters 1.0 μm, 0.65 μm, and 0.45 μm (two lots and respective duplicates), no cells were detected in any of the filtered wines analysed. The PES filters showed high *B. bruxellensis* removal efficacy in all micron rating tested. Similar efficacy was achieved for X grade borosilicate glass microfibre filter material. Polypropylene and V grade borosilicate glass microfibre was also shown to offer less retention. It should be noted that any filtration is only a transitory action, and the process will not protect the wine against subsequent *Brettanomyces* contamination and development of viable cells.

Further treatment is necessary to reduce the levels of related volatile phenols. Ugarte et al. obtained up to a 77% reduction in the total ethyl phenols (4-ethylguaiacol and 4-ethylphenol) to what was considered to be acceptable levels [41]. The three-hour process comprised a two-stage integrated process, a membrane with cross-flow filtration and a hydrophobic absorbent resin (Amberlite XAD-16 HP). However, a reduction in some (desired) aromatic compounds, including ethyl and methyl vanillate and other esters was also reported.

16.6.3 Removal of Smoke Taint

Grapes and wines may suffer smoke taint as a consequence of bush fires or controlled burning taking place near vineyards. This topic is discussed in detail in Chapter 12. In particular, Australia, South Africa, and California has had many instances of wines impacted by smoke taint in the last 10 years. Other countries, including Chile, have also had wines affected. Affected wines may smell of burnt ash, smoked salmon, or medicinal. As with *Brettanomyces*, high concentrations of volatile phenols are present in tainted wines. In this case, the main compounds are guaiacol and 4-methyl-guaiacol. Wines may contain low concentrations of these as a result of oak maturation or treatments.

Studies, tests, and commercial applications in attempting to remove smoke taints by RO have been recently taking place. The first study to examine reducing smoke taint in wines using RO and solid-phase adsorption was carried out by Fudge et al. [42]. Three samples of only one variety (Pinot Noir) was used for the evaluation. The study concluded that the process reduced the level of volatile phenols derived from smoke, and improved the aromas and taste. However, the taints were found to return after a period, and the authors put this down to hydrolysis of glycoconjugate precursors which the treatment process did not remove. The process of reduction by RO would be helped by rapid measures of smoke-related compounds (e.g. UV 275 nm phenolics measures) as they go through the RO process to assist in determining when appropriate reductions have been achieved. An improved understanding of the full range of smoke taint compounds is also required to assist in the design of specific membranes to filter these compounds selectively. Guaiacol and some other smoke-related compounds are relatively small molecules so they can be removed, but other larger compounds that are not removed may dominate and become more evident in the aroma. Although there is potential for using this technique when the smoke taint is low in highly tainted wines, it would take many passes through the system to reduce levels to below the sensory threshold concentrations of smoke taint compounds. There is also concern that some fruit characters may be removed and the smoke characters come-to-the-fore.

16.7 Wine Stabilisation and pH Adjustment by Electrodialysis

Membrane electrodialysis technique is widely used in large wineries for tartrate stabilisation and pH adjustment in wines. Electrodialysis is an electrochemical process in which ions dissolved in a solution move to another solution, employing an electrical potential as the driving force [11]. Ion-selective semi-permeable membranes distinguish the charged ions.

16.7.1 Tartrate Stabilisation

Tartrate crystals in the bottle or glass are *not* a wine fault and are very often visible in wines of the very highest quality. The crystals are usually potassium bitartrate (potassium hydrogen tartrate [KHT]) or occasionally calcium tartrate. This topic is discussed in detail in Chapter 15.

The conventional way of stabilising tartrates, particularly potassium bitartrate, is by a cold stabilisation, or contact process. Using cold stabilisation, the wine must be fined to negate the protective effect on micro-crystals. It is then chilled to just above its freezing point, e.g. chilled to −4.5 °C (24 °F) for a wine containing 12% abv or chilled to −8.5 °C (16.7 °F) in the case of a fortified wines containing 20% abv and is held in insulated tanks at this temperature for at least 10 days and often considerably more. The process is inefficient and expensive in terms of capital outlay and the running costs of refrigeration units [3]. A more effective, quicker, and cheaper system is the contact process, which involves chilling wine to chilling wine to −3 °C (27 °F) to +4 °C (39 °F) and adding 1–2 g/l of purified tetrahydrated crystals of potassium bitartrate. A vigorous stirring follows this for at least four hours to keep the crystals in suspension and maximise the contact area. This method usually referred to as 'the contact process', will take six days or so.

Ion exchange, replacing potassium and calcium ions in wine with hydrogen ions, is now permitted in the EU for this purpose. The addition of metatartaric acid is permitted in the EU up to 100 mg/l, but it is only effective at protecting from tartrate precipitation for a limited amount of time. The addition of carboxymethylcellulose (CMC), is also permitted for white wines, and potassium polyaspartate and/or mannoproteins for all wines, but these only effective for stabilising potassium bitartrate.

A far more effective method of tartrate stabilisation is by the use of electrodialysis, which uses selective membranes allowing the passages of potassium, calcium, and tartrate ions under the influence of an electrical charge. The machine operation can be monitored by continuous measurement of the wine being treated. The removal of tartrates is fast (a single cross-flow pass), efficient and reliable, and is adjusted to the characteristics of the wine being processed. Minimal pre-treatment of the wine is required. It is generally accepted that there is no effect on the organoleptic characteristics of wine treated.

In 2010 the AWRI undertook an evaluation of an electrodialysis system to remove KHT from wine [43]. The evaluation was conducted in association with Memstar (now VAF Memstar), specialists in membrane technology for the wine industry (and other industries) and Orlando Wines, owned by Pernod Ricard. Highlights of the evaluation, which considered several parameters, are shown as Table 16.1.

The speed of the electrodialysis process makes the system particularly attractive to producers who wish to get a wine such as a new season's Sauvignon Blanc to market early. There is also less wine loss, perhaps 1%, compared with up to 3.5% with cold stabilisation [44]. As the process takes place without refrigeration, there is a considerable reduction in energy usage and costs, up to 95% compared with the 10+ days for cold stabilisation detailed

Table 16.1 Comparison of electrodialysis vs cold stabilisation for potassium tartrate stability.

Quantity	Electrodialysis	Cold stabilisation
Volume of wine processed (l)	29 100	29 100
Power consumption (kWh)	77	1761–1968
Water consumption (l)	7683[a)]	3606
Waste water	7683	1581
Wine losses (l)	136	424
Labour (h)	17	9
Time taken for process(h)	17	384

a) The membranes needed a special clean, which increased the water
    consumption and waste.
Source: Based on Forsyth [43]. AWRI Report Comparison between electrodialysis and cold treatment as a method to produce potassium tartrate stable wine. Glen Osmond, the Australian Wine Research Institute.

above [44]. Fok reported that using electrodialysis process to stabilise, a wine consumed 8 watt hours (Wh) of electricity per (US) gallon compared with 22 Wh for wine cold stabilised in insulated tanks, and 1200 Wh if the tanks are uninsulated [45]. Although Hestekin claims that the process is commonly used in the wine industry [46], the capital cost of equipment is high, making the machines unaffordable for most small wineries. However, in some regions, mobile machines and operators can be hired.

16.7.2   pH and Acidity Adjustment

The pH of must or wine can be reduced by using bipolar membrane electrodialysis. The process removes potassium and replaces it with hydrogen. The pH is reduced, but the titratable acidity remains less changed, as does the level of sugar and phenolic compounds, and in the case of wine, alcohol. The process achieves very accurate pH reductions, a typical drop after a single pass being by approximately 0.1–0.2 pH units, the legal maximum according to International Oenological Codex is 0.3 pH units [47]. Brightness, colour, aromas, and flavours are unaffected, which is not the case when tartaric acid is added, a process that is commonplace (especially for white wines) produced in New World countries and the hotter parts of southern Europe. As an extra benefit, the by-product of the process is suitable for use in wineries as a (most desirably chlorine-free) cleaning agent.

16.8   Final Reflections

Despite, or perhaps because of, the developments in the last 60 years of wine technology in both the vineyard and winery, the move today, particularly amongst smaller quality-conscious producers , the so-called 'boutique wineries', is very much towards wines produced with minimum intervention. The media and the consumer perceive these to be more natural, although the term 'natural wine' has its own, still somewhat ill-defined, meaning. Although the individual effects might be small, in every stage of handling off wine, there is the potential for a loss of quality [3]. Whilst few would doubt that the use of modern technology has improved the average quality and consistency of wines, it is generally accepted that its use is no guarantee of making good wine.

Many growers have converted to organic or even biodynamic practices in the vineyard, and the 'natural wine' movement is rapidly gaining momentum as illustrated by the holding of annual major 'natural' wine fairs in London, Berlin, New York, Miami, and Los Angeles. To the wine lover, wine is the result of a unique combination of vineyard climate (including meso- and micro-climate) soil, aspect, the combination of which may be termed 'terroir', the weather in a particular year, grape variety(ies) and sensitive production techniques in both the vineyard and winery. However, bulk wine production is rapidly moving into a parallel world of 'the new oenology' as detailed by Fischer [48]. Here, grapes are merely a commodity, a source for valuable sugars, acids, flavours, tannins, and pigments. In this world, grape processing obtains a defined must in a reproducible manner. Wine corrections are made by RO or spinning cone columns to achieve specifications. Ethanol, volatile acidity, aromas, flavours, off-flavours, tannins, and coloured pigments may be fractionalised and recombined to produce a wine in the required style [48] .

The capital costs of membrane equipment are high, well beyond the means of many boutique wineries, as the equipment is lying idle for the vast majority of the time. However, for top-quality producers, such as the premier cru classés of Bordeaux, any techniques and equipment that help their wine achieve the highest perceived quality status is worth the cost. Interestingly, when I have visited the cuveries and chais of several grand cru classés the chief winemaker has readily shown the latest purchases optical grape selection equipment, flotation pumps, and nitrogen generators. However, the membrane equipment, particularly the RO machine, have always remained well hidden. The producers of many top wines in both the old and New Worlds have stated that they do not use any membrane equipment in the winery. The development of cheaper and longer-lasting membranes will make processing equipment more attractive to smaller wineries and less well-off wineries. The appearance of mobile RO and electrodialysis machines is beginning the make the technology available to those to whom capital outlay is the main concern.

Perhaps the most exciting area for development of membrane technologies will be the arena of the removal of fault compounds. *Brettanomyces* and smoke taint and have been discussed previously and in Chapters 4 and 12. Without a doubt, the most serious of taint compounds that may affect wines is contamination by haloanisoles, discussed in Chapter 3. Although haloanisoles are not removed by traditional filtration or RO, it would appear that progress is being made towards effective removal with other membrane technologies. So-called ladybird taint, as discussed in Chapter 13, is a growing and serious problem in some wine regions. Of all fining treatments, only charcoal is effective at removal of this taint, but unfortunately, it also removes desirable aromas and flavours. This is certainly another area where developments in membrane technology may provide an urgently needed solution.

Advances have been made in the knowledge of filtration membrane fouling resulting in considerable technological improvements. However, fouling remains an issue to be resolved. El Rayes et al. [49] noted that the large-scale development of CFMF is still hampered by the barriers induced by membrane fouling – it might be said to be the major block to the more widespread use of membranes in the wine industry.

The pace of advancement in technology leaves legislation trying to catch up. The OIV details conditions and limitations of the usage of winemaking techniques in the *International Code of Oenological Practices* [27]. The International Oenological Codex gathers descriptions of the main chemical, organic, and gas products used in the making and keeping of wines [47]. For wines produced or marketed in the European Union, the EU Regulations detail permitted practices. These are now in line with the International Code of Oenological Practices. Whilst derogations may be given in the EU for experiments unless a practice is in the regulations, no wine for commercial sale may be made using such a practice. However, it can be argued that such restrictions delay and limit future developments in the oenological use of membrane technology.

References

1 Anderson, K., Nelgen, S., and Pinilla, V. (2017). *Global Wine Markets, 1860 to 2016: A Statistical Compendium*. Adelaide: University of Adelaide Press. https://doi.org/10.20851/global-wine-markets License: CC-BY 4.0.

2 Urkiage, A., de las Fuentes, L., Acilu, M., and Uriarte, J. (2002). Membrane comparison for wine clarification by microfiltration. *Desalination* 148: 115–120. https://doi.org/10.1016/S0011-9164(02)00663-X.

3 Bird, D. (2010). *Understanding Wine Technology*, 3e. Newark: DBQA Publishing.

4 Peynaud, E. and Allard, J.J. (1970). Concentration des moûts de raisin par osmose inverse. *CR Académie d'Agriculture de France* 56 (18): 1476–1478.

5 Rankine, B. (2004). *Making Good Wine*, revised Macmillan edn. Sydney: Pan Macmillan Australia Pty Limited.

6 Ribéreau-Gayon, P., Glories, Y., Maujean, A., and Dubourdieu, D. (2006). *Handbook of Enology: The Chemistry of Wine: Stabilization and Treatments*, vol. 2. Chichester: Wiley.

7 Jackson, R.S. (2020). *Wine Science: Principles and Applications*, 5e. San Diego, CA: Academic Press.

8 Starbard, N. (2008). *Beverage Industry Microfiltration*. Iowa: Wiley-Blackwell.

9 Ribéreau-Gayon, J., Peynaud, E., Ribéreau-Gayon, P., and Sudraud, P. (1976). *Sciences et Techniques du Vin: Vinification – Transformation du Vin*, vol. III. Paris: Dunod.

10 Lipnizki, F. (2010). Cross-flow membrane applications in the food industry. In: *Membrane Technology: Membranes for Food Applications*, vol. 3 (eds. K.-V. Peinemann, S.P. Nunes and L. Giorno), 1–24. Weinheim: Wiley-VCH.

11 El Rayess, Y. and Mietton-Peuchot, M. (2016). Membrane technologies in wine industry: an overview. *Critical Reviews in Food Science and Nutrition* 56 (12): 2005–2020. https://doi.org/10.1080/10408398.2013.809566.

12 Cutler, L. (2007). Industry Round Table: Filtration. Winemakers discuss the truth, fiction and consequence of filtration in winemaking, *Wine Business Monthly*, July. https://www.winebusiness.com/wbm/?go=getArticleSignIn&dataId=49276 (paywall) (accessed 5 July 2020).

13 Rektor, A., Pap, N., Kókai, Z. et al. (2004). Application of membrane filtration methods for must processing and preservation. *Desalination* 162: 271–277. https://doi.org/10.1016/S0011-9164(04)00051-7.

14 Arriagada-Carrazana, J.P., Sáez-Navarrete, C., and Bordeu, E. (2005). Membrane filtration effects on aromatic and phenolic quality of Cabernet Sauvignon wines. *Journal of Food Engineering* 3: 363–368. https://doi.org/10.1016/j.jfoodeng.2004.06.011.

15 De Pinho, M.N. (2010). Membrane processes in must and wine industries. In: *Membrane Technology, Membranes for Food Applications: Membranes for Food Applications*, vol. 3 (eds. K.-V. Peinemann, S.P. Nunes and L. Giorno), 105–118. Weinheim: Wiley-VCH.

16 Boulton, R.B., Singleton, B.F., Bisson, L.L., and Kunkee, R.E. (1999). *Principles and Practices of Winemaking*. New York: Aspen Publishers, Inc.

17 Vernhet, A., Pellerin, M.P., Belleville, J. et al. (1999). Relative impact of major wine polysaccharides on the performances of an organic microfiltration membrane. *American Journal of Enology and Viticulture* 50 (1): 51–56.

18 Koehler, J.A., Ulbricht, M., and Belfort, G. (2000). Intermolecular forces between a protein and a hydrophilic modified polysulfone film with relevance to filtration. *Langmuir* 16: 10419–10427. https://doi.org/10.1021/la000593r.

19 Vernhet, A. and Moutounet, M. (2002). Fouling of organic microfiltration membranes by wine constituents: importance, relative impact of wine polysaccharides and polyphenols and incidence of membrane properties. *Journal of Membrane Science* 201 (1–2): 103–122.

20 Vernhet, A., Cartalade, D., and Moutounet, M. (2003). Contribution to the understanding of fouling build-up during microfiltration of wines. *Journal of Membrane Science* 211 (2): 357–370. https://doi.org/10.1016/S0376-7388(02)00432-5.

21 Ulbricht, M., Ansorge, W., Danielzik, I. et al. (2009). Fouling in microfiltration of wine: the influence of the membrane polymer on adsorption of polyphenols and polysaccharides. *Separation and Purification Technology* 68 (3): 335–342. https://doi.org/10.1016/j.seppur.2009.06.004.

22 Schneider, I. (2011). The influence of colloidal treatment media on wine filterability. *Der Winzer* (3): 16–17.

23 Hernández, E., Raventós, M., Auleda, J.M., and Ibarz, A. (2009). Freeze concentration of must in a pilot plant falling film cryoconcentrator. *Innovative Food Science and Emerging Technologies* 11 (1): 130–136. https://doi.org/10.1016/j.ifset.2009.08.014.

24 Guimberteau, G., Gaillard, M., and Wajsfelner, R. (1989). Observations récentes sur l'utilisation de l'osmose inverse en vinification. *Connaissance de Vigne et du Vin* 23 (2): 95–118.

25 Mietton-Peuchot, M., Milisic, V., and Noilet, P. (2002). Grape must concentration by using reverse osmosis. Comparison with chaptalization. *Desalination* 148: 125–129. https://doi.org/10.1016/S0011-9164(02)00665-3.

26 Ribéreau-Gayon, P., Dubourdieu, D., Donèche, B., and Lonvaud, A. (2006). *Handbook of Enology: The Microbiology of Wine and Vinifications*, vol. 1. Chichester: Wiley.

27 International Organisation of Vine and Wine (OIV) (2020). *International Code of Oenological Practices*. Paris: OIV. https://www.oiv.int/en/technical-standards-and-documents/oenological-practices.

28 Grainger, K. and Tattersall, H. (2015). *Wine Production and Quailty*, 2e. Chichester: Wiley Blackwell.

29 EUR-Lex (2019). Commission Delegated Regulation (EU) 2019/934 of 12 March 2019 supplementing Regulation (EU) No 1308/2013 of the European Parliament and of the Council as regards wine-growing areas where the alcoholic strength may be increased, authorised oenological practices and restrictions applicable to the production and con-servation of grapevine products, the minimum percentage of alcohol for by-products and their disposal, and publication of OIV files. https://eur-lex.europa.eu/legal-content/GA/TXT/?uri=CELEX:32019R0934 (accessed 25 May 2020).

30 Massot, A., Mietton-Peuchot, M., Peuchot, C., and Milisic, V. (2008). Nanofiltration and reverse osmosis in winemaking. *Desalination* 231: 283–289. https://doi.org/10.1016/j.desal.2007.10.032.

31 Mietton-Peuchot, M. (2010). New applications for membrane technologies in enol-ogy. In: *Membrane Technology: Membranes for Food Applications*, vol. 3 (eds. K.-V. Peinemann, S.P. Nunes and L. Giorno). Weinheim: Wiley-VCH.

32 Rankine, B. (1989). *Making Good Wine*, 1e. Sydney: The Macmillan Company of Aus-tralia.

33 Godden, P. and Gishen, M. (2005). Trends in the composition of Australian wine. *Wine Industry Journal* 20 (5): 21–46.

34 Liguori, L., Attanasio, G., Albanese, D., and di Matteo, M. (2010). Aglianico wine dealcoholization tests. In: *20th European Symposium on Computer Aided Process Engi-neering – ESCAPE20*, 325–330. Amsterdam: Elsevier.

35 Catarino, M. and Mendes, A. (2011). Dealcoholizing wine by membrane separation pro-cesses. *Innovative Food Science & Emerging Technologies* 12 (3): 330–337. https://doi.org/10.1016/j.ifset.2011.03.006.

36 Agence Nationale de La Recherche (2009). Quality wines with reduced alcohol content (VDQA). ANR-05-PNRA-011.

37 Umiker, N.L., Descenzo, R.A., Lee, J., and Edwards, C.G. (2012). Removal of *Bret-tanomyces bruxellensis* from red wine using membrane filtration. *Journal of Food Pro-cessing and Preservation* 37: 799–805. https://doi.org/10.1111/j.1745-4549.2012.00702.x.

38 Suárez, R., Suárez-Lepe, J.A., Morata, A., and Calderón, F. (2007). The production of ethylphenols in wine by yeasts of the genera *Brettanomyces* and *Dekkera*: a review. *Food Chemistry* 102 (1): 10–21. https://doi.org/10.1016/j.foodchem.2006.03.030.

39 Calderón, F., Morata, A., Uthurry, C., and Suárez, J.A. (2004). Aplicaciones de la ultra-filtración en la industria enológica – Ultimos avances tecnológicos. *Tecnología del Vino* 16: 49–54.

40 Duarte, F.L., Coimbra, L., and Baleiras-Couto, M. (2017). Filter media comparison for the removal of *Brettanomyces bruxellensis* from wine. *American Journal of Enology and Viticulture* 68: 504–508. https://doi.org/10.5344/ajev.2017.17003.

41 Ugarte, P., Agosin, E., Bordeu, E., and Villalobos, J.I. (2005). Reduction of 4-ethylphenol and 4-ethylguaiacol concentration in red wines using reverse osmosis and adsorption. *American Journal of Enology and Viticulture* 56: 30–36.

42 Fudge, A.L., Ristic, R., Wollan, D., and Wilkinson, K.L. (2011). Amelioration of smoke taint in wine by reverse osmosis and solid phase adsorption. *Australian Journal of Grape and Wine Research* 17 (2): S41–S48. https://doi.org/10.1111/j.1755-0238.2011.00148.x.

43 Forsyth, K. (2010). *AWRI Report Comparison Between Electrodialysis and Cold Treatment as a Method to Produce Potassium Tartrate Stable Wine*. Glen Osmond: The Australian Wine Research Institute. www.awri.com.au/wp-content/uploads/report_forsyth_PCS10004.pdf.

44 Thurston, C. (2008). Electrodialysis catches on for acid reduction. Tartrate removal system uses less 'juice' to stabilize wine and can save wineries thousands of dollars in energy costs. *Wine Business Monthly*. August 15.

45 Fok, P.E. (2008). PG&E studies electrodialysis for cold stability. *Practical Winery & Vineyard Journal*: 1–4.

46 Hestekin, J., Ho, T., and Potts, T. (2010). Electrodialysis in the food industry. In: *Membrane Technology: Membranes for Food Applications*, vol. 3 (eds. K.-V. Peinemann, S.P. Nunes and L. Giorno), 75–104. Weinheim: Wiley-VCH.

47 International Organisation of Vine and Wine (OIV) (2020). *International Oenological Codex*. Paris: OIV. http://www.oiv.int/en/technical-standards-and-documents/oenological-products/international-oenological-codex.

48 Fischer, U. (2006). Application of new technologies in the field of international oenology. *XIX Giornata Internazionale Vitivinicola*, 20 May 2006, Siena.

49 El Rayes, Y., Albasi, C., Bacchin, P. et al. (2011). Cross-flow microfiltration applied to oenology: a review. *Journal of Membrane Science* 382: 1–19.

17

The Impact of Container and Closure Upon Wine Faults

This chapter provides an overview of the impact of packaging materials on wine quality and aspects that may be implicated in wine faults. The topics covered include glass bottles, particularly with regard to the colour of the material, and polyethylene terephthalate (PET) bottles, bag-in-box, cans, natural cork stoppers, technical corks, screw caps, and synthetic closures. On each material, permeation, migration, and scalping are briefly considered. Cartons, including Tetrapaks are not discussed: these have a tiny market share.

17.1 Introduction

The impact of the container and closure upon packaged wine has been the subject of much discussion and research during the last 20 years. Many issues surround perceptions – both from consumers and journalists, writers, and wine critics. The popular wine press has featured numerous articles and comment on the topic of closures, their effects on wine development and preservation and their implication in possible faults and taints, but there has been less discussion about packaging containers, the effect of various colours of glass bottles, and the consequences of storing wine in materials other than glass. The choice of packaging design and materials is of particular pertinence to supermarket and other large trade buyers who will detail product parameters, a particular design and specification of container, and, in the case of bottled wine, the type of closure to be utilised.

In terms of product quality, there are three major issues to consider with regard to the impact of packaging upon the preservation of the quality of the wine at the time of bottling, its development in bottle, and its protection from faults and taints:

- Permeation
- Migration
- Scalping

Permeation is the extent to which substances present in the external environment, such as water, oxygen, other gasses, and microorganisms can penetrate through the packaging or closure material. Migration is the desorption of substances, including gasses, from the packaging material. Scalping is the sorption of aroma compounds and flavour constituents of the wine by packaging materials. The following sections will briefly consider each with regard to the type of container and, in the case of bottled wine, closure.

Wine Faults and Flaws: A Practical Guide, First Edition. Keith Grainger.
© 2021 John Wiley & Sons Ltd. Published 2021 by John Wiley & Sons Ltd.

17.2 Glass Bottles

17.2.1 Overview

The vast majority of wine in all major markets is packaged in glass bottles. It may seem incredible that some 300 years or so after glass bottles became the container of choice, it still is. Some 87% of wine packaging units are glass bottles [1]. Other forms of packaging including bag-in-box, cans, and PET bottles have made some inroads into the total market, but as one begins to move up the wine quality scale, glass bottles fast become the only container acceptable to both producers and consumers. Glass is subject to 0% gas and liquid permeation, 0% migration, and 0% flavour scalping. However, depending upon the thickness and colour of the bottles, there is a degree of penetration of ultraviolet (UV) light and visible light in the blue end of the spectrum that can have negative impact upon wine quality. Far more often than is generally realised, this can result in the fault of 'lightstrike', which is discussed in Chapter 6, or general product deterioration. Glass boasts considerable environmental credentials: it is 100% recyclable and can be reused, although sadly reuse has long since ceased in the wine industry. However, there are environmental downsides too. Glass manufacturing is among the most energy-intensive industries – in the United States in 2010 the energy usage in manufacturing glass containers (of all types) accounted for over 50 trillion British thermal units (Btus) [2]. Glass is heavy: a typical 75 cl bottle weighs between 450 and 550 g, making it the heaviest of all wine packaging containers, with a consequential greater environmental impact with regard to transportation.

17.2.2 A Brief History of Glass Bottles for Wine

There is some dispute as to when and where the first 'blown' glass bottles were produced, but they were certainly in existence by the first century BC (BCE). Mould-blown glass containers developed from the first century AD (CE), but production was a costly operation. Inexpensive glass bottles were 'invented' by an Englishman, Kenelm Digby in 1632 [3]. Prior to the 1630s furnaces in glass foundries had been fired by wood. Digby used coal firing together with fans in his manufacturing process, resulting in bottles that were much stronger than had previously been possible. His bottles had a rimmed neck too and could be sealed with a cork, although at this time wine would be transported in bulk. It would be dispensed from barrels in hostelries, and in large country and town houses bottled in the cellar directly from the barrel by the bottler (later to become known as the butler). Between 1660 and 1730 wine bottles were bulbous in shape, but by the end of the eighteenth century, and with the beginning of the concept of ageing of wines, the move to a cylindrical shape was well underway, which was almost complete by 1840. Of course, other shapes exist even today, such as the flagon-shaped 'bocksbeutel' commonly used for the wines from Franken and parts of Baden in Germany. However, it is perhaps a consequence of globalisation and economic reality that the wonderful array of shapes and sizes that once filled the shelves of wine merchants have mostly been replaced by the high-shoulder 'Bordeaux' bottle and round-shoulder 'Burgundy' bottle.

17.2.3   The Impact of Bottle Size Upon Wine Development and Quality

Today the 'standard' bottle size for wines, in most countries, is 75 cl. Prior to the 1970s individual regions and countries often had slightly different sizes. For example, Bordeaux wine was commonly bottled in 73 cl and German wine in 70 or 71 cl capacity. There are, of course, many small capacities such as 18.75 cl, 25 cl, 37.5 cl (half bottle), and 50 cl. Wines in bottle sizes of less than 75 cl generally mature faster and deteriorate quicker, and such small formats are regarded as less appropriate for long-term ageing. Wine matures more evenly in larger format bottles. The magnum of 1.5 cl is a size sought after by many collectors of fine wines, having the advantage of greater size and yet not being too unwieldly. The neck size of the magnum bottle is the same as for a standard bottle, and thus proportionately the volume of wine in contact with the closure is less. Magnums often attract over twice the price of standard bottles, even at the time of initial sale. Fine wines are occasionally bottled in even larger bottle formats. Much prized by collectors, the Jeroboam, whose the size varies according to the type of wine it contains (three litres of Champagne or Burgundy, but five litres of Bordeaux), is perhaps the largest that can be reasonably capably handled. These large formats have a slightly larger neck size and require a closure (invariably cork) wider in diameter.

17.2.4   The Impacts of Bottle Colour and Glass Thickness/Weight

Globally, green is bar far the most common wine bottle colour, although red wines from some countries, particularly Italy, may be bottled in brown glass. White wines, particularly sweet white, and rosé wines are often bottled in clear glass. Other colours are occasionally encountered, including blue that is sometimes used by producers in Germany.

In certain locations, including the shelves of supermarket/retailers, restaurants, and even at home wine can be stored for relatively long periods exposed to damaging UV light, and visible light. The thickness/weight of the glass, and particularly its colour, have a considerable impact on the amount of ultraviolet light that reaches the wine. Light damage can include unwanted changes in colour, the production of off-odours resulting from lightstrike, discussed in more detail in Chapter 6. Light can also deplete sulfur dioxide (SO_2) levels, resulting in a shortening of the life of the wine [4]. Sparkling wines together with white and rosé wines can be particularly badly affected – the phenolics in red wines can give some protection. Red wines stored in green or brown bottles do not suffer significantly from light damage [5].

Glass wine bottles do not transmit ultraviolet light at wavelengths below 300 nm. The most damaging wavelengths lie between this at 520 nm, with the range of 325–450 nm particularly implicated. Ultraviolet has wavelengths below 400 nm and is invisible to most humans; visible light wavelengths are from 400 to 760 nm. Critical wavelengths are 340, 380, and 440 nm (440 nm is in the visible region) [6]. Ultraviolet light resistant bottles can prevent the transmission at these wavelengths. Clear glass filters just 10% of UV light at wavelengths below 520 nm, green between 50% and 90%. and brown from 90% up to 99%. There are some variances and, particularly with clear glass, 'thicker' bottles filter out more ultraviolet light. There is no doubt that, standing on the retailers' shelves, some wines only look attractive in bottles made of clear glass, and marketing departments will dictate the precise bottle to be used. Rosé wines in particular are made to look pretty, but look anything but that when bottled in green glass. In the case of green bottles 70% or so of ultraviolet light at 370 nm, passes through. Brown is by far the most effective ultraviolet screen, and it is perhaps surprising that it is not more widely used by producers of premium and super-premium wines, especially as they may stay on retailers' shelves for relatively long periods. There has been a growing movement to bottle wines in lighter-weight glass – this has been primarily on account of environmental concerns, particularly with regard to the impact of transporting additional weight.

17.3 Bottle Closures

The main functions of a wine bottle closure are to ensure a good seal and prevent sensory deterioration of the wine, and to provide barriers to moisture, oxygen, carbon dioxide, and other gasses [7].

17.3.1 Types of Closure

The range of types of closures for wine has increased has increased dramatically in the last 30 years and may now be summarised as follows:

- Cork: Natural

             Technical: Micro-agglomerated cork
                               $1 + 1$, TwinTop
                               sparkling wine cork
                               Diam®

- Synthetic: Co-extruded: Brands: Nomacorc, Tasz Neo
                               Injection moulded – once important, but not today
                               Multi-component: Brand: ArdeaSeal®

- Screw cap: Liner: Saran-tin
                               Saranex
                               Liners with inbuilt oxygen transmission characteristics:
                               Stelvin® Inside, Oenoseal, VinPerfect, G3.

- Glass stopper: Brand: Vinolok

17.3.2 Dissolved Oxygen in Wine

17.3.2.1 Sources of Dissolved Oxygen and Total Package Oxygen in Wine

A sealed bottle will contain oxygen from several sources: dissolved oxygen (DO) in the wine at the time of bottling, oxygen that may remain in the headspace after any gas flushing or vacuum application at the time of closure application, and, particularly if a cellular structured closure (such as a natural cork) is used, any oxygen contained within the closure [8]. The sum of these is the Total Package Oxygen (TPO). Oxygen management at bottling time is a factor when considering the volume of dissolved oxygen in wine, whatever the closure type to be utilised TPO levels after bottling should be less than 1–1.25 mg/l for red wines and less than 0.5–0.6 mg/l for white wines. However, in practice, after bottling the initial amount of TPO in a bottle of wine, including the liquid and headspace can vary between 1 and 9 mg/l, depending upon the bottling process and closure [9]. Headspace oxygen alone can comprise up to 5 mg/l.

17.3.2.2 Consequences of Oxygen Uptake During the Bottling Operation

Oxygen uptake at bottling is the major source of oxygenation of bottled wine, regardless the type of closure. A TPO level of 2 mg/l in a 750 ml bottle can result in a rapid decrease of sulfur dioxide levels of 10 mg/l [10]. The amount of oxygen inserted during the bottling operation represents around 60% of the total amount of oxygen that will be present after 36 months of storage in bottles sealed with micro-agglomerated and natural corks. Oxygen contained in corks, and atmospheric oxygen entering through the closure-glass interface account for the remaining 40% or so. Bottling operations also contribute heavily to the total amount of oxygen entering bottles sealed under Saranex and Saran-tin-lined screw caps, representing, respectively, 60% and 85%, of the total amount of oxygen over 36 months [11]. Similar figures are likely for Stelvin Inside, Oenoseal or other 'variable transmission' liners, depending upon the liner-type gas permeability.

17.3.2.3 Oxygen Transmission Rates (OTRs), Permeation, Migration, and Scalping

Permeation Permeation is the extent to which substances present in the external environment, such as water, oxygen, other gasses, and microorganisms can penetrate through or around the closure into the packaged wine. Generally speaking, all closure types provide a barrier to liquids and microorganisms. However, gasses can transgress many closure types, oxygen being the gas to which bottles are constantly exposed. Accordingly, one of the key considerations with regard to the development or deterioration of bottled wine is that of the closure's Oxygen Transmission Rate (OTR), and this has become a prime consideration when producers make a closure selection decision. The Australian Wine Research Institute (AWRI) has undertaken trials of different closure types and their impact upon the preservation and development of bottled wine [12]. Very low OTR values can lead to the development of reductive characteristics in wine. The retention of fresh (primary) fruit characteristics are best achieved with a low OTR closure, whilst moderate OTRs leased to the development of stewed fruit aromas and flavours. Of course, high OTRs will lead to oxidation in the wine, something all producers would wish to avoid, even for wines that are deliberately oxidised in the production process, such as oloroso Sherries. The OTRs of the various closure types vary dramatically.

The latest developments in synthetic closures and screw caps have resulted in some manufacturers offering ranges of closures with differing OTRs, giving the wine producer the option to choose according to the type of wine and its anticipated shelf life. The amount of oxygen to permeate through the closure is generally the lowest with screw caps with a Saran-tin (or equivalent) liner, followed by those with a Saranex liner. Micro-agglomerated corks generally have the next lowest OTRs – the level increases fairly rapidly during the first six months post bottling, slows and after a year or so becomes almost linear at approximately 1.3–1.4 mg/l. Natural corks have the fourth lowest OTRs – the rate rises fairly rapidly during the first six months, slows, and then rises only slightly between one and five years, from approximately 2.6–2.8 mg/l. The OTR of co-extruded plastic stoppers rises fairly rapidly in the first six months post bottling, at a rate very similar to that of natural cork. In the case some co-extruded stoppers the uptake continues to rise rapidly and by five years after bottling may have reached some 5.4 mg/l, over 13 times that of the Saran-tin lined screw cap. However, there are other co-extruded closures that have an oxygen transmission of approximately 2.7 mg/l over five years.

The disastrous fault of contamination of wine with 2,4,6-trichloroanisole (TCA) or other haloanisoles was discussed in Chapter 3. Haloanisole contamination of wine storage sites, including producers and private cellars, can result in finished product becoming tainted via the closure. Bottles sealed with natural cork are most likely to prevent the wine absorbing TCA from such aerial contamination, due to the material's relative impermeability compared to Saranex-lined screw cap and some synthetic closures. The crown cap, as used for Champagne and many bottle fermented sparkling wines before disgorgement, is perhaps a particularly vulnerable closure in this regard. Over the last 10 years, I have noted a higher incidence of haloanisole related taints in sparkling wines than in other white or rosé wines, and consider the use of the crown cap may have had a part to play. Purely anecdotally, in 2018 I had in my house some 24 bottles of beer from nine different producers, sealed with Saranex-lined crown caps, that were stored in a wooden box in a slightly damp location. In the same box were eight bottles of artisan beer, sealed with a 'champagne'-type stopper and wire cage. Upon opening the beers I found that 22 of the 24 crown cap sealed beers reeked of haloanisle musty taint. All of the 'champagne'-stoppered beers were in good order. Upon examination, the storage box showed a film of growth of filamentous fungi, and my wife confirmed to me that domestic bleach had, on two occasions, been used to clean the box.

Migration Migration is the desorption of substances, mainly gasses, from the material of the closure, passing through the surface, into the packaged wine. In the case of natural cork there will also be diffusion or migration of oxygen from the cork itself into the wine. In fact the OTR from a natural cork stopper is mainly release of the oxygen contained within the stopper, not oxygen permeating through it. It is estimated that a typical cork of 44 mm length contains some 3.5 ml of oxygen [13]. When the cork is compressed during the bottling operation the pressure of the air contained within increases to between 88 and 132 lb/in.^{2} (6 and 9 atm). This pressure imbalance will be resolved mostly during the first six months post bottling by a gradual equalisation of gasses between cork and the headspace in the bottle. Cork also contains many phenolic, volatile and semi-volatile compounds, which can migrate into the wine.

Scalping Aroma and flavour scalping by closures may be a negative or, at times, positive characteristic. There is little evidence of desired volatile compounds being scalped by closures. However, several undesirable compounds may be scalped by some closure types. Naphthalene, regarded as undesirable, and 1,1,6-trimethyl-1,2-dihydronaphthalene (TDN), regarded by some tasters as undesirable (see Chapter 15) are scalped by both synthetic closures and to a lesser extent by natural corks [14]. The undesirable volatile sulfur compounds of hydrogen sulfide and dimethyl sulfide (see Chapter 6) may be considerably reduced in bottle due to scalping by natural corks, and to a less extent synthetic closures [15].

17.4 The Maintenance of Adequate Fee and Molecular SO_2 in Bottled Wine

17.4.1 The Adjustment of Free and Molecular SO_2 at Bottling

When adjusting the free SO_2 prior to bottling, consideration should be given to the TPO both at bottling time and during the anticipated shelf life of the wine. The amount of free SO_2 normally adjusted according to the policy and specification of the producer, with 30–35 mg/l being typically representative figures. Some 10 mg/l of free SO_2 should be present throughout the bottle life to maintain an adequate capacity to bind aldehydes and to protect against oxidation. For a wine to decline in free SO_2 from 30 to 10 mg/l, the amount of DO required is 5 mg/l. With regard to microbial stability, the level of molecular SO_2 is important (an absolute minimum of 0.5 mg/l might be considered). Allowance must be made for the OTR of the closure, including the amount of oxygen contained in and subsequently diffused from the cork, or the OTR of the alternative closure.

17.4.2 Retention of Adequate Free Sulfur Dioxide in Bottle

The retention of SO_2 post bottling of both white and red wines under different closure types has been measured in studies. Screw cap (Saran-tin) or Vinolok closures generally result in less loss of both free and total SO_2 than when wine is sealed with a natural cork or synthetic closures. A study by Skouroumounis et al. of the impact of type of closure and storage conditions on a Riesling and wooded Chardonnay showed, inter alia, that after storage the wines sealed with the synthetic closures were brown with aromas of oxidation and showed lower levels of both free and total SO_2 than those sealed with the other closure types [16]. A study by Paulo Lopes et al., using a Sauvignon Blanc, found that after two years, wines sealed with a natural or agglomerate cork were free from faults, but those sealed with synthetic closures with high oxygen permeability showed unwanted oxidative characters [17].

17.4.3 The Influence of Closure Type Upon Reductive Characters

The study by Lopes et al. also showed that wines sealed with a screw cap with a Saran-tin liner exhibited detrimental reductive thiol characters of struck flint and rubber [17]. The wines sealed with a natural cork showed negligible characteristics of reduction. More recently, certain bottle closures have also been shown to adsorb volatile sulfur compounds with varying efficiency [15]. There is increased removal of the reductive thiol compounds in wines bottled sealed with cork and technical closures compared with those bottled under screw cap with Saran-tin liners [15]. The accumulation and release of hydrogen sulfide, methanethiol, and ethanethiol complexes with copper and zinc ions may be favoured by low oxygen conditions [18]. Accordingly, wines sealed with screw cap (Saran-tin liner) are much more prone to development of 'reductive' character than wines sealed under closures with higher rate of oxygen transmission.

17.4.4 The Market for the Various Types of Closures

In 1999 it was estimated that cork stoppers, including 'technical' corks, had some 95% of the wine closure market, synthetic closures 2%, and screw cap 3%. Screw caps then entered a period of rapid growth spearheaded by initiatives in Clare Valley, Australia, and the New Zealand wine producers. By 2014 it was estimated screw caps had 29% of the market, synthetics 16% and cork had declined to 55%. Although the market had been growing in this period, the volume of cork stoppers sold had suffered serious decline. By 2019 cork products market share had increased to over 65%, largely at the expense of synthetic closures, which fell from 4.5 billion closures in 2006/7 to 2 billion closures in 2016 [19]. This decline of synthetic closures was largely due to the demise of injection-moulded plastic stoppers. According to APCOR, of the 18.5 billion bottles of wine produced each year, some 12 billion have cork closures. This figure includes technical and micro-agglomerated corks [20].

17.5 Cork Closures

Natural cork is, of course, the traditional way of sealing bottles of wine. It is highly elastic, impermeable to liquids and to a great extent impermeable to gasses. Other than one-piece natural cork, there are several other types of cork closure and, in fact the majority of the cork market today is in the form of multi-piece technical corks. A range of cork types is shown as Figure 17.1.

17.5.1 A Brief History of Cork Closures

It is often claimed that the use of cork as a wine stopper was discovered by the French monk, Dom Perignon, which is certainly not the case [21]. The year 1680 has been stated as when

Figure 17.1 Various types of cork closure.

the Benedictine monk first used cork for sealing wine [22]. Previously stoppers had generally been made of wood wrapped in hemp soaked in olive oil. Cork was probably first used to seal wine in the sixth century BC (BCE) in Toscana (Tuscany). Certainly, the Romans had used cork to seal amphorae and barrels containing wine, but such use was discontinued during the dark ages. Over the coming decades cork started to become more widely used to seal bottles, and its use was far from uncommon by 1680. This was not just for wine but also for other alcoholic drinks. John Worlidge noted in his *Vinetum Brittanicum – Treatise of Cider and such other wines and drinks that are extracted from all manner of fruits growing in this kingdom* in 1676 that defective corks could ruin drinks and 'therefore are glass stopples to be preferred'. A second edition of the work was published in 1678. The myth of being the first to use cork is but one of many spurious claims about Dom Perignon – perhaps the most widely spread misinformation is that he invented sparkling wine in or around 1697. In fact the production of sparkling wine by adding sugar or molasses to induce a second fermentation in the bottle was first described by an Englishman – Christopher Merret who in 1662 presented a paper to the Royal Society. At that time only the English were using coal-fired furnaces and able to produce bottles strong enough to resist the pressure of the CO_2 gas that would cause weaker bottles to explode. Bottles of sparkling wine in the 'Champagne' style and sealed with cork stoppers were first drunk in London during the 1660s [3].

At about the same time as Merret was working on his paper another English scientist, Robert Hooke of the Royal Society was examining a thin slice of cork under his microscope, and was amazed to find it 'much like a honeycomb'. Each of the compartments in the 'honeycomb' reminded Hooke of the cells in which monks slept, so he named them 'cells' – the first use of that term that subsequently was to be used for the building blocks of life [3]. For almost 300 years, from the end of the seventeenth century until the 1980s, wine in glass bottles was invariably sealed with cork. In the second part of the twentieth century, in the case of inexpensive wines, this might be an agglomerated cork closure. In the 1980's the growth in the use of alternative closure began, initially with injection-moulded plastic stoppers, but from the late 1990s co-extruded synthetic stoppers and screw caps. The issue of TCA and other haloanisoles that became a particular problem in the late twentieth century has been extensively discussed in Chapter 3.

17.5.2 The Origin of Cork and Production of Cork Stoppers

Natural cork comes from the bark of the cork oak: *Quercus suber*. The bark of the trees are stripped every nine to 12 years; it then regenerates. Figure 17.2 shows bark stripping taking place.

In the early life of the tree the bark is not thick enough or suitable for wine closures – it is only when the trees are at least 27 years old, and at the third stripping (called the second reproduction stripping) that the bark can be used. There are some 2.1 million hectares of cork oak forests in the world, of which some 1.44 million are in Europe. The largest producing country of corks is Portugal, with 737 000 ha of montado (cork forest) which produces over 100 000 tonnes of cork annually, representing approximately 50% of the world's total. The second largest producer is Spain with 61 500 tonnes. Within Portugal some 84% of the area of montado is situated in the region of Alentejo. Together these two countries account for over 80% of the world's wine cork production. Portugal produces 40 million cork stoppers

Figure 17.2 Cork bark stripping. Source: Courtesy of Amorim.

a day. Other countries where the cork oak thrives include (southern) France, Morocco, and China. Although cork is used in many products, wine corks account for some 72% of the raw material's usage, and an even higher percentage of the cork industry's product value.

The cork industry is keen to promote the product's environmental credentials. The cork oaks regenerate their bark and have a life of perhaps 150 years. There can be no doubt that the montados, the areas centred upon cork oak forests do provide a valuable environment. The average density of cork oaks is in the region of 80 trees per hectare. The forests are interspersed with pastures and just 5% or so of the total area may be used for growing cereals. As well as cork oak there are large areas of holm oak (*Quercus rotundifolia*), and some Pyrenean oak (*Quercus pyrenaica*). The area of montado has grown by 3% in the last 10 years.

The most important importing country for cork stoppers by value is France, which accounts for some 16.5% of value. Figure 17.3 shows cork stopper exports from Portugal to the six most important countries.

Cork is certainly one of the most wonderful natural materials with fascinating properties. The bark is made-up of 14 sided polyhedrons, which contain oxygen. In fact oxygen comprises some 80–90% the content of natural cork. The material is elastic, and offers a high level of insulation to heat, sound, and vibration. The non-gaseous components of cork comprises approximately 45% suberin (which gives elasticity), 33% polyphenols – mostly lignin

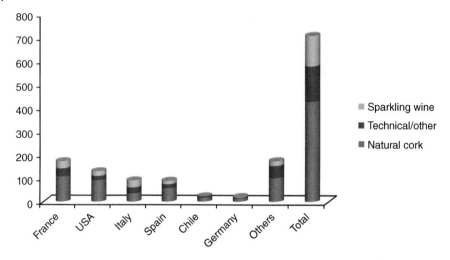

Figure 17.3 Value of Cork stoppers exported from Portugal in 2017 € Millions. Source: National Statistics Institute (INE) Portugal.

together with some tannins – and 12% polysaccharides, mainly cellulose and hemicellulose. The remainder of the composition includes waxes, terpenoids, salts, glycerine, and water.

Many in the wine industry believe that cork is an antiquated product with which to seal a bottle and that wine is a highly technical product that warrants more than a piece of tree bark as a closure. Areas of criticism include the risk of haloanisole contamination, unreliable OTRs, and incidences of random oxidation. It is perhaps interesting to note that the same people do not question the use of oak barrels for the maturation and storage of wines, which may also be subject to similar issues. Barrels have been used for wine storage since the first century BC (BCE), and the advantages of making them from oak and the positive impact this would have upon wine quality were realised not long thereafter [23]. Of course, the species of oak used for barrel manufacture, namely *Quercus petraea* and *Quercus robur*, are very different to the cork oak, *Quercus suber*.

17.5.3 The Cork Production Process

The world's largest cork producer is Amorim, based in Portugal but with finishing plants in several other countries. I will base my description of cork production upon the systems used by them.

The cork forests are mainly in the east and south of Portugal, and Amorim's first-stage processing plants are situated at Ponte-de-Sôr and Coruche. Cork is purchased from many farmers, with whom the company has long-term contracts. The bark is stripped from the trees manually during the summer. Each tree is stripped every 9 or 10 years; the outer bark is taken from the trunk (excluding the first metre or so close to the ground) and also from the lower part of the thick branches. Following stripping the bark is placed on metal pallets sited on concrete slabs for seasoning; the minimum period under Portuguese law is six months, although Amorim will season for at least one year. During this time the sap in the cells is washed out. A preliminary analysis the cork planks is made before they are despatched to

Figure 17.4   Cork bark awaiting processing at Amorim.

the nearby plant for initially processing. Figure 17.4 shows planks of cork bark awaiting processing.

Historically the harvested cork planks were boiled twice, once near the forests, and the second time at the company's main plant at Santa Maria de Lamas. As part of the recent changes in practices the second boil has been replaced by a system of steam evaporation. The boiling process called CONVEX, which lasts an hour, removes volatile organic compounds from the water. The factory uses a lot of steam, for the evaporation stage and also for the ROSA process discussed below. The cork industry is keen to expound its environmental credentials, and Amorim proudly state that 62% of the energy used in the plant is obtained by biofuel burning of the cork dust waste. The planks are then rested for two days to de-humidify.

The cork planks are inspected and cut into strips of bark in across the axis of the trunk or branch in preparation for punching. This process is undertaken either manually or by robots, the corks being punched along the axis. Figure 17.5 shows corks being punched manually.

One man can punch 14 000 corks in an 8 hour shift; in the same time a robot can punch 35 000 corks, or 105 000 in a 24 hour period (the robots can work continuously). The cork strips to be punched by robots need pre-selection to remove those with problem areas. However, the robots will still produce a higher percentage of corks that will be put to waste, for the human puncher can avoid areas of the bark with defects. Development of a new generation of robots is taking place that will be programmed to avoid defect areas.

Following punching, the corks are fed to a machine that carries out an analysis of the individual corks. They are subjected to a physical, air pressure test, to ensure that there are no passages in the cork though which wine might leak or oxygen could pass. Some corks will be rejected at this stage. Then a photo is made with both two- and three-dimensional

Figure 17.5 Hand punching of corks.

images that will reveal defects and the machine will remove suspect corks from the system. The corks next proceed to washing and 'bleaching'. No chlorine used at any stage of the process. Some customers and markets request 'white' corks and this is achieved by the use of peroxide. The corks are places in a large drum that can take a maximum of 90 000 corks where they will spend 60–90 minutes in the washing process. Historically the corks would just have been washed and dried, but now they are washed, steamed, dried, and then proceed to the ROSA process.

The ROSA system, developed by Amorim and introduced in 2003, is a method of extraction of volatiles, particularly TCA. ROSA is based on a steam distillation whereby steam and water under pressure force volatile compounds out from the cork cells. Initially the system was only suitable for cork granules as used for the manufacture of agglomerated or technical corks. However, subsequent generations ROSA 'Evolution' and the latest SUPER-ROSA treat whole, natural corks. Eighty per cent of any TCA content is removed by the process. Amorim admits that for a cork that contains 1 ppt of TCA the 80% removal figure is brilliant, but should a cork have 100 ppt of TCA, then following processing this would still leave 20 ppt, an unacceptable figure well above sensory detection threshold. Each batch is analysed for TCA, and those above threshold rejected. As discussed in Section 3.11.3.5, Amorim now offer a top-of-the-range natural cork called NDtech. The corks are individually analysed for traces of TCA using rapid gas chromatography. Any cork with more

0.5 ng/l of releasable TCA is removed during the production process. The company offers a *non-detectable* TCA guarantee on corks in the NDtech range.

The finishing of corks, including branding by ink, heat or laser and coating with paraffin or silicone to facilitate insertion and removal may take place in the destination country or at another Amorim plant at Santa Maria de Lamas. Amorim produces over 4 billion corks a year, with 1 billion of these being natural (1 piece) corks.

17.5.4 Natural Corks and Colmated Corks

17.5.4.1 Grades of Natural Cork

Natural corks are graded according to visual appearance, with the high grades attracting premium prices.

- Traditionally corks are classified into eight grades: Flower (Flor); Extra; Superior; 1st; 2nd; 3rd; 4th; 5th. The lower grades will have more and perhaps larger lenticels, and maybe vertical cracks, although these should not extend beyond 18% of the cork length.
- In the United States, the Cork Quality Council (CQC) has adopted a grading system which classifies corks in three grades: A, B, and C. Grade A corks have no holes or pores which exceed 2 mm, no cracks that exceed 11% of cork length, and no horizontal cracks.

17.5.4.2 Colmated Corks

Colmated corks are lower grade natural corks that have the lenticels on the surface filled cork dust obtained from the waste following the finishing of natural corks and sealed with a natural resin-based or water-based glue. This improves the cork's appearance, and perhaps also the seal between the cork/bottle interface.

17.6 Technical Corks and Agglomerated Corks

There are several types of technical cork, as briefly discussed below. Technical corks generally have lower OTRs than natural cork and some, particularly the Diam closure, are constructed in several formats with differing OTRs.

17.6.1 Micro-agglomerated Corks

Micro-agglomerated corks are made from small granules of cork that are ground from bark waste, including the area that is left following the punching of natural cork. The particles are then agglomerated using food-grade polyurethane glue and the closures are either individually moulded (for the higher quality) or extruded, the latter resulting in rods cut to the required length. Micro-agglomerated corks have granules of specified size and quality; the granules for agglomerated corks are less defined. The closures are suitable for wines that are to be drunk early, generally within six months of packaging. Micro-agglomerated corks are available with a non-detectable TCA guarantee.

17.6.2   Twin Top, 1 + 1 and Sparkling Wine Other Technical Corks

Twin Top (an Amorim brand) and other similar 1 + 1 closures comprise a thin solid disc of natural cork glued to each end of a body comprised of agglomerated granules. The discs, which are approximately 6.5 mm in thickness are selected and cleaned according to individual producers' methods. These include the use of steam, steam and alcohol, ethane solutions, hydrogen peroxide solutions, and ozonised water. The closure is type is suitable for wines that will be aged for up to two years in bottle. Twin Top corks are available with a non-detectable TCA guarantee. Sparkling wine corks have two thin discs glued to one end of an agglomerated body.

17.6.3   Helix

A new development is the Helix cork closure Helix is a technical cork that can be screwed out of (and back into) the specially designed and manufactured bottle without the need for a corkscrew. Helix has been developed by Amorim and Owens-Illinois, the world's largest glass container manufacturer.

17.6.4   ProCork

ProCork is a natural cork with an inert, semi-crystalline, oxygen-selective membrane that is heat-welded to each end. The membrane comprised five layers and is just 50 µm in thickness [24]. According to the manufacturer, large molecules including TCA and lignins are unable to pass through the membrane whilst small linear molecules can pass at a slow rate. This allows a slow and controlled OTR. ProCork claims that they have sold 500 million corks but have had 'less than ten TCA wines' [25].

17.7   Diam Cork Closure

17.7.1   History: From the Altec Disaster to Diam Success

The Altec closure, comprising ground cork granules with the lignin content replaced with polymer microspheres, all glued together binding agent, was launched in 1995 by Sabaté, then the second largest cork manufacturer in the world. The aim was to provide a low cost, largely TCA free alternative to natural cork. In 2001 Sabaté-Diosos and Sabaté USA sold 800 million Altec closures, with a total sales since the commencement of production topping 2.5 billion closures. However, in the last quarter of 2001 a crisis hit the company in the United States when press reports appeared claiming that some wineries were complaining of TCA contamination, and law suits quickly ensued (which were subsequently settled). In one suit Van Duzer Winery in the Willamette Valley, Oregon, claimed that Sabaté's Altec composite stoppers released TCA that ruined 1200 cases of 1999 Chardonnay with a retail value of more than US$ 240 000 [26]. Further, the 24-month results' of the AWRI closure trials found that each of the bottles sealed with Altec closures had a TCA-like aroma, and laboratory analysis revealed detectable levels of TCA in each of the Altec sealed wines [27]. The precise reason for the widespread contamination of the product has never been revealed, but the company had perhaps been victim of its own success, and the cork raw material compromised [3]. In July 2002 the company announced that it was developing new technology called *Diamant*. In June 2003 the Sabaté companies changed names to *Oeneo* and *Oeneo USA*. The new Diam closure, using the *Diamant* technology was launched in 2005. The market for the closures has expanded considerably over the last 15 years to a production level of over 2.4 billion Diam closures per annum by 2020 [28]. There are now over 10 000 customers in 68 countries [28]. Diam now offer a guarantee that each cork is free of releasable TCA, i.e. below the measurable limit of 0.3 ng/l of TCA.

17.7.2 The Technology of the Diam Closure

Ninety-five percent of the Diam closure component comprises processed de-lignified cork granules, with the other 5% being acrylate and polyurethane. Diam employs a supercritical carbon dioxide (sCO_2) cleansing process to remove TCA from the cork component. Supercritical carbon dioxide is a fluid carbon dioxide is that is hot and highly pressurised. The supercritical state is achieved at a temperature above the critical temperature of 31.1 °C (87.98 °F) and pressure above the critical pressure of 72.9 atm/bar or 7.39 MPa. As an extraction solvent it has many uses in the food, perfume, pharmaceutical, and dry cleaning industries. It is now frequently used for decaffeinating coffee. The system used for Diam was developed by Sabaté (the previous name of OENO) together with the Commissariat à l'Energie Atomique (CEA), a French government-funded research organisation. Acrylate particles when heated fill the gaps between the tiny fragments of cork, and the polyurethane acts as an adhesive that holds the components together. A mixture of paraffin and silicon is applied as a surface coating.

17.7.3 The Range of Diam Closures and OTRs

At the time of writing, Diam closures are available in a range of five different OTRs and suitabilities for various durations of bottle-ageing: Diam 1, 3, 5, 10, and 30. The Diam 1 has the highest OTR and is suitable for wines that will be consumed within a year, whilst the Diam 30 is designed for the ageing of fine wines: manufacturer offers a guarantee of no leakage over a period of 30 years, providing the wine is stored below 15 °C (59 °F). The Diam 5, 10, and 30 are also available as Origine by Diam that utilises a binder composed of 100% vegetable polyols and a beeswax emulsion coating. The OTRs of Diam closures are stated by the manufacturer to be much more consistent than that those of natural cork, who also claim that wines may be bottled with lower levels of free SO_2 – from 10 to 15 mg/l.

Many producers of high quality wines have changed closures from natural cork to Diam, especially for white wines. Following its in house research and testing, Château Margaux has moved to Diam for the Pavillon Blanc commencing with the 2016 vintage – the red wines are still closed with natural cork (Thomas Burke MS – Château Margaux, personal conversation 8 April 2018). Producers in Chablis who are using Diam include William Fèvre (since 2008), Benoît Droin (since 2011) and Demaine Leflaive (since 2014). The manufacturer claims that some 40% of Gand Cru Burgundies are now closed with Diams.

However, there are anecdotes of Diam imparting taints to wine, and several industry insiders report that this is their experience. Speaking at The Buyer's Closure debate in 2017,

Doug Wregg, sales, marketing and buying director at 'natural' wine specialist Les Cave de Pyrene noted that agglomerated corks like Diam impart unappealing off-flavours. Ben Stephenson, owner of award winning Manchester independent wine merchant 'Hangditch' agreed, noting that 'Diam corks are not neutral, even if they are popular' [29]. In 2016 a German wine merchant, Rolf Cordes, claimed that Diam closures introduce atypical bitterness (ATB) into wine [30]. This was immediately hotly disputed by the manufacturer [31], but some commentators agreed with Cordes.

17.8 Synthetic Closures

There are two basic methods for producing synthetic closures: the co-extrusion process, and injection moulding.

17.8.1 Co-extruded Closures

In this process, one extrusion forms an outer sleeve with a smooth surface and another extrusion fills the inner core. For example, Tasz Neo comprises a low density polyethylene (LDPE) core within a LDPE-based thermoplastic elastomer sleeve. The closures are available in three different OTRs.

17.8.1.1 Nomacorc
Nomacorc, manufactured by Viventions is, at the time of writing, the market leader in co-extruded closures. Nomacorc offer several ranges including 'Select Green' made from 60% to 80% sugar cane from Brazil, according, Vinventions. There are three different OTR options for the 'Select Green' series of closures [32]:

- *Select Green 100*: oxygen ingress **per bottle** is 0.40 mg of oxygen after 3 months, 0.7 mg after 6 months, 1.2 mg after 12 months, and 1.1 mg a year thereafter.
- *Select Green 300*: oxygen ingress is 1.6 mg after 3 months, 2.1 mg after 6 months, 2.8 mg after 12 months, and 1.1 mg a year thereafter.
- *Select Green 500*: 1.8 mg after 3 months, 2.3 mg after 6 months, 3.1 mg after a year, and 1.7 mg a year thereafter.

The company also produce a synthetic closure for wines destined for long ageing:

- *Reserva* has a claimed oxygen ingress of 0.3 mg after 3 months, 0.4 mg after 6 months, 0.7 mg after a year, and then 0.6 mg per year thereafter. It is designed for wines that can be aged for up to 25 years.

The company claim that natural corks are inconsistent and the OTR can range between 1 and 6 mg per year [33]. This figure is hotly disputed by the cork industry.

17.8.2 Injection-Moulded Closures

Injection corks are made of thermoplastic elastomer mixes. The product was popular in the years prior to 2010, but has waned since. The then market leader for injection-moulded closures, Supremecorq, ceased manufacture in 2011. There have been several criticisms of injection-moulded stoppers, including high OTRs and the difficulty of removal from the bottle, together with the near impossibility of re-insertion.

17.8.3   Flavour Scalping

One issue with synthetic closures that raises itself above the parapet every few years is that of possible flavour scalping. The 2003 study by Capone et al. found that after two years in bottle wines sealed with synthetic closures showed lower levels of (desirable) ethyl hexanoate, ethyl octanoate, and ethyl decanoate than those sealed with technical or natural corks [14]. It would perhaps appear that not much changed in 17 years, according to a study looking at the influence of 1 + 1 corks, micro-agglomerated corks and co-extruded synthetic closures over a four year storage period was undertaken at University of Porto by Oliveira et al., with results published in 2020. Wines sealed with 1 + 1 corks showed higher levels of ethyl hexanoate, ethyl octanoate, ethyl decanoate, and 2-phenylethyl acetate. These compounds have fruity and floral odour descriptors and may contribute to aroma quality, aroma intensity and balance. TDN was higher in the micro-agglomerated and 1 + 1 cork-sealed wines. The synthetic closures were associated with oxidised descriptors, and lower free and total sulfur dioxide [34].

17.9   Screw Caps

Screw caps are often referred to as 'twist-offs' in the United States. They are also called ROTE (roll-on tamper evident) and ROPP (roll-on pilfer proof) closures.

17.9.1   History

Although often regarded as a relatively new closure type, screw caps were developed in Burgundy in the 1960s by Stelvin, the first aluminium closure reaching its final form in 1964 and introduced in the US markets shortly afterwards. The closure was made possible by the discovery in 1933 by researchers at Dow Chemical of Polyvinylidene chloride (PVDC), subsequently branded as 'Saran' and 'Saranex' [35]. This was (and for the Saranex-lined closures remains) the main material in the liner that is in contact with wine. In the early 1970s Gallo was amongst the companies to market wines closed with screw cap in the United States, and at the same time the closure became popular in Scandinavia and Switzerland. Screw cap closures for wine were first introduced into the Australian market in 1976 [36]. Although the technical assets of the closure were advocated by several wine producers, the launch of the closure was regarded as a marketing failure [36]. Today screw caps account for over 25% of the wine closure market, and the worldwide sales of screw cap closures amount to some 4.5–5 billion units.

17.9.2   Screw Cap Liners

The seal for screwcap wines is not the metal closure itself, but the liner. The two most common liners are Saran-tin and Saranex. In 2013 Amcor, the owners of Stelvin, the biggest brand of screw caps, introduced four different types of liner, with four different OTRs. The products are marketed as Stelvin Inside. The range offered at the time of writing includes Stelvin 1 0_2 (OTR of less than 0.0005 cc per day), Stelvin 3 0_2 (0.0005 cc), Stelvin

5 0$_2$, (0.005 cc), and Stelvin 7 0$_2$ (0.05 cc per day). The liners consist of layers of expanded polyethylene foam (EPE), aluminium, polyethylene (PE), polyethylene terephthalate (PET) and, in the case of Stelvin 5 0$_2$, silicon oxide.

A new screw cap liner, ALKOvin has been designed to adsorb volatile sulfur compounds (VSCs) in bottled wine. The structure is complex with several layers comprising PE foam, paper, aluminium, an 'active layer' and PE white. The liner has been shown to substantially reduce levels of methanethiol and ethanethiol in heavily spiked wines, although there was a slight increase in hydrogen sulfide (H$_2$S) and a 22.3% increase in dimethyl sulfide. An expert sensory panel assessed a 'spiked' Chardonnay wine as exhibiting less reduced aromas, such as flint and vegetal aromas following 12 months bottler storage, compared with and identical wine seals with a Saran-tin-lined screw cap [37].

The VinPerfect is a closure that has been developed to avoid so-called post-bottling reduction. It uses a 30 × 60 Stelvin aluminium screw cap with, where the top of the cap has microscopic holes that allows a predictable and consistent OTR. The liner comprises a multi-layer laminate which has two PET films together with a partial aluminium coating. The bottle seal is provided by a 2 mm high density polyethylene foam which has elasticity. The manufacturer claims OTRs for the closure range from 0. 000 15 cc of oxygen per day for the 'light' to 0.0003 cc for the 'medium', and 0.0007 cc per day for the 'medium plus' [38].

17.9.3 Post-Bottling Reduction

The topic of reduction in wines, including so-called 'post-bottling reduction' was discussed in Chapter 6. The involvement of each type of closure, its OTR or range of OTRs, upon wine development and the formation of reductive characteristics has been the subject of considerable research, particularly by the AWRI, but remains much discussed by serious wine lovers. So-called 'post-bottling reduction' was largely unheard of until the early years of this century, and the problem has mostly been associated with wines sealed with screw cap closures with Saran-tin liners. There can be little doubt that many winemakers have adjusted their winemaking processes, including SO$_2$ management and the use of copper to avoid potential problems. Of course, it might be argued that a closure should be adjusted to the wine, and not the wine to the closure. Marcelo Papa, the head winemaker of major Chilean producer Concha y Toro notes ' with Saran-tin we saw some reduction in reds. [Now] we manage wines to arrive at the bottling line without potential redox, and we changed the liner to Saranex for reds' [39]. However, Saranex liners are generally regarded as only suitable for wines that will not be aged for more than three years [33].

17.9.4 Damage to Screw Caps

Screw caps are at risk of physical damage in the event of careless bottling operations, and during the storage, distribution, and retail cycles. The level of damage can vary from the cosmetic to such severe damage that compromises the integrity of the liner and seal. A survey to determine the level of screw cap damage leading to changes in wine chemistry was undertaken by Alison Eisermann-Ctercteko for her Master of Wine dissertation, and an abstract published in 2013 [40]. Ten thousand bottles were examined in 22 retail outlets in Australia, and a further 1500 in the United Kingdom. Staggeringly, some 26.1% of the closures were found to be damaged, with the damage severe enough in 8.2% of the total to cause significant chemical changes to the wine. Eisermann-Ctercteko noted that this figure is greater than that of the previous industry-wine problem of cork taint. Incorrect or faulty screw cap application was found to be responsible for cap damage of 7.4% of the total bottles (28.4% of the damaged caps), but this was mostly of a cosmetic nature and unlikely to affect the wine quality.

17.10 Vinolok

Often simply called the glass stopper, Vinolok was launched 2003, having been 'invented' less than two years earlier by Dr Karl Matheis. Of course, glass stoppers were not really new – they had been in relatively common usage in pharmacies and laboratories for centuries. The crystal producing company Preciosa are now the owners of the brand. The seal for the wine bottles is not a glass on glass, which would be ineffective, but an ethylene vinyl acetate co-polymer gasket is fitted between the underside of the stopper cap, and the top of the neck of the bottle. The OTR of Vinolok is in the range of 0.0026–0.0031 cc/day, similar to that of technical corks [41].

17.11 Some Advantages and Disadvantages of Various Closure Types

The search for the perfect wine closure is perhaps ongoing, with each type possessing strengths and weaknesses. The cork industry is making a concerted effort to stem the rise of alternative closures. Post-bottling reductivity is not generally a problem for wines sealed with natural corks, but is not unknown with technical corks. OTRs can be variable, and in cases of severe permeation wines can suffer random oxidation. Speaking at The Buyer's Closures Debate (held in association with the stopper supplier Vinventions) in 2018, Douglas Blyde, wine consultant, sommelier and drinks editor for The Evening Standard noted that 'that it was the inconsistency of traditional corks that was particularly difficult to work with and it is quite possible to find variations within a six-bottle case of wine' [29]. Membrane barriers, added to the ends of corks help provide a consistent OTR. The sides of the cork can also be coated, to reduce oxygen transmission between the neck and the bottle. ProCork uses this technology, and the product is available in a range of OTRs.

Although incidences of haloanisole taints have been considerable reduced in the last 15 years or so, the issue remains in producers minds with regard to the choice of closure, and other taints including bitter phenols or guaiacol may also migrate from cork into wine.

Natural corks can break or crumble upon opening, especially if the wine is old or the bottles has been stored incorrectly. It is still generally accepted that cork-sealed bottles should always be stored laying on their sides, but this adage has recently been challenged, particularly by Dr Miguel Cabral, director of R&D at Amorim. Speaking to *The Drinks Business* in June 2018, he stated, 'The cork will never dry out with almost 100% humidity in the headspace, so it is a myth that you need to store a bottle on its side' [42]. This conclusion, has either not reached or not been accepted by many industry players. Jon Tracey, sales director of English wine producer Bolney Estate states that one of the reasons for its 2018 switch from cork to screw cap was they 'wanted to remove the risk of upright storage affecting cork moisture levels, resulting in oxidation of the product' [43].

The retention in bottle of adequate levels of free SO_2 is important to reduce the risk of oxidation. A 2015 study by Liu et al. showed that bottles sealed with cork retained higher free and total SO_2 levels than synthetic closures after four years storage [44] and the 2020 study by Oliveira et al. reach a similar conclusion [34]. However, Saran-tin-lined screw caps retain the highest free SO_2 levels. A 2019 study by Meng-Qi Ling et al., looking at closure types for white and rosé wines found that rosé wines 'sealed with natural corks and screw caps exhibited no significant difference after an 18-month bottle aging, except for in the "berry" aroma description' [45]. However, the synthetic closure indicated a greater loss of aromas in rosé wines during ageing. Screw caps behaved better than any other closures in maintaining the aroma compounds in white wines [45].

Penfolds Grange is one of the finest Australian red wines, with recent vintages selling (at the time of writing) for in excess of £400 a bottle. Peter Gago, since 2002 Penfolds chief winemaker, says the reason that screw caps have not been used 'is because screw caps don't show heat damage, rather than because of concern about how the wine evolves' [46]. He says, 'A red wine under screw cap looks perfect. It might have been at 50 °C for four weeks but you would never know' [46]. It is well known that if a bottled wine has been subjected, albeit briefly, to hot conditions this may be evidenced by wine seeping around, and sometimes even through, the cork, and possibly the cork and capsule may be raised. Such a situation would immediately raise alarm bells with auctioneers, collectors and lovers of fine wine.

17.12 The Bottling Operation

Wine bottle manufacturers will supply the winery/bottling company with detailed drawings and specifications of bottles supplied. The drawings will include a suggested fill height (point), measured from the top of the bottle. In the case of standard 750 ml bottle this is usually approximately 64 mm from the top. It should be noted that this fill point is the appropriate point for wine that is bottled at 20 °C (68 °F), and should be adjusted according to the actual temperature of the wine at bottling. The level should be increased by 0.55 mm for every 0.55 °C (1 °F) the wine temperature is below 20 °C, and reduced by 0.55 mm for every 0.55 °C it is above 20 °C [47]. However, bottling at temperatures above 20 °C is not recommended. The bottle internal pressure should be adjusted according to the temperature of the wine to equate to a pressure of 2.0 psi at 20 °C. If bottles are sealed with a natural or technical cork they should be stored neck up for at least five minutes after corking, as it takes this time for a compressed cork to expand to 90% of its eventual expansion in the bottle [47].

17.13 PET Bottles

The image of wine in and PET bottles has always been low, and in most markets only a small percentage of inexpensive wine is sold in these containers, mostly inexpensive wines and small format bottles. Improvements in technology have in the consumer's eyes been negated by environmental considerations. Perhaps Laura Parker put it succinctly in in August 2019 in one of her several articles on the topic in *National Geographic*: 'The evolution of the plastic bottle from amazing to scourge of the land and sea has played out inside a generation' [48]. There are, of course, some advantages in the material: it is lightweight (perhaps its only environmental credential) and is largely unbreakable. Many events, e.g. outdoor concerts, sports events, prohibit glass containers being on site for security and safety reasons. However, from a product quality perspective there remain several major disadvantages. The material is not entirely impervious to oxygen, leading to bottlers generally using increase levels of sulfur dioxide (SO_2). Oxygen ingress results in reduced product shelf life, and there are many reported instances of oxidised wines. The acids in wine are known to attack PET, and some compounds migrate into wine. Flavour scalping by PET is another known issue. The material allows greater passage of UV light than glass. PET and antimony, the catalyst used in PET production, have been shown to be a source of phthalates especially if stored at high temperatures. On the positive side PET bottles are now available with oxygen scavenger additives (MXD6 or Oxyclear®) which help limit oxidation [49].

It is unlikely that many consumers buy wine in a PET bottle with the intention off keeping it for long before consumption. However, they almost certainly will not know how long it is since the bottle has been filled. The *Institut des Sciences de la Vigne at du Vin*, which is based in Bordeaux (with some 250 researchers), studied development of both white and red wines packaged in both single and multi-layered PET bottles. The report, which was widely publicised in both the general and trade press, found that after a period of six months since filling white wines had significant colour changes, and increase in oxygen, decrease in sulfur dioxide and had developed a 'rotten fruit flavour'. The changes in red wines were much less significant.

The majority of PET bottles are manufactured from virgin material that produces a food-grade resin. In a world that is questioning the place of single use plastic materials for packaging, is would seem probable that use of the PET bottle for a non-essential product such as wine will decline during the coming years.

17.14 Cans

Widespread packaging wine in cans has only been undertaken since the relatively recent past, but people have been buying food and drinks products in cans since the technology was introduced in the 1930s. A historic problem was reaction between wine, with its relatively high acid content, and the metal of the can but advances in coating/liner technology in the last 25 years have largely overcome this issue, although cans are still not suitable for long-term storage of wines. In 1996 the patented Vinsafe can coating system was introduced in Australia and this technology is now widely used in Australia, New Zealand, Europe, and Asia. However, as of 2019 the system was not licensed for use in the United States where an alternative system in which the metal is coated with an epoxy resin is in common use. During the canning operation oxygen is evacuated from the package which is then filled. A shot of liquid nitrogen is added, the lid applied and sealed.

The figures quoted for the global wine in can market vary considerably. Grand View research report a global figure amounting to US$ 70.3 for 2019 [50], whilst the US-published Wine Spectator Magazine quoted US$ 183.6 million over the 52-week period ending 11 July 2020 [51]. In some markets have has been a rapid growth in the last few years, albeit from a very low base: in the United States, from a modest US$ 6.4 million in 2015, wine in can sales increased to US$ 16 million in 2016 and then US$ 19.1 million in 2018 [50]. However, in 2019, less than 0.5% of retail sales value in the United States was in can format. Four of the world's top five wine producers (Constellation Brands, Treasury Wine Estates, E&J Gallo and The Wine Group) package wine in cans, in addition to bottles and other types of packaging [52]. Most canned wine is in a small format.

There are numerous advantages to packaging wine in cans, including

- 0% oxygen permeability;
- A complete absence of penetration of ultraviolet light;
- Largely unbreakable;
- Very light;
- Recyclable.

From a marketing point of view, a myriad of dynamic designs can be printed on the can. Cans are 100% recyclable, and lightweight. Cans also aid the consumer who wishes to practice portion control. They are particularly useful when several people prefer different wines; they are easily chilled and easily disposed of for recycling.

17.15   Bag-in-Box

Of all the packaging alternatives to glass bottles, only bag-in-box has made a significant impact in the market, particularly in some markets, in larger formats and for so-called 'entry level' wines. The packaging has been given various official and unofficial names in some countries, including 'bladder packs', goon bags and casks.

17.15.1   History of Bag-in-Box

In many ways the original wine bag was an animal skin, this being taken from various parts of the animal, or even the whole animal probably since wine was first made. The advantages included convenience, lightness, strength, flexibility, a barrier to oxygen ingress, and the container was not easily breakable. Of course there were significant downsides, including bacterial growth, and odours migrating into the wine. However, the above advantages are those of today's bag-in-box, although the issues of oxygen permeation, flavour scalping, and low perceived image are perhaps the major drawbacks. One major advantage to the small wine producer today, as that a basic bag filling machine is inexpensive.

The invention of the modern bag-in-box for wine is credited to Australian wine producer Tom Angove who obtained a patent in 1965 [23, 53]. Angove's bag was lacking a tap – the wine was obtained by cutting off a corner of the bag and resealing with a clothes peg. This weakness was addressed just two years later when the large wine producer Penfolds (today part of Treasury Wine Estates) marketed wine in their version [23]. Since then, there have been marked improvements in design, materials and manufacture of the bag, and the outer structural carton.

17.15.2   The Manufacture of the Bag

The bag comprises several layers of plastic and metal films, to provide structure, and/or a barrier to oxygen ingress. PET is one of the main materials and several layers of this may be incorporated in the manufacture. However, this material does not provide an effective barrier to oxygen and other gasses, so a layer or layers of an appropriate barrier, either metal or a resin co-polymer of ethylene vinyl alcohol and MXD6 nylon may be bonded to the PET. In this case alcohol and aroma losses through the bag are reduced, but the tap remains an area of vulnerability for oxygen ingression.

17.15.3   Filling Bag-in-Box Wines

The bag can be sterilised by ethylene oxide. Filling is a simple operation, and basic filling machines are inexpensive. However, the uptake of oxygen when filling can be high, which may promote product deterioration. A 2014 study by Ana Catarino et al. showed that oxygen uptake in filling red wine in bag-in-box was 2.22 mg/l, and white wine 2.22 mg/l [54]. These figures are alarmingly high.

17.16   Final Reflections

It's October 2019 and I'm tutoring a New Zealand wine tasting for the Cumbria Wine Society at Armathwaite Hall Hotel at Bassenthwaite, in England's beautiful Lake District. This was a fairly large audience and five bottles of each wine are required. I'm showing several flights of Sauvignon Blanc wines, including one from Hawke's Bay: the 2018 vintage of Trinity Hill. Hawke's Bay Sauvignon Blancs are quite different in style to those from Marlborough, from where stems the great majority of New Zealand's production. Marlborough has built its reputation on fresh, exotic, tropical fruit-driven Sauvignons, usually with overt aromas of gooseberry, and asparagus. Those from Hawke's Bay are somewhat fleshier, with an array of hedgerow flowers. The Trinity Hill is a delightful example. It is lees aged, and a small percentage of the wine is matured in barrels, giving a creamy texture without a hint of any aggressive characteristics. As always, after opening and before presenting to the tasters, I check each wine carefully for condition. Four of the five bottles are exactly as I would expect, but the fifth is a little different, with slightly less fruitiness, but showing some notes of hay, honey, and a hint of nuttiness. I examine the closure, a Saran-tin-lined screw cap, and note some damage on both the cap and liner: clearly the cause of the issue. The strange thing is I actually prefer the wine from this bottle – the little oxidation I have described has given an extra dimension. I have no idea when the cap was damaged, but I am sure that had the bottle been kept for a few more months, it would have been totally out of condition.

References

1 Pereira, B., Lopes, P., Marques, J. et al. (2013). Sealing effectiveness of different types of closures towards volatile phenols and haloanisoles. *Journal International des Sciences de la Vigne et du Vin* 47 (2): 145–157. https://doi.org/10.20870/oeno-one.2013.47.2.1545.

2 EIA (2013). *Glass Manufacturing is an Energy-Intensive Industry Mainly Fueled by Natural Gas*. US Energy Information Administration https://www.eia.gov/todayinenergy/detail.php?id=12631 (accessed 16 August 2920).

3 Taber, G.M. (2007). *To Cork or Not to Cork*. New York: Scribner, a division of Simon and Schuster Inc.

4 Grant-Preece, P., Barril, C., Leigh, M. et al. (2017). Light-induced changes in bottled white wine and underlying photochemical mechanisms. *Critical Reviews in Food Science and Nutrition* 57 (4): 743–754. https://doi.org/10.1080/10408398.2014.919246.

5 Guerrini, L., Pantani, O., Politi, S. et al. (2019). Does bottle color protect red wine from photo-oxidation? *Packaging Technology and Science* 2433 https://doi.org/10.1002/pts.2433.

6 Hartley, A. (2008). The effect of ultraviolet light on wine quality. www.wrap.org.uk/sites/files/wrap/UV%20&%20wine%20quality%20May%2708.pdf WRAP The Waste and Resource Action Programme.

7 Risch, S. (2009). Food packaging history and innovations. *Journal of Agricultural and Food Chemistry* 57: 8089–8092. https://doi.org/10.1021/jf900040r.

8 Robertson, G.L. (2013). *Food Packaging Principles and Practice*, 3e. Boca Raton, FL: CRC Press.

9 Ugliano, M., Dieval, J.B., Dimkou, E. et al. (2013). Controlling oxygen at bottling to optimize post-bottling development of wine. *Practical Winery & Vineyard Journal* 34: 44–50.

10 Marin, A.B. and Durham, C.A. (2007). Effects of wine bottle closure type on consumer purchase intent and price expectation. *The American Journal of Enology and Viticulture* 58: 192–201.

11 Lopes, P., Roseira, I., Cabral, M. et al. (2012). Impact of different closures on intrinsic sensory wine quality and consumer preferences. *Wine & Viticulture Journal* March/April, 34–41.

12 Francis, L., Field, J., Gishen, M. et al. (2003). The AWRI closure trial: sensory evaluation data 36 months after bottling. *Australian and New Zealand Grapegrower and Winemaker* 475: 59–64.

13 Cork Quality Council (2019). *Oxygen Ingress by Closure Type*. Forestville, CA: Cork Quality Council. https://www.corkqc.com/products/otr-ingress-by-closure-type.

14 Capone, D., Sefton, M.A., Prestorius, I., and Hoj, P. (2003). Flavour "scalping" by wine bottle closures. *Wine Industry Journal* 5 (18): 16–20.

15 Silva, M.A., Jourdes, M., Darriet, P., and Teissedre, P.-L. (2012). Scalping of light volatile sulfur compounds by wine closures. *Journal of Agricultural and Food Chemistry* 60 (44): 10952–10956. https://doi.org/10.1021/jf303120s.

16 Skouroumounis, G.K., Kwiatkowski, M.J., Francis, I.L. et al. (2005). The impact of closure type and storage conditions on the composition, colour and flavour properties of

a Riesling and a wooded Chardonnay wine during five years' storage. *Australian Journal of Grape and Wine Research* 11: 369–377. https://doi.org/10.1111/j.1755-0238.2005.tb00036.x.

17 Lopes, P., Silva, M.A., Pons, A. et al. (2009). Impact of oxygen dissolved at bottling and transmitted through closures on the composition and sensory properties of a Sauvignon Blanc wine during bottle storage. *Journal of Agricultural and Food Chemistry* 57: 10261–10270. https://doi.org/10.1021/jf9023257.

18 Ferreira, V., Bueno, M., Franco-Luesma, E. et al. (2014). Key changes in wine aroma active compounds during bottle storage of Spanish red wine under different oxygen levels. *Journal of Agricultural and Food Chemistry* 62: 10015–10027. https://doi.org/10.1021/jf503089u.

19 Schmitt, P. (2017). Wine closures: the facts. *The Drinks Business* 27 February. https://www.thedrinksbusiness.com/2017/02/wine-closures-the-facts (accessed 8 March 2019).

20 Apcor (2018, 2018). *Cork Yearbook 18/19*. Santa Maria de Lamas: Portuguese Cork Association http://www.apcor.pt/wp-content/uploads/2018/12/Anuario_APCOR_2018.pdf (accessed 17 January 2019).

21 Johnson, H. (1989). *Vintage: The Story of Wine*. London: Simon and Schuster.

22 Silva, S.P., Sabino, M.A., Fernandes, S.M. et al. (2005). Cork: properties, capabilities and applications. *International Materials Reviews* 50 (6): 345–365. https://doi.org/10.1179/174328005X41168.

23 Work, H.H. (2019). *The Shape of Wine – Its Packaging Evolution*. Abingdon: Routledge.

24 Mullen, T. (2019). This ProCork membrane can add consistency to wine quality. Forbes 2 December. https://www.forbes.com/sites/tmullen/2019/12/02/this-membrane-can-add-consistency-to-wine-quality/#480c1c42e0b9 (accessed 17 August 2020).

25 Procork (2020). Traditional Natural Cork and Technical Cork with OTR Technology. Webpage. https://www.procorktech.com/technical-science (accessed 17 August 2020).

26 Sogg, D. (2001). Oregon winery files lawsuit against maker of Altec corks. *Wine Spectator* 18 October 2001. https://www.winespectator.com/webfeature/show/id/Oregon-Winery-Files-Lawsuit-Against-Maker-of-Altec-Corks_21048 (accessed 21 January 2020).

27 Goode, J. (2002). Grain of truth. http://www.wineanorak.com/sabate_altectrial.htm (accessed 23 August 2020).

28 Diam. (2020). Press Kit. Dam Bouchage. https://www.diam-closures.com/medias/brochure/3250-dossier-de-presse.pdf (accessed 23 August 20200).

29 The Buyer. (2017) The Buyer's Closure Debate. http://www.the-buyer.net/wp-content/uploads/2017/06/The-buyer_Vinventions.pdf (accessed 23 August 2020).

30 Campbell, B. (2016). Diam closures flawed? *The Real Review*. 1 February. https://www.therealreview.com/2016/02/01/diam-closures-flawed (accessed 23 August 2020).

31 Millar, R. (2016). Diam hits back over 'bitterness' claim. *The Drinks Business* 5 February. https://www.thedrinksbusiness.com/2016/02/diam-hits-back-over-bitterness-claim (accessed 23 August 2020).

32 Nomacorc (2020). *Select Green – Green Line Solution for Premium Still Wines*. Product specification sheet. Vinventions https://www.vinventions.com/assets/e9772a31-d2e1-4741-b149-54b0989cee68/sellsheet-nomacorc-selectgreen-en-eu.pdf (accessed 24 August 2020).

33 Theron, C. (2016). The choice of bottle closures. *Wineland Media*, 1 May 2016. www.wineland.co.za/the-choice-of-bottle-closures (accessed 8 March 2019).

34 Oliveira, A.S., Furtado, I., de Lourdes Bastos, M. et al. (2020). The influence of different closures on volatile composition of a white wine. *Food Packaging and Shelf Life* 23: 100465. https://doi.org/10.1016/j.fpsl.2020.100465.

35 Work, H. (2005). Bottling under screwcaps. Quality control issues for premium wines. *Practical Winery & Vineyard Magazine*. July/August.

36 Wilson, D. (2010). Case study on the introduction of the screwcap to the Australian wine market. *5th International Academy of Wine Business Research Conference*, Auckland, New Zealand (8–10 February 2010). http://academyofwinebusiness.com/wp-content/uploads/2010/04/Wilson-Introduction-of-the-screwcap-Case-Study.pdf (accessed 30 June 2020).

37 Schneider, V., Schmitt, M., and Kroeger, R. (2017). Wine screwcap closures: the next generation. *Grapegrower & Winemaker* 638: 50–52.

38 Adams, A. (2015). Product focus: screwcaps with variable OTR. *Wines and Vines* August. https://winesvinesanalytics.com/features/article/155284/Product-Focus-Screwcaps-with-Variable-OTR (accessed 24 August 2020).

39 Easton, S. (2015). Are we reaching closure on the closure debate? http://www.winewisdom.com/articles/closures/are-we-reaching-closure-on-the-closure-debate (accessed 10 January 2019).

40 Eisermann-Ctercteko, A. (2013). Let's not be screwed! Screwcap damage levels greater than cork taint: implications for producers, the retail sector and consumers. *Wine and Viticulture Journal* September–October 2013, pp. 38–44.

41 Scrimgeour, N. (2014). *Vinolok (VINOSEAL) Closure Evaluation Stage 1: Fundamental Performance Assessment*. AWRI https://vinolok.com/wp-content/uploads/2019/09/awri-report.pdf (accessed 24 August 2020).

42 Schmitt, P. (2018). Storing wine on its side is nonsense, says scientist. *The Drinks Business*, 11th June 2018. https://www.thedrinksbusiness.com/2018/06/storing-wine-on-its-side-is-bullsht-says-scientist (accessed 9 January 2019).

43 Riley, L. (2019). Opening minds on closures. *Harpers Wine & Spirit* (February 2019): 44–46.

44 Liu, N., Song, Y.-Y., Dang, G.-F. et al. (2015). Effect of wine closures on the aroma properties of Chardonnay wines after four years of storage. *South African Journal of Enology and Viticulture* 36 (3): 296–303. https://doi.org/10.21548/36-3-963.

45 Ling, M.-Q., Xie, H., Hua, Y.-B. et al. (2019). Flavor profile evolution of bottle aged rosé and white wines sealed with different closures. *Molecules* 24 (5): 836. https://doi.org/10.3390/molecules24050836.

46 Atkin, T. (2019). *Finding Closure in the Cork-vs-Cap Conundrum*, 15 February. Harpers Wine & Spirit.

47 Cork Quality Council (2019). *Bottling Handbook for Proper Closures*. Forestville, CA: Cork Quality Council.

48 Parker, L. (2019). How the plastic bottle went from convenience to curse. National Geographic. www.nationalgeographic.co.uk/environment-and-conservation/2019/08/how-plastic-bottle-went-miracle-container-despised-villain (accessed 16 August 2020).

49 Sängerlaub, S. and Müller, K. (2017). Long-time performance of bottles made of PET blended with various concentrations of oxygen scavenger additive stored at different temperatures. *Packaging Technology and Science* 30 (1–2): 45–58. https://doi.org/10.1002/pts.2283.

50 Anon (2020). *Canned Wines Market Size, Share & Trends Analysis Report By Product (Sparkling, Fortified), By Distribution Channel (On-trade, Off-trade, Online Retail), By Region, And Segment Forecasts, 2020–2027*. San Francisco, CA: Global View research.

51 Weed, A. (2020). Canned wine sales are bursting at the seams. *Wine Spectator* 3 August.

52 Williams, H.A., Williams, R., and Bauman, B. (2018). *Growth of the Wine-in-a-Can Market*. Texas Wine Marketing Research Institute. http://www.depts.ttu.edu/hs/texaswine/docs/Wine_in_Can_Industry_Report.pdf.

53 Lower, G. (2010). Thomas Angove, king of the cask, dead at 92. *The Australian* 31 March.

54 Catarino, A., Alves, S., and Mira, H. (2014). Influence of technological operations in the dissolved oxygen content of wines. *Journal of Chemical Engineering* 8: 390–394.

18

Best Practice for Fault and Flaw Prevention

This Chapter comprises a general discussion on how faults and flaws in wine might be prevented. Topics briefly considered include Hazard Analysis Critical Control Point (HACCP), Standard Operating Procedures (SOPs), and traceability. The benefits of good winery design and layout are noted. Methods of effective cleaning and sanitation of the winery and equipment are explained. Winemaking practices and procedures that may lead to, or prevent faults, are considered. The use of additives and processing aids, including sulfur dioxide, dimethyl dicarbonate, lysozyme, and chitosan, are revisited.

18.1   The Wine Industry

In many ways, the wine industry has detached itself from the systems, procedures, and rigours that are undertaken daily in the food industry generally. Even though wine may be subject to spoilage during the process of production, the general absence of microbiological pathogens that are harmful to human health has resulted in a state of affairs in many wineries that would be regarded as unsatisfactory elsewhere in food manufacture. The implementation of HACCP systems and SOPs are not always applied with sufficient rigour in the wine industry. Indeed in some wineries, they are not applied at all. However, the conceptualisation of Risk Assessments, Risk Management Systems, Good Manufacturing Practice (GMP) and SOPs are important not only to avoid faults, flaws, and taints but to generally ensure a quality product.

18.2   HACCP

The concept of the HACCP system began in the early 1960s by the Pillsbury Company and the National Aeronautics and Space Administration (NASA) [1]. Its basis was the concept in engineering of Failure Mode and Effect Analysis (FMEA). This considers each stage of an operation and what can potentially go wrong at that stage, what the effect could be, and putting effective control mechanisms in place. The concept developed into a microbiological safety system to ensure food safety for astronauts. From these beginnings, HACCP has grown into an internationally accepted system for the prevention of foodborne disease. Accordingly, HACCP systems are in place in most food manufacturing and processing

Wine Faults and Flaws: A Practical Guide, First Edition. Keith Grainger.

plants. However, the production of wine is regarded as a low-risk food processing activity, and HACCP systems are generally regarded as not essential (although trade purchasers and retail consortia may demand that they are in place). For example, the New Zealand Food Safety Authority (NZFSA) has accepted that no critical control point was identified for the production of wine [2]. However, there are numerous physical, chemical, and microbiological hazards in the wine production process. By way of simple examples, excess additions of chemicals could result in faults, illegality of the finished product, or both. A mistake in calculating or weighing sulfur dioxide or dimethyl dicarbonate (DMDC) before bottling could result in excessive SO_2 or methanol, respectively. Contamination with haloanisoles, or *Brettanomyces*, would likely result in unsaleable wine. Accordingly, although not critical from a food safety perspective, there are control points that should be put in place to avoid the occurrence of faults, of loss of quality.

18.3 Standard Operating Procedures (SOPs)

SOPs are an important component of a quality system because they ensure consistency in operations. SOPs should contain detailed, written instructions for operations. Some examples of SOPs include labelling chemicals, additives, and processing aids; storing equipment; methods of racking; instructions for tank topping/gas blanketing; the preparation, checking, and additions of processing aids; and record keeping. Particularly important are cleaning and sanitation procedures, which are known as Sanitation Standard Operating Procedures (SSOPs).

18.4 Traceability

Traceability may be defined as the ability to track materials, from one point to another in the purchasing, production, and supply chains. A system should be designed, integrated, and implemented for:

- *Upstream traceability*: Goods that come into the production facility, including grapes, additives, processing aids and packaging materials;
- *Internal traceability*: The progress of grapes, must and wine through the production process, including the use of processing aids and additives;
- *Downstream traceability*: The tracking of the product after it leaves the production facility. This is vital should it be necessary to have a product recall.

18.5 Winery Design

The days when wines were mostly produced in timeless buildings constructed of stone or cellars whose walls were line with *Torula* or other fungi are, for better or worse, fast disappearing. Winery buildings of contemporary design using advanced structures, both natural and synthetic materials, are now commonplace at producers great and small. Illustrious estates wish to make a statement that exudes culture, art, and science. In recent years in the Haut–Médoc district of Bordeaux, many *Cru Classés* have made a massive investment in new cuveries and chais. Notable examples are Château Pichon Longueville Comtesse de Lalande, Château Montrose, Château Talbot, Château Cos d'Estournel, Château Beychevelle, and the striking, yet sympathetic, Norman Foster-designed cuverie at Château Margaux. At the time of writing a three-year total renewal programme is nearing completion at Château Lynch Bages in Pauillac.

Of course, when designing production facilities, there are numerous design and construction considerations to be addressed, and often these will conflict. Planning officers know little about wine production and may have very different considerations to property owners and winemakers. All producers hope that the visiting journalist, tourist, wine lover, or client will leave as an ambassador to the property. In a world where image is paramount, and beauty is associated with quality, having striking and appealing buildings is hugely desirable. Nobody wants to accept that fine wine is produced in a factory. Fundamental to the design are considerations as to achieving and maintaining the highest wine quality and eliminating possible sources of infection and contamination. In the last 20 years, environmental considerations have become most important, including achieving net zero carbon emissions which helps not only the planet but also the public relations (PRs) machine.

From a practical perspective, spacious, well-ventilated structures that are designed for minimal handling are ideal. A 'gravity-feed' winery minimises pumping (and oxygen uptake). Having simplicity and ease of must and wine flow through the buildings are needed for practicality and quality. The choice of construction materials requires careful consideration, and the avoidance of materials that might result to contamination is essential. As discussed in Chapter 3, any constructional and all internal timbers should be untreated or treated with compounds that do not pose any risk of contamination by chloroanisoles or bromoanisoles. Surfaces conducive to easy cleaning are essential. An adequate and efficient drainage system is vital – effluent should always be free-flowing. Ambient temperature and humidity control essential. Figure 18.1 shows a congested winery/barrel store that is difficult to operate and clean. Figure 18.2 shows the stylish and immaculate new chai at Château Talbot in Saint–Julien, Bordeaux.

18.6   Cleaning and Sanitation

The importance of a rigorous cleaning and sanitation regime cannot be overstated. All winery surfaces should be regularly cleaned and sanitised to inhibit microbial growth. Pressure washing is a relatively fast means of general basic cleaning, but many operatives place an overreliance on its effectiveness. By definition, debris is blasted against surfaces, often into containment areas or dirt traps.

18.6.1   Winery Water

It is vital that all water used in the winery for cleaning and other purposes, such as preparing processing aids, is chlorine-free. Otherwise, the risk of equipment, and even sections of the winery, becoming contaminated with one or more of the chloroanisoles is very real (see Chapter 3). Although contamination by 2,4,6-trichloroanisole (TCA) is

Figure 18.1 A somewhat congested winery.

the most widely documented; 2,3,4,6-tetrachloroanisole, 2-chloroanisole, 4-chloroanisole, 2,4-dichloroanisole, 2,6-dichloroanisole, and pentachloroanisole are also to be feared. Municipal water may well be treated with chlorine. Water should be filtered before use for cleaning filters, hoses, and bottling lines. Water stored in tanks is at an increases risk of microbial growth. In cases of doubt, water should be sterilised, e.g. with ultra-violet (UV) light.

18.6.2 Cleaning Agents

With the rise in popularity for craft beers, several small-scale wine producers have also entered the craft-brewing business. Hypochlorite bleaches are commonly used as a cleaning agent in breweries, and operatives must be trained in the dangers of introducing such products into the winery area. As discussed in Chapter 3, chlorine dioxide (ClO_2) is a safe sanitiser to use in wineries, as chlorine is not involved in any of the reactions. It is a cleaner and sanitiser that works by oxidation. It must be manufactured on-site, but it is stable in a dilute solution in the absence of light [3].

18.6.3 Cleaning of Equipment

All equipment including sorting tables/machines, de-stemmers, crushers, pumps, hoses, vats, filters, should be thoroughly cleaned after every use, and cleaned again and sanitised before subsequent use. Depending on the nature of the equipment, this may require partial

Figure 18.2 The immaculate new chai at Château Talbot.

or complete de-assembly. Dirt and bacteria are trapped on/in places sometimes missed or not even considered by cleaning personnel, including inside valves, threads, seals, sight tubes, sensors, and air locks. Cleaning in place (CIP) systems are most valuable from the point of view of efficiency and labour and time-saving. However, regular checks must be made on the effectiveness of the daily CIP operations, paying particular attention to the cleaning of 'hard to reach' areas. Cartridge membrane filters can be sanitised by flushing with water at 80 °C (182 °F), or by draining the housing and applying steam.

Steam cleaning is effective and environmentally friendly. However, care must be taken that steam does not contact surfaces and materials that may be damaged. These include many plastics, such as polypropylene linings of valves. Adequate post-cleaning ventilation of steamed surfaces is essential. Otherwise, the moist environment presents an ideal medium for the growth of moulds and bacteria. However, a steam generator can quickly repay its investment even in the smallest of wineries.

18.6.4 Suggested Sequence for Cleaning Equipment

The operative should wear Personal Protective Equipment (PPE) as required, including bodysuits, gloves, and goggles.

1. Disassemble as necessary and thoroughly clean with water and steam, using physical aids, such as brushes, as required. Avoid scratching surfaces with metal.

2. Use a mixture of sodium hydroxide (caustic soda) and water to clean all surfaces. CAU-TION: sodium hydroxide (alkaline) is highly caustic, and the utmost care must be taken to avoid contact with skin, eyes, hair, and every part of the body. The sodium hydroxide concentration can range from 2% to 5%, depending upon the level and nature of the soiling of the surfaces.
3. The equipment should be rinsed with clean water.
4. To neutralise the alkaline properties of the sodium hydroxide, the equipment should be rewashed with a solution of citric acid and water, in a similar concentration as was the sodium hydroxide solution previously used.
5. The equipment should be rinsed with a solution of clean water. All traces of citric acid must be removed. Citric acid can produce acetic acid in wine, resulting in excess volatile acidity (see Chapter 7).

18.6.5 Barrel Cleaning

Of all the equipment in a winery, it is the barrels, and by extension the entire barrel cellar(s) that are the most exposed to contamination. Haloanisoles, *Brettanomyces*, and acetic acid bacteria are just some of the faults that may result from lack of hygiene in this area.

It is strongly recommended that all barrels are purchased new directly from a reputable cooper, and commissioned according to the cooper's instructions. Purchasing previously used barrels poses a risk of contaminants entering the winery. Barrels are difficult to sterilise completely. It must be remembered that just one bad barrel poses a risk of affecting an entire batch of wine if the faulty wine it contains is not detected before blending. Depending upon the style and quality of the wines being produced, the use of second, third, fourth, and even fifth fill barrels is commonplace, but ideally, these should be from the winery's stock. After five years or so, the barrels will impart very little by way of oak flavours to a wine, and the risk of unwanted woody, or other aromas and flavours being absorbed increases.

A systematic and rigorous regime of cleaning between each use is required. Specialist barrel cleaning equipment is highly desirable for this purpose. Several barrel cleaning methods were detailed in Section 4.8.6.3. One generally effective method is thoroughly rinsing with cold water followed by three 70 °C (158 °F) water rinses and sanitising with SO_2 gas or filling with aqueous SO_2 at a concentration of 200 mg/l. The burning of a sulfur 'candle' in each barrel (as recommended by Pasteur) to disinfect the surfaces was de rigueur. This system is still used in many wineries. Today, the candle is usually a small disk with a hole (resembling a POLO® mint), which is suspended from a wire into the centre of the barrel. None of the solids from the candle should fall into the barrel, otherwise there is a risk of H_2S or other volatile sulfur compounds developing. In practice, the amount of SO_2 gas released is equivalent to the weight of the candle.

18.7 Good Practice Winemaking Procedures to Avoid Spoilage, Faults, and Flaws

The implementation of good practice winemaking procedures to avoid faults, flaws, and taints have been discussed throughout this book, but it is perhaps pertinent to revisit some key areas.

18.7.1   Harvesting

Pre-harvesting, it is important to withhold vineyard chemicals, including pesticides and fungicides, for four to six weeks before picking. All equipment used, including picking bins, grape scissors, harvesting machines should be scrupulously clean. Immediate washing after each use is essential. Minimal damage to the clusters and berries reduces the presence of vineyard flies, including spotted winged drosophila (*Drosophila suzukii*) and common fruit fly (*Drosophila melanogaster*). These flies can be responsible for the onset of sour rot, imparting a distinct vinegar smell to the grapes and even the wine. Damaged berries also commence the oxidation process.

Harvesting should ideally take place in cool conditions, e.g. early morning, and the filled bins kept as cool as possible

18.7.2   Transportation of Harvested Fruit

Grapes should be transported in small containers, e.g. 12–16 kg lug boxes and should ideally reach the winery and commence processing within an hour of picking. Steps should be taken to protect grapes from oxygen, and bacterial growth – these might include the covering of lug boxes with dry ice, or potassium metabisulfite (PMBS), or by blast-chilling. Although there are downsides to adding PMBS to grapes, it inactivates polyphenol oxidase and thus helps prevent enzymatic oxidation, as discussed in Chapter 5.

Figure 18.3 shows PMBS added to harvested grapes in picking bins.

Figure 18.3   Potassium metabisulfite added to harvested grapes.

18.7.3   Sorting

Any diseased or damaged fruit and materials other than grapes (MOGS) should be removed by sorting. This may be a multistage operation with initial sorting in the vineyard, bunch sorting before de-stemming and berry sorting before crushing. Vibrating tables work well provided the operatives are skilled. Optical sorters are very efficient and remove the risk of 'human-failure' from the process.

18.7.4   De-stemming, Crushing

This can be the dirtiest operation in the winemaking process. The crushing station is a magnet for flies, particularly *Drosophila*, who are vectors of bacteria, including the dreaded *Acetobacter*. Every effort should be made to exclude these. Discarded stems should be immediately removed and transported to at least 100 m from the winery for processing. Oxygen uptake at crushing can range from 5 to 10 mg/l. The addition of sulfur dioxide at crushing is usually the first stage of microbiological control in the winemaking process.

18.7.5   Pressing (for White or Rosé Wines)

Pressing should be gentle but rapid. Tank presses, which surround the grapes in an atmosphere of inert gas (which is recycled), prevent the oxidation of must and help limit bacterial growth at this stage. High phenolic fractions from the press should be separated. These are likely to come from later pressings and will generally have high pH. A stronger fining may be used for these.

18.7.6   Yeast Nutrients

The addition of yeast nutrients should be limited, as any excess may also provide nutrients for *Brettanomyces*. However, to avoid a sluggish or incomplete fermentation (and to reduce the risk of the formation of volatile sulfur compounds), sufficient yeast assimilable nitrogen (YAN) must be present in the must. Yeasts need approximately 110 mg/l of YAN for satisfactory red wine fermentations, and 165 mg/l for whites. The use of diammonium phosphate (DAP) should be avoided unless the musts are nitrogen deficient. Although DAP minimises yeast stress and helps a vigorous fermentation without the production of sulfides, such ferments may retain micronutrients that may subsequently be used by spoilage yeasts as a food source.

18.7.7   Oxygenating During Fermentation

During the yeast growth phase during fermentation, some 6–10 mg/l of oxygen are needed. Depending upon handling and unless protected, grape musts will contain some dissolved oxygen (DO). The addition of 2 mg/l per day may be undertaken early in the fermentation cycle. Additions should not be made near the end of the alcoholic fermentation as yeasts no longer require oxygen. The necessary supply of oxygen for fermenting musts may be undertaken by several methods, including aerated pump-overs, splashing the discharging juice into a vessel and/or retuning juice over the cap. Air may also be introduced in line during pump-overs.

18.7.8 Use of Inert Gasses and Topping of Containers

The use of inert gasses during winery operations is a most effective tool in minimising uptake of oxygen, avoiding oxidation, and limiting the growth of aerobic yeasts and bacteria. During racking operations, air should be displaced from both the racking and receiving tanks/barrels. The receiving vessel should be purged with three to seven volumes of gas, and the wine in the racking vessel blanketed too. Hoses and the pump should also be gas flushed. After racking, the DO in the wine should be checked, and the wine sparged if necessary.

All storage tanks and barrels should be topped regularly to avoid air space. If wine is stored on ullage in a tank, the headspace must be flushed, and a blanket maintained over the stored wine. The concentration of oxygen in the headspace should be regularly monitored and kept below 0.5%. The use of tanks with 'floating' lids held in place by an inflatable plastic tyre is not satisfactory for anything other than short term storage. If the tanks are constructed of light-gauge steel, the metal is likely to distort, the lids often do not sit directly on the entire wine surface, and loss of pressure in the tyre is not uncommon. Storing wine under carbon dioxide will result in an increase in its concentration in the wine. If the tank is 98% full, the increase is in the region of 7%, but if the tank is only half full, there will be a 297% increase [4]. For wines stored under nitrogen, there will be a decrease in the concentration of carbon dioxide of 1.5% in a 98% full tank, and 49.6% in a 50% full tank. The concentration of CO_2 in white wines should generally be in the range of 500–700 mg/l, and for red wines 200–400 mg/l. Tannic and solidly structured reds should be at the bottom end of this range. There are several methods for reducing CO_2 in wines, including fine nitrogen bubbling into a stream of wine at a temperature of approximately 18 °C (65 °F), and allowing degassing in a shallow container.

18.7.9 Bottling

The bottling line is the last production contamination source of wine. However, wines can become contaminated in bottle, e.g. with TCA (see Chapter 3). The bottling line should be sanitised before the bottling run by flushing with water at over 80 °C for at least 20 minutes [5], or the use of steam with caustic cleaning agents. There must be no 'cold spots' in the bottling unit. Biofilms can be resistant to hot water sterilisation, and the bottling line should be rinsed at the end of each day. Regular measurements of DO and headspace oxygen (HSO) should be made during bottling [6]. Carefully inspected closures and preventive maintenance are also essential components of efficient bottling.

18.8 The Use of Oenological Additions and Processing Aids

18.8.1 Sulfur Dioxide

18.8.1.1 Properties of Sulfur Dioxide
The importance of maintaining appropriate free and molecular sulfur levels at the various stages of winemaking cannot be overstated. Sulfur dioxide (SO_2) has four important properties:

- Antimicrobial;
- Antioxidant;

- Antioxidase/antienzymatic;
- Binder of carbonyl compounds, including acetaldehyde.

Although SO_2 can have a bleaching effect, it actually helps the release of anthocyanins during the red wine maceration process, improve colour, and helps in clarification by promoting the coagulation of colloids and increasing lees precipitation [7].

18.8.1.2 Total, Bound, Free, and Molecular Sulfur Dioxide

A large portion of the SO_2 added to wine immediately becomes reversibly 'bound' to various compounds, including phenols, carbonyl components, aldehydes (particularly acetaldehyde), and sugars. Thus there is no antimicrobial or antioxidant effect. The remaining 'unbound' portion is known as free SO_2, and although it is widely believed that this entire portion has antimicrobial properties, this is not the case. Another consideration is that for every 10 mg/l SO_2 added to the must, the bound SO_2 levels in the final wine will increase by 3–7 mg/l.

$$SO_2 + H_2O_2 \quad \rightarrow \quad SO_4 + 2H$$

Sulfur dioxide + Hydrogen peroxide $\quad \rightarrow \quad$ Sulfate + Hydrogen

Free Sulfur Dioxide It is generally accepted that a minimum level of 10 mg/l of free SO_2 is needed to be maintained in bottled wine for it to be an effective antioxidant. However, the actual critical threshold concentration of free SO_2 will vary according to the matrix of volatile carbonyl compounds present in the wine [8].

Free sulfur dioxide (SO_2) in either gaseous or aqueous solutions is present in three forms:

- Bisulfite – HSO_3;
- Molecular SO_2 – $SO_2 \cdot H_2O$;
- Sulfite – SO_3^{2-}
- Bisulfite: At the pH of wine, between 92% and 99% of SO2 present will be in the form of bisulfite [9]. This has antioxidant properties but does not have any antimicrobial effect.
- Molecular SO_2: Only a molecular portion of SO_2 has antimicrobial properties, and the percentage of total SO_2 that this comprises in wine is largely pH-dependent. The lower the pH, the higher the percentage of molecular SO_2. Sulfur dioxide is toxic to many bacteria, and to *Kloeckera apiculata* and other non-*Saccharomyces* yeasts. Molecular SO_2 can pass through the yeast cell membrane, and once inside will bind with proteins and enzymes, resulting in eventual cell death. The level at which non-*Saccharomyces* yeasts are inhibited is species and strain-dependent: a level above 0.4 mg/l of molecular SO_2 will be effective in most cases, but some strains of *Brettanomyces* may be tolerant up to 0.625 mg/l, others even higher, as discussed in Chapter 4. *Saccharomyces cerevisiae* is generally tolerant to a level of 0.8 mg/l of molecular SO_2.

18.8.1.3 Health and Legal Considerations

Some asthmatics are allergic to SO_2, with around 5% suffering from wheezing, chest tightness, coughing [10], or other severe reactions. Thankfully, anaphylaxis is very rare. Perhaps some 1% or so of the general population have an intolerance or sensitivity to sulfites. The Food and Agriculture Organization of the United Nations (FAO) and the World Health Organisation (WHO) sets the acceptable daily intake (ADI) of sulfites as 0.7 mg/kg of body weight [11]. For a person weighing 11 stones (70 kg), this would represent a maximum daily intake of 49 mg. Assuming no other sulfates from any source were ingested, this would equate to a maximum consumption of a third of a litre or less than half a standard 75 cl bottle of a wine containing 150 mg/l of sulfites.

There are legal limits at to the amount of Total SO_2 that may be present in wine when offered for sale, in most major markets. These may vary according to wine colour and style. The limits for the EU, the United Kingdom, USA, Canada, Australia, and New Zealand, at the time of writing, are detailed in Table 18.1.

Table 18.1 Legal limits on total SO_2 content of wine in the EU, UK, USA, Canada, Australia, and New Zealand.

Territory	Wine type	Limit of total SO_2 in wine (mg/l)	Legal reference
Canada	All	350[a]	Canadian Food & Drug Reg. (C.R.C., c. 870) B.02.100
USA	All	350	27 CFR 4.22(b)(1)
Australia	Less than 35 g/l RS	250	ANZFSC 4.5.1: Clause 5(5)(a)
	More than 35 g/l RS	300	
New Zealand	Less than 35 g/l RS	250[b]	ANZFSC 4.5.1: Clause 5(5)(a)
	More than 35 g/l RS	400[b]	
EU[c]	White/rosé		
	Less than 5 g/l RS	200	(EU) 2019/934 Annex I B
	More than 5 g/l RS	250	
	Specific wines e.g. Spätlese	300	
	Specific wines e.g. Auslese	350	
	Specific wines e.g. Sauternes	400	
	Red		
	Less than 5 g/l RS	150	
	More than 5 g/l RS	200	
	Organic red wines	100	EU No. 203/2012, Annex VIII a
	Organic white wines	150	

a) Free SO_2 is limited to 70 mg/l.
b) Expressed as mg/kg.
c) The limits for the United Kingdom are expected to remain in line with those of the EU.

18.8.1.4   The Use of Sulfur Dioxide in Winemaking

Sulfur dioxide may be added to grapes, must, or wine in one or more of three forms:

- Potassium metabisulfite (PMBS) – molecular formula: $K_2S_2O_5$;
- Gaseous;
- Aqueous.

Potassium Metabisulfite (PMBS)   This is a stable powder that in theory, yields 57.6% by weight of SO_2, i.e. 100 g of PMBS added 1 hl to must or wine would increase the **total** SO_2 by 57.6 mg/l. In fact, in commercial oenological products, the SO_2 content is approximately 50% [7].

The compound also makes an effective sanitising agent for equipment: as such it should be added to cold water (NB: water should never be added to PMBS as this results in a rapid release of SO_2 gas). A few grammes of citric acid added to each 10 l of solution further increase the effectiveness of PMBS as a sanitising agent. Sodium metabisulfite, which is less expensive, may be used as an alternative sanitising agent, but is not generally permitted as an additive in winemaking.

When filling barrels post malolactic fermentation (MLF) or maintaining the SO_2 levels of wines in barrique or other small barrels, the use of effervescent tablets comprising a mixture of 67% PMBS and 33% potassium bicarbonate (33%) is very convenient. These are simple, and give a regulated amount of SO_2 without the need for weighing. The potassium bicarbonate produces an effervescence that facilitates the combing the SO_2 released into the wine. The tablets are particularly useful for modest additions of sulfite during storage. Two popular brands are OENOSTERYL® EFFERVESCENT manufactured by Laffort, and Inodose made by Institut Oenologique de Champagne. The tablets are available in 2 and 5 g formats. Adding a 2 g tablet to a 225 l barrique of wine will increase the total SO_2 level by 8 mg/l, and a 5 g tablet by 22 mg/l.

Aqueous SO$_2$ Aqueous SO_2 may be made by bubbling SO_2 gas into water, or by adding PMBS to water, making a 5 or 10% solution. The use of aqueous SO_2 is recommended for pre-bottling additions, as this eliminates any risk of precipitation of potassium salts a might happen if PMBS is used.

The stages of the winemaking process at which sulfur dioxide should be added are:

- *Crushing*: The amount of SO_2 added at crushing will depend on the colour and style of wine being made, and the sanitary condition of the fruit. For healthy black grapes for making red wine, the addition might be some 40 mg/l. For healthy grapes for white wine, some 60 mg/l is recommended. If the fruit is damaged or is suffering rot or mildews, a larger addition is recommended, perhaps an additional 20 mg/l;
- *Post MLF*: A level of 20–40 mg/l of free SO_2, depending on the type of wine and pH, may be considered as appropriate to protect the wine against oxidation;
- *Racking*: A level of 0.4–0.8 mg/l of molecular SO_2, depending on the type of wine and pH, may be considered as appropriate to protect the wine against microbial growth;
- *Pre-bottling*: Free SO_2 should be adjusted to 25–35 mg/l, according to the wine type and pH. The total SO_2 concentration should be analysed to ensure that the wine will be legal at the point of sale (see Table 18.1). Levels of molecular SO_2 will fall over time in packaged wine, the amount of fall depending on several factors including the wine composition, type of container, and if in bottle, the closure. White wine might lose between 20% and 50% of molecular SO_2 during the first year in bottle.

18.8.2 Dimethyl Dicarbonate (Velcorin)

DMDC is a microbial control agent marketed under the trade name Velcorin™. It is a steri-lant that works by inactivating microbial enzymes, and its use may prove particularly valu-able to producers who make wines with no added SO_2 [12]. It is effective against most yeasts, including of the genera *Saccharomyces*, *Brettanomyces*, and *Schizosaccharomyces* and acetic and lactic acid bacteria (LAB) [13]. DMDC may be added shortly before or during the bot-tling process to obtain microbiological stability for bottled wines containing fermentable sugars and to prevent the development of unwanted yeast and LAB. The manufacturer of Velcorin (LANXESS) states that the usual dosage is between 125 mg/l and 200 mg/l [14]. The maximum addition according to International Organisation of Vine and Wine (OIV) International Code of Oenological Practices is 200 mg/l, and the wine must not be marketed whilst its presence is detectable [15]. It is generally accepted that the use of DMDC does not result in any organoleptic impact. Viable yeast or bacteria concentration should always be reduced to below 500 CFU/ml before the addition of the compound. The use of DMDC as a preservative (E242) is permitted for wine in the EU under Regulation 2019/934 [16] (which amends previous regulations). It is approved as an additive for wine in the USA by FDA 21 – Code of Federal Regulations, providing 'the viable microbial load has been reduced to 500 microorganisms/ml (500 CFU/ml) or less by current good manufacturing practices such as heat treatment, filtration, or other technologies prior to the use of dimethyl dicar-bonate' [17]. It is also approved as a *processing aid* in wine in Australia and New Zealand under Food Standard Code 1.3.3. [18]. DMDC should always be used in synergy with sulfur dioxide. Velcorin must be completely homogenised into wine, so requires a special dosing apparatus. It is recommended that Velcorin is added post-final filtration before bottling as illustrated in Figure 18.4.

18.8.3 Lysozyme

The addition of lysozyme may effectively inhibit the growth of LAB. This is a low molec-ular weight protein derived from egg whites that brings about lysis of the cell wall of Gram-positive bacteria, including LAB species. It is another product that is useful to producers who wish to make wines with no added SO_2 [13]. The OIV permits its use up to a maximum dose of 500 mg/l [15]. As the dosing equipment is expensive, it is generally out of reach of small producers, unless mobile facilities are available. Except in the case producers making 'no-added sulfur' wines, lysozyme should always be used in association with sulfur dioxide.

18.8.4 Chitosan

Chitosan is a polysaccharide derived from chitin and isolated from *Aspergillus niger* mycelium which can prevent the growth of *Brettanomyces*, together with that of some bacteria. Only the fungal form derived from *A. niger* using a process patented by the company *KitoZyme* is permitted by EU and OIV for use in wine [15, 16]. The product is marketed by Lallemand as 'No Brett Inside'™. The legal limit of addition, for this purpose, under EU (and OIV) Regulations is 10 g/hl (0.1 g/l) (100 mg/l). The usual rate of application is 4 g/hl, and this is best added incrementally. As the yeasts may grow again after treatment, racking an immediate addition of SO_2, followed by regular monitoring and maintaining an adequate molecular SO_2 level, is essential [19].

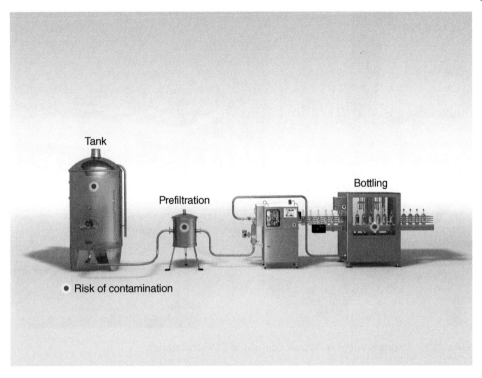

Figure 18.4 Velcorin dosing unit post filtration in bottling line. Source: Courtesy of LANXESS.

18.9   Routine Wine Analysis

During the individual chapters of this book, I have discussed methods of analysis for individual faults. Some of the equipment is expensive, even a basic gas chromatograph and mass spectrometer run into several thousands of £/$/€. Some are hugely expensive, and equipment such as GC/MS/MS (QqQ) ('Triple Quadrupole') is likely to remain mostly within the confines of professional laboratories. State-of-the-art laboratories, such as California's ETS have achieved sensitivity, accuracy, and speed of analysis to levels that could only be imagined 20 years ago. The laboratory developed Scorpions™ assays, which 'based on specific genetic targets, detect the full range of wine and juice spoilage organisms. This genetic analysis method detects microbial populations directly in wine or juice. Results are routinely reported within two business days, giving winemakers the ability to address problems before wine defects occur [20]. An image of Scorpions analyses is shown in Figure 18.5.

At the other end of the scale, the small winery, without trained laboratory staff can now have instant analysis results for many parameters, by using inexpensive spectrophotometers such as the CDR WineLab® or Megazyme MegaQuant Wave. These have a good degree of accuracy for most of their available parameters. Simple mini titrators for SO_2 and titratable acidity are also inexpensive very easy to use.

However, the traditional 'wet' laboratory still has its place, and the accuracy for some parameters, e.g. free and total SO_2 using aeration oxidation is unbeatable. The analysis methods in the OIV Compendium of International Methods of Wine and Must Analysis

Figure 18.5 Scorpions analyses. Source: Courtesy of ETS Laboratories.

(many of these are still based on traditional wet techniques) should be used for all legally required analyses [21]. Important analyses to be regularly undertaken during production and maturation include pH, titratable acidity, volatile acidity and free and total SO_2. Table 18.2 lists parameters and methods that are used for the analysis of wine.

18.10   Final, Final Reflections

Crucial to the quality of a wine is its balance: the relationship between the individual structural components and the interplay between the compounds that contribute aromas and flavours. However, there is another type of balance, the balance between technical perfection and excitement. As I noted in the Introduction, one wine can be analysed chemically and microbiologically and be declared technically very good and free from flaws, yet may taste distinctly uninteresting. Another may show technical weaknesses or even flaws, yet when tasted ,it can be so full of character and true to its origin that it sends a shiver down the spine, and should be regarded as the very highest quality. Sadly in the wine world today, there are far too many wines that fit the former description.

Lambic beers, produced in Belgium's Pajottenland region and Brussels, are like no other. They may be described as being at a crossroads where the differences between wine, beer, and cider merge. They are natural ferments in which a variety of yeasts are involved, including *Hanseniaspora uvarum*, *Pichia anomala*, and *Brettanomyces bruxellensis*. They are sharp. They are sour and sometimes have more than a whiff of vinegar, having high levels of lactic and acetic acids, the latter up to 1.6 g/l. They can smell of stables. They may

Table 18.2 Parameters and methods used for the analysis of wine.

Analyte	Method
2,3,4,6-Tetrachloroanisole (ng/l)	GC-MS
2,4,6-Tribromoanisole (ng/l)	GC-MS
2,4,6-Trichloroanisole (ng/l)	GC-MS
4-Ethylphenol (µg/l)	GC-MS
Acetaldehyde (mg/l)	Gas chromatography with flame ionisation detector (GC-FID)/spectrophotometer
Alcohol (vol% at 20 °C)[a]	Ebulliometry/alcohol meter/spectrophotometer
Arsenic (mg/l)[a]	Ion chromatography/trace elemental analysis
Ascorbic acid (mg/l)[a]	High performance liquid chromatography (HPLC)/spectrophotometry (spectrofluorometer)
Bacteria (CFU/ml)	Plate cultures/fluorescence microscopy/quantitative polymerase chain reaction (qPCR)
Benzoic acid (mg/l)	HPLC
Brettanomyces spp.	Plate cultures/fluorescence microscopy/qPCR
Calcium (mg/l)[a]	Flame atomic absorption spectrometry (FAAS)
Calorific value (kJ/100 ml [kCal/100 ml])	Calculation
Carbon dioxide (g/l)	Orbisphere
Cold stability[a]	Chill test
Copper (mg/l)[a]	FAAS
Density (g/l)	Densitometry
Dissolved oxygen (mg/l)	Dissolved oxygen meter
Dry extract (g/l)	Calculation
Ethyl acetate (mg/l)	GC-FID
Ethyl carbamate (µg/l)	LC-MS/MS
Filterability	Trial
Free sulfur dioxide (mg/l)[a]	Aspiration oxidation/Ripper/Platinum electrode
Fructose (g/l)	Sequential analyser
Glucose (g/l)	Sequential analyser
Glycerol (g/l)	HPLC
Haze (formazine turbidity unit)	Heat stability
Hydrogen sulfide (µg/l)	GC/sulfur chemiluminescent detection (SCD)
Iron (mg/l)[a]	FAAS
Lactic acid (g/l)	Spectrophotometer/sequential analyser
Lead (mg/l)[a]	Graphite furnace atomic absorption spectroscopy (GFAAS)
Magnesium (mg/l)	Inductively coupled plasma mass spectrometry (ICP-MS)

Table 18.2 (Continued)

Analyte	Method
Malic acid (g/l)	Paper chromatography/photospectrometer
Mould (CFU/ml)	Plate cultures/fluorescence microscopy/qPCR
Ochratoxin A (μg/l)	HPLC
Pentachloroanisole (μg/l)	GC-MS
pH[a]	pH meter
Potassium (mg/l)[a]	Flame photometry
Protein stability[a]	Heat test
Reducing (fermentable) sugar (g/l)	Rebelein
Residual (all forms of) sugar (g/l)[a]	HPLC/reaction-titration/spectrophotometer
Silver (mg/l)	ICP-MS
Sodium (mg/l)[a]	Flame atomic emission spectrometry (FAES)
Sorbic acid (mg/l)[a]	HPLC
Specific gravity[a]	Hydrometer/densitometry
Sugar-free dry extract (g/l)[a]	Calculation: sugar-free extract is the difference between the total dry extract and the total sugars
Tartrates	Conductivity/cold stability
Total (titratable) acidity (g/l)[a]	Indicator titration/spectrophotometer
Total dry extract (g/l)	Oven/calculation
Total phenolics (mg/l)	Spectrophotometer
Total sulfur dioxide (mg/l)[a]	Ripper/Platinum electrode
Volatile acidity (g/l [acetic acid])[a]	Cash still/HPLC/spectrophotometer
Yeast (CFU/ml)	Plate cultures/fluorescence microscopy/qPCR

a) A certificate showing analysis results for these items is commonly required by United Kingdom supermarkets.

be cloudy. They age for years. Many beer drinkers really dislike the styles, but to lovers, they offer a breadth of aromas and flavours like no other. Perhaps they offer a glimpse into a bygone world of wine production and are a million miles from standardised and globalised modern wines, that wine-producing technicians spend their lives ensuring that there is no hint of a fault or a flaw, or excitement.

As I write this, sales of still wines in many key markets, including the UK and USA, are in decline. Millennials are unexcited by the product. In an otherwise flat beer market sales of craft beers are growing: up 6% by value in the USA in 2019. Craft beers and real ales are perhaps the 'new wine', and some top restaurants now have lists of these that include food matching suggestions. Sales of 'artisan' gins, many using weird and wonderful botanicals and selling for two or three times the price of the familiar proprietary brands, have gone through the roof, particularly in the UK. The excitement is there, and consumers are reaching out for it. Perhaps wine production has become too obsessed with the science and lost sight of the art.

References

1 Mortimore, S. and Wallace, C. (2001). *HACCP*. Oxford: Blackwell Science.

2 New Zealand Food Safety Authority. Generic HACCP Application: Production of Grape Wine. https://www.mpi.govt.nz/dmsdocument/869/direct (accessed 10 August 2019).

3 UC Davis Viticultre and Enology. (2019). Chlorine dioxide. Webpage. https://wineserver.ucdavis.edu/industry-info/enology/methods-and-techniques/common-chemical-reagents/chlorine-dioxide (accessed 6 July 2019).

4 Lonvaud-Funel, A. (1976). Recherches sur le gaz carbonique du vin. Thèse de Doctorat Oenologie. Univerité de Bordeaux II.

5 Tran, T., Wilkes, E., and Johnson, D. (2015). Microbiological stability of wine packaging in Australia and New Zealand. *Wine and Viticulture Journal*, March/April 2015: 46–49.

6 Letaief, H. (2016). Key points of the bottling process. *Wines & Vines*. https://winesvinesanalytics.com/features/article/168227/Key-Points-of-the-Bottling-Process (accessed 30 April 2020).

7 Giacosa, S., Segade, S.R., Cagnasso, E. et al. (2019). SO_2 in wines: rational use and possible alternatives. In: *Red Wine Technology* (ed. A. Morata), 309–321. London: Academic Press.

8 Barril, C., Rutledge, D.N., Scollary, G.R., and Clark, A.C. (2016). Ascorbic acid and white wine production: a review of beneficial versus detrimental impacts. *Australian Journal of Grape and Wine Research* 22: 169–181. https://doi.org/10.1111/ajgw.12207.

9 Wilkes, E. (2018). Practical measurement of total SO_2. *Wine & Viticulture Journal* 33 (4): 32–34. www.awri.com.au/wp-content/uploads/2018/12/s2037.pdf (accessed 26 April 2020).

10 ASCIA (2019). Sulfite sensitivity. Information for patients, consumers and carers. Australian Society of Chemical Immunology and Allergy. www.allergy.org.au/images/pcc/ASCIA_PCC_Sulfite_sensitivity_2019.pdf (accesed 27 April 2020).

11 Food and Agriculture Organization of the United Nations/World Health Organization. (2008). Joint FAO/WHO Expert Committee on Food Additives. Sixty-ninth meeting, Rome 17-26 June 2008 Summary and Conclusions. http://www.fao.org/3/a-at870e.pdf (accessed 30 December 2020).

12 Fugelsang, K.C. and Edwards, C.G. (2007). *Wine Microbiology – Practical Applications and Procedures*, 2e. New York: Springer.

13 Lisanti, M.T., Blaiotta, G., Nioi, C., and Moio, L. (2019). Alternative methods to SO_2 for microbiological stabilization of wine. *Comprehensive Reviews in Food Science and Food Safety* 18 (2): 455–479. https://doi.org/10.1111/1541-4337.12422.

14 LANXESS. (2020). Velcorin® The best choice for wines. Webpage. https://velcorin.com/velcorin/spectrum-of-use/velcorin-wine (accessed 20 July 2020).

15 International Organisation of Vine and Wine (OIV) (2021). *International Code of Oenological Practices*. Paris: OIV. https://www.oiv.int/public/medias/7713/en-oiv-code-2021.pdf.

16 EUR-Lex. (2019) Commission Delegated Regulation (EU) 2019/934 of 12 March 2019 supplementing Regulation (EU) No 1308/2013 of the European Parliament and of the Council as regards wine-growing areas where the alcoholic strength may be increased, authorised oenological practices and restrictions applicable to the production and conservation of grapevine products, the minimum percentage of alcohol for by-products and their disposal, and publication of OIV files. https://eur-lex.europa.eu/legal-content/GA/TXT/?uri=CELEX:32019R0934 (accessed 25 May 2020).

17 e-CFR. Code of Federal Regulations, Title 27, Chapter I, Subchapter A, Part 24. https://www.ecfr.gov/cgi-bin/text-idx?c=ecfr&sid=506cf0c03546efff958847134c5527d3&rgn=div5&view=text&node=27:1.0.1.1.19&idno=27 (accessed 14 June 2020).

18 Australia and New Zealand Food Standard Code 1.3.3. www.foodstandards.gov.au/code/Documents/1.3.3%20Processing%20aids%20v157.pdf (accessed 8 April 2020).

19 Petrova, B., Cartwright, Z.M., and Edwards, C.G. (2016). Effectiveness of chitosan preparations against *Brettanomyces bruxellensis* grown in culture media and red wines. *Journal International des sciences de la vigne et du vin* 50 (1) https://doi.org/10.20870/oeno-one.2016.50.1.54.

20 ETS Laboratories (2015). *Scorpion Genetic Testing for Spoilage Organisms.* ETS https://www.etslabs.com/library/27 (accessed 20 August 2020).

21 OIV. (2016). Compendium of International Methods of Wine and Must Analysis. Paris: OIV. http://www.oiv.int/public/medias/2624/compendium-2016-en-vol1.pdf (accessed 28 March 2020).

Appendix A

Levels of Free SO$_2$ Required to Give 0.5, 0.625, and 0.8 mg/l of Molecular SO$_2$ for Differing Wine pH Values

Desired molecular SO$_2$	0.5 mg/l	0.625 mg/l	0.8 mg/l
pH of wine		Free SO$_2$ required (mg/l)	
2.9	7	8	11
3.0	8	10	13
3.1	10	13	16
3.2	13	16	21
3.3	16	20	26
3.4	20	25	32
3.5	25	31	40
3.6	31	39	50
3.7	39	49	63
3.8	49	62	79
3.9	62	78	99
4.0	78	97	125

NB: A free SO$_2$ level of over 50 mg/l will probably be detected upon the nose of a wine, and would be regarded as excessive and render the wine flawed.

Wine Faults and Flaws: A Practical Guide, First Edition. Keith Grainger.
© 2021 John Wiley & Sons Ltd. Published 2021 by John Wiley & Sons Ltd.

Further Reading

Easily Readable Books

Bird, D. (2010). *Understanding Wine Technology*, 3e. Nottingham: DBQA Publishing.

Fugelsang, K.C. and Edwards, C.G. (2007). *Wine Microbiology: Practical Applications and Procedures*, 2e. New York: Springer.

Grainger, K. and Tattersall, H. (2016). *Wine Production and Quality*, 2e. Chichester: Wiley-Blackwell.

Smith, C. (2014). *Postmodern Winemaking*. Berkeley: University of California Press.

Books with a More Scientific Approach

Bakker, J. and Clarke, R.J. (2012). *Wine Flavour Chemistry*, 2e. Oxford: Wiley-Blackwell.

Considine, J.A. and Frankish, E. (2014). *Quality in Small-Scale Winemaking*. Kidlington: Academic Press.

Flamini, R. and Traldi, P. (2010). *Mass Spectrometry in Grape and Wine Chemistry*. Hoboken, NJ: Wiley.

Jackson, R. (2020). *Wine Science – Principles and Applications*, 5e. London: Academic Press.

Jacobson, J.L. (2006). *Introduction to Wine Laboratory Practices and Procedures*. New York: Springer.

Lundanes, E., Reubsaet, L., and Greibrokk, T. (2013). *Chromatography: Basic Principles, Sample Preparations and Related Methods*. Weinheim: Wiley-VCH.

Morata, A. (ed.) (2019). *Red Wine Technology*. London: Academic Press.

Moreno-Arribas, M.V. and Sualdea, B.B. (eds.) (2016). *Wine Safety, Consumer Preference, and Human Health*. Cham: Springer.

Reynolds, A.G. (ed.) (2021). *Managing Wine Quality: Oenology and Wine Quality*, 2e, vol. II. Cambridge: Woodhead Publishing.

Ribéreau-Gayon, P., Dubourdieu, D., Donèche, B., and Lonvaud, A. (2006). *Handbook of Enology: The Microbiology of Wine and Vinifications*, 2e, vol. 1. Chichester: Wiley.

Ribéreau-Gayon, P., Glories, Y., Maujean, A., and Dubourdieu, D. (2006). *Handbook of Enology: The Chemistry of Wine: Stabilization and Treatment*, 2e, vol. 2. Chichester: Wiley.

Tamine, A.Y. (ed.) (2013). *Membrane Processing – Dairy and Beverage Applications*. Chichester: Wiley-Blackwell.

Waterhouse, A.L., Sacks, G.L., and Jeffery, D.W. (2016). *Understanding Wine Chemistry*. Chichester: Wiley.

Glossary

Aldehyde A compound which contains a carbonyl group (a carbon and oxygen atom covalently connected by a double bond) with at least one hydrogen atom attached.

Amino acid A class of biomolecules that contain both amino groups ($-NH^{3+}$) and carboxylate groups ($-COO^-$).

Anaerobiosis Life in the absence of atmospheric or free oxygen.

Anamorph The asexual reproductive stage of a fungus.

Anion A negatively charged ion, with more electrons than protons.

Anosmic Lacking in a sense of smell.

Anthropogenic Produced by human activity.

Appassimento The process of partially dehydrating grapes to make a particular style of rich, concentrated wine, e.g. Amarone Valpolicella.

ATP Adenosine triphosphate (ATP) is a compound that provides energy to drive processes in a living cell by capturing metabolic energy in the form of high-energy phosphate bonds and transporting them to sites within the cell where energy is required to drive a biochemical reaction.

Autolysis A chemical process by which wine breaks down dead yeast cells, which release mannoproteins, polysaccharides, and other products which impart textural changes sometimes 'bread-like' aromas and flavours.

Barrique A wooden cask of 225 l capacity, used in Bordeaux, but also in common usage elsewhere.

Bentonite A natural volcanic mineral comprising hydrated aluminium silicate. The material is used for fining wines, and particularly for protein removal. There are two basic forms used in winemaking: sodium bentonite and calcium bentonite.

Bound sulfur dioxide The inactive proportion of SO_2.

British thermal unit (Btu) The amount of heat required to raise the temperature of one pound of water by one degree Fahrenheit.

Butt A barrel, capacity 550 l, used for ageing Sherry.

Carbonyl A carbon and oxygen atom covalently connected by a double bond. Aldehydes, ketones and carboxylic acids are examples of carbonyls present in wines.

Chai A wine maturation cellar (with or without production or bottling facilities) situated at ground level, as is usually the case in the Bordeaux region.

Charcoal, activated A form of carbon that has been processed to have micro or meso pores, thus increasing the surface area available for chemical reactions or adsorption.

Wine Faults and Flaws: A Practical Guide, First Edition. Keith Grainger.
© 2021 John Wiley & Sons Ltd. Published 2021 by John Wiley & Sons Ltd.

Charm analysis See Gas Chromatography Olfactometry (GCO).

Chelating A type of bonding of ions or molecules to metal ions.

Chemical compound A substance comprising of atoms of two or more chemical elements bonded together in a fixed ratio. A simple example is H_2O (water) which comprises two atoms of Hydrogen (H) and one atom of oxygen (O).

Chemotroph An organism that manufactures its own food by oxidising chemical compounds.

Cleanskin A wine that is sold under an alternative label that does not name the producer. The concept is strong in Australasia, but is also used elsewhere, including by top Bordeaux châteaux.

Cliqueage Named after the sound of an activated solenoid, cliquegae is the instantaneous injection of oxygen to a tank or in barrel, rather than the slow, and continuous infusion that comprises micro-oxygenation (q.v.).

CODEX An international directory compiled by OIV (q.v.) detailing the chemical organic and gas products whose use is permitted for the making and maturation of wines.

Cold soak(ing) Low-temperature pre-fermentation maceration of juice and grape solids.

Colony forming unit (CFU) A measurement, normally per millilitre (in the case of liquid food products) of the number of viable cells of microorganisms (bacteria, yeast, and fungi). NB: A single living cell can reproduce through binary fission.

Coolship A large shallow (18–24″ [46–61 cm]) vat historically used in beer production to cool wort before fermentation, used today by producers of Lambic beers and a small but increasing numbers of craft beer producers for specialist ales.

Crabtree effect The production by some genera of yeasts, including *Saccharomyces* spp. and *Brettanomyces* spp., of alcohol (ethanol), rather than a large volume of biomass, under aerobic conditions and high glucose/fructose concentrations. Respiration, one energy source, is repressed and replaced by fermentation, an alternative energy source. Named after Herbert Grace Crabtree, an English biochemist.

Crossflow filtration A type filtration using membranes of less than 1 µm where the majority of wine is diverted across rather than through the membrane, which reduces fouling. The majority of the wine is returned to the feed tank (retentate), but a small percentage of the flow passes through the membrane (permeate).

Cru Classé An elite Bordeaux Château, officially classified as a producer of superior wine. The most famous classification was conducted in 1855 and remains largely unchanged.

Custer effect (sometimes called Custers effect or negative Pasteur effect). The inhibition of alcoholic fermentation in an environment without oxygen (anaerobic conditions). The activities *Brettanomyces* yeasts provide an example of the Custer effect.

Cuverie The part of a winery where alcoholic fermentation takes place.

Derivatization The modification of a compound by which a polar group in a molecule (e.g. a carboxyl or hydroxyl group) is chemically converted to a non-polar group in order to volatise the molecule so that it can be analysed by Gas Chromatography/Mass Spectrometry or High Performance Liquid Chromatography .

Diatomaceous earth A soft rock comprising crushed fossils or marine creatures that is used for wine filtration. Such use is decreasing. Also known as *Kieselguhr*

Diploid Containing two complete sets of chromosomes

En primeur A marketing and sales system practised particularly in Bordeaux, but also in several other regions, by which fine wines are sold "as futures" part way through their period of barrel maturation.

Ester An aromatic compound produced by a reaction between an acid and an alcohol. Produced during the alcoholic fermentation by yeasts making by-products that react with alcohol. The aromas of esters are classified as secondary aromas.

Esterase An enzyme that splits esters in a reaction with water (hydrolysis).

Facultative Heterofermentative (LAB) Bacteria that produce different end products, e.g. lactic acid, acetic acid, or only produce lactic acid depending upon the type of sugars utilised.

Fenton reaction A catalytic process that converts hydrogen peroxide into a hydroxyl free radical.

Filtration The process of passing the wine through a medium to remove bacteria or solids. A process that should be undertaken with care so as not to remove desirable aromas and flavours.

Fining Removal of microscopic troublesome matter (colloids) from the wine, generally undertaken before bottling but sometimes at intermediate stages of winemaking. Materials that may be used for fining include bentonite, albumen (egg whites), gelatine and isinglass.

Fino A type of dry Sherry, kept on ullage in butts for three to seven years, and undergoing microbiological ageing due to the influence of the growth of "flor" surface yeasts, q.v.

Flash détente A technique whereby must is heated to 85 °C (185 °F) and sent to a high-pressure vacuum chamber where the liquid is vaporised. The grape skins are deconstructed, and the chamber cooled rapidly to 32 °C (90 °F). The process gives highly-coloured, fruity red wines with soft tannins.

Flavanols a group of flavonoids, including catechins and procyanidins, present in high in grapes and red wines.

Flexitank A flexible bag that can be fitted into a standard 6.25 m (20 ft) container. They are generally single use. The best flexitanks for wine comprise a polyethylene bag with a barrier of ethylene vinyl alcohol (EVOH) copolymer.

Flor Lit. "flower" A yeast film, mostly of the species *Saccharomyces beticus*, that grows on the surface of fino Sherry which is deliberately matured on ullage in butts.

Free sulfur dioxide The proportion of SO_2 that acts as an antioxidant and protective.

Gas chromatography olfactometry (GCO) Gas chromatography separates Individual aromatics from a compound and shows the concentration of each. Each of the odorants is then nosed as it leaves the gas chromatograph and the human response is measured.

Genome An organism's complete set of genes and DNA.

Genotype The set of genes that make up an organism.

Glycoside A natural substance with a carbohydrate component, comprising one or more sugars, combined with a hydroxy compound (which may be a derivative of phenol or alcohol or other carbohydrate).

Gran Reserva A Spanish term for a wine that has undergone lengthy ageing, some of which is in barrel. The minimum ageing period varies according to the origin and colour of the wine. In the case of red Rioja, the minimum is two years in barrel, and the wine may not leave the winery until the sixth year after the vintage.

Haemolymph A blood-like fluid present in most invertebrates.

Hexose A simple sugar, the molecules of which contain six carbon atoms. Glucose and fructose are hexoses.

High Pressure Processing (HPP) A technique of cold 'pasteurisation'. Food or drink sealed the flexible container which it is to be sold is subjected to pressure up to 600 MPa, transmitted by water. Unlike conventional pasteurisation, the aromatics and flavours remain almost unchanged.

Hydroxy A prefix in chemistry that shows the presence of a (–OH) functional group.

Hyperoxygenation Also referred to as hyperoxidation. This is a process of deliberate oxidation of must to increase resistance to subsequent oxidation.

International Organisation of Vine and Wine (OIV) Intergovernmental organisation that is regarded as the scientific and technical reference of the vine and wine world. Its member states account for over 85% of world wine production, but exclude USA, Canada, Mexico and China.

Ion An atom or molecule that has a net (positive or negative) electrical charge.

ISO tank A 26 000 l stainless-steel tanks that fits can be transported by sea, rail or road, fitting on a standard truck. Unlike flexitanks ISO tanks are re-usable.

Ketone An organic compound belonging to a class that is characterised by having a carbonyl group (a carbon and oxygen atom covalently connected by a double bond).

Kieselguhr A coarse-grade diatomaceous earth powder used as a filtration adjuvant. Kieselguhr filtration will remove gross solids, including lees.

Killer yeast A yeast that secretes toxins lethal to some other yeasts and filamentous fungi.

Lees Sediment, including dead yeast cells, which settles at the bottom of the vat or barrel following the end of the fermentation process. 'Gross' lees are the initial coarse sediments, but lighter 'fine' lees may settle after each racking.

Lenticel A (usually) elliptical raised pore in cork (and other plant material).

Liposoluble Soluble in lipids, which are long chains of carbon and hydrogen molecules.

Lug box A stackable, ideally food-grade plastic, storage container as might be used for transporting and holding grapes prior to processing.

Lysis The disintegration of a cell by rupture of the cell wall or membrane.

Macro-oxygenation The intentional and controlled addition of measured amounts of oxygen to the wine, in larger amounts than is the case with micro-oxygenation (q.v.) using specialist equipment. The process usually takes place over a short period.

Maillard reaction A chemical reaction between an amino acid and a reducing sugar that can result in browning.

Membrane filtration Separation of dissolved materials (solutes) or particulates using microporous barriers of polymeric, ceramic or metallic materials.

Mercaptans Group of volatile, foul-smelling organic compounds similar to alcohols but containing a sulfur atom in place of the oxygen atom (–SH). Found in reduced wines.

Micro-oxygenation The intentional and controlled addition of minute, measured amounts of oxygen to wine using specialist equipment, usually taking place over several weeks or months.

Minimal media Growth media containing the minimum nutrients necessary for colony growth, but usually not containing amino acids.

Molecular sulfur dioxide The proportion of the free SO_2 that acts as antimicrobial protective.

Morphology The study of, and the shape, size and structure of living organisms, and the relationship between their structures.

Must adjustment The addition of various substances before fermentation to ensure the desired chemical balance. For example, these may include additions of tartaric acid (acidification), often practised in hot climates. Deacidification may be required in cool climates.

Must concentrators Machines used to remove water from the juice of grapes. They may prove particularly valuable following a wet vintage.

Must enrichment Process that may be undertaken before or in the early stages of fermentation whereby the sugar content is increased to raise the potential alcohol level of the wine. It may be undertaken in cool climates where grapes struggle to ripen.

Must Unfermented grape juice and grape solids, which may include skins and seeds.

Nanogramme One billionth (one thousand-millionth of a gram).

Necrosis The death of tissue.

Norisoprenoid A diverse class of aromatic compounds, originating in carotenoid molecules in grapes, whose aromas are released after fermentation or, in some cases, extended ageing.

Obligate aerobe An organism that requires oxygen to grow, metabolising sugars or fats to produce energy.

Pasteurisation Process of heating wine in order to ensure stability by destruction of microorganisms.

Pathogen A microorganism, bacterium or virus that can cause disease.

Pectolytic enzymes Proteins used to break down and destroy pectin haze to improve the clarification in wine.

Pentose A monosaccharide (simple sugar) with five carbon atoms.

Peptide A short chain of amino acids, linked by peptide bonds.

Petri dish A shallow glass or plastic plate, with lid, used for culturing cells, particularly bacteria.

pH A measure of the hydrogen ion (H^+) concentration in a polar liquid, and hence the acidity of the liquid. pH is measured on a scale from 0 to 14, where 7 is neutral, below 7 is acidic, and above 7 is alkaline.

Picogramme One trillionth of a gram.

Pied de cuve A yeast starter prepared in advance with a small quantity of grape must and desired yeast cultures. When fermenting vigorously, this is added to the bulk of the must to help ensure a rapid onset of fermentation.

Pipe Cask of 534 or 550 l capacity, used in the Douro Valley in Portugal for Port production.

Ploidy The number of complete sets of chromosomes in a cell.

Polyvinylpolypyrrolidone (PVPP) Synthetic, high-molecular-weight clarifying agent used in fining wines as a polyphenol adsorbent. It binds and removes small phenolic compounds such as catechins and anthocyanins.

Potassium metabisulfite A white crystalline powder (E224), molecular formula $K_2S_2O_5$ that releases approximately 58% by weight of sulfur dioxide.

Precision viticulture A set of methodologies, analysis and processes for a site-specific vine and grape management.

Primeurs week (Bordeaux) An event that usually takes place annually in early April each year that gives the wine trade and press the opportunity to taste wines from the previous vintage, whilst they are still undergoing cellar maturation.

Pyrazines Chemical compounds that give green, peppery aromas and a herbaceous character in some wines. Cabernet Sauvignon, especially if slightly under-ripe and Sauvignon Blanc are two of the grape varieties that can show pyrazine characteristics, both in aroma and on the palate.

Pyrolysis Derived from the Greek "pyro" (fire) and "lysis" (separating), pyrolysis is the irreversible thermal decomposition of materials at elevated temperatures.

Quinone One of a series of aromatic compounds derived from benzene by the replacement of two atoms of hydrogen with two atoms of oxygen.

Racemic Composed of dextrorotatory (D) and laevorotatory (L) forms of a compound in equal proportions.

Reverse osmosis A technology of purification of liquids, using a pressure-driven separation processes that employs a semipermeable membrane.

Scalping (flavour) Absorption by the package or closure of volatile aromas and flavours generally resulting in a loss of quality of the packaged product.

Sheet filtration Sometimes called plate and frame filtration. This system employs a series of filtration sheets usually made of cellulose or perlite. The wine only passes through one of the sheets, which are manufactured in various porosities and accordingly can perform tough, polishing or even from a basic polishing up to total sterile filtration.

Spinning cone column A device used for dealcoholising wines. A stainless steel column contains a series of inverted cones alternatively attached to a rotating central shaft and the (stationary) column body. Wine passes downwards over the cones twice through rising vapour: the first pass removes aromatics (which are later re-introduced), and the second pass removes alcohol.

Strain A genetic variant of a microorganism.

Strecker degradation A reaction that, in the presence of a carbonyl compound, converts α-amino acids into aldehydes by the removal of an amino group and carboxyl group.

Stuck fermentation A fermentation that has unwantedly ceased to be active before all the sugars have been consumed by the yeasts.

Substrate The surface on which an organism grows and lives.

Sulfur dioxide (SO_2) Compound used widely in winemaking as an antimicrobial and antioxidant protective.

Tannins A loose term encompassing polyphenols that bind and precipitate proteins. Present in grape skins, stalks and seeds (condensed tannins) and also oak wood and products (hydrolysable tannins). Tannins have an astringent feel in the mouth, particularly on the gums.

Teleomorph The sexually reproductive stage of a fungus.

Terroir A French term bringing together the notion of soil, microclimate, landscape and environment within a particular vineyard site and the resultant effect on the vines.

Thermo détente A pre-fermentation extraction process that involves the heating of grapes to 75 °C (167 °F), which are then pressurised with compressed gas, the pressure then being rapidly released. Useful in wet vintages.

Thiol An aromatic organic compound similar to an alcohol but containing a sulfur atom in place of the oxygen atom.

Tinctorial Relating to dyeing, staining or colouring.

Total sulfur dioxide The combination of free and bound SO_2. It is the total SO_2 that is subject to legal limits, although the majority of this is usually bound, and thus inactive.

Traditional method (sparkling wines) A method of making high-quality sparkling wines by the undertaking of a second fermentation in the bottle in which the wine will be sold. The yeasty sediment that is formed is removed by process of riddling and disgorging.

Transfer method (sparkling wines) A method of making sparkling wines by the undertaking of a second fermentation in a cellar bottle. The bottles will then be decanted under pressure into a vat, the dead yeast is removed by filtration, and the wine is re-bottled under pressure.

Trehalose A disaccharide sugar formed from two glucose units.

Triploid Containing three complete sets of chromosomes.

Ullage The air space between the wine and the top of the tank or barrel, or (in bottle) the wine and closure.

Veraison The beginning of grape ripening, when the skin softens and the colour starts to change. Following veraison, the growth in the size of berry is mainly due to the expansion rather than the reproduction of cells. As ripening continues, sugars increase and acidity levels fall.

Vitis vinifera European species of the *Vitis* genus, and from which nearly all of the world's wine is made. There are over 5000 known varieties of the species, e.g. Chardonnay and Cabernet Sauvignon.

Yeast assimilable nitrogen (YAN) Nitrogen available for yeasts, comprising primary amino acids, ammonia and ammonium ions.

Useful Websites

This list contains a selection of websites that the author considers to be well constructed and contain valuable information for the reader. It is, by definition, selective. Unless otherwise stated, the sites are in English.

Although all the sites were 'live' on 12 August 2020, by the nature of the internet websites come and go, and thus some may no longer be available. Should the reader find any sites to be defunct, the author would be grateful if they are brought to his attention.

Adams & Chittenden Scientific Glass Coop: www.adamschittenden.com/

Based in Berkeley, California, Adams & Chittenden are manufacturers of laboratory glassware and tools, including distillation extraction and chromatography glass.

Aroxa: www.aroxa.com

Aroxa manufactures a range of certified flavour standards, including many of the fault compounds detailed in this book. These can be used for training winery staff, including technicians and quality control personnel.

Association of Wine Educators: www.wineeducators.com

The Association of Wine Educators is a professional association whose members are involved in the field of wine education. The website includes a directory of members, many of whom are specialists in different aspects of wine production and assessment.

The Australian Wine Research Institute: www.awri.com.au

The Australian Wine Research Institute, whose aim is to advance the competitive edge of the Australian Wine Industry, researches the composition and sensory characteristic of wines, undertakes an analytical service and provides industry development and support. The website contains a good deal of free information.

Bordeaux Wines: Conseil Interprofessionnel du Vin de Bordeaux https://www.bordeaux .com/gb

The CIVB represents, advises and controls the wine industry of Bordeaux, which is the largest fine wine region in the world. This is a lively website with much useful information about Bordeaux and its wines.

Circle of Wine Writers Circle of Wine Writers: www.circleofwinewriters.org

The Circle of Wine Writers is an association of wine writers and communicators with members all around the world.

Wine Faults and Flaws: A Practical Guide, First Edition. Keith Grainger.
© 2021 John Wiley & Sons Ltd. Published 2021 by John Wiley & Sons Ltd.

ETS Laboratories. www.etslabs.com

ETS was founded in St. Helena, California in 1978 by Gordon and Marjorie Burns. They were pioneers in wine analysis for the fast-growing Californian wine industry. The company now has three sites in California, one in Oregon and one in Washington State. The Library section of the website contains much useful information on wine faults and wine quality.

Grainger, Keith www.keithgrainger.com

This is the website of the author of this book.

Institut des Sciences de la Vigne et du Vin University of Bordeaux: http://www.isvv.u-bordeaux.fr/en

The centre of excellence for studies of oenology in France.

Institut Français de la Vigne et du Vin:http://www.vignevin.com/ (in French). The institute provides technical information for growers and winemakers in France. The website (in French) includes details of their publications.

Lallemand: https://www.lallemandwine.com/en/north-america/

Lallemand are producers of winemaking yeasts, bacteria, nutrients and other products. The website contains useful winemaking tools.

LANXESS/Velcorin: www.velcorin.com This is the website of the manufacturer of the wine 'sterilant' Velcorin, who also market the necessary dosing equipment. The website gives technical information.

Organisation Internationale de la Vigne et du Vin (OIV): www.oiv.int

The OIV is an intergovernmental organisation and is regarded as the science and technical reference of the vine and wine world. Member states account for over 85% of world wine production but exclude the UK, USA, Canada, Mexico and China. The website provides up-to-date news and many useful statistics.

UC Davis: http://wineserver.ucdavis.edu

The Department of Viticulture and Enology at Davis has been at the forefront of research for 40 years.

Wiley: www.wiley.com

The website of this book's publishers, which includes details of publications and links to numerous resources.

Wine Australia: https://www.wineaustralia.com/

Wine Australia is an Australian Government statutory authority which helps foster and encourage Australian wine grape and wine businesses. Its activities include investing in research and development. The website, particularly the Growing and Making Page, contains much useful information on current issues.

Winetitles: www.winetitles.com.au

Winetitles is an Australian publisher of wine books, journals and seminar papers.

Vinolab https://www.vinolab.hr/calculator/en4

Vinolab is a wine laboratory based in Croatia. The linked page is to over 25 pages of instant calculators, e.g. additions required for molecular SO_2 concentrations or hydrometer temperature correction.

Vintessential Laboratories: www.vintessential.com.au

Vintissential has five laboratories in Australia undertaking wine analysis, and are manufacturers of enzymatic test kits, and market oenological supplies and laboratory equipment.

The website contains many useful factsheets and discussions about wine faults and wine-making generally.

WeinPlus Glossary: https://glossary.wein-plus.eu/

This site is 'the largest wine encyclopaedia in the world' with over 23 000 key words and over 154 000 cross-references. It is auto-translated into English.

Wineland Media: www.wineland.co.za

Wineland is published in South Africa. The website includes a link to some excellent technical articles written in an easy-to-understand manner.

Index

Wine Faults and Flaws: A Practical Guide, First Edition. Keith Grainger.
© 2021 John Wiley & Sons Ltd. Published 2021 by John Wiley & Sons Ltd.

bag-in-box 188, 405, 426
  filling 427
  history 426
  manufacture 427
  oxygen permeability 188
  taint from 358
balance (vine) 387
balance (wine) 23–24, 26, 42, 47, 50–53, 122,
    176, 352, 361–362, 392
  acidity (impact on) 362–363
  flavour concentration 364
  residual sugar 364
  tannins 363–364
banana (aroma) 37, 39, 48, 129, 271
BAND-AID 36, 42, 117, 128, 159
Banyuls 176
Bardolino 46
bark, cork 71, 75–77, 84, 104, 106, 412–415
Barolo 35, 45, 258
barrel(s)
  ageing (*see* barrel maturation)
  Brettanomyces in 121, 133, 135–136, 142,
    149, 159
  cellars and stacking 13, 61–62, 87, 96–97,
    144, 434
  cleaning and sanitising 95, 142–144, 276,
    437
  maturation 29–31, 173, 186–188, 202–203,
    223, 227, 258, 268, 277–278, 348, 363
  TCA in 89–90
  toasting 267–268, 277
  topping up 144, 187, 440
Bartoshuk, Linda 24, 43
bâtonnage 37, 136, 187, 191, 193, 321
Beaujolais 38, 51, 55, 124
beer 58, 122, 241, 243, 307, 352, 409
  bottle-conditioned 297
  Brettanomyces, role in 119, 121–122, 127,
    297
  craft 121–122, 435, 448
  haloanisoles, contamination with 58–59,
    70, 95
  Lambic 121–122, 446, 448, 455
beeswax (aroma) 173, 175, 189
beetroot (aroma) 36, 105, 359
bentonite 107, 189, 228, 236, 273, 281,
    311–312, 314, 359–360
  alternatives 313

calcium 311, 373
  contamination of 1, 6, 73, 96–97, 102, 358
  fining 232, 246, 249, 280, 299, 309, 311,
    348, 354–355, 363, 373
  loss of aromas 311–312
  fining trials 312
  haloanisole trap (use as) 97
  ochratoxin A, treatment for 354–355
  protein removal by fining 298, 311
  sodium 311
  traps 90
  types 311
  wine loss 311–312
Bentotest 311
benzaldehyde 174
benzenemethanethiol (benzene mercaptan)
    23, 41, 228–229
biofilms 95, 138, 143, 272, 304, 440
biogenic amines 4–5, 119, 273, 318, 352,
    356–357, 364
biomethylation 71–72, 74, 94
biotin 140, 220
bisulfite 185, 222, 242, 441
Bisson, Linda 66, 124
bitterness 24, 31, 42, 44, 46, 50, 53, 157, 172,
    290, 292, 323, 364, 393, 420
bleach, hypochlorite 59, 71, 76, 79, 95–96,
    195, 409, 435
bleeding 392
blood-like odour 5, 356
blue fining 235, 314
Bocksin 237
body (tasting term) 29, 42–43, 46–47, 50,
    148, 209, 239, 362–363
boiled sweet (aroma) 37
Bordeaux 35, 60, 75, 147, 287, 363, 374, 387,
    392, 399, 406, 434
  Brettanomyces in 123–125, 128, 131, 136,
    144
  haloanisole contaminated cellars (historic)
    62–64
  mixture 220, 313
  premox in 189–190, 192, 195
Botrytis cinerea 13, 48, 75, 106, 133, 198–199,
    221–222, 258, 263, 269, 272, 353, 359
bottle ageing 38, 52, 136, 188, 195, 243, 246,
    345, 348
bottle fermentation, unwanted 296–298, 300

Printed and bound by CPI Group (UK) Ltd, Croydon, CR0 4YY

16/04/2025

14658382-0003